Edited by
Claus Daniel and Jürgen O. Besenhard

Handbook of Battery Materials

Related Titles

Lui, R.-S., Zhang, L., Sun, X., Lui, H., Zhang, J., (eds.)

Electrochemical Technologies for Energy Storage and Conversion

2011
ISBN: 978-3-527-32869-7

Garcia-Martinez, J. (ed.)

Nanotechnology for the Energy Challenge

2010
ISBN: 978-3-527-32401-9

Stolten, D. (ed.)

Hydrogen and Fuel Cells

Fundamentals, Technologies and Applications

2010
ISBN: 978-3-527-32711-9

Aifantis, K. E., Hackney, S. A., Kumar, R. V. (eds.)

High Energy Density Lithium Batteries

Materials, Engineering, Applications

2010
ISBN: 978-3-527-32407-1

Hirscher, M. (ed.)

Handbook of Hydrogen Storage

New Materials for Future Energy Storage

2010
ISBN: 978-3-527-32273-2

Liu, H., Zhang, J. (eds.)

Electrocatalysis of Direct Methanol Fuel Cells

From Fundamentals to Applications

2009
ISBN: 978-3-527-32377-7

Ozawa, K. (ed.)

Lithium Ion Rechargeable Batteries

Materials, Technology, and New Applications

2009
ISBN: 978-3-527-31983-1

Vielstich, W., Gasteiger, H. A., Yokokawa, H. (eds.)

Handbook of Fuel Cells

Advances in Electrocatalysis, Materials, Diagnostics and Durability, Volumes 5 & 6

2009
ISBN: 978-0-470-72311-1

Mitsos, A., Barton, P. I. (eds.)

Microfabricated Power Generation Devices

Design and Technology

2009
ISBN: 978-3-527-32081-3

Edited by Claus Daniel and Jürgen O. Besenhard

Handbook of Battery Materials

Second, Completely Revised and Enlarged Edition

WILEY-VCH Verlag GmbH & Co. KGaA

The Editor

Dr.-Ing. Claus Daniel
Oak Ridge National Laboratory
MS6083
P.O. Box 2008
Oak Ridge, TN 37831-6083
USA

1st Reprint 2012

All books published by **Wiley-VCH** are carefully produced. Nevertheless, authors, editors, and publisher do not warrant the information contained in these books, including this book, to be free of errors. Readers are advised to keep in mind that statements, data, illustrations, procedural details or other items may inadvertently be inaccurate.

Library of Congress Card No.: applied for

British Library Cataloguing-in-Publication Data
A catalogue record for this book is available from the British Library.

Bibliographic information published by the Deutsche Nationalbibliothek
The Deutsche Nationalbibliothek lists this publication in the Deutsche Nationalbibliografie; detailed bibliographic data are available on the Internet at <http://dnb.d-nb.de>.

© 2011 Wiley-VCH Verlag & Co. KGaA, Boschstr. 12, 69469 Weinheim, Germany

All rights reserved (including those of translation into other languages). No part of this book may be reproduced in any form – by photoprinting, microfilm, or any other means – nor transmitted or translated into a machine language without written permission from the publishers. Registered names, trademarks, etc. used in this book, even when not specifically marked as such, are not to be considered unprotected by law.

Typesetting   Laserwords Private Limited, Chennai
Printing and Binding   betz-druck GmbH, Darmstadt
Cover Design   Formgeber, Eppelheim

Printed in the Federal Republic of Germany
Printed on acid-free paper

Print ISBN: 978-3-527-32695-2

Dedication: Jürgen O. Besenhard (1944–2006)

The first edition of the "Handbook of Battery Materials" was edited by Professor Jürgen Otto Besenhard. Jürgen Besenhard began his scientific career at the time, when the era of lithium batteries came up. With his strong background in chemistry and his outstanding ability to interpret and understand the complex phenomena behind the many exploratory research findings on lithium batteries in the late 1960s and early to mid 1970s, Jürgen Besenhard was able to attribute "performance" to material properties. His early work is evidence for this:

1) Understanding of reversible alkali metal ion intercalation into graphite anodes (*J. Electroanal. Chem.*, **53** (1974) 329 and *Carbon*, **14** (1976) 111)
2) Understanding of reversible alkali metal ion insertion into oxide materials for cathodes (*Mat. Res. Bull.*, **11** (1976) 83 and *J. Power Sources*, **1** (1976/1977) 267)
3) First reviews on lithium batteries (*J. Electroanal. Chem.*, **68** (1976) 1 and *J. Electroanal. Chem.*, **72** (1976) 1)
4) Preparation of lithium alloys with defined stoichiometry in organic electrolytes at ambient temperature (*Electrochim. Acta*, **20** (1975) 513).

Jürgen Besenhard's research interests were almost unlimited. After he received a Full Professorship at the University of Münster (Germany) in 1986 and especially after 1993, when he assumed the position as head of the Institute of Chemistry and Technology at Graz University of Technology in Austria, he expanded his activities to countless topics in the field of applied electrochemistry. But his favorite topic, "his dedication," has always been "Battery Materials."

Jürgen Otto Besenhard was an exceptional and devoted scientist and he leaves behind an enduring record of achievements. He was considered as a leading authority in the field of lithium battery materials. His works will always assure him a highly prominent position in the history of battery technology.

Prof. Besenhard was also a highly respected teacher inside and outside the university. Consequently, it was only natural that he edited a book, which attempted to give explanations, rather than only summarizing figures and facts. The "Handbook of Battery Materials" was one of Jürgen Otto Besenhard's favorite projects. He knew that materials are the key to batteries. It is the merit of Claus Daniel and the publisher Wiley, that this project will be continued.

May this new edition of the "Handbook of Battery Materials" be a useful guide into the complex and rapidly growing field of battery materials. Beyond that, it is my personal wish and hope that the readers of this book may also take the chance to review Prof. Besenhard's work. Jürgen Otto Besenhard has been truly one of the fathers of lithium batteries and lithium ion batteries.

Münster, Germany, July, 2011 *Martin Winter*

Contents

Preface *XXVII*
List of Contributors *XXIX*

Part I   Fundamentals and General Aspects of Electrochemical Energy Storage   *1*

1   **Thermodynamics and Mechanistics**   *3*
Karsten Pinkwart and Jens Tübke
1.1   Electrochemical Power Sources   *3*
1.2   Electrochemical Fundamentals   *6*
1.2.1   Electrochemical Cell   *6*
1.2.2   Electrochemical Series of Metals   *9*
1.2.3   Discharging   *11*
1.2.4   Charging   *12*
1.3   Thermodynamics   *13*
1.3.1   Electrode Processes at Equilibrium   *13*
1.3.2   Reaction Free Energy ΔG and Equilibrium Cell Voltage $\Delta \varepsilon_{00}$ *14*
1.3.3   Concentration Dependence of the Equilibrium Cell Voltage   *15*
1.3.4   Temperature Dependence of the Equilibrium Cell Voltage   *16*
1.3.5   Pressure Dependence of the Equilibrium Cell Voltage   *18*
1.3.6   Overpotential of Half Cells and Internal Resistance   *19*
1.4   Criteria for the Judgment of Batteries   *21*
1.4.1   Terminal Voltage   *21*
1.4.2   Current–Voltage Diagram   *21*
1.4.3   Discharge Characteristic   *22*
1.4.4   Characteristic Line of Charge   *22*
1.4.5   Overcharge Reactions   *23*
1.4.6   Coulometric Efficiency and Energy Efficiency   *24*
1.4.7   Cycle Life and Shelf Life   *24*
1.4.8   Specific Energy and Energy Density   *25*
1.4.9   Safety   *25*
1.4.10   Costs per Stored Watt Hour   *26*
References   *26*

2	**Practical Batteries** 27
	Koji Nishio and Nobuhiro Furukawa
2.1	Introduction 27
2.2	Alkaline-Manganese Batteries 27
2.3	Nickel–Cadmium Batteries 30
2.4	Nickel–MH Batteries 36
2.5	Lithium Primary Batteries 43
2.5.1	Lithium–Manganese Dioxide Batteries 43
2.5.2	Lithium–Carbon Monofluoride Batteries 52
2.5.3	Lithium–Thionyl Chloride Batteries 54
2.6	Coin-Type Lithium Secondary Batteries 55
2.6.1	Secondary Lithium–Manganese Dioxide Batteries 55
2.6.2	Lithium–Vanadium Oxide Secondary Batteries 60
2.6.3	Lithium–Polyaniline Batteries 62
2.6.4	Secondary Lithium–Carbon Batteries 62
2.6.5	Secondary Li-LGH–Vanadium Oxide Batteries 63
2.6.6	Secondary Lithium–Polyacene Batteries 64
2.6.7	Secondary Niobium Oxide–Vanadium Oxide Batteries 64
2.6.8	Secondary Titanium Oxide–Manganese Oxide Batteries 65
2.7	Lithium-Ion Batteries 66
2.7.1	Positive Electrode Materials 66
2.7.2	Negative Electrode Materials 70
2.7.3	Battery Performances 75
2.8	Secondary Lithium Batteries with Metal Anodes 78
	References 80
	Further Reading 84

Part II	**Materials for Aqueous Electrolyte Batteries** 87
3	**Structural Chemistry of Manganese Dioxide and Related Compounds** 89
	Jörg H. Albering
3.1	Introduction 89
3.2	Tunnel Structures 90
3.2.1	β-MnO_2 90
3.2.2	Ramsdellite 91
3.2.3	γ-MnO_2 and ε-MnO_2 93
3.2.4	α-MnO_2 100
3.2.5	Romanèchite, Todorokite, and Related Compounds 102
3.3	Layer Structures 104
3.3.1	Mn_5O_8 and Similar Compounds 105
3.3.2	Lithiophorite 106
3.3.3	Chalcophanite 108
3.3.4	δ-MnO_2 Materials 109
3.3.5	10 Å Phyllomanganates of the Buserite Type 114

3.4	Reduced Manganese Oxides	116
3.4.1	Compounds of Composition MnOOH	116
3.4.1.1	Manganite (γ-MnOOH)	116
3.4.1.2	Groutite (α-MnOOH)	116
3.4.1.3	δ-MnOOH	118
3.4.1.4	Feitknechtite β-MnOOH	118
3.4.2	Spinel-Type Compounds Mn_3O_4 and γ-Mn_2O_3	119
3.4.3	Pyrochroite, $Mn(OH)_2$	119
3.5	Conclusion	120
	References	120
	Further Reading	122
4	**Electrochemistry of Manganese Oxides**	**125**
	Akiya Kozawa, Kohei Yamamoto, and Masaki Yoshio	
4.1	Introduction	125
4.2	Electrochemical Properties of EMD	126
4.2.1	Discharge Curves and Electrochemical Reactions	126
4.2.2	Modification of Discharge Behavior of EMD with $Bi(OH)_3$	128
4.2.3	Factors which Influence MnO_2 Potential	129
4.2.3.1	Surface Condition of MnO_2	129
4.2.3.2	Standard Potential of MnO_2 in 1 mol L^{-1} KOH	131
4.2.4	Three Types of Polarization for MnO_2	131
4.2.5	Discharge Tests for Battery Materials	134
4.3	Physical Properties and Chemical Composition of EMD	137
4.3.1	Cross-Section of the Pores	137
4.3.2	Closed Pores	139
4.3.3	Effective Volume Measurement	140
4.4	Conversion of EMD to $LiMnO_2$ or $LiMn_2O_4$ for Rechargeable Li Batteries	140
4.4.1	Melt-Impregnation (M–I) Method for EMD	142
4.4.2	Preparation of $Li_{0.3}MnO_2$ from EMD	143
4.4.3	Preparation of $LiMn_2O_4$ from EMD	145
4.5	Discharge Curves of EMD Alkaline Cells (AA and AAA Cells)	147
	References	147
	Further Reading	148
5	**Nickel Hydroxides**	**149**
	James McBreen	
5.1	Introduction	149
5.2	Nickel Hydroxide Battery Electrodes	150
5.3	Solid-State Chemistry of Nickel Hydroxides	151
5.3.1	Hydrous Nickel Oxides	151
5.3.1.1	β-$Ni(OH)_2$	151
5.3.1.2	α-$Ni(OH)_2$	154
5.3.1.3	β-NiOOH	157

5.3.1.4	γ-NiOOH 158
5.3.1.5	Relevance of Model Compounds to Electrode Materials 158
5.3.2	Pyroaurite-Type Nickel Hydroxides 159
5.4	Electrochemical Reactions 161
5.4.1	Overall Reaction and Thermodynamics of the Ni(OH)$_2$/NiOOH Couple 161
5.4.2	Nature of the Ni(OH)$_2$/NiOOH Reaction 162
5.4.3	Nickel Oxidation State 164
5.4.4	Oxygen Evolution 164
5.4.5	Hydrogen Oxidation 164
	References 165

6	**Lead Oxides** 169
	Dietrich Berndt
6.1	Introduction 169
6.2	Lead/Oxygen Compounds 170
6.2.1	Lead Oxide (PbO) 170
6.2.2	Minium (Pb$_3$O$_4$) 171
6.2.3	Lead Dioxide (PbO$_2$) 171
6.2.4	Nonstoichiometric PbO$_x$ Phases 172
6.2.5	Basic Sulfates 172
6.2.6	Physical and Chemical Properties 172
6.3	The Thermodynamic Situation 173
6.3.1	Water Decomposition 174
6.3.2	Oxidation of Lead 175
6.3.3	The Thermodynamic Situation in Lead–Acid Batteries 177
6.3.4	Thermodynamic Data 180
6.4	PbO$_2$ as Active Material in Lead–Acid Batteries 181
6.4.1	Planté Plates 182
6.4.2	Pasted Plates 184
6.4.2.1	Manufacture of the Active Material 184
6.4.2.2	Tank Formation 187
6.4.2.3	Container Formation 187
6.4.3	Tubular Plates 187
6.5	Passivation of Lead by Its Oxides 189
6.5.1	Disintegration of the Oxide Layer at Open-Circuit Voltage 191
6.5.2	Charge Preservation in Negative Electrodes by a PbO Layer 192
6.6	Ageing Effects 192
6.6.1	The Influence of Antimony, Tin, and Phosphoric Acid 193
	References 194
	Further Reading 196

7	**Bromine-Storage Materials** 197
	Christoph Fabjan and Josef Drobits
7.1	Introduction 197

7.2	Possibilities for Bromine Storage	199
7.2.1	General Aspects	199
7.2.2	Quaternary Ammonium-Polybromide Complexes	200
7.3	Physical Properties of the Bromine Storage Phase	204
7.3.1	Conductivity	204
7.3.2	Viscosity and Specific Weight	207
7.3.3	Diffusion Coefficients	208
7.3.4	State of Aggregation	209
7.4	Analytical Study of a Battery Charge Cycle	210
7.5	Safety, Physiological Aspects, and Recycling	212
7.5.1	Safety	212
7.5.2	Physiological Aspects	214
7.5.3	Recycling	214
	References	214
8	**Metallic Negatives**	**219**
	Leo Binder	
8.1	Introduction	219
8.2	Overview	219
8.3	Battery Anodes ('Negatives')	220
8.3.1	Aluminum (Al)	220
8.3.2	Cadmium (Cd)	221
8.3.3	Iron (Fe)	222
8.3.4	Lead (Pb)	223
8.3.5	Lithium (Li)	224
8.3.6	Magnesium (Mg)	224
8.3.7	Zinc (Zn)	225
8.3.7.1	Zinc Electrodes for 'Acidic' (Neutral) Primaries	226
8.3.7.2	Zinc Electrodes for Alkaline Primaries	226
8.3.7.3	Zinc Electrodes for Alkaline Storage Batteries	229
8.3.7.4	Zinc Electrodes for Alkaline 'Low-Cost' Reusables	230
8.3.7.5	Zinc Electrodes for Zinc-Flow Batteries	232
8.3.7.6	Zinc Electrodes for Printed Thin-Layer Batteries	233
	References	234
9	**Metal Hydride Electrodes**	**239**
	James J. Reilly	
9.1	Introduction	239
9.2	Theory and Basic Principles	239
9.2.1	Thermodynamics	240
9.2.2	Electronic Properties	242
9.2.3	Reaction Rules and Predictive Theories	243
9.3	Metal Hydride–Nickel Batteries	244
9.3.1	Alloy Activation	245
9.3.2	AB_5 Electrodes	246

9.3.2.1	Chemical Properties of AB_5 Hydrides 246
9.3.2.2	Preparation of AB_5 Electrodes 249
9.3.2.3	Effect of Temperature 250
9.3.3	Electrode Corrosion and Storage Capacity 250
9.3.4	Corrosion and Composition 251
9.3.4.1	Effect of Cerium 254
9.3.4.2	Effect of Cobalt 257
9.3.4.3	Effect of Aluminum 257
9.3.4.4	Effect of Manganese 258
9.4	Super-Stoichiometric AB_{5+x} Alloys 259
9.5	AB_2 Hydride Electrodes 261
9.6	XAS Studies of Alloy Electrode Materials 264
9.7	Summary 265
	Acknowledgment 266
	References 266
	Further Reading 268

10   Carbons 269
Kimio Kinoshita

10.1	Introduction 269
10.2	Physicochemical Properties of Carbon Materials 269
10.2.1	Physical Properties 269
10.2.2	Chemical Properties 273
10.3	Electrochemical Behavior 274
10.3.1	Potential 274
10.3.2	Conductive Matrix 275
10.3.3	Electrochemical Properties 277
10.3.4	Electrochemical Oxidation 277
10.3.5	Electrocatalysis 278
10.3.6	Intercalation 282
10.4	Concluding Remarks 283
	References 283

11   Separators 285
Werner Böhnstedt

11.1	General Principles 285
11.1.1	Basic Functions of the Separators 285
11.1.2	Characterizing Properties 286
11.1.2.1	Backweb, Ribs, and Overall Thickness 286
11.1.2.2	Porosity, Pore Size, and Pore Shape 287
11.1.2.3	Electrical Resistance 289
11.1.3	Battery and Battery Separator Markets 291
11.2	Separators for Lead–Acid Storage Batteries 293
11.2.1	Development History 293
11.2.1.1	Historical Beginnings 293

11.2.1.2	Starter Battery Separators	294
11.2.1.3	Industrial Battery Separators	296
11.2.1.3.1	Stationary Battery Separators	296
11.2.1.3.2	Traction Battery Separators	298
11.2.1.3.3	Electrical Vehicle Battery Separators	299
11.2.2	Separators for Starter Batteries	300
11.2.2.1	Polyethylene Pocket Separators	300
11.2.2.1.1	Production Process	300
11.2.2.1.2	Mixing and Extrusion	301
11.2.2.1.3	Properties	301
11.2.2.1.4	Profiles	304
11.2.2.1.5	Product Comparison	306
11.2.2.2	Leaf Separators	306
11.2.2.2.1	Sintered PVC Separators	307
11.2.2.2.2	Cellulosic Separators	309
11.2.2.2.3	Glass Fiber Leaf Separators	310
11.2.2.2.4	Leaf Separators with Attached Glass Mat	311
11.2.2.2.5	'Japanese' Separators	311
11.2.2.2.6	Microfiber Glass Separators	312
11.2.2.3	Comparative Evaluation of Starter Battery Separators	313
11.2.3	Separators for Industrial Batteries	316
11.2.3.1	Separators for Traction Batteries	316
11.2.3.1.1	Polyethylene Separators	316
11.2.3.1.2	Rubber Separators	319
11.2.3.1.3	Phenol–Formaldehyde–Resorcinol Separators (DARAK 5000)	320
11.2.3.1.4	Microporous PVC Separators	320
11.2.3.1.5	Comparative Evaluation of the Traction Battery Separators	321
11.2.3.2	Separators for Open Stationary Batteries	321
11.2.3.2.1	Polyethylene Separators	322
11.2.3.2.2	Phenol–Formaldehyde Resin–Resorcinol Separators (DARAK 2000/5005)	322
11.2.3.2.3	Microporous PVC Separators	322
11.2.3.2.4	Sintered PVC Separators	322
11.2.3.2.5	Comparative Evaluation of Separators for Open Stationary Batteries	323
11.2.3.3	Separators for Valve Regulated Lead–Acid (VRLA) Batteries	324
11.2.3.3.1	Batteries with Absorptive Glass Mat	324
11.2.3.3.2	Batteries with Gelled Electrolyte	325
11.3	Separators for Alkaline Storage Batteries	328
11.3.1	General	328
11.3.2	Primary Cells	329
11.3.3	Nickel Systems	329
11.3.3.1	Nickel–Cadmium Batteries	329
11.3.3.1.1	Vented Construction	329
11.3.3.1.2	Sealed Construction	330

11.3.3.2	Nickel–Metal Hydride Batteries *331*
11.3.4	Zinc Systems *331*
11.3.4.1	Nickel–Zinc Storage Batteries *331*
11.3.4.2	Zinc–Manganese Dioxide Secondary Cells *332*
11.3.4.3	Zinc–Air Batteries *332*
11.3.4.4	Zinc-Bromine Batteries *333*
11.3.4.5	Zinc–Silver Oxide Storage Batteries *333*
11.3.5	Separator Materials for Alkaline Batteries *334*
	Acknowledgments *337*
	References *337*

Part III   Materials for Alkali Metal Batteries *341*

12   Lithium Intercalation Cathode Materials for Lithium-Ion Batteries *343*
Arumugam Manthiram and Theivanayagam Muraliganth

12.1	Introduction *343*
12.2	History of Lithium-Ion Batteries *343*
12.3	Lithium-Ion Battery Electrodes *345*
12.4	Layered Metal Oxide Cathodes *347*
12.5	Layered $LiCoO_2$ *348*
12.6	Layered $LiNiO_2$ *350*
12.7	Layered $LiMnO_2$ *352*
12.8	$Li[Li_{1/3}Mn_{2/3}]O_2$ - $LiMO_2$ Solid Solutions *352*
12.9	Other Layered Oxides *354*
12.10	Spinel Oxide Cathodes *355*
12.11	Spinel $LiMn_2O_4$ *355*
12.12	5 V Spinel Oxides *359*
12.13	Other Spinel Oxides *361*
12.14	Polyanion-containing Cathodes *362*
12.15	Phospho-Olivine $LiMPO_4$ *363*
12.16	Silicate Li_2MSiO_4 *369*
12.17	Other Polyanion-containing Cathodes *370*
12.18	Summary *370*
	Acknowledgments *371*
	References *371*

13   Rechargeable Lithium Anodes *377*
Jun-ichi Yamaki and Shin-ichi Tobishima

13.1	Introduction *377*
13.2	Surface of Uncycled Lithium Foil *379*
13.3	Surface of Lithium Coupled with Electrolytes *380*
13.4	Cycling Efficiency of Lithium Anode *381*
13.4.1	Measurement Methods *381*
13.4.2	Reasons for the Decrease in Lithium Cycling Efficiency *382*
13.5	Morphology of Deposited Lithium *382*

13.6	The Amount of Dead Lithium and Cell Performance	385
13.7	Improvement in the Cycling Efficiency of a Lithium Anode	385
13.7.1	Electrolytes	386
13.7.2	Electrolyte Additives	387
13.7.2.1	Stable Additives Limiting Chemical Reaction between the Electrolyte and Lithium	387
13.7.2.2	Additives Modifying the State of Solvation of Lithium Ions	389
13.7.2.3	Reactive Additives Used to Make a Better Protective Film	391
13.7.3	Stack Pressure on Electrodes	396
13.7.4	Composite Lithium Anode	396
13.7.5	Influence of Cathode on Lithium Surface Film	397
13.7.6	An Alternative to the Lithium-Metal Anode (Lithium-Ion Inserted Anodes)	397
13.8	Safety of Rechargeable Lithium Metal Cells	398
13.8.1	Considerations Regarding Cell Safety	399
13.8.2	Safety Test Results	400
13.8.2.1	External Short	400
13.8.2.2	Overcharge	400
13.8.2.3	Nail Penetration	400
13.8.2.4	Crush	400
13.8.2.5	Heating	400
13.9	Conclusion	400
	References	401
	Further Reading	404
14	**Lithium Alloy Anodes**	**405**
	Robert A. Huggins	
14.1	Introduction	405
14.2	Problems with the Rechargeability of Elemental Electrodes	406
14.3	Lithium Alloys as an Alternative	407
14.4	Alloys Formed *In situ* from Convertible Oxides	409
14.5	Thermodynamic Basis for Electrode Potentials and Capacities under Conditions in which Complete Equilibrium can be Assumed	409
14.6	Crystallographic Aspects and the Possibility of Selective Equilibrium	412
14.7	Kinetic Aspects	413
14.8	Examples of Lithium Alloy Systems	414
14.8.1	Lithium–Aluminum System	414
14.8.2	Lithium–Silicon System	415
14.8.3	Lithium–Tin System	417
14.9	Lithium Alloys at Lower Temperatures	419
14.10	The Mixed-Conductor Matrix Concept	423
14.11	Solid Electrolyte Matrix Electrode Structures	427

14.12	What about the Future?	429
	References	429
	Further Reading	431

15 Lithiated Carbons 433
Martin Winter and Jürgen Otto Besenhard

15.1	Introduction	433
15.1.1	Why Lithiated Carbons?	436
15.1.2	Electrochemical Formation of Lithiated Carbons	437
15.2	Graphitic and Nongraphitic Carbons	437
15.2.1	Carbons: Classification, Synthesis, and Structures	438
15.2.2	Lithiated Graphitic Carbons (Li_xC_n)	441
15.2.2.1	In-Plane Structures	441
15.2.2.2	Stage Formation	442
15.2.2.3	Reversible and Irreversible Specific Charge	444
15.2.3	Li_xC_6 vs $Li_x(solv)_yC_n$	447
15.2.4	Lithiated Nongraphitic Carbons	452
15.2.5	Lithiated Carbons Containing Heteroatoms	461
15.2.6	Lithiated Fullerenes	462
15.3	Lithiated Carbons vs Competing Anode Materials	462
15.4	Summary	466
	Acknowledgments	466
	References	466
	Further Reading	478

16 The Anode/Electrolyte Interface 479
Emanuel Peled, Diane Golodnitsky, and Jack Penciner

16.1	Introduction	479
16.2	SEI Formation, Chemical Composition, and Morphology	480
16.2.1	SEI Formation Processes	480
16.2.2	Chemical Composition and Morphology of the SEI	483
16.2.2.1	Ether-Based Liquid Electrolytes	483
16.2.2.1.1	Fresh Lithium Surface	483
16.2.2.1.2	Lithium Covered by Native Film	484
16.2.2.2	Carbonate-Based Liquid Electrolyte	485
16.2.2.2.1	Fresh Lithium Surface	485
16.2.2.2.2	Lithium Covered by Native Film	485
16.2.2.3	Polymer (PE), Composite Polymer (CPE), and Gelled Electrolytes	486
16.2.3	Reactivity of e^-_{sol} with Electrolyte Components – a Tool for the Selection of Electrolyte Materials	487
16.3	SEI Formation on Carbonaceous Electrodes	490
16.3.1	Surface Structure and Chemistry of Carbon and Graphite	490
16.3.2	The First Intercalation Step in Carbonaceous Anodes	493
16.3.3	Parameters Affecting Q_{IR}	499

16.3.4	Graphite Modification by Mild Oxidation and Chemically Bonded (CB) SEI 500	
16.3.5	Chemical Composition and Morphology of the SEI 503	
16.3.5.1	Carbons and Graphites 503	
16.3.5.2	HOPG 505	
16.3.6	SEI Formation on Alloys 508	
16.4	Models for SEI Electrodes 508	
16.4.1	Liquid Electrolytes 508	
16.4.2	Polymer Electrolytes 511	
16.4.3	Effect of Electrolyte Composition on SEI Properties 513	
16.4.3.1	Lithium Electrode 513	
16.4.3.2	Li_xC_6 Electrode 517	
16.5	Summary and Conclusions 518	
	References 519	
	Further Reading 523	

17 Liquid Nonaqueous Electrolytes 525
Heiner Jakob Gores, Josef Barthel, Sandra Zugmann, Dominik Moosbauer, Marius Amereller, Robert Hartl, and Alexander Maurer

17.1	Introduction 525	
17.2	Components of the Liquid Electrolyte 526	
17.2.1	The Solvents 526	
17.2.2	The Salts 530	
17.2.2.1	Lithium Perchlorate 530	
17.2.2.2	Lithium Hexafluoroarsenate 531	
17.2.2.3	Lithium Hexafluorophosphate 531	
17.2.2.4	Lithium Tetrafluoroborate 532	
17.2.2.5	Lithium Fluoroalkylphosphates 532	
17.2.2.6	Lithium Bis(oxalato)borate 532	
17.2.2.7	Lithium Difluoro(oxalato)borate and Lithium bis(fluorosulfonyl)imide 533	
17.2.3	Ionic Liquids 537	
17.2.3.1	Physical Chemical Properties 538	
17.2.3.1.1	Viscosity 538	
17.2.3.1.2	Conductivity 538	
17.2.3.1.3	Diffusion Coefficient 539	
17.2.3.1.4	Electrochemical Stability 539	
17.2.3.1.5	Thermal Stability 539	
17.2.3.2	Crystallization and Melting Points 539	
17.2.3.3	Applications of ILs in Lithium-Ion Batteries 539	
17.2.4	Purification of Electrolytes 548	
17.2.5	Hydrolysis of Salts 549	
17.3	Intrinsic Properties 550	
17.3.1	Chemical Models of Electrolytes 551	

17.3.2	Ion-Pair Association Constants	551
17.3.3	Triple-Ion Association Constants	554
17.3.3.1	Bilateral Triple-Ion Formation	554
17.3.3.2	Unilateral Triple Ion Formation	555
17.3.3.3	Selective Solvation of Ions and Competition between Solvation and Ion Association	558
17.4	Bulk Properties	560
17.4.1	Electrochemical Stability Range	560
17.4.2	Computational Determination of Electrochemical Stability	565
17.4.3	Passivation and Corrosion Abilities of Lithium Salt Electrolytes	569
17.4.4	Chemical Stability of Electrolytes with Lithium and Lithiated Carbon	573
17.4.5	Conductivity of Concentrated Solutions	579
17.4.5.1	Introduction	579
17.4.5.2	Conductivity-Determining Parameters	586
17.4.5.3	The Walden Rule and the Haven Ratio	588
17.4.5.4	The Role of Solvent Viscosity, Ionic Radii, and Solvation	589
17.4.5.5	The Role of Ion Association	591
17.4.5.6	Application of the Effects of Selective Solvation and Competition between Solvation and Ion Association	592
17.4.5.7	Conductivity Optimization of Electrolytes by Use of Simplex Algorithm	595
17.4.6	Transference Numbers	598
17.4.6.1	Introduction	598
17.4.6.2	Hittorf Method	600
17.4.6.3	Electromotive Force (emf) Method	601
17.4.6.4	Potentiostatic Polarization Method	602
17.4.6.5	Conductivity Measurement	603
17.4.6.6	Galvanostatic Polarization Method	603
17.4.6.7	Transference numbers from NMR-diffusion coefficients	605
17.4.6.8	Impedance Measurements	605
17.4.7	Diffusion Coefficients in Liquids	606
17.4.7.1	Introduction	606
17.4.7.2	Pfg-NMR	607
17.4.7.3	Moiré Pattern	608
17.4.7.4	Microelectrodes	608
17.4.7.5	Restricted Diffusion	608
17.5	Additives	609
	References	611
18	**Polymer Electrolytes**	**627**
	Fiona Gray and Michel Armand	
18.1	Introduction	627
18.2	Solvent-Free Polymer Electrolytes	629
18.2.1	Technology	629

18.2.2	The Fundamentals of a Polymer Electrolyte	630
18.2.3	Conductivity, Structure, and Morphology	632
18.2.4	Second-Generation Polymer Electrolytes	632
18.2.5	Structure and Ionic Motion	635
18.2.6	Mechanisms of Ionic Motion	637
18.2.7	An Analysis of Ionic Species	639
18.2.8	Cation-Transport Properties	639
18.3	Hybrid Electrolytes	643
18.3.1	Gel Electrolytes	644
18.3.2	Batteries	647
18.3.3	Enhancing Cation Mobility	649
18.3.4	Mixed-Phase Electrolytes	650
18.4	Looking to the Future	652
	References	652
	Further Reading	656
19	**Solid Electrolytes**	**657**
	Peter Birke and Werner Weppner	
19.1	Introduction	657
19.2	Fundamental Aspects of Solid Electrolytes	658
19.2.1	Structural Defects	658
19.2.2	Migration and Diffusion of Charge Carriers in Solids	666
19.3	Applicable Solid Electrolytes for Batteries	668
19.3.1	General Aspects	668
19.3.2	Lithium-, Sodium-, and Potassium-Ion Conductors	669
19.3.3	Capacity and Energy Density Aspects	671
19.4	Design Aspects of Solid Electrolytes	674
19.5	Preparation of Solid Electrolytes	676
19.5.1	Monolithic Samples	676
19.5.1.1	Solid-State Reactions	676
19.5.1.2	The Pechini Method	677
19.5.1.3	Wet Chemical Methods	677
19.5.1.4	Combustion Synthesis and Explosion Methods	678
19.5.1.5	Composites	678
19.5.1.6	Sintering Processes	679
19.5.2	Thick-Film Solid Electrolytes	679
19.5.2.1	Screen Printing	679
19.5.2.2	Tape Casting	679
19.5.3	Thin-Film Solid Electrolytes	680
19.5.3.1	Sputtering	680
19.5.3.2	Evaporation	680
19.5.3.3	Spin-On Coating and Spray Pyrolysis	681
19.6	Experimental Techniques for the Determination of the Properties of Solid Electrolytes	681
19.6.1	Partial Ionic Conductivity	681

19.6.1.1	Direct-Current (DC) Measurements	681
19.6.1.2	Impedance Analysis	682
19.6.1.3	Determination of the Activation Energy	683
19.6.2	Partial Electronic Conductivity	683
19.6.2.1	Determination of the Transference Number	685
19.6.2.2	The Hebb–Wagner Method	685
19.6.2.3	Mobility of Electrons and Holes	686
19.6.2.4	Concentration of Electrons and Holes	686
19.6.3	Stability Window	688
19.6.4	Determination of the Ionics Conduction Mechanism and Related Types of Defects	689
	Acknowledgment	690
	References	690
	Further Reading	691

20 Separators for Lithium-Ion Batteries 693
Robert Spotnitz

20.1	Introduction	693
20.2	Market	694
20.3	How a Battery Separator Is Used in Cell Fabrication	697
20.4	Microporous Separator Materials	700
20.5	Gel Electrolyte Separators	707
20.6	Polymer Electrolytes	708
20.7	Characterization of Separators	708
20.8	Mathematical Modeling of Separators	712
20.9	Conclusions	714
	References	714

21 Materials for High-Temperature Batteries 719
H. Böhm

21.1	Introduction	719
21.2	The ZEBRA System	720
21.2.1	The ZEBRA Cell	720
21.2.2	Properties of ZEBRA Cells	721
21.2.3	Internal Resistance of ZEBRA Cells	723
21.2.4	The ZEBRA Battery	726
21.3	The Sodium/Sulfur Battery	728
21.3.1	The Na–S System	728
21.3.2	The Na/S Cell	729
21.3.3	The Na/S Battery	731
21.3.4	Corrosion-Resistant Materials for Sodium/Sulfur Cells	733
21.3.4.1	Glass Seal	733
21.3.4.2	Cathode and Anode Seal	733
21.3.4.3	Current Collector for the Sulfur Electrode	734
21.4	Components for High-Temperature Batteries	735

21.4.1	The Ceramic Electrolyte β''-Alumina	735
21.4.1.1	Doping of β''-Al$_2$O$_3$	735
21.4.1.2	Manufacture of β''-Alumina Electrolyte Tubes	736
21.4.1.3	Properties of β''-Alumina Tubes	740
21.4.1.4	Stability of β-Alumina and β''-Alumina	742
21.4.2	The Second Electrolyte NaAlCl$_4$ and the NaCl–AlCl$_3$ System	742
21.4.2.1	Phase Diagram	742
21.4.2.2	Vapor Pressure	743
21.4.2.3	Density	744
21.4.2.4	Viscosity	744
21.4.2.5	Dissociation	745
21.4.2.6	Ionic Conductivity	746
21.4.2.7	Solubility of Nickel Chloride in Sodium Aluminum Chloride	746
21.4.3	Nickel Chloride NiCl$_2$ and the NiCl$_2$–NaCl System	748
21.4.3.1	Relevant Properties of NiCl$_2$	748
21.4.3.2	NiCl$_2$–NaCl System	748
21.4.4	Materials for Thermal Insulation	749
21.4.4.1	Multifoil Insulation	750
21.4.4.2	Glass Fiber Boards	750
21.4.4.3	Microporous Insulation	751
21.4.4.4	Comparison of Thermal Insulation Materials	751
21.4.5	Data for Cell Materials	754
21.4.5.1	Nickel	754
21.4.5.2	Liquid Sodium	754
21.4.5.3	NaCl	754
21.4.5.4	Sulfur and Sodium Polysulfides	754
	References	755
	Further Reading	756

Part IV	**New Emerging Technologies**	**757**
22	**Metal–Air Batteries**	**759**
	Ji-Guang Zhang, Peter G. Bruce, and X. Gregory Zhang	
22.1	General Characteristics	759
22.1.1	The Pros	760
22.1.1.1	High Specific Capacity and Energy	760
22.1.1.2	Low-Cost Cathode	763
22.1.2	The Cons	763
22.1.2.1	Power Limitations	763
22.1.2.2	Electrolyte Evaporation and Flooding	763
22.1.2.3	Side Reactions	763
22.1.2.4	Solid Discharge Products	764
22.2	Air Electrode	764
22.2.1	Catalyst	765
22.2.2	Carbon Sources	766

22.3	Zinc–Air Batteries 767
22.3.1	Primary Zinc–Air Batteries 768
22.3.2	Rechargeable Zinc–Air Batteries 770
22.3.2.1	Electrically Rechargeable Zinc–Air Batteries 770
22.3.2.2	Mechanically rechargeable Zinc–Air Batteries 772
22.3.3	Hydraulically Rechargeable Zinc–Air Batteries 772
22.4	Lithium–Air Batteries 773
22.4.1	Lithium–Air Batteries Using a Nonaqueous Electrolyte 775
22.4.2	Lithium–Air Batteries Using Protected Lithium Electrodes 781
22.4.3	Lithium–Air Batteries Using an Ionic Liquid Electrolyte 783
22.4.4	Lithium–Air Batteries Using Solid Electrolytes 784
22.4.5	Rechargeable Lithium–Air Batteries 785
22.5	Other–Air Batteries 789
22.6	Conclusions 792
	Acknowledgment 792
	References 792
23	**Catalysts and Membranes for New Batteries** 797
	Chaitanya K. Narula
23.1	Introduction 797
23.2	Catalysts 798
23.2.1	Catalysts in Metal–Air Batteries 798
23.2.2	Catalysts in Lithium–Thionyl Chloride Batteries 800
23.2.3	Catalysts in Other Batteries 800
23.3	Separators 802
23.3.1	Separator Types 802
23.3.2	Separators for Batteries Based on Nonaqueous Electrolytes 803
23.3.2.1	Primary Batteries Based on Lithium 803
23.3.2.2	Secondary Batteries Based on Lithium 804
23.3.2.2.1	The Lithium-Ion Battery 804
23.3.2.2.2	Lithium Polymer Battery 805
23.3.2.2.3	Lithium-Ion Gel Polymer Battery 805
23.3.3	Separators for Batteries Based on Aqueous Electrolytes 806
23.4	Future Directions 807
	References 808
24	**Lithium–Sulfur Batteries** 811
	Zengcai Liu, Wujun Fu, and Chengdu Liang
24.1	Introduction 811
24.2	Polysulfide Shuttle and Capacity-Fading Mechanisms 812
24.2.1	Origin of Polysulfide Shuttle 813
24.2.2	Influence of Polysulfide Shuttle on Charge Profile 814
24.2.3	Effect of Polylsulfide Shuttle on Charge–Discharge Capacities 815
24.2.4	Capacit-Fading Mechanism 816

24.3	Cell Configuration	816
24.4	Positive Electrode Materials	818
24.4.1	Carbon–Sulfur Composites	818
24.4.2	Polymer–Sulfur Composites	827
24.4.3	Organosulfides	828
24.4.4	Other Positive Electrode Materials	828
24.4.5	Binders	829
24.5	Electrolytes	830
24.5.1	Liquid Electrolytes	830
24.5.2	Solid Electrolytes	832
24.6	Negative Electrode Materials	833
24.7	Conclusions and Prospects	835
	Acknowledgments	835
	References	836

Part V Performance and Technology Development for Batteries *841*

25 Modeling and Simulation of Battery Systems *843*
Partha P. Mukherjee, Sreekanth Pannala, and John A. Turner

25.1	Introduction	843
25.2	Macroscopic Model	848
25.2.1	Direct Numerical Simulation	850
25.2.1.1	Conservation of Charge	850
25.2.1.2	Conservation of Species	850
25.2.1.3	Energy Balance	851
25.2.1.4	Electrode Kinetics	852
25.2.2	Volume-Averaged Formulation	853
25.2.2.1	Species Conservation	854
25.2.2.2	Charge Conservation	855
25.2.2.3	Thermal Energy Conservation	855
25.2.2.4	Electrode Kinetics	856
25.2.2.5	Boundary and Initial Conditions	856
25.2.3	Pseudo-2D Model	857
25.2.4	Single Particle Model	858
25.2.5	Simplifications to Solid Phase Diffusion	858
25.3	Aging Model	859
25.4	Stress Model	861
25.5	Abuse Model	863
25.6	Life Prediction Model	865
25.7	Other Battery Technologies	866
25.7.1	Li–Air Battery	866
25.7.2	Redox Flow Battery	868
25.8	Summary and Outlook	869
	Acknowledgments	870

Nomenclature 870
References 871

26 Mechanics of Battery Cells and Materials 877
Xiangchun Zhang, Myoungdo Chung, HyonCheol Kim, Chia-Wei Wang, and Ann Marie Sastry

26.1 Mechanical Failure Analysis of Battery Cells and Materials: Significance and Challenges 877
26.1.1 Introduction 877
26.1.2 Complications Associated with Analysis 878
26.1.2.1 Stochastic Microstructure of Electrode Materials 878
26.1.2.2 Multiple Physicochemical Processes 879
26.1.2.3 Multiple Length Scales and Time Scales 880
26.1.2.4 Mechanical Loads on Battery Materials 881
26.2 Key Studies in the Mechanical Analysis of Battery Materials 883
26.2.1 Identified Stresses in Battery Materials 883
26.2.1.1 Compaction/Residual Stresses due to the Manufacturing Process 883
26.2.1.2 Intercalation-Induced Stress 884
26.2.1.3 Thermal Stress 885
26.2.1.4 Compaction due to Packaging Constraints 885
26.2.2 Modeling and Experimental Analysis of Single Electrode Particles 886
26.2.2.1 Modeling of Intercalation-Induced Stress 886
26.2.2.2 Experimental Studies of Electrode Particles 889
26.2.3 Mechanical Stress and Electrochemical Cycling Coupling in Carbon Fiber Electrodes 892
26.2.4 Battery Cell Modeling with Stress 893
26.2.5 Comparison of the Magnitude of Various Stresses 896
26.3 Key Issues Remaining to be Addressed 897
26.3.1 Modeling and Simulations 897
26.3.1.1 From Stress Calculation to Fracture Analysis 897
26.3.1.2 How to Account for Real Stochastic Geometry 897
26.3.1.3 Reduced-Order Models 898
26.3.2 Experimental Approaches 898
26.3.3 Remediation of Stresses 899
26.4 Outlook for the Future 900
References 901

27 Battery Safety and Abuse Tolerance 905
Daniel H. Doughty

27.1 Introduction 905
27.2 Evaluation Techniques for Batteries and Battery Materials 907
27.2.1 Electrochemical Characterization 907
27.2.2 Thermal Characterization 907

27.2.2.1	Differential Scanning Calorimetry (DSC)	907
27.2.2.2	Accelerating-Rate Calorimeter (ARC)	908
27.2.2.3	Thermal Ramp Test	908
27.2.2.4	Large-Scale Calorimetry	908
27.2.3	Standardized Safety and Abuse Tolerance Test Procedures	909
27.3	Typical Failure Modes	910
27.3.1	Gas Generation	915
27.3.2	Physical Damage	916
27.3.3	Charge and Discharge Failures	916
27.3.4	Short Circuit	917
27.4	Safety Devices	918
27.5	Discussion of Safety and Abuse Response for Battery Chemistries	919
27.5.1	Zinc–Carbon Batteries (Leclanché Cells)	919
27.5.2	Alkaline Batteries	920
27.5.3	Lead–Acid Batteries	921
27.5.4	Nickel–Cadmium Batteries	922
27.5.5	Nickel–Metal Hydride Batteries	923
27.5.6	Lithium Primary Batteries	924
27.5.7	Lithium–Manganese Dioxide (Li–MnO_2)	924
27.5.8	Lithium–Carbon Monofluoride (Li–$(CF)_x$)	925
27.5.9	Lithium–Thionyl Chloride (Li–$SOCl_2$)	925
27.5.10	Lithium–Sulfur Dioxide (Li–SO_2)	926
27.5.11	Lithium Secondary Batteries	927
27.5.12	Lithium Metal Secondary Batteries	928
27.5.13	Lithium-Ion Batteries	929
27.5.14	Separators in Lithium-Ion Batteries	932
27.5.15	Electrolytes in Lithium-Ion Batteries	933
27.5.16	Lithium Polymer Batteries	934
27.5.17	Summary	935
	Acknowledgments	935
	References	936
28	**Cathode Manufacturing for Lithium-Ion Batteries**	**939**
	Jianlin Li, Claus Daniel, and David L. Wood III	
28.1	Introduction	939
28.2	Electrode Manufacturing	940
28.2.1	Slurry Processing	940
28.2.1.1	Casting	941
28.2.1.1.1	Tape Casting	941
28.2.1.1.2	Slot-Die Coating	942
28.2.1.2	Printing	943
28.2.1.2.1	Screen Printing	944
28.2.1.2.2	Ink-jet Printing	944
28.2.1.3	Spin Coating	945

28.2.2	Vacuum Techniques 947	
28.2.2.1	Chemical Vapor Deposition 947	
28.2.2.2	Electrostatic Spray Deposition 949	
28.2.2.3	Pulsed Laser Deposition 952	
28.2.2.4	Radio Frequency (RF) Sputtering 953	
28.2.3	Other Processing – Molten Carbonate Method 955	
28.3	Summary 955	
	References 957	

Index 961

Preface to the Second Edition of the Handbook of Battery Materials

For Kijan and Stina

> The language of experiment is more authoritative than any reasoning, facts can destroy our ratiocination – not vice versa.
>
> Count Alessandro Volta, 1745–1827
> Inventor of the Battery

You are looking at the second edition of the Handbook of Battery Materials. It has been 12 years since the first edition edited by Prof. Jürgen Besenhard was published.

This second edition is dedicated in memory of world renowned Prof. Jürgen Besenhard who was a pioneer in the field of electrochemical energy storage and lithium batteries. As a young scientist in the field of electrochemical energy storage, I am humbled to inherit this handbook from him.

Over the last decade, driven by consumer electronics, power tools, and recently automotive and renewable energy storage, electrochemical energy storage chemistries and devices have been developed at a never before seen pace. New chemistries have been discovered, and continued performance increases to established chemistries are under way. With these developments, we decided to update the handbook from 1999. The new edition is completely revised and expanded to almost double its original content.

Due to the fast pace of the market and very quick developments on large scale energy storage, we removed the chapter on *Global Competition*. It might be outdated by the time this book actually hits the shelves. Chapters from Parts I and II from the first edition on *Fundamentals and General Aspects of Electrochemical Energy Storage, Practical Batteries*, and *Materials for Aqueous Electrolyte Batteries* have been revised for the new edition to reflect the work in the past decade. Part III on *Materials for Alkali Metal Batteries* has been expanded in view of the many research efforts on lithium ion and other alkali metal ion batteries. In addition, we added new Parts IV on *New Emerging Technologies* and V on *Performance and Technology Development* with chapters on *Metal Air, Catalysts, and Membranes, Sulfur, System Level Modeling, Mechanics of Battery Materials*, and *Electrode Manufacturing*.

In our effort, we strongly held on to Prof. Besenhard's goal to *"fill the gap"* between fundamental electrochemistry and application of batteries in order to provide a *"comprehensive source of detailed information"* for *"graduate or higher level"* students and *"those who are doing research in the field of materials for energy storage."*

I would like to thank all authors who contributed to this book; Craig Blue, Ray Boeman, and David Howell who made me apply my experience and knowledge from a different area to the field of electrochemical energy storage; and Nancy Dudney who continues to be a resourceful expert advisor to me.

Finally, I thank my wife Isabell and my family for the many sacrifices they make and support they give me in my daily work.

Oak Ridge, TN, July 2011   *Claus Daniel*

List of Contributors

Jörg H. Albering
Graz University of Technology
Institute for Chemical
Technology of Inorganic
Materials
Stremeyrgasse 16/III
8010 Graz
Austria

Marius Amereller
University of Regensburg
Institute of Physical and
Theoretical Chemistry
Universitätsstr. 31
93053 Regensburg
Germany

Michel Armand
Université de Montréal
Département de Chimie
C.P. 6128
Succursale Centre-Ville
Montréal
Québec H3C 3J7
Canada

Josef Barthel
University of Regensburg
Institute of Physical and
Theoretical Chemistry
Universitätsstr. 31
93053 Regensburg
Germany

Dietrich Berndt
Am Weissen Berg 3
61476 Kronberg
Germany

Jürgen Otto Besenhard[†]
Graz University of Technology of
Inorganic Materials
Stremayrgasse 16/III
8010 Graz
Austria

Leo Binder
Graz University of Technology
Institute for Inorganic Chemistry
Stremayrgasse 9
8010 Graz
Austria

Peter Birke
Christian-Albrechts University
Technical Faculty
Chair for Sensors and
Solid State Ionics
Kaiserstr. 2
24143 Kiel
Germany

H. Böhm
AEG Anglo Batteries GmbH
Söfliger Str. 100
889077 Ulm
Germany

Werner Böhnstedt
Daramic Inc.
Erlengang 31
22844 Norderstedt
Germany

Peter G. Bruce
University of St. Andrews
School of Chemistry
North Haugh
St. Andrews, KY16 9ST
Scotland

Myoungdo Chung
Sakti3
Incorporated
Ann Arbor, MI
USA

Claus Daniel
Oak Ridge National Laboratory
Oak Ridge, TN 37831-6083
USA

and

University of Tennessee
Department of Materials Science
and Engineering
Knoxville, TN 37996
USA

Daniel H. Doughty
Battery Safety Consulting Inc.
139 Big Horn Ridge Dr. NE
Albuquerque, NM 87122
USA

Josef Drobits
Technische Universität Wien
Institut für Technische
Elektrochemie
Getreidemarkt 9/158
1060 Wien
Austria

Christoph Fabjan
Technische Universität Wien
Institut für Technische
Elektrochemie
Getreidemarkt 9/158
1060 Wien
Austria

Wujun Fu
Center for Nanophase Materials
Sciences
Oak Ridge National Laboratory
Oak Ridge
TN 37831
USA

Nobuhiro Furukawa
Sanyo Electric Co., Ltd.
Electrochemistry Department
New Materials Research Center
1-18-3 Hashiridani
Hirahata City
Osaka 573-8534
Japan

Diane Golodnitsky
Tel Aviv University
Department of Chemistry
Tel Aviv 69978
Israel

Heiner Jakob Gores
University of Regensburg
Institute of Physical and
Theoretical Chemistry
Universitätsstr. 31
93053 Regensburg
Germany

and

WWU Münster (Westfälische
Wilhelms-Universität Münster)
MEET - Münster Electrochemical
Energy Technology
Corrensstraße 46
48149 Münster
Germany

Fiona Gray
University of St Andrews
School of Chemistry
The Purdie Building
North Haugh
St Andrews
Fife KY16 9ST
UK

Robert Hartl
University of Regensburg
Institute of Physical and
Theoretical Chemistry
Universitätsstr. 31
93053 Regensburg
Germany

Robert A. Huggins
Christian-Albrechts-University
Kaiserstr. 2
24143 Kiel
Germany

HyonCheol Kim
Sakti3
1490 Eisenhower Place
Building 4
Ann Arbor, MI 48108

Kimio Kinoshita
Lawrence Berkeley Laboratory
Environmental Energy
Technology
Berkeley, CA 94720
USA

Akiya Kozawa
ITE Battery Research Institute
39 Youke, Ukino
Chiaki-cho
Ichinomiyashi
Aichi-ken 491
Japan

Jianlin Li
Oak Ridge National Laboratory
Materials Science and Technology
Division
Oak Ridge, TN 37831-6083
USA

Chengdu Liang
Center for Nanophase Materials
Sciences
Oak Ridge National Laboratory
Oak Ridge, TN 37831
USA

Zengcai Liu
Center for Nanophase Materials
Sciences
Oak Ridge National Laboratory
Oak Ridge, TN 37831
USA

Arumugam Manthiram
The University of Texas at Austin
Materials Science and
Engineering Program
Austin, TX 78712
USA

Alexander Maurer
University of Regensburg
Institute of Physical and
Theoretical Chemistry
Universitätsstr. 31
93053 Regensburg
Germany

James McBreen
Brookhaven National Laboratory
Department of Applied Science
Upton, NY 1973
USA

Dominik Moosbauer
University of Regensburg
Institute of Physical and
Theoretical Chemistry
Universitätsstr. 31
93053 Regensburg
Germany

and

Gamry Instruments, Inc.
734 Louis Drive
Warminster, PA 18974
USA

Partha P. Mukherjee
Oak Ridge National Laboratory
Computer Science and
Mathematics Division
Oak Ridge, TN 37831
USA

Theivanayagam Muraliganth
The University of Texas at Austin
Materials Science and
Engineering Program
Austin, TX 78712
USA

Chaitanya K. Narula
University of Tennessee
Materials Science and Technology
Division
Physical Chemistry of Materials
Oak Ridge, TN 37831
USA

and

Department of Materials Science
and Engineering
Knoxville, TN 37966
USA

Koji Nishio
Sanyo Electric Co., Ltd.
Electrochemistry Department
New Materials Research Center
1-18-3 Hashiridani
Hirahata City
Osaka 573-8534
Japan

Sreekanth Pannala
Oak Ridge National Laboratory
Computer Science and
Mathematics Division
Oak Ridge, TN 37831
USA

Emanuel Peled
Tel Aviv University
Department of Chemistry
Tel Aviv 69978
Israel

Jack Penciner
Tel Aviv University
Department of Chemistry
Tel Aviv 69978
Israel

Karsten Pinkwart
Fraunhofer Institut für
Chemische Technologie (ICT)
Angewandte Elektrochemie
Josef-von-Fraunhofer-Str. 7
76327 Pfinztal
Germany

James J. Reilly
Brookhaven National Laboratory
Department of Sustainable
Energy Technologies
Upton, NY 11973
USA

Ann Marie Sastry
Sakti3
1490 Eisenhower Place
Building 4
Ann Arbor, MI 48108

and

University of Michigan
Departments of Mechanical
Biomedical and Materials
Science and Engineering
2300 Hayward Street
Ann Arbor, MI 48109
USA

Robert Spotnitz
Battery Design LLC
2277 Delucchi Drive
Pleasanton, CA 94588
USA

Shin-ichi Tobishima
Gunma University
Department of Chemistry and
Biochemistry
Faculty of Engineering
1-5-1 Tenjin-cho
Kiryu, Gunma, 376-8515
Japan

Jens Tübke
Fraunhofer Institut für
Chemische Technologie
Angewandte Elektrochemie
Josef-von-Fraunhofer-Str. 7
76327 Pfinztal
Germany

John A. Turner
Oak Ridge National Laboratory
Computer Science and
Mathematics Division
Oak Ridge, TN 37831
USA

Chia-Wei Wang
Sakti3
1490 Eisenhower Place
Building 4
Ann Arbor, MI 48108

Werner Weppner
Christian-Albrechts University
Technical Faculty
Chair for Sensors and
Solid State Ionics
Kaiserstr. 2
24143 Kiel
Germany

Martin Winter
Graz University of Technology of
Inorganic Materials
Stremayrgasse 16/III
8010 Graz
Austria

David L. Wood III
Oak Ridge National Laboratory
Materials Science and
Technology Division
Oak Ridge, TN 37831-6083
USA

Jun-ichi Yamaki
Kyushu University
Institute for Materials Chemistry
and Engineering
6-1 Kasuga-koen
Kasuka-shi 816-8508
Japan

Kohei Yamamoto
Fuji Electrochemical Co.
Washizu, Kosai-shi
Shizuoka-ken 431
Japan

Masaki Yoshio
Saga University
Faculty of Science and
Engineering
Department of Applied Chemistry
1 Honjo
Saga 8408502
Japan

Ji-Guang Zhang
Pacific Northwest
National Laboratory
Energy and Environment
Directory
Richland, WA 99352
USA

X. Gregory Zhang
Independent consultant
3 Weatherell Street
Ontario M6S 1S6
Canada

Xiangchun Zhang
Sakti3
Incorporated
Ann Arbor, MI
USA

Sandra Zugmann
University of Regensburg
Institute of Physical and
Theoretical Chemistry
Universitätsstr. 31
93053 Regensburg
Germany

Part I
Fundamentals and General Aspects of Electrochemical Energy Storage

1
Thermodynamics and Mechanistics

Karsten Pinkwart and Jens Tübke

1.1
Electrochemical Power Sources

Electrochemical power sources convert chemical energy into electrical energy (see Figure 1.1). At least two reaction partners undergo a chemical process during this operation. The energy of this reaction is available as electric current at a defined voltage and time [1].

Electrochemical power sources differ from others such as thermal power plants in the fact that the energy conversion occurs without any intermediate steps; for example, in the case of thermal power plants, fuel is first converted into thermal energy (in furnaces or combustion chambers), then into mechanical energy, and finally into electric power by means of generators. In the case of electrochemical power sources, this multistep process is replaced by one step only. As a consequence, electrochemical systems show some advantages such as high energy efficiency.

The existing types of electrochemical storage systems vary according to the nature of the chemical reaction, structural features, and design. This reflects the large number of possible applications.

The simplest system consists of one electrochemical cell – the so-called galvanic element [1]. This supplies a comparatively low cell voltage of 0.5–5 V. To obtain a higher voltage the cell can be connected in series with others, and for a higher capacity it is necessary to link them in parallel. In both cases the resulting ensemble is called a *battery*.

Depending on the principle of operation, cells are classified as follows:

1) *Primary cells* are nonrechargeable cells in which the electrochemical reaction is irreversible. They contain only a fixed amount of the reacting compounds and can be discharged only once. The reacting compounds are consumed by discharging, and the cell cannot be used again. A well-known example of a primary cell is the Daniell element (Figure 1.2), consisting of zinc and copper as the electrode materials.

Handbook of Battery Materials, Second Edition. Edited by Claus Daniel and Jürgen O. Besenhard.
© 2011 Wiley-VCH Verlag GmbH & Co. KGaA. Published 2011 by Wiley-VCH Verlag GmbH & Co. KGaA.

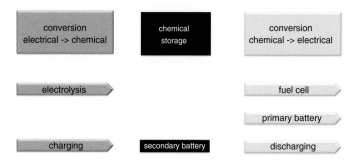

Figure 1.1 Chemical and electrical energy conversion and possibilities of storage.

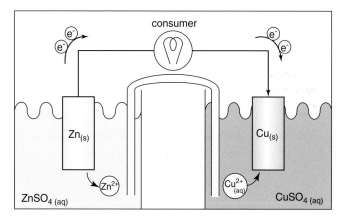

Figure 1.2 Daniell element.

2) *Secondary cells* are rechargeable several times [1]. Only reversible electrochemical reactions offer such a possibility. After the cell is discharged, an externally applied electrical energy forces a reversal of the electrochemical process; as a consequence the reactants are restored to their original form, and the stored electrochemical energy can be used once again by a consumer. The process can be reversed hundreds or even thousands of times, so that the lifetime of the cell can be extended. This is a fundamental advantage, especially as the cost of a secondary cell is normally much higher than that of a primary cell. Furthermore, the resulting environmental friendliness should be taken into account.

3) *Fuel cells* [2]: In contrast to the cells so far considered, fuel cells operate in a continuous process. The reactants – often hydrogen and oxygen – are fed continuously to the cell from outside. Fuel cells are not reversible systems.

Typical fields of application for electrochemical energy storage systems are in portable systems such as cellular phones, notebooks, cordless power tools,

Table 1.1 Comparison of cell parameters of different cells [4].

Cell reaction	Standard potential	Terminal voltage	Capacity (Ah kg^{-1})	Specific energy (Wh kg^{-1})
Zn + CuSO$_4$ → ZnSO$_4$+Cu	$\varepsilon_{00}(Zn/Zn^{2+}) = -0.76$ V $\varepsilon_{00}(Cu/Cu^{2+}) = 0.34$ V	$\Delta\varepsilon_{00} = 1.10$ V	238.2	262
Cd + 2NiOOH + 2H$_2$O → Cd(OH)$_2$ + 2Ni(OH)$_2$	$\varepsilon_{00}(Cd/Cd^{2+}) = -0.81$ V $\varepsilon_{00}(Ni^{2+}/Ni^{3+}) = 0.49$ V	$\Delta\varepsilon_{00} = 1.30$ V	161.5	210
Li + MnO$_2$ → LiMnO$_2$	$\varepsilon_{00}(Li/Li^{+}) = -3.04$ V $\varepsilon_{00}(Mn^{3+}/Mn^{4+}) = 0.16$ V	$\Delta\varepsilon_{00} = 3.20$ V	285.4	856.3

Table 1.2 Comparison of the Efficiencies.

System	Coulometric efficiency	Energy efficiency
Lead–acid accumulator	0.80	0.65–0.70
Nickel–cadmium accumulator	0.65–0.70	0.55–0.65
Nickel–metal hydride accumulator	0.65–0.70	0.55–0.65

Table 1.3 Comparison of Primary and Secondary Battery Systems.

System	Specific energy (theoretical) (Wh kg^{-1})	Specific energy (practical) (Wh kg^{-1})	Energy density (practical) (Wh L^{-1})
Alkaline (zinc)–manganese cell	336	50–80	120–150
Zinc–carbon	358	60–90	140–200
Lead–acid	170	35	90
Nickel–cadmium	209	50	90
Nickel–metal hydride	380	60	80
Lithium-ion–metal oxide	500–550	150	220

SLI (starter-light-ignition) batteries for cars, and electrically powered vehicles. There are also a growing number of stationary applications such as devices for emergency current and energy storage systems for renewable energy sources (wind, solar). Especially for portable applications the batteries should have a low weight and volume, a large storage capacity, and a high specific energy density. Most of the applications mentioned could be covered by primary batteries, but economical and ecological considerations lead to the use of secondary systems.

Apart from the improvement and scaling up of known systems such as the lead–acid battery, the nickel–cadmium, and the nickel–metal hydride batteries, new types of cells have been developed, such as the lithium-ion system. The latter seems to be the most promising system, as will be apparent from the following sections [3].

To judge which battery systems are likely to be suitable for a given potential application, a good understanding of the principles of functioning and of the various materials utilized is necessary (see Table 1.1).

The development of high-performance primary and secondary batteries for different applications has proved to be an extremely challenging task because of the need to simultaneously meet multiple battery performance requirements such as high energy (watt-hours per unit battery mass or volume), high power (watts per unit battery mass or volume), long life (5–10 years and some hundreds of charge-discharge cycles), low cost (measured per unit battery capacity), resistance to abuse and operating temperature extremes, near-perfect safety, and minimal environmental impact (see Table 1.2 and Table 1.3). Despite years of intensive worldwide R&D, no battery can meet all of these goals.

The following sections therefore present a short introduction to this topic and to the basic mechanisms of batteries [4]. Finally, a first overview of the important criteria used in comparing different systems is given.

1.2
Electrochemical Fundamentals

1.2.1
Electrochemical Cell

The characteristic feature of an electrochemical cell is that the electronic current, which is the movement of electrons in the external circuit, is generated by the electrochemical processes at the electrodes. In contrast to the electric current in the external system, the transportation of the charge between the positive and the negative electrode within the electrolyte is performed by ions. Generally the current in the electrolyte consists of the movement of negative and positive ions.

A simplified picture of the electrode processes is shown in Figure 1.3. Starting with an open circuit, metal A is dipped in the solution, whereupon it partly dissolves. Electrons remain at the electrode until a characteristic electron density is built up. For metal B, which is more noble than A (see Section 1.2.2), the same process takes place, but the amount of dissolution and therefore the resulting electron density is lower.

If these two electrodes are connected by an electrical conductor, an electron flow starts from the negative electrode with the higher electron density to the positive electrode. The system electrode/electrolyte tries to keep the electron density constant. As a consequence additional metal A dissolves at the negative electrode forming A^+ in solution and electrons e^-, which are located at the surface

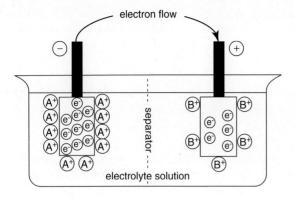

Figure 1.3 Electrochemical cell with negative and positive electrodes.

of metal A.

$$A \rightarrow A^+ + e^-$$

At the positive electrode the electronic current results in an increasing electron density. The system electrode B/electrolyte compensates this process by the consumption of electrons for the deposition of B^+-ions:

$$B^+ + e^- \rightarrow B$$

The electronic current stops if one of the following conditions is fulfilled:

- the base metal A is completely dissolved
- all B^+-ions are precipitated.

As a consequence, it is necessary to add a soluble salt to the positive electrode compartment to maintain the current for a longer period. This salt consists of B^+-ions and corresponding negative ions. The two electrode compartments are divided by an appropriate separator to avoid the migration and the deposition of B^+-ions at the negative electrode A. Since this separator blocks the exchange of positive ions, only the negative ions are responsible for the charge transport in the cell. This means that for each electron flowing in the outer circuit from the negative to the positive electrode, a negative ion in the electrolyte diffuses to the negative electrode compartment.

Generally, the limiting factor for the electronic current flow is the transport of these ions. Therefore the electrolyte solution should have a low resistance.

An electrolyte is characterized by its *specific resistance* $\rho(\Omega\,\text{cm})$, which is defined as the resistance of the solution between two electrodes each with an area of 1 cm^2 and at a distance apart of 1 cm. The reciprocal of this value is known as the specific conductivity κ ($\Omega^{-1}\,\text{cm}^{-1}$) [5]. For comparison, the values for different materials are given in Figure 1.4.

The conductivity of different electrolyte solutions varies widely. The selection of a suitable, highly conductive electrolyte solution for an electrochemical cell

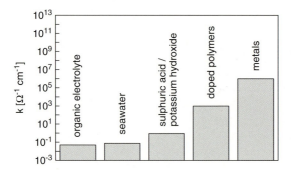

Figure 1.4 Comparison of the specific conductivity of different materials.

depends on its compatibility with the other components, particularly the positive and negative electrodes.

From the chemical viewpoint, a galvanic cell is a current source in which a local separation of oxidation and reduction process exists. In the following, this is explained using the example of the Daniell element.

Here the galvanic cell contains copper as the positive electrode, zinc as the negative electrode, and the sulfates of these metals as the electrolyte.

A salt bridge serves as an ion-conducting connection between the two half cells. On closing the external circuit, the oxidation reaction starts with the dissolution of the zinc electrode and the formation of zinc ions in half cell I. In half cell II copper ions are reduced and metallic copper is deposited. The sulfate ions remain unchanged in the solution. The overall cell reaction consists of an electron transfer between zinc and copper ions:

$$Zn \rightarrow Zn^{2+} + 2e^- \quad \text{oxidation/half cell I}$$
$$Cu^{2+} + 2e^- \rightarrow Cu \quad \text{reduction/half cell II}$$
$$Zn + Cu^{2+} \rightarrow Zn^{2+} + Cu \quad \text{overall cell reaction}$$

A typical feature of a redox reaction is an exchange of electrons between at least two reaction partners. It is characterized by the fact that oxidation and reduction always occur at the same time. For the Daniell element, the copper ions are the oxidizing agent and the zinc ions the reducing agent. Both together form the corresponding redox pair:

$$Red_1 + Ox_2 \rightarrow Ox_1 + Red_2$$
$$CuSO_4 + Zn \rightarrow ZnSO_4 + Cu$$

The electrode at which the oxidation dominates during discharge is named the anode (negative pole), and the other, where the reduction dominates, is the cathode (positive pole). This nomenclature is valid only for the discharging reaction; for the charging reaction the names are reversed.

1.2.2
Electrochemical Series of Metals

The question arises, which metal is dissolved and which one is deposited when they are combined in an electrochemical cell. The electrochemical series indicates how easily a metal is oxidized or its ions are reduced, that is, converted into positive charged ions or metal atoms respectively. To compare different metals we use the standard potential, which is described below.

In Galvanic cells it is only possible to determine the potential difference as a voltage between two half cells, but not the absolute potential of the single electrode. For the measurement of the potential difference it has to be ensured that an electrochemical equilibrium exists at the phase boundaries (electrode/electrolyte). At least it is required that there is no flux of current in the external and internal circuit.

To compare the potentials of half cells a reference had to be defined. For this reason it was decided arbitrarily that the potential of the hydrogen electrode in a 1 M acidic solution should be equal to 0 V at a temperature of 25 °C and a pressure of 101.3 kPa. These conditions are called *standard conditions* [6].

The reaction of hydrogen in acidic solution is a half-cell reaction and can therefore be handled like the system metal/metal salt solution.

$$H_2 + 2\,H_2O \rightarrow 2\,H_3O^+ + 2\,e^-$$

An experimental setup for the hydrogen half cell is illustrated in Figure 1.5.

The potentials of the metals in their 1 M salt solution are all related to the standard or normal hydrogen electrode (NHE). To measure the potential of such a system, the hydrogen half cell is combined with another half cell to form a Galvanic cell. The measured voltage is called the *normal potential* or *standard electrode potential* ε_{00} of the metal. If the metals are arranged in the order of their normal potentials, the resulting order is named the electrochemical series of the metals (Figure 1.6). Depending on their position in this potential series, they are called *base* ($\varepsilon_{00} < 0$) or *noble* ($\varepsilon_{00} > 0$) metals.

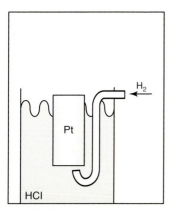

Figure 1.5 Hydrogen electrode with hydrogen-saturated platinum electrode in hydrochloric acid.

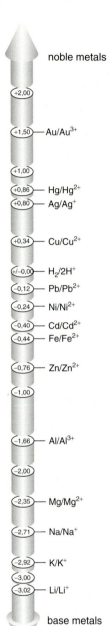

Figure 1.6 Electrochemical series of metals and the standard potential in volt (measured against NHE).

$Zn \rightarrow Zn^{2+} + 2\,e^-$, normal potential $\varepsilon_{00} = -0.76\ V_{NHE}$

$Cu \rightarrow Cu^{2+} + 2\,e^-$, normal potential $\varepsilon_{00} = +0.34\ V_{NHE}$

For the Daniell element in Figure 1.2 the following potential difference is obtained:

$$\Delta\varepsilon_{00} = \varepsilon_{00,\,Cu/Cu^{2+}} - \varepsilon_{00,\,Zn/Zn^{2+}} \tag{1.1}$$

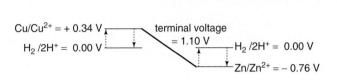

Under equilibrium conditions the potential difference $\Delta\varepsilon_0$ corresponds to the terminal voltage of the cell.

If there are no standard conditions or if it should not be possible to measure the standard potential, the value can be determined by thermodynamic calculations (see Section 1.4.1).

For the arrangement of a galvanic cell for use as a power source the half cells are chosen such that their potentials $\varepsilon_{I,II}$ are as far apart as possible. Therefore, it is obvious why alkaline metals, especially lithium or sodium, are interesting as new materials for the negative electrode. As they have a strong negative standard potential and a comparatively low density, a high specific energy can be realized by combination with a positive electrode.

The following examples, the Daniell element, nickel-cadmium cells, and lithium-manganese dioxide cells, show the influences of the electrode materials on different cell parameters.

1.2.3 Discharging

During the discharge process, electrons are released at the anode from the electrochemically active material, which is oxidized. At the same time, cathodic substances are reduced by receiving electrons. The transport of the electrons occurs through an external circuit (the consumer).

Looking at first at the anode, there is a relationship between the electronic current I and the mass m of the substance which donates electrons, and this is known as the *first Faraday law* [7]:

$$m = \frac{M}{z \cdot F} \cdot I \cdot t \tag{1.2}$$

m = active mass
M = molar mass
z = number of electrons exchanged
F = Faraday constant: $96\,485\ C\ mol^{-1} = 26.8\ Ah\ mol^{-1}$.
t = time

The Faraday constant is the product of the elementary charge e (1.602×10^{-19} C) and the Avogadro constant N_A (6.023×10^{23} mol^{-1}).

$$F = \frac{I \cdot t}{n} = \frac{Q}{n} = N_A \cdot e \qquad (1.3)$$

Q = quantity of electricity, electric charge
n = number of moles of electrons exchanged.

For the Daniell-element the electron-donating reaction is the oxidation of zinc. In the following the active mass m which is necessary to deliver a capacity of 1 Ah, is calculated.

$$Zn \rightarrow Zn^{2+} + 2\,e^-$$

$M = 65.4$ g mol^{-1}, $z = 2$, $F = 26.8$ Ah mol^{-1}, $Q = 1$ Ah

$$m = \frac{M}{z \cdot F} \cdot Q$$
$$m = 1.22 \text{ g}$$

Of course, Faraday's first law applies to cathodic processes. Therefore, the deposition of 1 Ah copper ions results in an increase in the electrode mass of $m = 1.18$ g.

In addition, Faraday recognized that for different electrode reactions and the same amount of charge the proportion of the reacting masses is equal to the proportion of the equivalent masses

$$\frac{m_A}{m_B} = \frac{M_A \cdot |z_B|}{M_B \cdot |z_A|} \qquad (1.4)$$

Equation 1.4 expresses the fact that 1 mol electrons discharges

- 1 mol monovalent ions,
- 1/2 mol bivalent ions, or
- 1/z mol z-valent ions.

1.2.4
Charging

The charging process can only be applied to secondary cells, because, in contrast to primary cells, the electrochemical reactions are reversible. If primary cells are charged, this may lead to electrochemical side reactions, for example, the decomposition of the electrolyte solution with dangerous follow-up reactions leading to explosions [8].

While charging, ions are generally reduced at the negative electrode and an oxidation process takes place at the positive electrode. The voltage source must be at least equivalent to the difference $\Delta\varepsilon_{00}$ between the equilibrium potentials of the two half cells. Generally the charge voltage is higher.

1.3
Thermodynamics

1.3.1
Electrode Processes at Equilibrium

Corresponding to chemical reactions, it is possible to treat electrochemical reactions in equilibrium with the help of thermodynamics.

As well as determining the potential at standard conditions by means of measurement, it is possible to calculate this value from thermodynamic data [9]. In addition, one can determine the influence of changing pressure, temperature, concentration, and so on.

During the determination of standard electrode potentials an electrochemical equilibrium must always exist at the phase boundaries, for example, electrode/electrolyte. From a macroscopic viewpoint, no external current flows and no reaction takes place. From a microscopic viewpoint or on a molecular scale, however, a continuous exchange of charges occurs at the phase boundaries. Figure 1.7 demonstrates this fact at the anode of the Daniell element.

The exchange of charge carriers in the molecular sphere at the phase boundary zinc/electrolyte solution corresponds to an anodic and an equal cathodic current. These compensate each other in the case of equilibrium.

Three kinds of equilibrium potentials are distinguishable:

1) A *metal ion potential* exists if a metal and its ions are present in balanced phases, for example, zinc and zinc ions at the anode of the Daniell element.
2) A *redox potential* exists if both phases exchange electrons and the electron exchange is in equilibrium; for example, the normal hydrogen half cell with an electron transfer between hydrogen and protons at the platinum electrode.
3) If two different ions are present, only one of which can cross the phase boundary, which may exist at a semi-permeable membrane, one gets a so-called

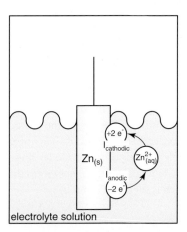

Figure 1.7 Phase boundary metal (zinc)/electrolyte solution (zinc sulfate) in equilibrium.

membrane potential. Well-known examples are the sodium/potassium ion pumps in human cells.

1.3.2
Reaction Free Energy ΔG and Equilibrium Cell Voltage $\Delta \varepsilon_{00}$

Instead of measuring the equilibrium cell voltage $\Delta \varepsilon_{00}$ at standard conditions directly, this can be calculated from the reaction free energy ΔG for one formula conversion. In this context one of the fundamental equations is the GIBBS–HELMHOLTZ relation [7].

$$\Delta G = \Delta H - T \cdot \Delta S \tag{1.5}$$

For the electrochemical cell reaction, the reaction free energy ΔG is the utilizable electric energy. The reaction enthalpy ΔH is the theoretical available energy, which is increased or reduced by $T \cdot \Delta S$. The product of the temperature and the entropy describes the reversible amount of heat consumed or released during the reaction. With tabular values for the enthalpy and the entropy it is possible to obtain ΔG.

Using the reaction free energy, ΔG, the cell voltage $\Delta \varepsilon_0$ can be calculated. First, the number n of exchanged moles of electrons during an electrode reaction must be determined from the cell reaction. For the Daniell element (see example below), 2 mol of electrons are released or received, respectively.

$$1 \text{ mol Zn} \rightarrow 1 \text{ mol Zn}^{2+} + 2 \text{ mol e}^-$$
$$1 \text{ mol Cu} \rightarrow 1 \text{ mol Cu}^{2+} + 2 \text{ mol e}^-$$

With the definition of the Faraday constant (Equation 1.3), the amount of charge of the cell reaction for one formula conversion is given by the following equation:

$$Q = I \cdot t = n \cdot F \tag{1.6}$$

With this quantity of charge, the electrical energy is

$$\Delta \varepsilon_{00} \cdot Q = \Delta \varepsilon_{00} \cdot n \cdot F \tag{1.7}$$

The thermodynamic treatment requires that during one formula conversion the cell reaction is reversible. This means that all partial processes in a cell must remain in equilibrium. The current is kept infinitely small, so that the cell voltage ε and the equilibrium cell voltage $\Delta \varepsilon_{00}$ are equal. Furthermore, inside the cell no concentration gradient should exist in the electrolyte; that is, the zinc and copper ion concentrations must be constant in the whole Daniell element. Under these conditions, the utilizable electric energy, $\Delta \varepsilon_{00} \times z \times F$ per mol, corresponds to the reaction free energy ΔG of the Galvanic cell, which is therefore given by

$$\Delta G = -z \cdot F \cdot \Delta \varepsilon_{00} \tag{1.8}$$

For the Daniell element under standard conditions $T = 298$ K

$Zn + CuSO_4 \longrightarrow ZnSO_4 + Cu$

Reaction enthalpy	$\Delta H = -210.1$ kJ mol^{-1}
Entropy	$\Delta S = -7.2$ J K^{-1} mol^{-1}
Reaction free energy	$\Delta G = \Delta H - T \cdot \Delta S$
	$\Delta G = -208$ kJ mol^{-1}
Faraday constant	$F = 96\,485$ C mol^{-1}
Number of exchanged electrons	$z = 2$
Cell voltage	$\Delta \varepsilon_{00} = -\frac{\Delta G}{z \cdot F} \left[\frac{\text{kJ mol}^{-1}}{\text{C mol}^{-1}} \right]$
	$\Delta \varepsilon_{00} = 1.1$ V

1.3.3
Concentration Dependence of the Equilibrium Cell Voltage

It is established from the chemical thermodynamics that the sum of the chemical potentials μ_i of the substances v_i involved in the gross reaction is equal to the reaction free energy.

$$\Delta G = \sum v_i \cdot \mu_i \tag{1.9}$$

Here v_i are the stoichiometric factors of the compounds used in the equation for the cell reaction, having a plus sign for the substances formed and a negative sign for the consumed compounds.

As a result of the combination of Equations 1.8 and 1.9, the free reaction enthalpy ΔG and the equilibrium cell voltage $\Delta \varepsilon_{00}$ under standard conditions are related to the sum of the chemical potentials μ_i of the involved substances.

$$-\frac{\Delta G}{z \cdot F} = \Delta \varepsilon_{00} = \frac{1}{z \cdot F} \sum v_i \cdot \mu_i \tag{1.10}$$

Earlier it was shown that the equilibrium cell voltage $\Delta \varepsilon_{00}$ is equal to the difference of the equilibrium potentials of its half cells, for example, for the Daniell element:

$$\Delta \varepsilon_{00} = \varepsilon_{00,\,Cu/Cu^{2+}} - \varepsilon_{00,\,Zn/Zn}^{2+} \tag{1.11}$$

The chemical potential of one half cell depends on the concentrations c_i of the compounds, which react at the electrode:

$$\mu_i = \mu_{i,0} + R \cdot T \cdot \ln c_i \tag{1.12}$$

R = universal gas constant: 8.3 J·mol^{-1}·K^{-1}.

As a consequence, the equilibrium potential of the single half cell also depends on the concentrations of the compounds. The NERNST equation (Equation 1.13), which is one of the most important electrochemical relations, expresses this [10]. It results if Equation 1.12 is inserted into Equation 1.10 with regard to one half cell:

$$\Delta \varepsilon_0 = \Delta \varepsilon_{00} + \frac{R \cdot T}{z \cdot F} \cdot \sum v_i \cdot \ln c_i \tag{1.13}$$

For a metal-ion electrode the NERNST equation is

$$\Delta \varepsilon_0 = \Delta \varepsilon_{00} + \frac{R \cdot T}{z \cdot F} \cdot \ln \frac{c_{Me^{z+}}}{c_{Me}} \tag{1.14}$$

and this is used in the following example for the calculation of the concentration dependence of the zinc electrode.

For one half cell of the Daniell element at a temperature of $T = 298$ K

$$Zn \longrightarrow Zn^{2+} + 2\,e^-$$

with the concentration	$c_{Zn^{2+}} = 0.1$ mol L^{-1}
universal gas constant	$R = 8.3$ J $\cdot$ mol^{-1} $\cdot$ K^{-1}
Faraday constant	$F = 96\,485$ C $\cdot$ mol^{-1}
number of exchanged electrons	$z = 2$
standard potential vs NHE,	$\Delta \varepsilon_{00}(Zn/Zn^{2+}) = -0.76$ V

$$\Delta \varepsilon_0 = \Delta \varepsilon_{00} + \frac{R \cdot T}{z \cdot F} \cdot \ln \frac{c_{Zn^{2+}}}{c_{Zn}}$$

$$\Delta \varepsilon_0 = -0.79 \text{ V}$$

The variation of the concentration from 1 mol L^{-1} (standard condition) to 0.1 mol L^{-1} is related to a change in the potential of -0.03 V.

If the concentrations of the copper and zinc ions within a Daniell element are known, the following cell voltage $\Delta \varepsilon_0$ results:

$$\Delta \varepsilon_{00} = \varepsilon_{0,\,Cu/Cu^{2+}} - \varepsilon_{0,\,Zn/Zn^{2+}} \tag{1.15}$$

1.3.4
Temperature Dependence of the Equilibrium Cell Voltage

The temperature dependence of the equilibrium cell voltage forms the basis to determine the thermodynamic variables ΔG, ΔH, and ΔS. The values of the equilibrium cell voltage $\Delta \varepsilon_{00}$ and the temperature coefficient $d\Delta \varepsilon_{00}/dT$, which are necessary for the calculation, can be measured exactly in experiments.

The temperature dependence of the cell voltage $\Delta \varepsilon_0$ results from Equation 1.10 by partial differentiation at a constant cell pressure.

$$\left(\frac{\partial \Delta \varepsilon_0}{\partial T}\right)_p = -\frac{1}{z \cdot F} \cdot \left(\frac{\partial \Delta G}{\partial T}\right)_p \tag{1.16}$$

For the temperature coefficient of the reaction free energy follows, because of thermodynamic relations [7], by partial differentiation of Equation 1.5:

$$\left(\frac{\partial \Delta G}{\partial T}\right)_p = -\Delta S \tag{1.17}$$

$$\left(\frac{\partial \Delta \varepsilon_0}{\partial T}\right)_p = -\frac{1}{z \cdot F} \cdot (-\Delta S)_p \tag{1.18}$$

The *reversible reaction heat of the cell* is defined as the reaction entropy multiplied by the temperature (Equation 1.5). For an electrochemical cell this is also called the *PELTIER effect* and can be described by the difference between the reaction enthalpy ΔH and the reaction free energy ΔG. If the difference between the reaction free energy ΔG and the reaction enthalpy ΔH is less than zero, the cell becomes warmer. On the other hand, for a difference greater than zero, it cools down. The reversible heat of formation W of the electrochemical cell is therefore:

$$W = \Delta G - \Delta H \tag{1.19}$$
$$W = -T \cdot \Delta S$$

For the Daniell element at standard conditions, $T = 298$ K

$$Zn + CuSO_4 \longrightarrow ZnSO_4 + Cu$$

reaction enthalpy	$\Delta H = -210.1$ kJ mol^{-1}
reaction free energy	$\Delta G = -208$ kJ mol^{-1}
Heat	$W = \Delta G - \Delta H$
	$W = 2.1$ kJ mol^{-1}

The reversible amount of heat of 2.1 kJ·mol^{-1} is consumed by charging and released by discharging.

The relationship between free reaction enthalpy, temperature, cell voltage, and reversible heat in a Galvanic cell is reflected by the GIBBS–HELMHOLTZ equation (Equation 1.20).

$$\Delta H = \Delta G - T \cdot \left(\frac{\partial \Delta G}{\partial T}\right)_p \tag{1.20}$$

Insertion of Equation 1.8 for ΔG results in

$$\Delta H = z \cdot F \cdot \left[\Delta \varepsilon_{00} + T \cdot \left(\frac{\partial \Delta \varepsilon_0}{\partial T}\right)_p\right] \tag{1.21}$$

Earlier it was deduced that for ΔS and ΔG:

$$\Delta S = z \cdot F \cdot \left(\frac{\partial \Delta \varepsilon_0}{\partial T}\right)_p \tag{1.22}$$

$$\Delta G = -z \cdot F \cdot \Delta \varepsilon_0 \tag{1.23}$$

From experiments it is possible to obtain the temperature coefficient for the Daniell element, $\Delta\varepsilon_0/T = -3.6 \times 10^{-5}$ V K^{-1}:

temperature	$T = 298$ K
equilibrium cell voltage	$\Delta\varepsilon_{00} = 1.1$ V
Faraday constant	$F = 96\,485$ C mol^{-1}
number of exchanged electrons	$z = 2$
reaction enthalpy	$\Delta H = z \cdot F \cdot \left[\Delta\varepsilon_{00} + T \cdot \left(\frac{\partial \Delta\varepsilon_0}{\partial T}\right)_p\right]$
	$\Delta H = 212.2$ kJ mol^{-1}
reaction entropy	$\Delta S = z \cdot F \cdot \left(\frac{\partial \Delta\varepsilon_0}{\partial T}\right)_p$
	$\Delta S = -2.1$ kJ K^{-1}
free reaction enthalpy	$\Delta G = -z \cdot F \cdot \Delta\varepsilon_0$
	$\Delta G = -208$ kJ mol^{-1}

The calculation of the free reaction enthalpy is possible with Equation 1.8, and the determination of the reaction entropy ΔS follows from Equation 1.22.

1.3.5
Pressure Dependence of the Equilibrium Cell Voltage

It is obvious that the cell voltage is nearly independent of the pressure if the reaction takes place between solid and liquid phases where the change in volume is negligibly low. On the other hand, in reactions involving the evolution or disappearance of gases, this effect has to be considered [11].

The pressure dependence of the reaction free energy is equal to the volume change associated with one formula conversion.

$$\left(\frac{\partial \Delta G}{\partial p}\right)_T = \Delta V \tag{1.24}$$

With $\Delta G = -n \times F \times \Delta\varepsilon_0$ and $\Delta V = -RT/p$ we have

$$\left(\frac{\partial \Delta\varepsilon_0}{\partial p}\right)_T = -\frac{R \cdot T}{n \cdot F} \cdot \frac{1}{p} \tag{1.25}$$

By integration, the equilibrium cell voltage as a function of the partial pressure of the solved gas (with the integration constant K equivalent to $\Delta\varepsilon_{00}$ [10]) is obtained:

$$\Delta\varepsilon_0 = K - \frac{R \cdot T}{n \cdot F} \ln p \tag{1.26}$$

The following example of a hydrogen/oxygen fuel cell illustrates this relationship.

> For a hydrogen/oxygen fuel cell at standard conditions, $T = 298\,\text{K}$ and $p = 101.3\,\text{kPa}$, where
>
> | | |
> |---|---|
> | cell reaction is | $2\,H_2 + O_2 \rightarrow 2\,H_2O$ |
> | standard potential (oxygen) | $\varepsilon_{00} = +1.23\,\text{V}$ |
> | standard potential (hydrogen) | $\varepsilon_{00} = 0\,\text{V}$ |
> | standard cell voltage | $\Delta\varepsilon_{00} = +1.23\,\text{V}$ vs NHE |
> | For the anode | $\varepsilon_0 = \varepsilon_{00} + \frac{R \cdot T}{n \cdot F} \ln p_{O_2} = 1.23 + 0.03\,\text{V}$ |
> | | $\varepsilon_0 = 1.26\,\text{V}$ |
> | For the cathode | $\varepsilon_0 = \varepsilon_{00} - \frac{R \cdot T}{n \cdot F} \ln p_{H_2}^2 = 0 - 0.06\,\text{V}$ |
> | | $\varepsilon_0 = -0.06\,\text{V}$ |
> | | $\Delta\varepsilon_0 = 1.26\,\text{V} - (-0.06\,\text{V}) = 1.32\,\text{V}$ |
>
> an increase in the pressure to $1013\,\text{kPa}$ results in an increase in the standard cell voltage of $0.09\,\text{V}$.

1.3.6
Overpotential of Half Cells and Internal Resistance

The potential of the electrode surface is determined using the Nernst equation introduced in Section 1.3.3. In equilibrium, the currents in anodic and cathodic direction are equal. If they are related to an electrode area, they are called *exchange current densities* j_0.

$$j_a = |j_c| = j_0 \tag{1.27}$$

$j_{a,c}$ represents anodic, cathodic current density ($A\,cm^{-2}$).

If a current flows, for example, while discharging a battery, a shift in the potential of the single half cell is measured. This deviation is called *overpotential*, η [12]. Thus, the real potential $\Delta\varepsilon_{\text{real}}$ has to be calculated using the following equation:

$$\Delta\varepsilon_{\text{real}} = \Delta\varepsilon_0 - \sum |\eta| \tag{1.28}$$

It is obvious that for a half cell the sum of the overpotentials should be as low as possible. Depending on their origin, a distinction has to be made between:

- *Charge transfer overpotential*: The charge transfer overpotential is caused by the fact that the speed of the charge transfer through the phase-boundary electrode/electrolyte is limited. It generally depends on the kind of substances that are reacting, the conditions in the electrolyte, and the characteristic of the electrode (for example, what kind of metal). The formulae which deal with this form of overpotential are called the *Butler–Volmer equation* and the Tafel equation [10].

- *Diffusion overpotential:* When high current densities j at electrodes (at the boundary to the electrolyte) exist, depletion of the reacting substances is possible, resulting in a concentration polarization. In this case the reaction kinetics is determined only by diffusion processes through this zone, the so-called Nernst layer. Without dealing with its derivation in detail, the following formula is obtained for the occurring diffusion overpotential (j_{limit} being the maximum current density):

$$\eta_{\text{diff}} = \left| \frac{RT}{zF} \cdot \ln\left(1 - \frac{j}{j_{\text{limit}}}\right) \right| \tag{1.29}$$

As expected, the value of η_{diff} increases with increasing current densities.

- *Reaction overpotential:* Both the overpotentials mentioned above are normally of greater importance than the reaction overpotential. But sometimes it may happen that other phenomena which occur in the electrolyte or during electrode processes such as adsorption and desorption are the rate-limiting factors.
- *Crystallization overpotential:* This can occur as a result of the inhibited intercalation of metal ions into their lattice. This process is of fundamental importance when secondary batteries are charged, especially during the metal deposition at the negative side.

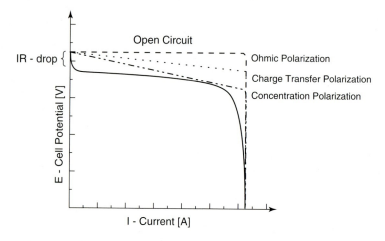

Figure 1.8 Cell polarization as a function of operating current.

Corresponding to the change in the potential of the single electrodes, which is related to their different overpotentials, a shift in the overall cell voltage is observed (see Figure 1.8). Moreover, an increasing cell temperature can be noticed. Besides joulic heat, caused by voltage losses due to the internal resistance R_i (electrolyte, contact to the electrodes, etc.) of the cell, thermal losses W_K (related to overpotentials) are the reason for this phenomenon.

$$W_J = I^2 \cdot R_i \cdot t \tag{1.30}$$

$$W_K = I \cdot \sum \eta_i \cdot t \tag{1.31}$$

1.4
Criteria for the Judgment of Batteries

The need to operate electrically powered tools or devices independently of stationary power sources has led to the development of a variety of different battery systems, the preference for any particular system depending on the field of application. In the case of a occasional use, for example, for electric torches in the household or for long-term applications with low current consumption such as watches or pacemaker, primary cells (zinc–carbon, alkaline manganese, or lithium–iodide cells) are chosen. For many other applications such as notebooks, MP3-players, cellular phones, or starter batteries in cars only rechargeable battery systems, for example, lithium-ion batteries or lead–acid batteries, can be considered from the point of view of cost and the environment.

The wide variety of applications has led to an immense number of configurations and sizes, for example, small round cells for hearing aids or large prismatic cells like lead–acid batteries for use in trucks. Here the great variety of demands has the consequence that nowadays no battery system is able to cope with all of them. The choice of the 'right' battery system for a single application is therefore often a compromise.

The external set-up of different battery systems is generally simple and in principle differs only a little from one system to another. A mechanically stable cell case carries the positive and negative electrodes, which are kept apart by means of a membrane and are connected to electrically conducting terminals. Conduction of the ions between the electrodes takes place in a fluid or gel-like electrolyte [13].

To assess the different battery systems, their most important features need to be compared.

1.4.1
Terminal Voltage

During charging and discharging of the cell the terminal voltage U between the poles is measured. Also, it should be possible to calculate the theoretical thermodynamic terminal voltage from the thermodynamic data of the cell reaction. This value often differs slightly from the voltage measured between the poles of the cell because of an inhibited equilibrium state or side reactions.

1.4.2
Current–Voltage Diagram

An important experimentally available feature is the current–voltage characteristic. This gives the terminal voltage provided by the electrochemical cell as a function of the discharge current (see Figure 1.9). The product of current and the accompanying terminal voltage is the electric power delivered by the battery system at any time.

$$P = I \cdot U_{\text{terminal voltage}} \tag{1.32}$$

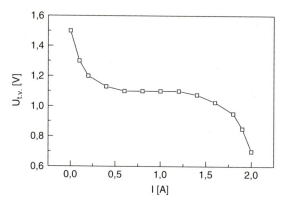

Figure 1.9 Current-voltage characteristic of a Leclanché element.

The power as a function of the battery weight is known as the power density P_s of the element in watts/kilogram. Figure 1.9 shows the current-voltage characteristic of a Leclanché element.

1.4.3
Discharge Characteristic

The discharge curve is another important feature of battery systems. Here, the terminal voltage is plotted against the discharge capacity. For an ideal battery the terminal voltage drops to zero in a single step when the whole of the stored energy is consumed.

The discharge rate C is defined by the discharge current and the nominal capacity of the secondary cell. It is equal to the reciprocal value of the discharging time:

$$C = \frac{\text{discharge current}}{\text{nominal capacity}} \qquad (1.33)$$

The nominal capacity of every system is defined by a specific value of C; for example, for th nickel–cadmium system, it is $\frac{1}{20}$C. By discharging at a higher current, the final capacity obtainable becomes lower because the IR losses and the polarization effects increase (see Figure 1.10).

The mode of the discharge (for example, at constant current, constant load, or constant power) can also have a significant effect on the performance of the battery. It is advisable that the mode of discharge used in a test or evaluation setup should be the same as the one used in the application.

1.4.4
Characteristic Line of Charge

During charging, the secondary cell receives the same amount of electric energy as that previously released, and this is stored in the form of chemical energy (see Figure 1.11 for nickel–cadmium system). Terminal voltage, charging time, number

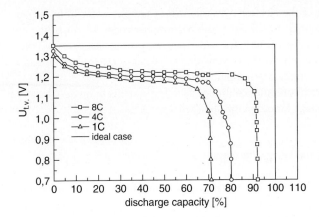

Figure 1.10 Ideal discharge characteristic of a nickel–cadmium system.

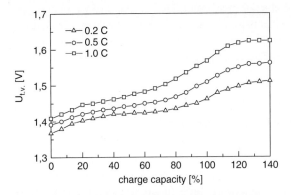

Figure 1.11 Dependence of the cell voltage on the charge capacity for three different currents in the nickel–cadmium system: $Cd + 2\, NiOOH + 2\, H_2O \rightarrow Cd(OH)_2 + 2\, Ni(OH)_2$.

of cycles, and other parameters are influenced by the charging procedure in a single battery system.

1.4.5
Overcharge Reactions

Nearly all electric consumers demand a high voltage, which is realized by connecting cells in series. Since the single cells have different capacities, it is impossible to maintain the optimal charge voltage in the weakest cell at the end of the charge process, while to charge, current passes through all the serially connected cells. As a consequence, the cell voltage increases, and, as well as the main charging reaction, chemical or electrochemical side reactions are possible. A well-known problem is the decomposition of the electrolyte solution (for example, water to hydrogen at the negative electrode or to oxygen at the positive electrode). In some battery systems these evolved gases react back with formation of educts. For example, in the nickel–cadmium cell oxygen is formed at the positive electrode and reacts back at the negative electrode, warming up the cell [8].

To avoid this problem, computer-controlled charging systems in modern battery stacks regulate the voltage for each individual cell.

1.4.6
Coulometric Efficiency and Energy Efficiency

The *efficiency during an energy conversion* is defined as the ratio of the energy converted to the energy consumed. This parameter is only decisive for secondary systems. The charge (Q_{charge}) necessary to load a secondary cell, is always higher than the charge ($Q_{discharge}$) released during discharge. This is caused by an incomplete conversion of the charging current into utilizable reaction products. Useless side reactions with heat production may occur. Here, numerous parameters are important such as the current density, the temperature, the thickness, the porosity of the separator, and the age of the cell.

There are two possible ways to describe the efficiency of batteries – the coulometric efficiency and the energy efficiency.

- *Coulometric efficiency*:

$$q_{Ah} = \frac{Q_{discharge}}{Q_{charge}} \tag{1.34}$$

The reciprocal value $f = \frac{1}{q_{Ah}}$ of the coulometric efficiency is called the *charging factor*. The coulometric efficiency for electrochemical energy conversion is about 70–90% for nickel–cadmium and nearly 100% for lithium-ion batteries [14].
- *Energy efficiency*:

$$q_{Wh} = q_{Ah} \cdot \frac{\overline{U}_{discharge}}{\overline{U}_{charge}} \tag{1.35}$$

Here, $\overline{U}_{discharge}$ and $\overline{U}_{charge}$ are the average terminal voltages during charge and discharge. The discharge voltage is normally lower than the charge voltage because of the internal resistance and overpotentials. For this reason the coulometric efficiency is always higher than the energy efficiency. It is influenced by the same terms as the charge efficiency but in addition by the discharge current and the charging procedure.

1.4.7
Cycle Life and Shelf Life

Another important parameter to describe a secondary electrochemical cell is the achievable number of cycles or the lifetime. For economic and ecological reasons, systems with a high cycle life are preferred. The number of cycles illustrates how often a secondary battery can be charged and discharged repeatedly before a lower limit (defined as a failure) of the capacity is reached. This value is often set at 80% of the nominal capacity. To compare different battery systems, the depth of discharge has to be quoted as well as the number of cycles.

Additionally, batteries deteriorate as a result of chemical side reactions that proceed during charging and discharging, but also during storage. Cell design, temperature, the electrochemical system, and the charge state affect the shelf life and the charge retention of the battery.

1.4.8
Specific Energy and Energy Density

With respect to the specific energy (the electric energy per unit mass) of today's battery systems, there is a major difference between the performance of aqueous systems and that of nonaqueous systems [15]. Apart from batteries for some special applications, there are

- Aqueous batteries with about 140 Wh kg^{-1} for primary and about 80 Wh kg^{-1} for secondary systems
- Nonaqueous batteries with about 400 Wh kg^{-1} for primary and about 180 Wh kg^{-1} for secondary systems
- For comparison: the utilizable electric or mechanic energy of a gasoline engine is 3000 Wh per 1 kg gasoline.

The zinc–carbon and alkaline manganese cells are primary battery systems, while lead, nickel–cadmium, and nickel–metal hydride batteries are secondary batteries with aqueous electrolyte solutions. The aqueous battery systems generally show only a limited performance at low temperatures. Because of the decomposition of the water, the voltage of a single cell is limited. For this reason lithium-ion batteries are of great interest when using organic or polymer electrolytes, allowing cell potentials of up to 4.5 V to be achieved.

1.4.9
Safety

Batteries are sources of energy and deliver their energy in a safe way when they are properly used. Therefore it is of crucial importance to choose the right electrochemical system in combination with the correct charge, discharge, and storage conditions to assure optimum, reliable, and safe operation.

There are instances when a battery may vent, rupture, or even explode if it is abused. To avoid this, a cell and/or a battery should include protective devices to avoid

- application of too high charge or discharge rates
- improper charge or discharge voltage or voltage reversal

- short-circuiting
- charging primary batteries
- charging or discharging at too high or too low temperatures.

To ensure that the right operating conditions are used every time, a type of electronic control, the so-called battery management system, can be used. This is especially important for the lithium-ion battery, where a too low end of discharge voltage, a too high end of charge voltage, or a too high charge or discharge rate not only can affect the lifetime and the cycle life but also can amount to abuse of the equipment resulting in possible rupture or explosion of the cell.

1.4.10
Costs per Stored Watt Hour

The cost per watt hour delivered from a primary battery is the ratio between the price of the battery and its capacity. For a secondary battery the cost of the battery installation has to be taken into consideration as well as the ratio of the charging cost to the delivered energy.

References

1. Linden, D. and Reddy, T.B. (2002) *Handbook of Batteries*, 3rd edn, McGraw-Hill, Inc.
2. Kordesch, K. and Simader, G. (1996) *Fuel Cells and Their Applications*, Wiley-VCH Verlag GmbH, Weinheim.
3. Kiehne, H.A. (2003) *Battery Technology Handbook*, Marcel Dekker, New York.
4. Jaksch, H.D. (1993) *Batterielexikon*, Pflaum Verlag, München.
5. Hamann, C.H. and Vielstich, W. (2005) *Elektrochemie*, Wiley-VCH Verlag GmbH, Weinheim.
6. Sawyer, D.T., Sobkowiak, A., and Roberts, J.L. (1995) *Electrochemistry for Chemists*, John Wiley & Sons, Inc.
7. Alberty, R.A. and Silbey, R.J. (1996) *Physical Chemistry*, John Wiley & Sons, Inc.
8. Levy, S.C. and Bro, P. (1994) *Battery Hazards and Accident Prevention*, Plenum Press, New York, London.
9. Lide, D.R. (2009) *Handbook of Chemistry and Physics*, 90th edn, CRC Press, Boca Raton, FL, Ann Arbor, MI, Boston, MA.
10. Bard, A.J. and Faulkner, L.R. (2001) *Electrochemical Methods: Fundamentals and Applications*, John Wiley & Sons, Inc., London.
11. Bockris, J.O.M., Reddy, A.K.N. and Gamboa-Adeco, M.E. (2006) *Modern Electrochemistry 1, 2A and 2B*, Springer Verlag GmbH, Berlin.
12. Southampton Electrochemistry Group (2002) *Instrumental Methods in Electrochemistry*, Ellis Horwood Limited.
13. Munshi, M.Z.A. (1995) *Handbook of Solid State Batteries and Capacitors*, World Scientific Publishing Co. Pte. Ltd., Singapore.
14. Gabano, J.-P. (1983) *Lithium Batteries*, Academic Press, London.
15. Barsukov, V. and Beck, F. (1996) *New Promising Electrochemical Systems for Rechargeable Batteries*, Kluwer Academic Publishers, Dordrecht.

2
Practical Batteries
Koji Nishio and Nobuhiro Furukawa

2.1
Introduction

Batteries can be roughly divided into primary and secondary batteries. Primary batteries cannot be electrically charged, but they have high energy density and good storage characteristics. Lithium primary batteries, which were commercialized about 20 years ago, exist in many forms, for example, lithium–manganese dioxide, lithium–carbon monofluoride, and lithium–thionyl chloride batteries. Other batteries include carbon–zinc, alkaline-manganese, zinc–air, and silver oxide–zinc batteries.

Secondary batteries can be electrically charged, and this can offer savings in costs and resources. Recently, lithium-ion and nickel–metal hydride (MH) batteries have been developed, and are used with the other secondary batteries, such as nickel–cadmium, lead–acid, and coin-type lithium secondary batteries.

The variety of practical batteries has increased during the last 20 years. Applications for traditional and new practical battery systems are increasing, and the market for lithium-ion batteries and nickel–MH batteries has grown remarkably. This chapter deals with consumer-type batteries, which have developed relatively recently.

2.2
Alkaline-Manganese Batteries

Batteries using an alkaline solution as the electrolyte are commonly called *alkaline batteries*. They are high-power owing to the high conductivity of the alkaline solution. Alkaline batteries include primary batteries, typical of which are alkaline-manganese batteries, and secondary batteries, typical of which are nickel–cadmium and nickel–MH batteries. These batteries are widely used.

The dry cell was invented by Leclanché in the 1860s, and this type of battery was developed in the nineteenth century. In the 1940s, Rube1 achieved significant

Handbook of Battery Materials, Second Edition. Edited by Claus Daniel and Jürgen O. Besenhard.
© 2011 Wiley-VCH Verlag GmbH & Co. KGaA. Published 2011 by Wiley-VCH Verlag GmbH & Co. KGaA.

progress in alkaline-zinc batteries and manufactured zinc powder with high surface area to prevent zinc passivation.

The discharge of alkaline-manganese batteries comes from the electrochemical reactions at the anode and cathode. During discharge, the negative electrode material, zinc, is oxidized, forming zinc oxide; at the same time, MnO_2 in the positive electrode is reduced (MnOOH):

Cathode reaction:
$$2MnO_2 + H_2O + 2e^- \rightarrow 2MnOOH + 2OH^- \quad 0.12 \text{ V vs NHE} \tag{2.1}$$
Anode reaction:
$$Zn + 2OH^- \rightarrow ZnO + H_2O + 2e^- \quad -1.33 \text{ V vs NHE} \tag{2.2}$$
Overall reaction:
$$Zn + 2MnO_2 \rightarrow ZnO + 2MnOOH \quad 1.45 \text{ V} \tag{2.3}$$

The initial voltage of an alkaline-manganese dioxide battery is about 1.5 V. Alkaline-manganese batteries use a concentrated alkaline aqueous solution (typically in the range of 30–45% potassium hydroxide) for electrolyte. In this concentrated electrolyte, the zinc electrode reaction proceeds, but if the concentration of the alkaline solution is low, then the zinc tends to passivate.

The cell construction of an alkaline-manganese battery is shown in Figure 2.1. The steel can serves as a current collector for the manganese dioxide electrode. Inside the can is a cathode containing manganese dioxide and graphite powder.

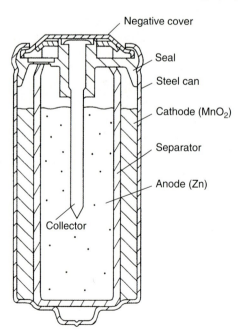

Figure 2.1 Cell construction of an alkaline-manganese battery.

Zinc powder is packed inside the separator together with the electrolyte solution and a gelling agent. An anode collector is inserted into the zinc powder. The battery is hermetically sealed, which contributes to its good shelf life.

Figure 2.2 shows a comparison of the discharge characteristics between alkaline-manganese batteries and Leclanché batteries. The capacity of the alkaline-manganese batteries is about three times that of the Leclanché batteries.

Amalgamated zinc powder has been used as the negative material to prevent zinc corrosion and zinc passivation. Recently, from the viewpoint of environmental problems, mercury-free alkaline-manganese batteries were developed by using zinc

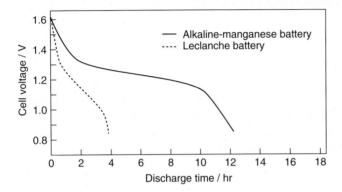

Figure 2.2 Comparison between the discharge characteristics of alkaline-manganese and Leclanché batteries (load 7.5 Ω; temperature 20 °C).

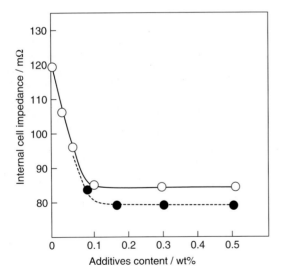

Figure 2.3 Variation of internal impedance of alkaline-manganese cells with the additives content of the zinc powder: ○, Hg additive; ●, In additive.

powder with indium, bismuth, and other additives [1–3]. Adding indium to zinc powder is the most effective way to improve the characteristics of the cells [2]. Figure 2.3 shows the variation in the internal impedance of the cells according to the additive content of the zinc powder.

Today's battery performance has greatly improved. The capacity of newly developed alkaline-manganese batteries is about 1.5 times higher than that of conventional batteries [4]. Figure 2.4 shows a comparison of the discharge characteristics of cells between newly developed and conventional types. Therefore, alkaline-manganese batteries have become more suitable than they once were when requiring a high discharge current.

2.3
Nickel–Cadmium Batteries

The nickel–cadmium battery [5] has a positive electrode made of nickel hydroxide and a negative electrode in which a cadmium compound is used as the active material. Potassium hydroxide is used as the electrolyte. During charge and discharge, the following reactions take place:

Positive electrode reaction:

$$\text{NiOOH} + \text{H}_2\text{O} + \text{e}^- \underset{\text{Charge}}{\overset{\text{Discharge}}{\rightleftarrows}} \text{Ni(OH)}_2 + \text{OH}^- \quad 0.52 \text{ V vs NHE} \quad (2.4)$$

Negative electrode reaction:

$$\text{Cd} + 2\text{OH}^- \underset{\text{Charge}}{\overset{\text{Discharge}}{\rightleftarrows}} \text{Cd(OH)}_2 + 2\text{e}^- \quad -0.80 \text{ V vs NHE} \quad (2.5)$$

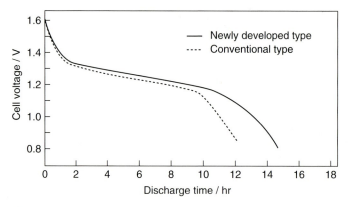

Figure 2.4 Comparison between the discharge characteristics of newly developed and conventional alkaline–manganese cells (load 7.5 Ω; temperature 20 °C).

Overall battery reaction:

$$2\text{NiOOH} + \text{Cd} + 2\text{H}_2\text{O} \underset{\text{Charge}}{\overset{\text{Discharge}}{\rightleftarrows}} 2\text{Ni(OH)}_2 + \text{Cd(OH)}_2 \quad 1.32\,\text{V} \quad (2.6)$$

Reactions take place at the positive electrode between nickel oxyhydroxide and nickel hydroxide, and at the negative electrode between cadmium metal and cadmium hydroxide. In addition, the H_2O molecules, which are generated during charging, are consumed during discharging. Therefore, variations in electrolyte concentration are insignificant. Because of this reaction, the nickel–cadmium battery excels in temperature characteristics, high-rate discharge characteristics, durability, and so on [6]. Most significant is the fact that the amount of electrolyte in the cell can be reduced enough to allow the manufacture of completely sealed cells.

The nickel–cadmium battery was invented by Jungner in 1899. The battery used nickel hydroxide for the positive electrode, cadmium hydroxide for the negative electrode, and an alkaline solution for the electrolyte. Jungner's nickel–cadmium battery has undergone various forms of the development using improved materials and manufacturing processes to achieve a superior level of performance.

In 1932, Shlecht and Ackermann invented the sintered plate. In those days, conventional plates involved a system in which the active materials were packed into a metal container called a *pocket* or *tube*. However, with the sintered-plate method, the active materials are placed inside a porous electrode formed of sintered nickel powder. In 1947, Neumann achieved a completely sealed structure. This idea of protection against overcharge and overdischarge by proper capacity balance is illustrated in Figure 2.5.

Focusing on the concept of the completely sealed system, the Sanyo Electric Co. developed sealed-type nickel–cadmium batteries in 1961. This type of battery enjoys a wide application range that is still expanding; a large variety of nickel–cadmium batteries has been developed to meet user needs ranging from low-current uses like emergency power sources and semiconductor memories to high-power applications such as cordless drills.

Figure 2.6 shows the typical structural design of a cylindrical nickel–cadmium battery. It has a safety vent, as illustrated in Figure 2.7, which automatically opens and releases excessive pressure when the internal gas pressure increases. Formation of hydrogen is avoided by 'extra' Cd(OH)_2; oxygen is removed by reaction with Cd.

Figure 2.8 shows the charge characteristics when charging is performed at a constant current. In nickel–cadmium batteries, characteristics such as cell voltage, internal gas pressure, and cell temperature vary during charging, depending

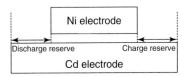

Figure 2.5 Electrode capacity balance of a sealed Ni–Cd battery.

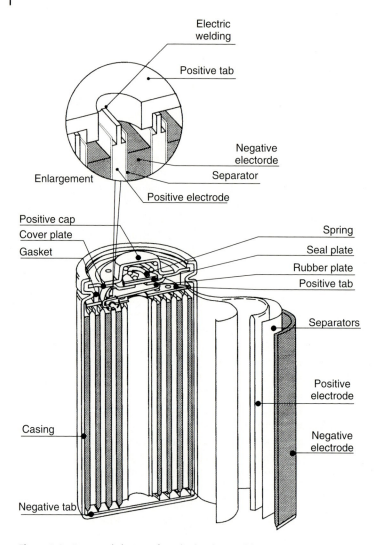

Figure 2.6 Structural design of a cylindrical Ni–Cd battery.

on the charge current and ambient temperature. Figure 2.9 shows the discharge characteristics at various discharge rates. The discharge capacity of the cell decreases as the discharge current increases. However, compared with other batteries, nickel–cadmium batteries have excellent high-current discharge characteristics. A continuous, high-current discharge at 4 C or, in some types, over 10 C is possible.

Figure 2.10 shows the charge–discharge cycle characteristics. As shown in this figure, nickel–cadmium batteries exhibit excellent cycle characteristics and no noticeable decline is observed after 1000 charge–discharge cycles.

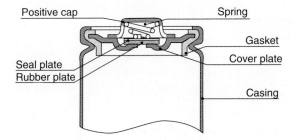

Figure 2.7 Safety vent of an Ni–Cd battery.

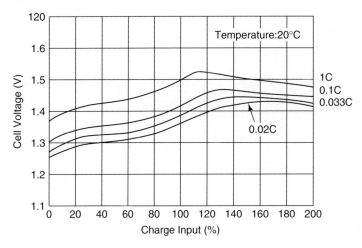

Figure 2.8 Charge characteristics of an Ni–Cd battery at a constant current (cell type 1200SC; temperature 20 °C).

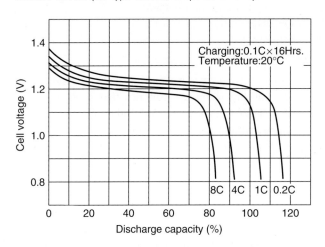

Figure 2.9 Discharge characteristics of an Ni–Cd battery at various discharge currents (cell type 1200SC).

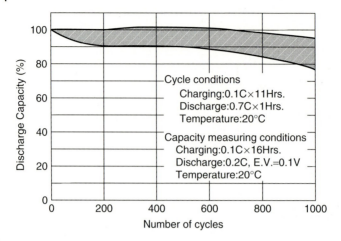

Figure 2.10 Charge–discharge cycle characteristics of an Ni–Cd battery (cell type 1200SC).

The significant features of nickel–cadmium batteries can be summarized as follows:

1) Outstanding economy and long service life, which can exceed 500 charge–discharge cycles.
2) Low internal resistance, which enables a high-rate of discharge, and a constant discharge voltage, which provides an excellent source of DC power for any battery-operated appliance.
3) A completely sealed construction which prevents the leakage of electrolyte and is maintenance-free. No restrictions on mounting direction enable use in any appliance.
4) Ability to withstand overcharge and overdischarge.
5) A long storage life without deterioration in performance and recovery of normal performance after recharging.
6) Wide operating-temperature range.

Recent advances in electronics technologies have accelerated the trend toward smaller and lighter devices. For the secondary batteries that serve as power supplies for these devices, there is also an increasing demand for the development of more compact, lighter batteries with high energy density and high performance. Improvements have been made possible mainly because of progress in the nickel electrode.

For many years, sintered-nickel electrodes have been used as the positive electrodes for sealed-type nickel–cadmium batteries. With an increase in the demand for high energy density, this type of electrode has been improved. Figure 2.11 shows an improved sintered substrate with high porosity. In addition, a new type of manufacturing process has been developed for a nickel electrode, which is made by pasting nickel hydroxide particles (Figure 2.12) into a three-dimensional nickel substrate (Figure 2.13). To increase the energy density of nickel electrode, it is

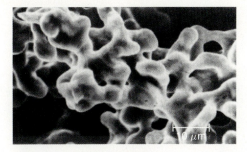

Figure 2.11 Improved sintered substrate with high porosity.

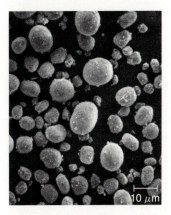

Figure 2.12 Nickel hydroxide particles for active materials.

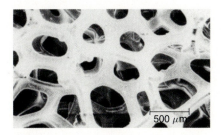

Figure 2.13 Three-dimensional nickel substrate.

important to put as many nickel hydroxide particles as possible into a given substrate, and improve its utilization. Such new electrodes are used for high-capacity nickel–cadmium batteries.

As mentioned above, nickel–cadmium batteries have excellent characteristics and are used in diverse fields. Special-purpose batteries (Figure 2.14) comply effectively with the requirements for improvement of various devices, for example, high-capacity, fast-charge, high-temperature, heat-resistant, memory backup.

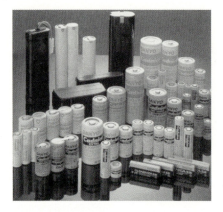

Figure 2.14 Various Ni–Cd batteries.

2.4
Nickel–MH Batteries

Nickel–MH batteries contain a nickel electrode similar to that used in nickel–cadmium batteries as the positive electrode, and a hydrogen-absorbing alloy for the negative electrode. This has made the development of a hydrogen-absorbing alloy electrode important.

Hydrogen-absorbing alloy can reversibly absorb and desorb a large amount of hydrogen. Hydrogen gas is rapidly absorbed in the gas phase, then desorbed on the alloy (gas-solid reaction). In the electrode reaction, the alloy electro-chemically absorbs and desorbs hydrogen in an alkaline solution (electrochemical reaction):

Positive electrode reaction:

$$NiOOH + H_2O + e^- \underset{\text{Charge}}{\overset{\text{Discharge}}{\rightleftarrows}} Ni(OH)_2 + OH^- \quad 0.52 \text{ V vs NHE} \quad (2.7)$$

Negative electrode reaction:

$$MH + OH^- \underset{\text{Charge}}{\overset{\text{Discharge}}{\rightleftarrows}} M + H_2O + e^- \quad -0.80 \text{ V vs NHE} \quad (2.8)$$

Overall battery reaction:

$$NiOOH + MH \underset{\text{Charge}}{\overset{\text{Discharge}}{\rightleftarrows}} Ni(OH)_2 + M \quad 1.32 \text{ V} \quad (2.9)$$

where M = hydrogen-absorbing alloy and MH = metal hydride.

Figure 2.15 shows a typical mechanism of the charge–discharge reaction. During charging, the electrolytic reaction of water causes the hydrogen, which is present in atomic form on the surface of the hydrogen-absorbing alloy in the negative electrode, to disperse into and be absorbed by the alloy (discharge reaction). During discharge, the absorbed hydrogen reacts with hydroxide ions at the surface of the hydrogen-absorbing alloy to become water once again (charge reaction). In other words, the active material of the electrode reaction is hydrogen, and the hydrogen-absorbing alloy acts as a storage medium for the active material.

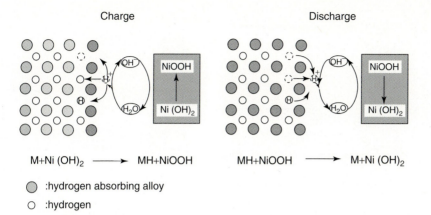

Figure 2.15 Reaction mechanism of the charging-discharging reaction of an MH electrode.

Hydrogen-absorbing alloys were discovered in the 1960s [7]. MH electrode materials were studied in the 1970s and 1980s [8–11]. To be suitable as the negative electrode material for a high-performance cell, a hydrogen-absorbing alloy must allow a large amount of hydrogen to be absorbed and desorbed in an alkaline solution, its reaction rate must be high, and it must have a long charge–discharge cycle life.

Much of this study was conducted on $LaNi_5$-based alloys [12–19] and $TiNi_x$-based alloys [20–22]. Sanyo Electric, Matsushita Battery, and most other battery manufacturers have been using $LaNi_5$-based rare earth–nickel-type alloys [23, 24]. Some manufacturers are using a $TiNi_x$-based alloy [22].

It was thought that rare earth–nickel-type alloys had a large exchange current density and that they absorbed a large amount of hydrogen, thereby enabling the construction of high-energy-density batteries. The first step in this development was to obtain a sufficient cycle life for their use as an electrode material.

Figure 2.16 shows the charge–discharge cycle characteristics of alloys in which part of the nickel component was replaced with cobalt. Misch metal (Mm), which is a mixture of rare earth elements such as lanthanum, cerium, praseodymium, and neodymium, was used in place of lanthanum. It was found that the partial replacement of nickel with cobalt and the substitution of the lanthanum content with Mm was very useful in improving the charge–discharge cycle life. However, such alloys have insufficient capacity, as shown in Figure 2.17 [18]. From study of the effect that their compositions had on the charge–discharge capacity, it was concluded that the best alloy elements were $Mm(Ni–Co–Al–Mn)_x$. This alloy led to the commercialization of sealed nickel–MH batteries. All the battery manufacturers who use a rare earth–nickel-type alloy for the negative electrode material employ similar alloys with slightly different compositions.

The nickel–MH battery comes in two shapes: cylindrical and prismatic. The internal structure of the cylindrical battery is shown in Figure 2.18. It consists of positive and negative electrode sheets wrapped within the battery, with separators between. Figure 2.19 shows the internal structure of the prismatic battery: it consists

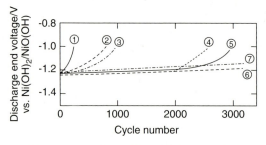

Hydrogen absorbing alloy : ① LaNi$_5$, ② LaNi$_4$Co, ③ LaNi$_3$Co$_2$, ④ LaNi$_2$Co$_3$, ⑤ La$_{0.8}$Ce$_{0.2}$Ni$_2$Co$_3$, ⑥ La$_{0.8}$Nd$_{0.2}$Ni$_2$Co$_3$, ⑦ MmNi$_2$Co$_3$

Figure 2.16 Charge–discharge cycle characteristics of various MH alloy electrodes.

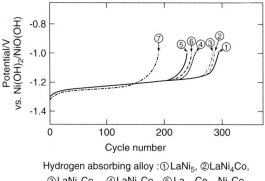

Hydrogen absorbing alloy : ① LaNi$_5$, ② LaNi$_4$Co, ③ LaNi$_3$Co$_2$, ④ LaNi$_2$Co$_3$, ⑤ La$_{0.8}$Ce$_{0.2}$Ni$_2$Co$_3$, ⑥ La$_{0.8}$Nd$_{0.2}$Ni$_2$Co$_3$, ⑦ MmNi$_2$Co$_3$

Figure 2.17 Discharge characteristics of various MH alloy electrodes.

of layered positive and negative electrode sheets, interlayered with separators. These structures are similar to that of the nickel–cadmium battery.

Figure 2.20 shows the charge–discharge characteristics of the AA-size nickel–MH battery in comparison with the nickel–cadmium battery produced by Sanyo Electric. Its capacity density is 1.5–1.8 higher than that of nickel–cadmium batteries.

Charging is the process of returning a discharged battery to a state in which it can be used again. The nickel–MH battery is normally charged with a constant current. This method has the advantage of allowing an easy calculation of the amount of charging based on the charging time. The standard for determining discharge capacity is a charging time of 16 h using a 0.1 C current at $20 \pm 5\,°C$. Battery voltage increases as the charging current increases, and decreases as the battery temperature increases. The general charging characteristics of a nickel–MH battery are shown in Figure 2.21. The battery voltage, gas pressure within the battery, and battery temperature change as time elapses under continued charging.

2.4 Nickel–MH Batteries

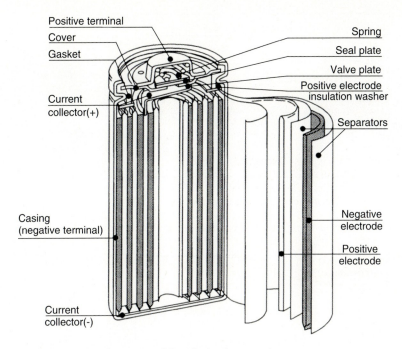

Figure 2.18 Internal structure of the cylindrical Ni–MH battery.

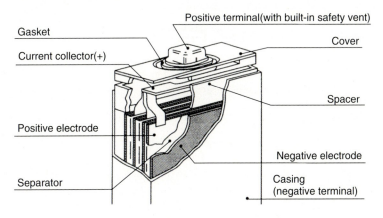

Figure 2.19 Internal structure of the prismatic Ni–MH battery.

The discharge voltage of nickel–MH batteries is almost the same as that of nickel–cadmium batteries.

Figure 2.22 shows the discharge characteristics at the 0.2, 1, and 3 C rate. The high-rate discharge characteristics of a nickel–MH battery compare unfavorably with those of a nickel–cadmium battery, because the specific surface area of the MH electrode is smaller than that of the cadmium electrode. Since the battery voltage drops dramatically if the discharge current exceeds 3 C, it is better to use a

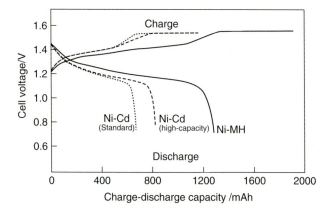

Figure 2.20 Charge–discharge characteristics of an Ni–MH battery (cell type AA).

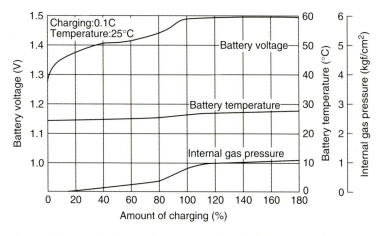

Figure 2.21 General charging characteristics of an Ni–MH battery (cell type 4/3 A).

current under 3 C. Figure 2.23 shows the charge–discharge characteristics. A life of 1000 cycles was obtained.

The outstanding characteristics of the nickel–MH battery are as follows:

1) A discharge capacity 80% higher than that of the standard nickel–cadmium battery;
2) A low internal resistance, which enables high-rate discharge;
3) A long charge–discharge cycle life, which can exceed 1000 cycle, and cell materials which are adaptable to the environment.

Since nickel–MH batteries were commercialized in 1990, they have become increasingly popular as a power source for computers, cellular phones, electric shavers, and other products.

The high capacity of the nickel–MH battery, which is approximately twice that of a standard nickel–cadmium battery, is possible because a hydrogen-absorbing alloy

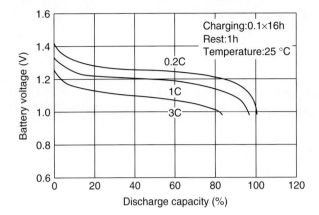

Figure 2.22 Discharge characteristics of an Ni–MH battery at various rates (cell type 4/3A).

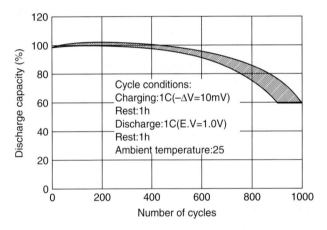

Figure 2.23 Charge–discharge characteristics of an Ni–MH battery (cell type 4/3 AA).

is used for the negative electrode. This alloy absorbs a large amount of hydrogen and features excellent reversibility of hydrogen absorption and desorption; thus the batteries' characteristics mainly depend on the physical and chemical properties of the hydrogen-absorbing alloy used for the negative electrode.

Improvement of Mm(Ni–Co–Al–Mn)$_x$ type alloys has been achieved in various ways. It was reported that alloys with a nonstoichiometric composition (Mm(Ni–Co–Mn–Al)$_x$: $4.5 \leq x \leq 4.8$) had a larger discharge capacity than those with stoichiometric alloys [25, 26]. Using X-ray diffraction analysis, it was found that the larger capacity is dependent on an increase in the unit cell volume of alloys with $x = 4.5$–4.8. It was also reported that annealing treatment improved the durability of this type of alloy.

The effects of both chemical compositional factors and the production process on the electrochemical properties of MH alloy electrodes were investigated

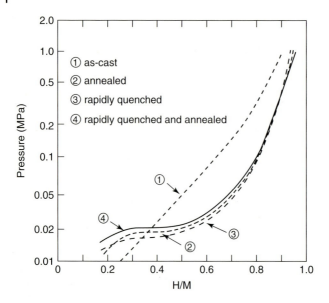

Figure 2.24 P–C isotherms of Mm(Ni–Co–Al–Mn)$_{4.76}$ alloys prepared through a rapid quenching and/or annealing process.

Figure 2.25 Nickel-metal hydride batteries manufactured by new technology.

[27]. Figure 2.24 shows the P–C isotherms of Mm(Ni–Co–Al–Mn)$_{4.76}$ alloys prepared by a rapid quenching and/or annealing process. The P–C isotherms of an induction-melted and as-cast alloy showed no plateau region, while the others, particularly the rapidly quenched and annealed alloy, showed clear plateau regions between 0.2 and 0.6 H/M, indicating that the rapid quenching and/or annealing process succeeded in homogenizing the microstructure. It was concluded that this process provides a larger hydrogen storage capacity in an alloy with a nonstoichiometric composition, AB$_{4.76}$.

Figure 2.25 shows nickel–MH batteries that have been improved by using the technique mentioned above.

These techniques are useful for improving cell characteristics such as cell capacity and charge–discharge cycle life.

2.5
Lithium Primary Batteries

The electrode potential of lithium is −3.01 V vs normal hydrogen electrode (NHE), which is the lowest value among all the metals. Lithium has the lowest density (0.54 g cm^{-3}) and the lowest electrochemical equivalent (0.259 g Ah^{-1}) of all solids. As a result of these physical properties, nonaqueous electrolyte batteries using lithium offer the possibility of high voltage and a high energy density. Organic and inorganic solvents which are stable with lithium are selected as the electrolytes for lithium batteries.

Primary lithium batteries offer these advantages as well as good low-temperature characteristics. There are many kinds of primary lithium batteries, with various cathode active materials; the main ones are lithium–manganese dioxide, lithium–carbon monofluoride, and lithium–thionyl chloride batteries [28].

2.5.1
Lithium–Manganese Dioxide Batteries

MnO_2 is used for the same purpose as the cathode active material in lithium–manganese dioxide (Li–MnO_2) batteries; it has been used for a long time in zinc–carbon and alkaline–manganese dioxide batteries, which are aqueous-electrolyte systems. In 1975, the Sanyo Electric Co. identified a novel reaction between lithium and MnO_2 and succeeded in exploiting this as the Li–MnO_2 battery. Sanyo has also granted the manufacturing technology for Li–MnO_2 batteries to major battery manufacturers around the world, and more than 15 companies are now producing it worldwide.

The following reaction mechanism is suggested to occur in Li–MnO_2:

$$\text{Anode reaction:} \quad Li \rightarrow Li^+ + e^- \quad (2.10)$$

$$\text{Cathode reaction:} \quad MnO_2 + Li^+ + e^- \rightarrow MnO_2^-(Li^+) \quad (2.11)$$

$$\text{Overall battery reaction:} \quad MnO_2 + Li \rightarrow MnO_2^-(Li^+) \quad (2.12)$$

where $MnO_2^-(Li^+)$ signifies that the lithium ion is introduced into the MnO_2 crystal lattice.

Figure 2.26 shows a schematic representation of the solid phase during the discharge of the MnO_2 crystal lattice, where tetravalent manganese is reduced to trivalent manganese.

In Li–MnO_2 batteries, lithium perchlorate ($LiClO_4$) or lithium trifluoromethanesulfonate ($LiCF_3SO_3$) is widely employed as an electrolytic solute, and mainly propylene carbonate (PC) and 1,2-dimethoxyethane (DME) are employed as a mixed solvent. The PC–DME–LiCLO$_4$ electrolyte shows high conductivity (>10^{-2} Ω^{-1} cm^{-1}) and low viscosity (<3 cP).

The requirements for the MnO_2 active material in Li–MnO_2 batteries are as follows:

1) It must be almost anhydrous.
2) It must have an optimized crystal structure suitable for the diffusion of Li$^+$ ions into the MnO_2 crystal lattice.

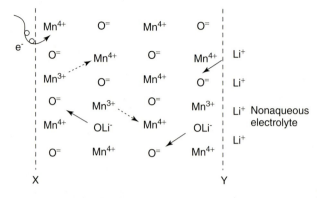

Figure 2.26 Schematic presentation of the solid phase during the discharge of MnO_2. The arrows show directions of movement of the electrons and lithium ions: →, lithium–ion movement; →, electron movement; X, MnO_2–electronic conductor interface; and Y, MnO_2–solution interface.

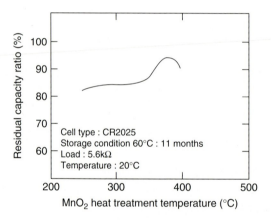

Figure 2.27 Relation of MnO_2 heat treatment temperature and residual capacity ratio after 11 months at 60 °C.

Although it is important that no water should exist in the cathode materials of nonaqueous batteries, the presence of a little water is unavoidable when MnO_2 is used as the active material. It is believed that this water is bound in the crystal structure, and that it has no effect on the storage characteristics, as shown in Figure 2.27, where the relationship of the MnO_2 heat-treatment temperature to the residual capacity ratio after 11 months of storage at 60 °C is plotted.

Figure 2.28 shows the discharge characteristics at a current density of 1.2 mA cm^{-2} of electrolytic MnO_2 heat-treated at various temperatures. From the characteristics shown, it may be concluded that the optimum heat-treatment temperature range for stable discharge is between 375 and 400 °C, which agrees with the data of Figure 2.27.

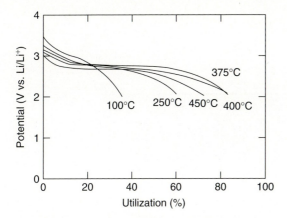

Figure 2.28 Discharge characteristics at a current density of 1.2 mA cm^{-2} of electrolytic MnO$_2$ heat-treated at various temperatures.

The general advantages of the Li–MnO$_2$ battery system are as follows:

1) **High voltage and high energy density.** Li–MnO$_2$ batteries are capable of maintaining a stable voltage of 3 V, which is about twice that of conventional dry-cell batteries. Because of this advantage, a single Li–MnO$_2$ battery can be used to replace two, and in practice even three, conventional dry-cell batteries.
2) **Excellent discharge characteristics.** Since Li–MnO$_2$ batteries are capable of maintaining stable voltage levels throughout long periods of discharge, a single battery can be used as the internal power source throughout the operational lifetime of a given item of equipment, eliminating the need for battery replacement. In addition, batteries using a crimp-sealed system with a spiral electrode can be used to provide high current discharge for a wide variety of applications.
3) **Superior leakage resistance.** The use of an organic solvent rather than an alkaline aqueous solution for the electrolyte results in significantly reduced corrosion and a much lower possibility of electrolyte leakage.
4) **Superior storage characteristics.** Li–MnO$_2$ batteries employing MnO$_2$, lithium, and a stable electrolyte exhibit a very low tendency toward self-discharge. The degree of self-discharge exhibited by Li–MnO$_2$ batteries stored at room temperature is as follows:
   a. Crimp-sealed batteries: 1% per annum
   b. Laser-sealed batteries: 0.5% per annum
5) **A wide operating-temperature range.** Because they use an organic electrolyte with a very low freezing point, lithium batteries operate at extremely low temperatures. Moreover, they demonstrate superior characteristics over a wide temperature range from cold to hot, as follows:
   a. Crimp-sealed batteries: -20 to $+70\,°C$
   b. Laser-sealed batteries: -40 to $+85\,°C$

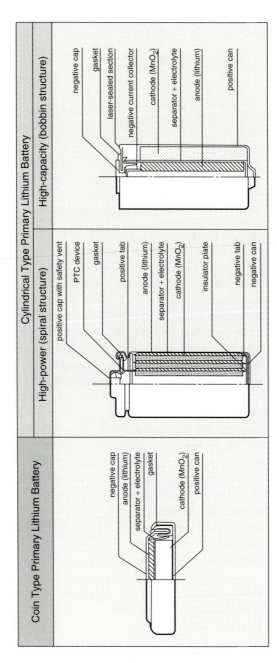

Figure 2.29 Shapes and construction of lithium–manganese dioxide batteries.

6) **A high degree of stability and safety.** Since Li–MnO$_2$ batteries do not contain toxic liquids or gases, they pose no major pollution problems.

Li–MnO$_2$ batteries are classified according to their shape and construction, which are shown in Figure 2.29.

The cathode of coin-type batteries consists of MnO$_2$ with the addition of a conductive material and binder. The anode is a disk made of lithium metal, which is pressed onto the stainless steel anode can. The separator is nonwoven cloth made of polypropylene, which is placed between the cathode and the anode.

Cylindrical batteries can be classified into two basic types: one with a spiral structure, and one with an inside-out structure. The former consists of a thin, wound cathode, and the lithium anode with a separator between them. The latter is constructed by pressing the cathode mixture into a high-density cylindrical form. Batteries with the spiral construction are suitable for high-rate drain, and those with the inside-out construction are suitable for high energy density.

The sealing system can also be classified into two types: crimp sealing and laser sealing. A comparison of these sealing methods is shown in Figure 2.30, the degree of airtightness with laser sealing being equivalent to a ceramic-based hermetic seal.

Figure 2.31 shows the construction of the 2CR5 Li–MnO$_2$ battery, which is used as the central power source for fully automatic cameras. The 2CR5 is composed of two CR15400 batteries connected in series. It is encapsulated in a plastic material and designed in shapes that will prevent misuse. The nominal voltage of the 2CR5 is 6 V.

When the 2CR5 is short-circuited, a thermal protector prevents the battery from overheating by substantially increasing the protector resistance. When the short circuit is removed, the 2CR5 operates normally. The thermal protector does not impede the ability of the 2CR5 to deliver high current. When the discharge current is depleted, the user can easily remove the 2CR5 from the camera and replace it with a new one.

Li–MnO$_2$ batteries are available in a variety of shapes and construction [29] in accordance with their particular use. Figure 2.32 shows various applications of lithium batteries based on their drain current. Coin-type batteries are generally

Name	Crimp-sealing	Laser-sealing
Structure	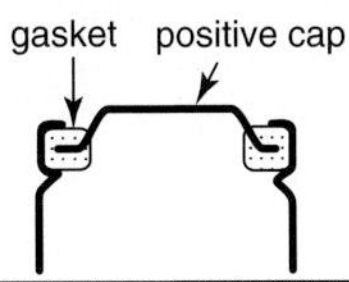 gasket positive cap	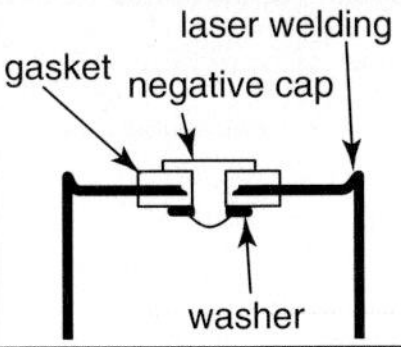 laser welding, gasket, negative cap, washer
Helium gas leakage (atm. cc/sec.)	10^{-7}	10^{-9}

Figure 2.30 The relationship of the seal type to the leak rate of helium for cylindrical lithium–manganese dioxide batteries.

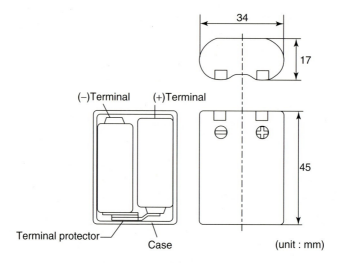

Figure 2.31 The construction, shape and dimensions of the 2CR5 lithium–manganese dioxide battery for fully automatic cameras.

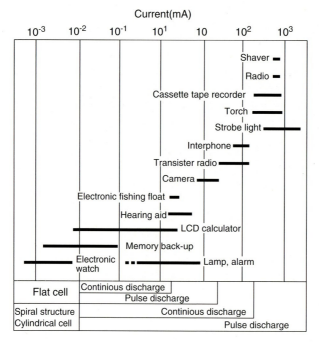

Figure 2.32 Various applications of lithium-manganese dioxide batteries, based on their drain currents.

used for low-rate drain. Cylindrical batteries with the inside-out construction can serve as a memory backup power source. Cylindrical batteries with the spiral construction, as mentioned above, are suitable for high-rate drains such as strobe light sources and camera autowiding systems, and will spread to various other fields, as high-power sources for cassette tape recorders, high-performance lights, 8 mm VTRs, LCD TVs, mobile telephones, transceivers, and other highly portable electronic equipment. Tables 2.1–2.4 show the specifications of coin-type, cylindrical inside-out construction, cylindrical spiral construction, and user-replaceable batteries, respectively.

Figure 2.33 shows the load characteristics of the coin-type CR2032. The cell voltage of the battery is approximately 3 V. Figure 2.34 shows the temperature characteristics of the CR2032. The battery discharges at a stable voltage over the wide temperature range of −20 to 70 °C.

Figure 2.35 shows the storage characteristics of the cylindrical inside-out construction CR17335SE. This battery demonstrates extremely good storage

Table 2.1 Specifications of coin-type manganese dioxide–lithium batteries.

Model	Nominal voltage (V)	Nominal capacity (mAh)	Standard discharge current (mA)	Discharge current (mA)		Dimensions (mm)		Weight (g)
				Maximum	Standard	Diameter	Height	
CR1220	3	35	0.1	2	10	12.5	2.0	0.8
CR1620	3	60	0.2	3	20	16.0	2.0	1.2
CK2016	3	80	0.3	5	50	20.0	1.6	1.7
CR2025	3	155	0.3	6	50	20.0	2.5	2.7
CR2032	3	220	0.3	4	40	20.0	3.2	3.2
CR2430	3	280	0.3	6	50	24.5	3.0	4.0
CR2450	3	560	0.3	3	50	24.5	5.0	6.2

Table 2.2 Specifications of cylindrical, inside-out construction, manganese dioxide-lithium batteries.

Model	Nominal voltage (V)	Nominal capacity (mAh)	Standard discharge current (mA)	Discharge current (mA)		Dimensions (mm)		Weight (g)
				Maximum	Standard	Diameter	Height	
CR4520SE	3	850	0.5	7	70	14.5	25.0	9
CR12600SE	3	1500	1.0	15	250	12.0	60.0	15
CR17335SE	3	1800	1.0	8	100	17.0	33.5	17
CR17450SE	3	2500	1.0	9	150	17.0	45.0	22
CR23500SE	3	5000	1.0	10	200	23.0	50.0	42

Table 2.3 Specifications of cylindrical, spiral construction, manganese dioxide–lithium batteries.

Model	Nominal voltage (V)	Nominal capacity (mAh)	Standard discharge current (mA)	Discharge current (mA)		Dimensions (mm)		Weight (g)
				Maximum	Standard	Diameter	Height	
CR–1/3N	3	160	2	60	80	11.6	10.8	3.3
2CR–/3N	6	160	2	60	80	13.0	25.2	9.1
CR 15270	3	750	10	1000	2500	15.5	27.0	11
CR 15400	3	1300	10	1500	3500	15.5	40.0	17
CR 17335	3	1300	10	1500	3500	17.0	33.8	16
CR17450E–R	3	2000	5	1000	2500	17.0	45.0	22

Table 2.4 Specifications of cylindrical, spiral construction, user-replaceable, manganese dioxide-lithium batteries.

Model	Nominal voltage (V)	Nominal capacity (mAh)	Standard discharge current (mA)	Discharge current (mA)		Dimensions (mm)		Weight (g)
				Maximum	Standard	Diameter	Height	
CR2	3	750	10	1000	2500	15.6	27.0	3.3
CR123A	3	1300	10	1500	3500	17.0	34.5	9.1
CR-P2	6	1300	10	1500	3500	$34.8(L) \times 19.5(W) \times 35.8(H)$	–	11
2CR5	3	1300	10	1500	3500	$34(L) \times 17(W) \times 45(H)$	–	17

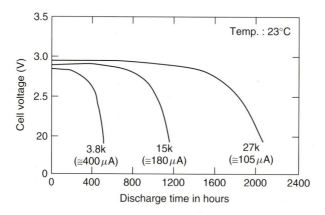

Figure 2.33 Load characteristics of the CR2032 lithium–manganese dioxide battery.

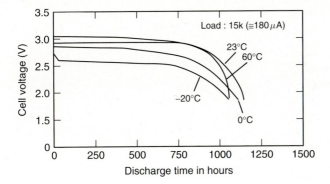

Figure 2.34 Temperature characteristics of the CR2032 lithium–manganese dioxide battery.

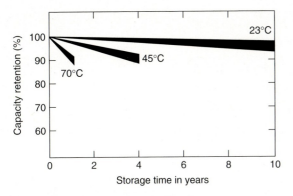

Figure 2.35 Self-discharge characteristics of the CR17335SE lithium–manganese dioxide battery.

characteristics; storage for 100 days at 70 °C is equivalent to 10 years at room temperature. Figure 2.36 shows the pulse discharge characteristics of the user-replaceable 2CR5. The operating voltage is stable over the wide temperature range of −20 to 60 °C. It can be used as a power source for tape recorders, LCD TVs, camera motors for film rewinding, and camera flash systems. Figure 2.37 shows practical tests of the 2CR5 in a fully automatic camera at 23 °C. When the shutter is released, the discharged current powers the exposure meter and the electromagnetic shutter, and it is also used for winding the film and charging the strobe light for the next photograph. Since the strobe light can be charged within 2 s, continuous photographs can be taken with the strobe light at short time intervals, as the figure shows. Continuous photographs can be taken with the strobe light even at −40 °C. Moreover, there is no voltage delay during the initial discharge stage, even at low temperatures at high pulse rates [30–34].

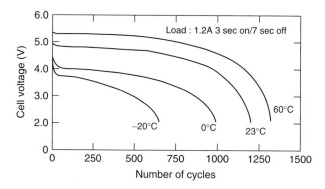

Figure 2.36 Pulse discharge characteristics of the 2CR5 lithium–manganese dioxide battery.

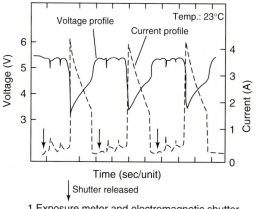

1 Exposure meter and electromagnetic shutter
2 Winding of film
3 Charge fo strobe light

Figure 2.37 Practical test results of a 2CR5 lithium–manganese dioxide battery in a fully automatic camera at 23 °C.

2.5.2
Lithium–Carbon Monofluoride Batteries

The world's first Li–(CF)$_n$ battery was developed by Matsushita Battery Industrial Co., and several types are being manufactured.

(CF)$_n$ is prepared by the reaction of carbon powder with fluorine gas at an elevated temperature. The properties of (CF)$_n$ are similar to those of polytetrafluoroethylene (PTFE) which is prepared by organic synthesis.

The discharge reaction of (CF)$_n$ is generally considered to be as follows:

$$\text{Anode reaction:} \quad n\text{Li} \rightarrow n\text{Li}^+ + ne^- \tag{2.13}$$

Cathode reaction: $(CF)_n + ne^- \rightarrow nC + nF^-$ (2.14)

Overall battery reaction: $nLi + (CF)_n \rightarrow nC + nLiF$ (2.15)

The general advantage of the Li–(CF)$_n$ batteries are the same as those of Li–MnO$_2$ batteries.

They may be classified by their structure, as coin, cylindrical, and pin types. Tables 2.5–2.7 respectively show their specifications. Applications of Li–(CF)$_n$ batteries as power sources are spreading from professional and business uses, such as in wireless transmitters and integrated circuit (IC) memory preservation, to consumer uses in electronic watches, cameras, calculators, and the like. Pin-type batteries are used for illumination-type fishing floats with a light-emitting diode. Coin-type batteries, which have a stable packing insulation, separator, and electrolyte for high-temperature usage, are applicable at temperatures as high as 150 °C. The packing insulation and separator are made of special-use

Table 2.5 Specifications of coin-type lithium–carbon monofluoride batteries.

Model	Nominal voltage (V)	Nominal capacity (mAh)	Discharge current (mA)		Dimensions (mm)		Weight (g)
			Maximum	Standard	Diameter	Height	
BR1216	3	25	5	0.03	12.5	1.60	0.6
BR1220	3	35	5	0.03	12.5	2.00	0.7
BR1225	3	48	5	0.03	12.5	2.50	0.8
BR1616	3	48	8	0.03	1.60	1.60	1.0
BR1632	3	120	8	0.03	16.0	3.20	1.5
BR2016	3	75	10	0.03	20.0	1.60	1.5
BR2020	3	100	10	0.03	20.0	2.00	2.0
BR2032	3	190	10	0.03	20.0	3.20	2.5
BR2320	3	110	10	0.03	23.0	2.00	2.5
BR2325	3	165	10	0.03	23.0	2.50	3.2
BR2330	3	255	10	0.03	23.0	3.00	3.2
BR3032	3	500	10	0.03	30.0	3.20	5.5

Table 2.6 Specifications of cylindrical lithium–carbon monofluoride batteries.

Model	Nominal voltage (V)	Nominal capacity (mAh)	Discharge current (mA)		Dimensions (mm)		Weight (g)
			Maximum	Standard	Diameter	Height	
BR–2/3A	3	1200	250	2.5	17.0	33.5	13.5
BR–A	3	1800	250	2.5	17.0	45.5	18.0
BR–C	3	5000	300	150.0	26.0	50.5	42.0

Table 2.7 Specifications of pin-type lithium–carbon monofluoride batteries.

Model	Nominal voltage (V)	Nominal capacity (mAh)	Discharge current (mA)		Dimensions (mm)		Weight (g)
			Maximum	Standard	Diameter	Height	
BR425	3	25	4	0.5	4.2	25.9	0.55
BR435	3	50	6	1.0	4.2	35.9	0.85

Table 2.8 Specification of coin-type lithium–carbon monofluoride batteries for high-temperature range.

Model	Nominal voltage (V)	Nominal capacity (mAh)	Dimensions (mm)		Operating temperature range (°C)
			Diameter	Height	
BR1225A	3	48	12.5	2.5	−40 to 150
BR1632A	3	120	16.0	3.0	−40 to 150

engineering plastics. Table 2.8 shows the specifications of coin-type batteries for high-temperature usage [35].

2.5.3
Lithium–Thionyl Chloride Batteries

The Li–SOCl$_2$ battery consists of a lithium-metal foil anode, a porous carbon cathode, a porous non-woven glass or polymeric separator between them, and an electrolyte containing thionyl chloride and a soluble salt, usually lithium tetrachloroaluminate. Thionyl chloride serves as both the cathode active material and the electrolytic solvent. The carbon cathode serves as a catalytic surface for the reduction of thionyl chloride and as a repository for the insoluble products of the discharge reaction.

Although the detailed mechanism for the reduction of thionyl chloride at the carbon surface is rather complicated and has been the subject of much controversy, the battery reactions are described as follows:

$$\text{Anode reaction:} \quad 4\text{Li} \rightarrow 4\text{Li}^+ + 4e^- \tag{2.16}$$

$$\text{Cathode reaction:} \quad 2\text{SOCl}_2 + 4\text{Li}^+ + 4e^- \rightarrow 4\text{LiCl} + \text{S} + \text{SO}_2 \tag{2.17}$$

$$\text{Overall battery reaction:} \quad 4\text{Li} + 2\text{SOCl}_2 \rightarrow 4\text{LiCl} + \text{S} + \text{SO}_2 \tag{2.18}$$

Sulfur dioxide is soluble in the electrolyte. Sulfur is soluble up to about 1 mol dm^{-3}, but it precipitates in the cathode pores near the end of discharge. Lithium

Table 2.9 Specifications of cylindrical lithium–thionyl chloride batteries.

Model	Nominal voltage (V)	Nominal capacity (mAh)	Dimensions (mm)		Weight (g)
			Diameter	Height	
ER3V P	3.6	1000	19.5	24.5	8.5
ER4V P	3.6	1200	19.5	24.5	10
ER6V P	3.6	2000	19.5	47.0	16
ER6LV P	3.6	1800	19.5	47.0	16
ER17330V P	3.6	1700	20.5	29.5	13
ER17500V P	3.6	2700	20.5	47.0	19

chloride is essentially insoluble and precipitates on the surfaces of the pores of the carbon cathode, forming an insulating layer which terminates the operation of cathode-limited cells [36].

The battery, which features a high (3.6 V) operating voltage and wide operating temperature range (−55 to 85 °C) can serve as a memory backup power source. Table 2.9 shows their specifications [37].

2.6 Coin-Type Lithium Secondary Batteries

2.6.1 Secondary Lithium–Manganese Dioxide Batteries

It has been [38] reported that MnO_2 has poor rechargeability [11, 12]. However, most investigations were on γ, γ/β, and β-MnO_2, which are similar to the MnO_2 used in primary Li–MnO_2 batteries. In γ/β-MnO_2, an expansion of the crystal lattice occurs when Li^+ ions are inserted into its crystal structure. However, the degree of expansion does not increase much after a large initial change at quite a low level of discharge. It was considered that, if MnO_2 contained a small amount of Li in its crystal structure beforehand, the reversibility of its crystal structure would be improved.

In order to improve the rechargeability of γ/β-MnO_2, two types of lithium-containing manganese oxides, spinel $LiMn_2O_4$ and heat-treated $LiOH_3$·MnO_2 (composite dimensional manganese oxide: CDMO), were prepared. First, the discharge and charge curves of γ/β-MnO_2, spinel $LiMn_2O_4$, and CDMO were measured. The cycle tests and discharge tests were carried out with flat–cells, with Li–Al alloy as the negative electrode; the electrolyte was 1 mol L^{-1} $LiClO_4$–PC/DME. The results are shown in Figure 2.38: when spinel $LiMn_2O_4$ capacity and CDMO were discharged to 2 V, both showed stable curves.

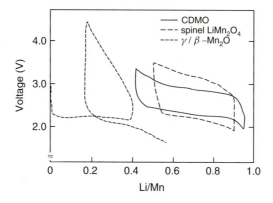

Figure 2.38 Discharge and charge curves for $\gamma/\beta-MnO_2$, spinel $LiMn_2O_4$, and CDMO electrodes.

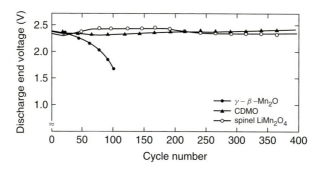

Figure 2.39 Cycling performance of various manganese oxide electrodes.

CDMO showed a 0.2 e/Mn larger capacity than spinel $LiMn_2O_4$, but γ/β-MnO_2 could not be fully charged to the 0.4 e/Mn level; in the second discharge, the discharge voltage of γ/β-MnO_2 was lower than that in the first discharge. Figure 2.39 shows the results of cycle tests on coin-type cells at a depth of 0.14 e/Mn. It was found that spinel $LiMn_2O_4$ and CDMO had better rechargeability than γ/β-MnO_2. No deterioration was observed in spinel $LiMn_2O_4$, or CDMO.

The crystal structure model of heat-treated $LiOH-MnO_2$ is considered to be as shown in Figure 2.40. It is composed of γ/β-MnO_2 which includes some Li, and Li_2MnO_3. γ/β-MnO_2 has one-dimensional channels, whereas Li_2MnO_3 has a structure in which Li atoms reside as layers, which accounts for its being named CDMO.

An Li–Al alloy was investigated for use as a negative electrode material for lithium secondary batteries. Figure 2.41 shows the cycle performance of an Li–Al electrode at 6% depth of discharge (DOD). The Li–Al alloy was prepared by an electrochemical method. The life of this electrode was only 250 cycles, and the Li–Al alloy was not adequate as a negative material for a practical lithium battery.

In order to clarify the reason for the deterioration in the Li–Al alloy electrode, morphological changes in it were investigated by scanning electron microscopy

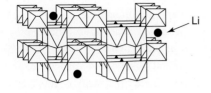

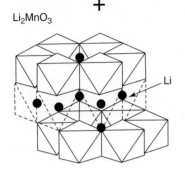

Figure 2.40 Proposed structure of CDMO.

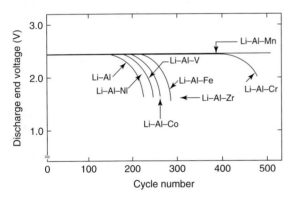

Figure 2.41 Cycling performance of several Li–Al alloy electrodes (discharge end 6% of total Li in Li–Al alloy; current density 1.1 mA cm^{-2}).

(SEM) after electrochemical alloying and cycling. The Li reacted with the Al nonuniformly during electrochemical alloying, and after the cycling fine particles were observed. It was thought that the pulverization resulted from the nonuniform reaction of Li and Al.

Several metal additives were investigated to improve this nonuniform reaction. Figure 2.41 shows the cycle performance of several Li–Al alloy electrodes. It was found that Li–Al–Mn and Li–Al–Cr alloys had better rechargeability than Li–Al alloy; in the Li–Al–Mn alloy, particularly no deterioration was observed even at the 500th cycle. It was confirmed by SEM that the Li–Al–Mn alloy did not turn to powder after cycling. Based on these results, Li–Al–Mn alloy was chosen as the negative electrode material for coin-type secondary lithium batteries.

Figure 2.42 shows the structure of the ML series of secondary lithium-manganese dioxide batteries, and Figure 2.43 shows the discharge curves of the ML2430 cell (diameter 24.5 mm, height 3.0 mm). The nominal voltage and capacity of the ML2430 are 3 V and 90 mAh, respectively, and the energy density is 160 Wh^{-1}. Figures 2.44 and 2.45 show the pulse characteristics and the dependence of discharge capacity on load; the discharge capacity is 90 mAh, even with a 1 kΩ load.

As regards the cycle performance, the ML2430 exhibits 3000 cycles at 5% depth and 500 cycles at a 20% depth of charge (Figure 2.46). It can be used over a wide range of temperatures, from −20 to 60 °C. The discharge capacity at −20 °C is 90% of the discharge capacity at 23 °C (Figure 2.47). The storage characteristics of the ML2430 were also measured (Figure 2.48); storage for 60 days at 60 °C is considered to be equivalent to storage for three years at room temperature. The loss of discharge capacity is less than 5% per year [39–44].

Finally, Table 2.10 shows the specifications of secondary lithium–manganese dioxide batteries. Recently, the use of these batteries as sources for memory backup has expanded remarkably [44].

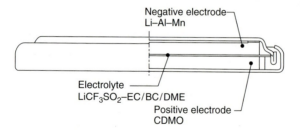

Figure 2.42 Cell structure of the Li–Al–CDMO cell (ML2430). EC, ethylene carbonate; BC, butylene carbonate.

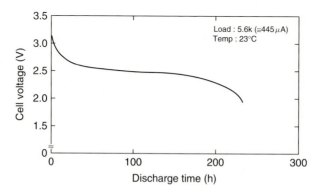

Figure 2.43 Discharge characteristics of the Li–Al–CDMO cell (ML2430).

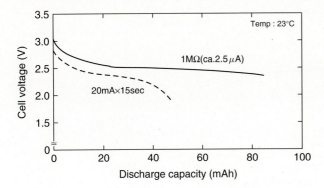

Figure 2.44 Pulse characteristics of the Li–Al–CDMO cell (ML2430).

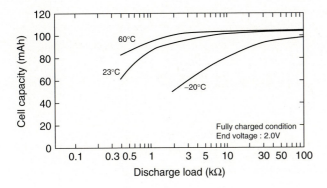

Figure 2.45 Dependence of discharge capacity on load (ML2430).

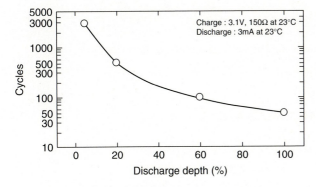

Figure 2.46 Cycling performance of the Li–Al–CDMO cell (ML2430). The number of 100% charge–discharge cycles is calculated until the capacity drops to 100% of the nominal value (end voltage 2.0 V). The number of 5, 20, and 60% charge–discharge cycles is calculated until an end voltage of 2.0 V.

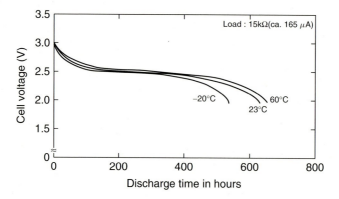

Figure 2.47 Discharge characteristics of the Li–Al–CDMO cell (ML2430) at several temperatures.

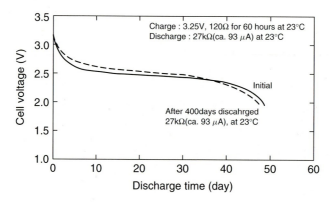

Figure 2.48 Storage characteristics of the Li–Al–CDMO cell (ML2430).

2.6.2
Lithium–Vanadium Oxide Secondary Batteries

Lithium–vanadium oxide rechargeable batteries were developed as memory backup power sources with high reliability and high energy density.

The active material of the positive electrode is vanadium pentoxide, and that of the negative electrode is a lithium–aluminum alloy. The electrolyte contains an organic solvent. The operating voltage is high, flat 3 V. The energy density is 100–140 Wh L^{-1}. The batteries have excellent overcharge-withstanding characteristics. They can serve as a memory backup power source, and they are applicable to various types of microcomputer equipment, because they can be installed in a small space. Table 2.11 shows the specifications of these batteries [45, 46].

2.6 Coin-Type Lithium Secondary Batteries

Table 2.10 Specifications of secondary lithium–manganese dioxide batteries.

Model	Nominal voltage (V)	Nominal capacity (mAh)	Standard charge–discharge current (mA)	Max discharge current (mA)		Charge–discharge cycle characteristics	Charging method	Dimensions (mm)		Weight (g)
				Continuous	Pulse			Diameter	Height	
ML1220	3	12	0.1	2	5	—	Constant-voltage charging at 3.1 ± 0.15 V	12.5	2.0	0.8
ML2016	3	25	0.3	8	20	3000 cycles (discharge depth of 0.5%)		20.0	1.6	1.8
ML2430	3	90	0.5	10	20	500 cycles (discharge depth of 20%)	2.95 ± 0.15 V (continuous charging at high temperature)	24.5	3.0	4.1

2 Practical Batteries

Table 2.11 Specifications of secondary lithium–vanadium oxide batteries.

Model	Nominal voltage (V)	Nominal capacity (mAh)	Discharge current (mA)		Dimensions (mm)		Weight (g)
			Maximum	Standard	Diameter	Height	
VL621	3	1.5	–	0.01	6.8	2.1	0.3
VL1261	3	5	–	0.03	12.5	1.6	0.7
VL1220	3	7	–	0.03	12.5	2.0	0.8
VL2020	3	20	–	0.07	20.0	2.0	2.2
VL2320	3	30	–	0.10	23.0	2.0	2.8
VL2330	3	50	–	0.10	23.0	3.0	3.7
VL3032	3	100	–	0.20	3.2	3.2	6.3

Table 2.12 Specifications of secondary lithium–polyaniline batteries.

Model	Nominal voltage (V)	Nominal capacity (mAh)	Standard discharge current (mA)	Cycling characteristics	Dimensions		Weight (g)
					Diameter	Height	
A1920	3	0.5 (3–2 V)	0.001–1	0.1 mAh > 1000 cycles	9.5	2.0	0.4
A12016	3	3 (3–2 V)	0.001–5	1 mAh > 1000 cycles	20.0	1.6	1.7
A12032	3	8 (3–2 V)	0.001–5	3 mAh > 1000 cycles	20.0	3.2	2.6

2.6.3 Lithium–Polyaniline Batteries

This battery is a completely new system with a conductive polymer of polyaniline for the positive electrode, a lithium alloy for the negative electrode and an organic solvent for the electrolyte. The battery features an operating voltage of 2–3 V. The energy density of the AL920 (diameter 9.5 mm, height 2.0 mm) is 11 Wh L^{-1}. It can serve as a memory backup power source. Table 2.12 shows the specifications of these batteries [47]. Chemically synthesized conductive polyaniline which is suitable for mass production has been investigated by Sanyo; conductive polymers of this type will be used as nonpollution materials in the future [48].

2.6.4 Secondary Lithium–Carbon Batteries

Some fusible alloys composed of Bi, Pb, Sn, and Cd exhibit good characteristics as material for the negative electrode of secondary lithium batteries. The alloy can absorb the lithium into the negative electrode during charge and it can release the absorbed lithium into the electrolyte as ions during discharge. Dendritic deposition does not occur and the coulombic efficiency is high, because lithium metal is not deposited.

The active material of the negative electrode is an alloy which contains 50% Bi, 25% Pb, 12.5% Sn, and 12.5% Cd. The active material of the positive electrode is an activated carbon in which the specific surface area is about $1000\,\mathrm{m^2\,g^{-1}}$, and it has an electrical capacity through the large electric double layer. Table 2.13 shows the specifications of the batteries [49]. The operating voltage is 2–3 V. The energy density of the CL2020 (diameter 20 mm, height 2 mm) is $4.0\,\mathrm{Wh\,L^{-1}}$. Long-term charge and discharge are possible. The batteries are used as a memory-backup supply for microcomputerized equipment and as maintenance-free power sources for solar-battery hybrid clocks, watches, and pocket calculators [50, 51].

2.6.5
Secondary Li-LGH–Vanadium Oxide Batteries

The active material of the negative electrode is a newly produced linear-graphite-hybrid (LGH) as the supporting carrier of lithium, and the active material of the positive electrode is made of amorphous V_2O_3–P_2O_5. By use of these active materials, the short cycle life of the charge–discharge characteristics due to lithium dendrite can be improved and the capacity decrease due to overdischarge can be reduced.

The battery features an operating voltage of 1.5–3 V. The energy density of the VG2025 (diameter 20.0 mm, height 2.5 mm) is $96\,\mathrm{Wh\,L^{-1}}$. It can serve as a memory backup power source. Table 2.14 shows the specifications of the batteries [52].

Table 2.13 Specifications of secondary lithium–carbon batteries.

Model	Nominal voltage (V)	Nominal capacity (mAh)	Dimensions (mm)		Weight (g)	Recommended discharge current
			Diameter	Height		
CL2020	3	1.0 (3–2 V)	19.7–20.0	2.0 ± 0.2	1.9	1 µA – 5 mA

Table 2.14 Specifications of secondary Li–LGH–amorphous V_2O_5 batteries.

Model	Nominal voltage (V)	Nominal capacity (mAh)	Dimensions (mm)		Weight (g)	Recommended discharge current
			Diameter	Height		
VG2025	3	25	20.0	2.5	2.5	−20 to 60 °C
VG2430	3	50	24.5	3.0	4.3	−20 to 60 °C

Table 2.15 Specifications of secondary lithium–polyacene batteries.

	Electrical characteristics (at room temperature)						Dimensions (mm)		
Model	Nominal voltage (V)	Nominal capacity (mAh)	Internal resistance (Ω)	Standard charging current (mA)	Standard charging method	Cycle time (min)	Diameter	Height	Weight
SL414	3	0.013	800	0.01–0.2	Constant-voltage charging	1000	4.8	1.4	0.06
SL614	3	0.07	160	0.001–0.5	–	–	6.8	1.4	0.16
SL621	3	0.15	190	0.001–1	–	–	6.8	2.1	0.2
SL920	3	0.30	90	0.001–3	–	–	9.5	2.1	0.4

2.6.6
Secondary Lithium–Polyacene Batteries

These batteries incorporate a polyacenic semiconductor (PAS) for the active material of the positive electrode, lithium for that of the negative electrode and an organic solvent for the electrolyte. PAS is essentially amorphous with a rather loose structure of molecular-size order with an interlayer distance of 4.0 Å, which is larger than the 3.35 Å of graphite [53, 54].

The batteries feature a high operating voltage of 2.0–3.3 V. The energy density of SL621 (diameter 6.8 mm, height 2.1 mm) is 6.5 Wh L^{-1}. It is applicable to various types of small, thin equipment requiring backup for memory and clock function. Table 2.15 shows the specifications of lithium–polyacene batteries [55].

2.6.7
Secondary Niobium Oxide–Vanadium Oxide Batteries

These batteries have vanadium oxide as the active material of the positive electrode, niobium oxide for the active material of the negative electrode, and an organic solvent for the electrolyte. Lithium ions enter the vanadium oxide from the niobium oxide during discharge, and lithium ions enter the niobium oxide from the vanadium oxide during charge.

The energy density of the VN1616 (diameter 16 mm) is 1.0–1.8 V. The charge–discharge cyclelife is in excess of 700. These batteries can be charged relatively fast and withstand over-discharging (0 V). They can serve as power sources for memory backup and for compact equipment in place of Ni–Cd button batteries. They are also applicable to medical equipment, solar clocks, solar radios, and pagers. Table 2.16 shows the specifications of the batteries [56].

Table 2.16 Specifications of secondary niobium oxide–vanadium oxide batteries.

Model	Nominal voltage (V)	Nominal capacity (mAh)	Dimensions (mm)		Weight (g)
			Diameter	Height	
VN621	1.5	1.2	6.8	2.1	0.3
VN1616	1.5	8	16	1.6	1.2

2.6.8
Secondary Titanium Oxide–Manganese Oxide Batteries

These batteries are new systems which use a lithium–manganese composite oxide for the active material of the positive electrode – a lithium–titanium oxide with a spinel structure for that of the negative electrode, and an organic solvent for the electrolyte. Lithium ions enter the manganese oxide from the titanium oxide during discharge and lithium ions enter the titanium oxide from the manganese oxide during charge.

The lithium–titanium oxides are prepared by heating a mixture of anatase (TiO_2) and LiOH at a high temperature. The product heated at 800–900 °C has a spinel structure of $Li_{4/3}Ti_{5/3}O_4$. When the charge and discharge cycles are performed between 2.5 and 0.5 V versus a lithium electrode, good cyclability (>400 cycles) is obtained in the plateau region. The open-circuit potential in the charge plateau is 1.58 V relative to lithium electrode and polarization during charge and discharge is small. The capacity density in the plateau is about 147 mAh g^{-1}, which corresponds to a 0.84-electron transfer for $Li_{4/3}Ti_{5/3}O_4$ [57, 58].

The batteries feature a 1.5 V operating voltage. The energy density of the MT920 (diameter 9.5 mm, height 2.0 mm) is 45 Wh L^{-1}. It is applicable to watches in which the power source is rechargeable with a solar battery, and it can serve as a memory backup power source. Table 2.17 shows the specifications of TiO_2–MnO_2 batteries [59].

Table 2.17 Specifications of secondary titanium oxide–manganese oxide batteries.

Model	Nominal voltage (V)	Nominal capacity (mAh)	Discharge current (mA)		Dimensions (mm)		Weight (g)
			Maximum	Standard	Diameter	Height	
MT621	1.5	1.2	–	0.1	6.8	2.1	0.3
MT920	1.5	3	–	0.2	9.5	2.0	0.3
MT1620	1.5	11	–	0.5	16.0	2.0	0.3

2.7
Lithium-Ion Batteries

Lithium-ion batteries are generally composed of lithium containing a transition-metal oxide as the positive electrode material and a carbon material as the negative electrode material. Figure 2.49 illustrates the principle of the lithium-ion battery. When the cell is constructed, it is in the discharged state. Then charged, lithium ions move from the positive electrode through the electrolyte and electrons also move from the positive electrode to the negative electrode through the external circuit with the charger. As the potential of the positive electrode rises and that of the negative electrode is lowered by charging, the voltage of the cell becomes higher. The cell is discharged by the connection of a load between the positive and negative electrodes. In this case, the lithium ions and electrons move in opposite directions while charging. Consequently, electrical energy is obtained.

2.7.1
Positive Electrode Materials

Many studies have been done on complex oxides of lithium and a transition metal, such as $LiCoO_2$, $LiNiO_2$, and $LiMn_2O_4$. $LiCoO_2$ and $LiNiO_2$ have α-$NaFeO_2$ structure. These materials are in space group $r3m$, in which the transition metal and lithium ions are located at octahedral 3(a) and 3(b) sites, respectively, and oxygen ions are at 6(c) sites. The oxygen ions form cubic close packing. This structure can be described as layered, giant, with alternating lithium-cation sheets and CoO_2/NiO_2-anion sheets.

In contrast, $LiMn_2O_4$ has a spinel structure. This material has the space group $Fd3m$ in which the transition-metal and lithium ions are located at octahedral 8(a) and tetrahedral 16(d) sites, respectively, and the oxygen ions are at 32(e) sites. There are octahedral 16(c) sites around the 8(a) sites and lithium ions can diffuse through

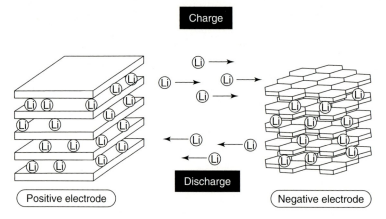

Figure 2.49 Principle of the lithium-ion battery.

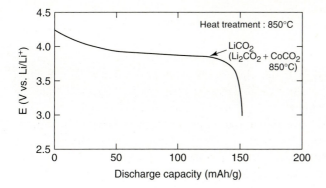

Figure 2.50 Discharge characteristics of LiCoO$_2$ (current density 0.25 mA cm^{-2}).

the 16(c) and 8(a) sites. As this structure contains a diffusion path for the lithium ions, these ions can be deintercalated and intercalated in these compositions.

The research on LiCoO$_2$ is more advanced because of the simplicity of sample preparation [60]. Figure 2.50 shows the first charge–discharge curves of LiCoO$_2$. The sample was prepared from Li$_2$CO$_3$ and CoCO$_3$. Lithium and cobalt salts were mixed well, and reacted at 850 °C for 20 h in air. The reaction conditions were such that the sample could show the maximum rechargeable capacity.

The electrolyte was a mixture of ethylene carbonate and diethyl carbonate containing 1 mol L^{-1} LiPF$_6$. In order to attain a high-voltage charge, an aluminum substrate was used. The data in Figure 2.50 were taken at the charge cutoff potential of 4.3 V (versus Li/Li^{-1}). The working voltage is extremely high, so an oxidation-resistant electrolyte is necessary in the development of 4 V secondary batteries.

As can be seen in Figure 2.50, the average working potential is about 3.6 V and re-chargeability is reasonably good. The capacity of LiCoO$_2$ was 150 mAh g^{-1}.

The conditions for synthesizing LiNiO$_2$ are said to be more complicated than those for LiCoO$_2$, but LiNiO$_2$ offers an advantage in terms of the availability of natural resources and cost [61–64]. Suitable conditions for synthesizing LiNiO$_2$, such as raw materials, heat-treating temperature, and atmosphere, have been investigated [65].

Lithium–nickel oxides form various lithium compounds, lithium hydroxides (LiOH), Li$_2$CO$_3$, nickel hydroxide (Ni(OH)$_2$), nickel carbonate (NiCO$_3$), and nickel oxide (NiO). Figure 2.51 shows the discharge characteristics of lithium–nickel oxides synthesized from these compounds. They were heat-treated at 850 °C for 20 h in air. Although the lithium–nickel oxides showed a smaller discharge capacity than that of LiCoO$_2$, LiOH and Ni(OH)$_2$ were considered to be appropriate raw materials.

Figure 2.52 shows the discharge characteristics of LiCoO$_2$ and lithium–nickel oxides prepared from LiOH and Ni(OH)$_2$ at 650, 750, and 850 °C. Lithium–nickel oxide heat-treated at 750 °C showed nearly the same discharge capacity as LiCoO$_2$ while the discharge potential was lower than that of LiCoO$_2$. Composition of these

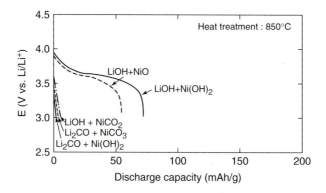

Figure 2.51 Discharge characteristics of some lithium–nickel oxides (current density 0.25 mA cm^{-2}).

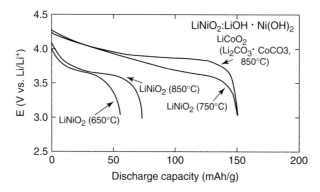

Figure 2.52 Discharge characteristics of some lithium–nickel oxides and LiCoO$_2$ (current density 0.25 mA cm^{-2}).

oxides was determined by chemical analysis. The compositions of lithium–cobalt oxide prepared at 850 °C and lithium–nickel oxides prepared at 650 and 750 °C were very close to LiCoO$_{2.0}$ and LiNiO$_{2.0}$, respectively. On the other hand, the composition of lithium–nickel oxides prepared at 850 °C was LiNiO$_{1.8}$, and the decrease in their discharge capacity was caused by oxygen defects in their structure.

In order to examine the influence of the heat-treatment atmosphere, LiCoO$_2$ and LiNiO$_2$ were synthesized in an oxygen atmosphere. As a result, LiNiO$_2$ heat-treated in oxygen showed much better discharge characteristics than that in the air or oxygen. LiNiO$_2$ heat-treated in oxygen showed a discharge capacity of more than 190 mAh g^{-1}, which was greater than that of LiCoO$_2$, as shown in Figure 2.53. From these results, LiOH and Ni(OH)$_2$ were found to be appropriate raw materials, and the most suitable conditions were 750 °C in oxygen, which produced a greater discharge capacity (more than 190 mAh g^{-1}) than LiCoO$_2$.

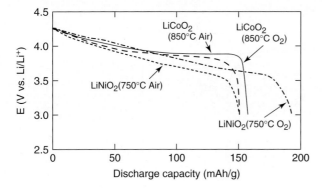

Figure 2.53 Discharge characteristics of LiNiO$_2$ and LiCoO$_2$ synthetized in air or oxygen (current density 0.25 mA cm^{-2}).

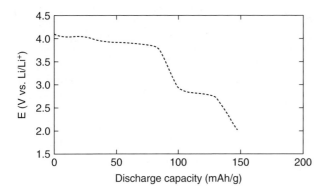

Figure 2.54 Discharge characteristics of LiMn$_2$O$_4$.

As LiMn$_2$O$_4$ offers an advantage in terms of the availability of natural resources and cost, many studies were made concerning charge–discharge characteristics and structure [57–68]. Figure 2.54 shows the discharge curve of LiMn$_2$O$_4$.

The operating voltage is extremely high, so an oxidation-resistant electrolyte is necessary for developing 4 V secondary batteries. As can be seen in Figure 2.54, the average operating potential is about 3.6 V and rechargeability is reasonably good. However, the discharge capacity of LiMn$_2$O$_4$ is less than 150 mAh g^{-1}. Consequently, the main feature of LiMn$_2$O$_4$ is its low cost, but the discharge capacity is also lower than LiCoO$_2$ and LiNiO$_2$.

LiCo$_{1-x}$Ni$_x$O$_2$ composite oxides consisting of LiNiO$_2$ and LiCoO$_2$ have also been studied; the influence of the Co/Ni ratio in these materials ($x = 0.1$–0.9) was examined. Figure 2.55 shows their discharge characteristics. The highest discharge capacity was obtained in the case of $x = 0.7$. The discharge capacity of LiCo$_{0.3}$Ni$_{0.7}$O$_2$ was more than 150 mAh g^{-1}; as it has almost the same capacity as LiCoO$_2$ and LiNiO$_2$, this material is desirable as the positive electrode material for lithium-ion batteries.

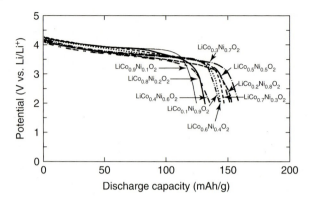

Figure 2.55 Discharge characteristics of $LiCo_xNi_{1-x}O_0$.

2.7.2
Negative Electrode Materials

Carbon materials which have the closest-packed hexagonal structures are used as the negative electrode for lithium-ion batteries; carbon atoms on the (0 0 2) plane are linked by conjugated bonds, and these planes (graphite planes) are layered. The layer interdistance is more than 3.35 Å and lithium ions can be intercalated and deintercalated. As the potential of carbon materials with intercalated lithium ions is low, many studies have been done on carbon negative electrodes [69–72].

There are many kinds of carbon materials, with different crystallinity. Their crystallinity generally develops due to heat-treatment in a gas atmosphere ('soft' carbon). However, there are some kinds of carbon ('hard' carbon) in which it is difficult to develop this crystallinity by the heat-treatment method. Both kinds of carbon materials are used as the negative electrode for lithium-ion batteries.

Soft carbon is also classified by its crystallinity. For example, acetylene black and carbon black are regarded as typical carbon materials with low crystallinity. Coke materials are carbon materials with intermediate crystallinity. It is easy to obtain these materials because they are made from petroleum and coal and they were actively studied in the 1980s. In contrast, there are some graphite materials which have high crystallinity; their capacity is greater than that of coke materials, and these materials have been studied more recently, in the 1990s [73–77].

Coke materials are generally made by heat-treatment of petroleum pitch or coal-tar pitch in an N_2 atmosphere. Coke made from petroleum is called '*petroleum coke*' and that from coal is called '*pitch coke*.' These materials have the closest-packed hexagonal structures. The crystallinity of coke materials is not so high as that of graphite. The crystallite size of coke along the c-axis (L_c) is small (about 10–20 Å) and the interlayer distance (d value; about 3.38–3.80 Å) is large.

Figure 2.56 shows the charge–discharge characteristics of coke materials such as petroleum coke and pitch coke in PC containing 1 mol L^{-1} LiPF$_6$. The discharge capacity of the coke electrodes was from 180 to 240 mAh g^{-1}. The initial efficiency (charge–discharge efficiency coulombic efficiency) of the coke electrodes was

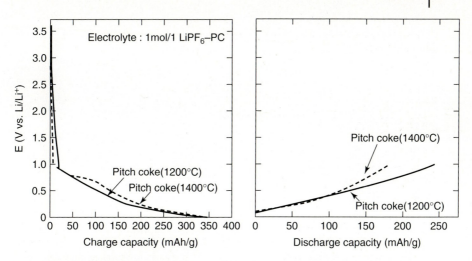

Figure 2.56 Charge–discharge characteristics of some carbon material electrodes (first cycle current density 0.2 mA cm^{-2}).

75–82%, and the efficiency after the second cycle was 100%. The charge–discharge characteristics in different electrolytes, such as butylene carbonate, γ-butyrolactone, sulfolane, and ethylene carbonate, were also tested. The results are almost the same as those for PC. It was found that the charge–discharge characteristics are not strongly influenced by the nature of the electrolyte.

The cycling characteristics of coke materials were also tested: the deterioration ratio of the charge–discharge efficiency after 500 cycles was small, and coke materials showed sufficiently good cycling performance to be used as negative electrode materials for lithium-ion batteries. The performance of coke materials does not depend very much on the electrolyte, but their disadvantage is low discharge capacity.

Graphite materials with high crystallinity are further classified by their production method. Graphite materials made by heat-treating coke materials at temperatures higher than about 2000 °C are called '*artificial graphite.*' On the other hand, there are also natural graphite materials which have the highest crystallinity of all carbon materials. These materials have the ideally closest-packed hexagonal structure. The L_c of natural graphite is more than 1000 Å and the d value is 3.354 Å, values which are close to the ideal graphite crystal structure [78, 79].

The crystallinity of artificial graphite can generally be controlled by the heat-treatment temperature, but it is lower than that of natural graphite. The L_c of artificial graphite is less than 1000 Å and the d value is more than 3.36 Å.

Figure 2.57 shows the charge–discharge characteristics of a natural graphite electrode in typical electrolytes such as PC and ethylene carbonate containing 1 mol L^{-1} LiPF$_6$. Natural graphite could not be charged in PC; the gas evolved during the attempt to charge was identified as propylene by gas chromatography, and it is

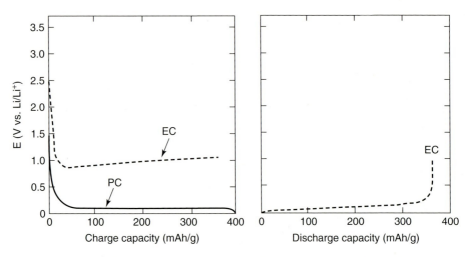

Figure 2.57 Charge–discharge characteristics of natural graphite in different electrolytes.

believed that its evolution was caused by the decomposition of the solvent. No gas evolution was observed in ethylene carbonate.

The discharge capacity of the natural graphite electrodes was 370 mAh g^{-1}. The initial efficiency of the coke electrodes was 92%, and the efficiency after the second cycle was 100%. The theoretical capacity of C$_6$Li is 372 mAh g^{-1}. These results suggest that C$_6$Li was produced by the electrochemical reduction of natural graphite, and the formation of C$_6$Li was confirmed. Figure 2.58 shows the X-ray diffraction pattern of natural graphite during charge: the peak was shifted to a lower angle by charging. In the case of full charging, the peak was 24°, which indicates the formation of C$_6$Li. The discharge capacity of natural graphite is close to that of C$_6$Li. Its charge–discharge curves are very flat and the charge–discharge potential is very low. These features are advantageous for lithium-ion batteries because it is anticipated that the voltage of a lithium-ion battery using natural graphite as the negative electrode is high and its charge–discharge curve is flat.

The charge–discharge characteristics of artificial graphite were also tested; artificial graphite could not be charged in PC for the same reason as natural graphite, but it could also undergo charge–discharge in ethylene carbonate. Figure 2.59 shows the charge–discharge characteristics of some graphite electrodes in ethylene carbonate containing 1 mol L^{-1} LiPF$_6$. Those of the artificial graphite electrode are also very flat, and the charge–discharge potential is also very low as for the natural graphite electrode.

However, the discharge capacity of artificial graphite is smaller than that of natural graphite, and depends on the heat-treatment temperature. Artificial graphite made by heat-treatment at a higher temperature showed a higher discharge capacity; as

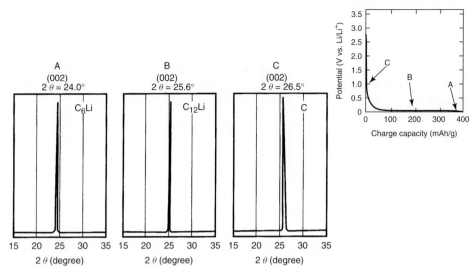

Figure 2.58 X-ray diffraction patterns of natural graphite.

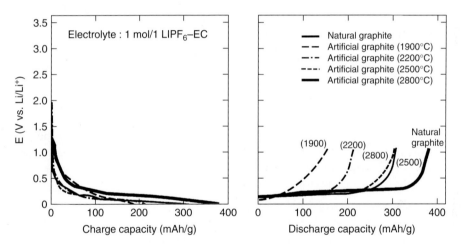

Figure 2.59 Charge–discharge characteristics of some graphite material electrodes (first cycle current density 0.2 mA cm^{-2}).

it has higher crystallinity, it is suggested that the discharge capacity of the graphite electrode may be related to the crystallinity of the graphite material.

The cycling characteristics of graphite electrodes were also tested. The deterioration ratio of the charge–discharge efficiency after 500 cycles was small and the graphite materials showed good cycling performance. The crystal structures of charged and discharged natural graphite electrodes at the 5th and 100th cycles were measured by the X-ray diffraction method; the results are shown in Figure 2.60. For both cycles, the peak shifted to a lower angle after charging, and 2θ of the peak

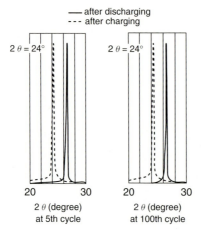

Figure 2.60 X-ray diffraction of natural graphite.

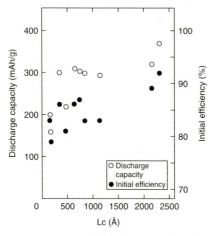

Figure 2.61 Relationship between discharge capacity, initial efficiency, and L_c of soft carbon materials.

was 24°, which indicates the formation of C_6Li. By discharging, the 2θ of the peak became 26.5°, which indicates the extraction of lithium. No change was observed in the crystal structure of natural graphite up to 100 cycles.

Figure 2.61 shows the relationship between the discharge capacity, the initial efficiency, and the L_c of some soft carbon materials when ethylene carbonate was used as a solvent. Figure 2.62 shows the relationship between the discharge capacity, the initial efficiency, and the d value in the same conditions. The carbon materials with longer L_c and smaller d values showed a higher discharge capacity and a higher initial charge–discharge efficiency. Natural graphite had the highest discharge capacity and the highest initial efficiency.

Both hard and soft carbons are used as negative electrode materials for lithium-ion batteries. Hard carbon is made by heat-treating organic polymer materials such

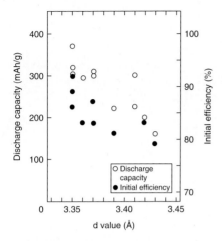

Figure 2.62 Relationship between discharge capacity, initial efficiency, and d value of soft carbon materials (○, discharge capacity; ●, initial efficiency).

as phenol resin. The heat-treatment temperature of these materials is the same as that of petroleum and coal when making coke materials.

Good charge–discharge characteristics have been reported [80], and the cycle characteristics were as good as those of soft carbon. The discharge capacity was strongly influenced by the charge–discharge conditions. There are some reports that the discharge capacity is larger than that of C_6Li as measured by the best charge–discharge method, but this method is difficult to use for practical lithium-ion batteries. The discharge capacity of hard carbon is expected to be smaller than that of soft carbon electrodes as measured by a practical charge–discharge method.

Polyacene is classified as a material which does not belong to either soft or hard carbons [81]. It is also made by heat-treatment of phenol resin. As the heat-treatment temperature is lower than about 1000 °C, polyacene contains hydrogen and oxygen atoms. It has a conjugated plane into which lithium ions are doped. It was reported that the discharge capacity of polyacene is more than 1000 mAh g^{-1}. However, there are no practical lithium-ion batteries using polyacene.

2.7.3
Battery Performances

Figure 2.63 shows the structure of a commercialized cylindrical-type lithium-ion battery. The lithium-ion battery is generally constructed with a spiral structure which serves as the separator between the positive and negative electrodes. An organic electrolyte containing lithium salts of which the conductivity is smaller than that of an aqueous electrolyte is used for this battery, but the short distance between the positive and negative electrodes and the large area of the electrode confer good characteristics.

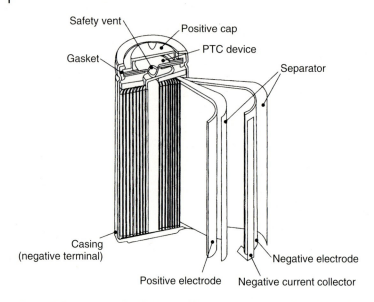

Figure 2.63 Structure of a lithium-ion battery. PTC, positive thermal coefficient device.

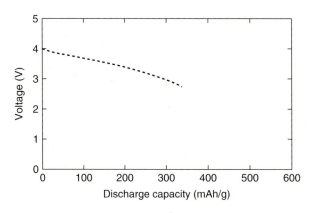

Figure 2.64 Discharge characteristic of an LiCoO$_2$/coke cell.

As mentioned above, the typical positive electrode material is LiCoO$_2$, and there are typically two types of negative electrode materials, such as coke and graphite. The characteristics of lithium-ion batteries constructed using these electrode materials are discussed below.

Figure 2.64 shows the discharge characteristics of a cylindrical type LiCoO$_2$–coke cell. The discharge capacity is 350 mAh, the average discharge voltage is 3.6 V, and the energy density of this battery is 164 Wh L^{-1} or 66 Wh kg^{-1}. The cell voltage of this battery decreased greatly during discharge. This feature is not favorable from the viewpoint of total energy density, but it is easy to determine the residual capacity from the cell voltage [82].

Figure 2.65 shows commercialized LiCoO$_2$-natural graphite cells and Table 2.18 shows the specification of these batteries. There are two types of batteries: cylindrical and prismatic. The cylindrical-type battery in Figure 2.65 is called *18650* because its diameter is 18 mm and its length is 65 mm.

Figure 2.66 shows the discharge characteristics of the 18650-type cell. The discharge capacity is 1350 mAh, the average discharge voltage is 3.6 V, and the

Figure 2.65 LiCoO$_2$-natural graphite cells.

Table 2.18 Specifications of LiCoO$_2$-natural graphite cells.

Model	Nominal voltage (V)	Nominal capacity (mAh)[a]	Standard charging method	Dimensions (mm)				Weight
				Diameter	Thickness	Width	Height	
Cylindrical	–	–	–	–	–	–	–	–
UR18650	3.6	1350	1 C to 4.1 V	18	–	–	6.5	~40
UR18500	3.6	900	Constant	18	–	–	50	~30
Rectangular	–	–	Current–	–	–	–	–	–
UF812248	–	550	Constant	–	8.1	225	48	~18
UF102248	3.6	750	Voltage	–	10.5	22.5	48	~24
UF611958	3.6	400	2.5/h	–	6.1	19.5	58	~15

[a]Guaranteed discharge capacity at 0.2 C (E_v = 2.7.5 V).
Operating temperature: charge at 0–40 °C, discharge at −20 to 60 °C.

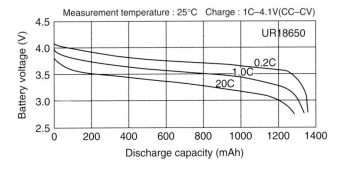

Figure 2.66 Discharge characteristics of the LiCoO$_2$-natural graphite cell.

energy density of this battery is 294 Wh L^{-1} or 122 Wh kg^{-1}. The energy density of this battery is higher than that of the LiCoO$_2$-coke cell, and its decrease in cell voltage was small during discharge, which is favorable from the viewpoint of total energy density.

These features are caused by the graphite negative electrode. The LiCoO$_2$-graphite system is superior to LiCoO$_2$-coke in energy density and charge–discharge characteristics [83].

As the cost of LiCoO$_2$ is high, other positive electrode materials will eventually take the place of LiCoO$_2$. LiNiO$_2$ and LiMnO$_2$ are often mentioned as positive electrode materials instead of LiCoO$_2$ [84]. LiNiO$_2$ is desirable because it offers a larger capacity and lower cost than LiCoO$_2$, and it is expected that a LiNiO$_2$-graphite cell will be commercialized in the near future.

2.8
Secondary Lithium Batteries with Metal Anodes

Secondary lithium-metal batteries which have a lithium-metal anode are attractive because their energy density is theoretically higher than that of lithium-ion batteries. Lithium–molybdenum disulfide batteries were the world's first secondary cylindrical lithium-metal batteries. However, the batteries were recalled in 1989 because of an overheating defect. Lithium–manganese dioxide batteries are the only secondary cylindrical lithium–metal batteries which are manufactured at present. Lithium–vanadium oxide batteries are being researched and developed. Furthermore, electrolytes, electrolyte additives, and lithium surface treatments are being studied to improve safety and rechargeability.

Li–MoS$_2$ batteries were developed by Moli Energy; lithium is intercalated into the positive MoS$_2$ material. The value of x can vary from about 0.2 for a fully charged battery to about 1.0 for a fully discharged battery in accordance with reaction:

$$x\text{Li} + \text{MoS}_2 \rightarrow \text{Li}_x\text{MoS}_2 \tag{2.19}$$

Products include an AA-size battery, a C-size battery, and a developmental butylene carbonate (BC)-size (diameter 66 mm, height 152 mm) battery with a nominal 65 Ah capacity. The features of these batteries are a long charge-retention time, a direct state-of-charge indicator based on a variable open-circuit voltage, a high energy density relative to that of other rechargeable batteries, and a high power density [85].

A new rechargeable Li-Li$_x$MnO$_2$ 3 V battery system was developed by Tadiran Ltd. The active material of the negative electrode is lithium metal, and that of the positive electrode is lithiated manganese dioxide. These batteries have an organic electrolyte and separator, and exhibit excellent performance and safe behavior. The capacity of the AA–size battery is 800–750 mAh, and the energy density is 125–145 Wh kg^{-1} or 280–315 Wh L^{-1}. At charging regimes around C/10, more than 350 cycles at 100% DOD could be obtained. An accumulated capacity of about 200 Ah can be achieved under cycling [86].

The system can prevent explosion, fire, and venting with fire under conditions of abuse. These batteries have a unique battery chemistry based on $LiAsF_6$/1,3-dioxolane/tributylamine electrolyte solutions which provide internal safety mechanism that protect the batteries from short-circuit, overcharge, and thermal runaway upon heating to 135 °C. This behavior is due to the fact that the electrolyte solution is stable at low-to-medium temperatures but polymerizes at a temperature over 125 °C [87].

The active material of the negative electrode is lithium metal, and that of the positive electrode is amorphous V_2O_5 (a–V_2O_5). A prototype AA-size battery has an energy of 2 Wh (900 mAh), an energy density of 110 Wh kg^{-1} or 250 Wh L^{-1}, and a life of 150–300 cycles depending on the discharge and charge currents.

One of the most important factors determining whether or not secondary lithium metal batteries become commercially viable is battery safety, which is affected by many factors: insufficient information is available about safety of practical secondary lithium metal batteries [88]. Vanadium compounds dissolve electrochemically and are deposited on the lithium anode during charge–discharge cycle. The low reactivity of the vanadium-deposited lithium anode has been observed by calorimetry; a chemical-state analysis and morphological investigation of the lithium anode suggest that the improvement in stability is primarily due to a passivation film [89].

Films on lithium play an important part in secondary lithium metal batteries. Electrolytes, electrolyte additives, and lithium surface treatments modify the lithium surface and change the morphology of the lithium and its current efficiency [90].

Various cyclic ethers are reported to be superior solvents for secondary lithium metal batteries. 1,3-Dioxolane [91, 92] and DME [92] show good cyclic characteristics. 1,3-Dioxolane-LiB$(CH_3)_4$ is highly conductive and has shown utility as an electrolyte in ambient temperature secondary lithium battery systems wherein a high rate of current drain is a desirable feature [93]. Researchers at Exxon used 1,3-dioxolane or DME-LiClO$_4$ or LiB$(C_6H_5)_4$ and 2-methyltetrahydrofuran-LiAsF$_6$ (2MeTHF) in a lithium-titanium disulfide system [94]. 1,3-Dioxolane-DME-Li$_2$B$_{10}$Cl$_{10}$ exhibited chemical stability toward the components of a lithium-titanium disulfide cell and showed promise as an electrolyte in such cells [95]. Among various systems composed of an ether-based solvent and a lithium salt, THF-LiAsF$_6$ was the least reactive to lithium at elevated temperature and gave the best cycling efficiency [96, 97]. THF-diethyl ether-LiAsF$_6$ afforded lithium electrode cycling efficiency in excess of 98% [98].

2MeTHF showed good cycling characteristics [99–101], and 2MeTHF–LiAsF$_6$ showed promise of yielding high energy density and cycle life [102]. In an investigation of THFs methylated in the α-position: 2MeTHF-LiAsF$_6$ and 2,5-dimethyltetrahydrofuran–LiAsF$_6$ showed good cycling characteristics [103].

Several solvents other than ethers have also been reported to be superior solvents for secondary lithium batteries. Ethylene carbonate showed good cycling characteristics [104, 105].

The addition of 2-methylfuran, thiophene, 2-methylthiophene, pyrrole, and 4-methylthiazole to PC-LiPF$_6$ or PC-THF-LiPF$_6$ improved the cycling efficiency

[106]. THF-2MeTHF-LiAsF$_6$ with an additional of 2-methylfuran showed the longest cycle life [107, 108].

The addition of some metal ions, such as Mg^{2+}, Zn^{2+}, In^{3+}, or Ga^{3+}, and some organic additives, such as 2-thiophene, 2-methylfuran, or benzene, to PC-LiClO$_4$ improved the coulombic efficiency for lithium cycling [109]. Lithium deposition on a lithium surface covered with a chemically stable, thin, and tight layer which was formed by the addition of HF to electrolyte can suppress the lithium dendrite formation in secondary lithium batteries [110].

The dendritic growth of lithium was suppressed on a lithium electrode surface modified by an ultrathin solid polymer electrolyte prepared from 1,1-difluoroethane by plasma polymerization [111].

A high rate discharge led to the recombination of isolated lithium which resulted in an increase in cycle life, and the cycle life decreased with an increase in the charge current density [112].

While the initial surface species formed on lithium in alkyl carbonates consist of ROCO$_2$Li compounds, these species react with water to form Li$_2$CO$_3$, CO$_2$, and ROH. This reaction gradually changes the composition of the surface films formed on lithium in these solvents, and Li$_2$CO$_3$ becomes the major component [113]. A film of Li$_2$CO$_3$ was formed on lithium by the direct reaction of PC with lithium [114]. Diethyl carbonate was found to react with lithium to form lithium ethyl carbonate [115]. The main reaction products in the surface film on lithium were CH$_3$OLi in DME, and C$_4$H$_9$OLi in THF [116]. A surface film which contained ROLi, ROCO$_2$Li, and Li$_2$CO, was formed on lithium in 1,3-dioxolane-LiClO$_4$ [117].

A lithium electrode is reported to show high rechargeability in solutions containing LiAsF$_6$. A brown film composed of an $(-O-As-O)_n$ polymer and LiF was formed on lithium in THF-LiAsF$_6$ [118]; elsewhere, a film on lithium was determined to be As$_2$O$_3$ and F$_2$AsOAsF$_2$ in THF–LiAsF$_6$ [119]. Another film had a probable two-layer structure consisting of Li$_2$O covered by an outer Li$_2$O–CO$_2$ adduct in 2MeTHF-LiAsF$_6$ [120]. A film of reduction product ROCO$_2$Li was formed on lithium in ethylene carbonate-LiAsF$_6$ or PC-LiAsF$_6$ [121].

As the cycling efficiency of metallic lithium is always significantly below 100% (~99%), the lithium anode has to be over-dimensioned (200–400%) in practical cells.

References

1. (1995) *Alkaline Manganese Battery Catalogue*, Sanyo Electric Co., Ltd.
2. Yano, M. and Nogami, M. (1997) *Denki Kagaku*, **65**, 154.
3. Miura, A. (1989) *Denki Kagaku*, **57** (6), 459.
4. Akazawa, T., Sekiguchi, W., and Nakagawa, J. (1987) Proceedings of 28th Battery Symposium, Tokyo.
5. (1988) *Engineering Handbook Of Sealed Type Nickel-cadmium Batteries*, Sanyo Electric Co., Ltd., Osaka.
6. Tuck, C.D.S. (1991) *Modern Battery Technology*, Sanyo Electric Co., Ltd, Osaka, pp. 244–289.
7. van Vucht, J.H.N., Kuijpers, F.A., and Brunning, H.C.A.M. (1970) *Philips Res. Rep.*, **25**, 133.

8. Gutjahr, M.A., Buchner, H., and Beccu, K. (1974) Proceedings of 8th International Symposium of a New Reversible Negative Electrode for Alkaline Storage Batteries Based on Metal Alloy Hydrides, p. 79.
9. Cohen, R.L. and Wernick, J.H. (1981) *Science*, **214**, 1081.
10. Suda, S. (1987) *Int. J. Hydrogen Energy*, **12**, 323.
11. Willimes, J.J.G. (1984) *Philips J. Res.*, **39**, 1.
12. van Beek, J.R., Donkersloot, H.C., and Willimes, J.J.G. (1985) *J. Power Sources*, **10**, 317.
13. Ikoma, M., Kawano, Y., Yanagihara, N., Ito, N., and Matsumoto, I. (1986) Proceedings of 27th Battery Symposium Osaka, p. 89.
14. Furukawa, N., Inoue, Y., and Matsumoto, T. (1987) Proceedings of 28th Battery Symposium, Tokyo, p. 107.
15. Sato, Y., Kanda, M., Yagasaki, E., and Kanno, K. (1987) Proceedings of 28th Battery Symposium Tokyo, p. 109.
16. Ikoma, M., Ito, Y., Kawano, H., Ikeyama, M., Iwasaki, K., and Matsumoto, I. (1987) Proceedings of 28th Battery Symposium Tokyo, p. 112.
17. Ogawa, H., Ikoma, M., Kawano, H., and Matsumoto, I. (1988) *J. Power Sources*, **12**, 393.
18. Yonezu, I., Nogami, M., Inoue, K., Matsumoto, T., Saito, T., and Furukawa, N. (1989) Proceedings of Hydrogen Energy System Society of Japan, Vol. 14-1, p. 21.
19. Nogami, M., Tadokoro, M., and Furukawa, N. (1989) 176th Meeting of Electrochemical Society Florida, p. 130.
20. Wakao, S., Sawa, H., Nakao, H., Chubachi, S., and Abe, M. (1987) *J. Less-Common Met.*, **131**, 311.
21. Fetcenko, M.A., Venkatesan, S., and Ovshinsky, S.R. (1990) Proceedings of 34th International Power Sources Symposium, New Jersey, p. 305.
22. Nagai, R., Wada, S., Hrista, H., Kajita, K., and Uetani, Y. (1991) Proceedings of 23rd Battery Symposium Kyoto, p. 175.
23. Nogami, M., Morioka, Y., Ishikura, Y., and Furukawa, N. (1993) *Denki Kagaku*, **61**, 997.
24. Nogami, M. and Furukawa, N. (1995) *J. Chem. Soc. Jpn.*, 1.
25. Nogami, M., Tadokoro, M., Kimoto, M., Chikano, Y., Ise, T., and Furukawa, N. (1993) *Denki Kagaku*, **61**, 1088.
26. Chikano, Y., Kimoto, M., Maeda, R., Nogami, M., Nishio, K., Saito, T., Nakahori, S., Murakami, S., and Furukawa, N. (1994) in *Hydrogen and Metal Hydride Batteries* (eds P.D. Bennett and T. Sakai), Proceedings of the Electrochemical Society, Publication 94–27, p. 403.
27. Yonezu, I. and Nogami, M. (1995) Proceedings of 7th Canadian Hydrogen Workshop, Quebec, p. 171.
28. Venkatasetty, H.V. (1990) in *Lithium Battery Technology* (ed. H.V. Venlakasetty), John Wiley & Sons, Inc., New York, p. 61.
29. (1997) *Catalogue of Lithium-Manganese Dioxide Batteries*, Sanyo Electric Co., Ltd.
30. Narukawa, S. and Furukawa, N. (1993) in *Modern Battery Technology*, (ed. C.V. Stuck), Ellis Horwood, New York, p. 348.
31. Ikeda, H. (1983) in *Lithium Batteries* (ed. J.P. Gabano), Academic Press, New York, p. 169.
32. Nohma, T., Yoshimura, S., Nishio, K., and Saito, T. (1994) in *Lithium Batteries* (ed. G. Pistoia), Elsevier, Amsterdam, p. 417.
33. Takahashi, M., Yoshimura, S., Nakane, I., Nohma, T., Nishio, K., Saito, T., Fujimoto, M., Narukawa, S., Hara, M., and Furukawa, N. (1993) *J. Power Sources*, **43–44**, 253.
34. Nishio, K., Yoshimura, S., and Saito, T. (1995) *J. Power Sources*, **55**, 115.
35. (1996) *Catalogue of Lithium-Carbon Monofluoride Batteries*, Matsushita Battery Industrial Co., Ltd.
36. Fukuda, M. and Iijima, T. (1983) in *Lithium Batteries* (ed. J.P. Gabano), Academic Press, New York, p. 211.
37. Gibbard, F. and Reddy, T.B. (1993) in *Modern Battery Technology* (ed. C.V. Stuck), Ellis Horwood, New York, p. 287.

38. (1996) *Catalogue of Lithium-Thionyl Chloride Batteries*, Toshiba Battery Co., Ltd.
39. Pistoia, G. (1982) *J. Electrochem. Soc.*, **129**, 1861.
40. Nohma, T., Yoshimura, S., Nishio, K., and Saito, T. (1994) in *Lithium Batteries* (ed. G. Pistoia), Elsevier, Amsterdam, p. 417.
41. Nohma, T., Saito, T., and Furukawa, N. (1989) *J. Power Sources*, **26**, 389.
42. Nohma, T., Yamamoto, Y., Nishio, K., Nakane, I., and Furukawa, N. (1990) *J. Power Sources*, **32**, 373.
43. Nohma, T., Yamamoto, Y., Nakane, I., and Furukawa, N. (1992) *J. Power Sources*, **39**, 51.
44. Watanabe, H., Nohma, T., Nakane, I., Yoshimura, S., Nishio, K., and Saito, T. (1990) *J. Power Sources*, **32**, 373.
45. (1997) *Catalogue of Secondary Lithium-Manganese Dioxide Batteries*, Sanyo Electric Co., Ltd.
46. Koshiba, N., Ikehata, T., and Takata, K. (1991) *Natl. Tech. Rep.*, **37** (1), 64.
47. (1996) *Catalogue of Lithium-Vanadium Oxide. Secondary Batteries*, Matsushita Battery Industrial Co., Ltd.
48. (1996) *Catalogue of Lithium-Polyaniline Batteries*, Seiko Instruments Inc.
49. Nishio, K., Fujimoto, M., Yoshinaga, N., Furukawa, N., Ando, O., Ono, H., and Murayama, T. (1991) *J. Power Sources*, **34**, 153.
50. (1996) *Catalogue of Secondary Lithium-Carbon Batteries*, Matsushita Battery Industrial Co., Ltd.
51. Koshiba, N., Hayakawa, H., and Momose, K. (1985) Proceedings of Symposium Battery Association Japan, p. 145.
52. Toyoguchi, Y., Yamaura, J., Matui, T., Koshiba, N., Shigematsu, T., and Ikehata, T. (1986) *Natl. Tech. Rep.*, **32** (5), 116.
53. (1996) *Catalogue of Secondary Li-Lgh-Vanadium Oxide Batteries*, Toshiba Battery Co., Ltd.
54. Yata, S. (1993) Proceedings of 60th Electrochemical Meeting, Japan, p. 184.
55. Yata, S., Hata, Y., Kinoshita, H., Ando, N., Hashimoto, T., Tanaka, K., and Yamabe, T. (1993) Proceedings of the Symposium on Battery Association of Japan, p. 63.
56. (1996) *Catalogue of Secondary Lithium-Polyacene Batteries*, Seiko Instruments Inc.
57. (1996) *Catalogue of Secondary Niobium Oxide-Vanadium Oxide Batteries*, Matsushita Battery Industrial Co., Ltd.
58. Koshba, N., Takata, K., Nakanishi, M., Asaka, E., and Takahara, Z. (1994) *Denki Kagaku*, **62**, 870.
59. Ohzuku, T. and Ueda, A. (1994) *Solid State Ionics*, **69**, 201.
60. (1996) *Catalogue of Secondary Titanium Oxide-Manganese Oxide Batteries*, Matsushita Battery Industrial Co., Ltd.
61. Reimers, J.N. and Dahn, J.R. (1992) *J. Electrochem. Soc.*, **139**, 2091.
62. Dahn, J.R. (1990) *Solid State Ionics*, **44**, 87.
63. Kanno, R., Kubo, H., and Kawamoto, Y. (1994) *J. Solid State Chem.*, **110**, 216.
64. Ohzuku, T., Ueda, A., and Nagayama, M. (1993) *J. Electrochem. Soc.*, **140**, 1862.
65. Ohzuku, T., Komori, H., Nagayama, M., Sawai, K., and Hirai, T. (1991) *Chem. Express*, **16**, 161.
66. Nohma, T., Kurokawa, H., Uehara, M., Takahashi, M., Nishio, K., and Saito, T. (1995) *J. Power Sources*, **54**, 522.
67. Momchilov, A., Manev, V., Nassalevska, A., and Kozawa, A. (1993) *J. Power Sources*, **41**, 305.
68. Ohzuku, T., Kitagawa, M., and Hirai, T. (1990) *J. Electrochem. Soc.*, **137**, 769.
69. Ohzuku, T., Komori, H., Sawai, K., and Hirai, T. (1991) *Chem. Express*, **5**, 733.
70. Imanishi, N., Ohashi, S., Ichikawa, T., Takeda, T., and Yamamoto, O. (1992) *J. Power Sources*, **39**, 185.
71. Kanno, R., Takeda, Y., Ichikawa, T., Nakanishi, K., and Yamamoto, O. (1989) *J. Power Sources*, **26**, 535.
72. Mohri, M., Yanagisawa, N., Tajima, Y., Tanaka, H., Mitate, T., Nakajima, S., Yoshida, M., Yoshimoto, Y., Suzuki, T., and Wada, H. (1989) *J. Power Sources*, **26**, 545.
73. Mori, Y., Iriyama, T., Hashimoto, T., Yamazaki, S., Fawakami, F., Shiroki, H., and Yamabe, T. (1995) *J. Power Sources*, **56**, 205.

74. Fujimoto, H., Mabuchi, A., Tokumitsu, K., and Kasuh, T. (1995) *J. Power Sources*, **54**, 440.
75. Tatsumi, K., Iwashita, N., Sakaebe, H., Shioyama, H., Higuchi, S., Mabuchi, A., and Fujimoto, H. (1995) *J. Electrochem. Soc.*, **142**, 716.
76. Takami, N., Satoh, A., Hara, M., and Ohsaki, T. (1995) *J. Electrochem. Soc*, **142**, 371.
77. Komatsu, S., Fukunaga, T., Terasaki, M., Mizutani, M., Yamachi, M., and Ohsaki, T. (1992) 33rd Battery Symposium in Japan, p. 89.
78. Tamaki, T. and Tamaki, M. (1992) Proceedingsof 36th Battery Symposium in Japan, p. 105.
79. Fujimoto, M., Ueno, K., Nouma, T., Takahashi, M., Nishio, K., and Saito, T. (1993) Proceedings of the Symposium on New Sealed Rechargeable Batteries and Supercapacitors, p. 280.
80. Fujimoto, M., Kida, Y., Nouma, T., Takahashi, M., Nishio, K., and Saito, T. (1996) *J. Power Sources*, **63**, 127.
81. Sonobe, N., Ishikawa, M., and Iwasaki, T. (1994) 35th Battery Symposium in Japan, p. 47.
82. Yata, S., Hato, Y., Kinoshita, H., Anekawa, A., Hashimoto, T., Tanaka, K., and Yamabe, T. (1994) 35th Battery Symposium in Japan, p. 59.
83. Ozawa, K. (1994) *Solid State Ionics*, **69**, 212.
84. Saitoh, T., Nohma, T., Takahashi, M., Fujitomo, M., and Nishio, K. (1993) Proceedings on New Sealed Rechargeable Batteries and Supercapacitors, p. 355.
85. Guyomard, D. and Tarascon, J.M. (1992) *J. Electrochem. Soc.*, **139**, 937.
86. Stiles, A.R. (1985) *New Mater. New Processes*, **3**, 89.
87. Dan, P., Mengeritski, E., Geronov, Y., Aurbach, D., and Weismann, I. (1995) *J. Power Sources*, **54**, 143.
88. Mengeritski, E., Dan, P., Weismann, I., Zaban, A., and Aurbach, D. (1996) *J. Electrochem. Soc.*, **143**, 2110.
89. Tobishima, S., Sakurai, Y., and Yamaki, J. (1996) 8th International Meeting on Lithium Batteries, p. 362.
90. Arakawa, M., Nemoto, Y., Tobishima, S., and Ichimura, M. (1993) *J. Power Sources*, **43–44**, 517.
91. Aurbach, D., Zahan, A., Gofes, Y., Ely, V.E., Weismann, I., Chusid, O., and Abranzon, O. (1994) 7th International Meeting on Lithium Batteries, p. 97.
92. Newmann, G.H. (1980) *Proceedings of the Workshop on Lithium Non-Aqueous Battery Electrochemistry*, vol. 80-7, The Electrochemical Society, p. 143.
93. Olmstead, W.N. (1981) *Proceedings of the Symposium on Lithium Batteries*, vol. 81-4, The Electrochemical Society, p. 150.
94. Klemann, L.O. and Newmann, G.H. (1981) *J. Electrochem. Soc.*, **128**, 13.
95. Abraham, K.M. (1981/2) *J. Power Sources*, **7**, 1.
96. Johnson, J.W. and Whittingham, M.S. (1980) *J. Electrochem. Soc.*, **127**, 1653.
97. Koch, V.R. and Young, J.H. (1978) *J. Electrochem. Soc.*, **125**, 1371.
98. Koch, V.R. (1979) *J. Electrochem. Soc.*, **126**, 181.
99. Koch, V.R., Goldman, J.L., Mattos, C.J., and Mulvaney, M. (1982) *J. Electrochem. Soc.*, **129**, 1.
100. Koch, V.R. (1978) US Patent 4, 118,550.
101. Abraham, K.M. (1981) *J. Electrochem. Soc.*, **128**, 2493.
102. Whittingham, M.S. (1981) *J. Electroanal. Chem.*, **118**, 229.
103. Yen, S.P.S., Shen, D., Vasquez, R.P., Grunthaner, F.J., and Somoano, R.B. (1981) *J. Electrochem. Soc.*, **128**, 1434.
104. Goldman, J.W., Mank, R.M., Young, J.H., and Koch, V.R. (1980) *J. Electrochem. Soc.*, **127**, 1461.
105. Watanabe, M., Kanba, M., Nagaoka, K., and Shinohara, I. (1982) *J. Appl. Polym. Sci.*, **27**, 4191.
106. Heermann, L. and Van Baelen, J. (1972) *Bull. Soc. Chim. Belg.*, **81**, 379.
107. Iizima, T. (1974) *Natl. Tech. Rep.*, **20** (1), 35.
108. Matsuda, Y. and Morita, M. (1988) *Prog. Batteries Sol. Cells*, **9**, 266.

109. Surampudi, S., Shen, D.H., Huang, C.K., Narayan, S.R., Attia, A., and Halper, G. (1993) *J. Power Sources*, **43–44**, 27.
110. Matsuda, Y. (1993) *J. Power Sources*, **43**, 1.
111. Takehara, Z. (1994) 7th International Meeting on Lithium Batteries, p. 37.
112. Takehara, Z., Ogumi, Z., Utimoto, Y., Yasuda, K., and Yoshida, H. (1993) *J. Power Sources*, **43–44**, 377.
113. Arakawa, M., Tobishima, S., Nemoto, Y., and Ichimura, M. (1993) *J. Power Sources*, **43–44**, 27.
114. Aurbach, D., Zaban, A., Ely, Y.E., Weissman, I., Chusid, O., and Abramzon, O. (1994) 7th International Meeting on Lithium Batteries, p. 97.
115. Fong, R., Reid, M.C., McMillan, R.S., and Dahn, J.R. (1987) *J. Electrochem. Soc.*, **134**, 516.
116. Aurbach, D., Daroux, M.L., Faguy, P.W., and Yeager, E. (1987) *J. Electrochem. Soc.*, **134**, 1611.
117. Aurbach, D., Daroux, M.L., Faguy, P.W., and Yeager, E. (1988) *J. Electrochem. Soc.*, **135**, 1863.
118. Gofer, Y., Ben-Zion, M., and Aurbach, D. (1992) *J. Power Sources*, **39**, 163.
119. Koch, A.V.R. (1979) *J. Electrochem. Soc.*, **39**, 163.
120. Odziemkowski, M., Krell, M., and Irish, D.E. (1992) *J. Electrochem. Soc.*, **139**, 3052.
121. Yen, S.P.S., Shen, D., Vasquez, R.P., Grunthaner, F.J., and Somoano, R.B. (1981) *J. Electrochem. Soc.*, **128**, 1434.

Further Reading

Aurbach, D. and Chusid, O. (1993) *J. Electrochem. Soc.*, **140**, L1.

Bartolozzi, M. (1990) *Resour. Conserv. Recycl.*, **4** (3), 233–240.

Brummer, S.B. (1989) in *Lithium Battery Technology* (ed. H.V. Venkatasetty), John Willey & Sons, Inc., New York, p. 159.

Chen, J. and Cheng, F.Y. (2009) *Acc. Chem. Res.*, **42** (6), 713–723.

Chen, D., Tang, L.H., and Li, J.H. (2010) *Chem. Soc. Rev.*, **39** (8), 3157–3180.

Chen, Y.H., Zhao, Y.M., An, X.N., Liu, J.M., Dong, Y.Z., and Chen, L. (2009) *Electrochim. Acta*, **54** (24), 5844–5850.

Chernova, N.A., Roppolo, M., Dillon, A.C., and Whittingham, M.S. (2009) *J. Mater. Chem.*, **19** (17), 2526–2552.

Dampier, D.W. (1974) *J. Electrochem. Soc.*, **121**, 656.

Deng, D., Kim, M.G., Lee, J.Y., and Cho, J. (2009) *Energy Environ. Sci.*, **2** (8), 818–837.

Ellis, B.L., Lee, K.T., and Nazar, L.F. (2010) *Chem. Mater.*, **22** (3), 691–714.

Fergus, J.W. (2010) *J. Power Sources*, **195** (4), 939–954.

Goodenough, J.B. and Kim, Y. (2010) *Chem. Mater*, **22** (3), 587–603.

Graham, R.W. (1978) Secondary Batteries.

Jindra, J. (1997) *J. Power Sources*, **66** (2), 15–25.

Kim, M.G. and Cho, J. (2009) *Adv. Funct. Mater.*, **19** (10), 1497–1514.

Li, H., Wang, Z.X., Chen, L.Q., and Huang, X.J. (2009) *Adv. Mater.*, **21** (45), 4593–4607.

Liu, C., Li, F., Ma, L.P., and Cheng, H.M. (2010) *Adv. Mater.*, **22** (8), E28.

Liu, J., Wang, J.W., Yan, X.D., Zhang, X.F., Yang, G.L., Jalbout, A.F., and Wang, R.S. (2009) *Electrochim. Acta*, **54** (24), 5656–5659.

Maher, K., Edstrom, K., Saadoune, I., Gustafsson, T., and Mansori, M. (2009) *Electrochim. Acta*, **54** (23), 5531–5536.

Mai, L.Q., Xu, L., Hu, B., and Gu, Y.H. (2010) *J. Mater. Res.*, **25** (8), 1413–1420.

McLarnon, F.R. and Cairns, E.J. (1991) *J. Electrochem. Soc.*, **138** (2), 645–664.

Park, C.M., Kim, J.H., Kim, H., and Sohn, H.J. (2010) *Chem. Soc Rev.*, **39** (8), 3115–3141.

Peng, B. and Chen, J. (2009) *Coord. Chem. Rev.*, **253** (23–24), 2805–2813.

Sakaguchi, H., Hatakeyama, K., Fujii, M., Inoue, H., Iwakura, C., and Esaka, T. (2006) *Res. Chem. Intermed.*, **32** (6), 473–481.

Santhanam, R. and Rambabu, B. (2010) *J. Power Sources*, **195** (17), 5442–5451.

Sayilgan, E., Kukrer, T., Civelekoglu, G., Ferella, F., Akcil, A., Veglio, F., and Kitis, M. (2009) *Hydrometallurgy*, **97** (4), 158–166.

Su, D.S. and Schlogl, R. (2010) *ChemSusChem*, **3** (2), 136–168.

Su, F.B., Zhou, Z.C., Guo, W.P., Liu, J.J., Tian, X.N., and Zhao, X.S. (2008) *Chem. Phys. Carbon*, **30**, 63–128.

Takeuchi, K.J., Marschilok, A.C., Davis, S.M., Leising, R.A., and Takeuchi, E.S. (2001) *Coord. Chem. Rev.*, **219**, 283–310.

Vijayamohanan, K., Balasubramanian, T.S. and Shukla, A.K. (1991) *J. Power Sources*, **34** (3), 269–285.

Yan, X.D., Yang, G.L., Liu, J., Ge, Y.C., Xie, H.M., Pan, X.M., and Wang, R.S. (2009) *Electrochim. Acta*, **54** (24), 5770–5774.

Yang, Z.G., Choi, D., Kerisit, S., Rosso, K.M., Wang, D.H., Zhang, J., Graff, G., and Liu, J. (2009) *J. Power Sources*, **192** (2), 588–598.

Yang, G., Ying, J.R., Gao, J., Jiang, C.Y., and Wan, C.R. (2008) *Rare Metal Mater. Eng.*, **37** (5), 936–940.

Yi, T.F., Zhu, Y.R., Zhu, X.D., Shu, J., Yue, C.B., and Zhou, A.N. (2009) *Ionics*, **15** (6), 779–784.

Yu, X.W. and Lich, S. (2007) *J. Power Sources*, **171** (2), 966–980.

Part II
Materials for Aqueous Electrolyte Batteries

3
Structural Chemistry of Manganese Dioxide and Related Compounds
Jörg H. Albering

3.1
Introduction

In this section the structural properties of the most common manganese dioxide modifications and closely related compounds will be presented and briefly compared. A huge number of compounds are designated as *'Braunstein'* (i.e., manganese dioxide). This term includes all natural and synthetic manganese oxides of composition $MnO_{1.4}$–$MnO_{2.0}$, regardless of the presence of foreign cations, hydroxide anions, or water molecules in the structure. All the known – and more or less structurally characterized – materials of that composition cannot be reviewed here. The main purpose is to give an overview of the variety of features in the chemistry of manganese oxides and to point out the structural correlations of various modifications within the large family of manganese oxide minerals and synthetic compounds. The occurrence of MnO_2 in natural ores and some methods of synthesizing the different modifications in the laboratory will be briefly described. Details of its electrochemical behavior in Leclanché cells and alkaline Zn/MnO_2 cells, and of its use as a lithium storage material in lithium-ion cells are given in Part II Chapter 4 and Part III Chapter 1, respectively.

In general, it is very difficult to obtain well-developed single crystals of any modification of MnO_2 in the laboratory [1]. Hence most of the structural data on manganese dioxide are obtained either from single crystals selected from natural ores (e.g., crystals up to 30 cm in length are reported for α-MnO_2 [2]) or by X-ray diffraction (XRD) or neutron powder techniques combined with Rietveld refinements [3]. Most manganese oxides are usually fine-grained and the crystallites may contain many defects, twin domains, superstructures, and partially occupied crystallographic sites. Therefore the XRD technique – although it is the most commonly used and successful method for the determination of the structural properties – is often limited by the poor crystallinity of the materials. Consequently, the effects of structural disorder (e.g., selective reflex broadening, preferred orientation, and the presence of indistinct, overlapping Bragg reflections) on the X-ray powder diffraction patterns can lead to incomplete or false interpretations of the diffraction data and subsequently to incorrect structural

Handbook of Battery Materials, Second Edition. Edited by Claus Daniel and Jürgen O. Besenhard.
© 2011 Wiley-VCH Verlag GmbH & Co. KGaA. Published 2011 by Wiley-VCH Verlag GmbH & Co. KGaA.

models. Thus other methods have to be applied to overcome this problem. The most powerful tool is high-resolution transmission electron microscopy (HRTEM) [4–6]. Other methods, for example, IR spectroscopy [7] and extended X-ray absorption fine structure (EXAFS) measurements [8], give additional information about the near-neighbor environment, the connection scheme of $Mn(O,OH)_n$ polyhedra, and the different oxidation states of the manganese atoms in the structure.

The most obvious structural feature in all oxides containing manganese in the oxidation states II, III, or IV is the more or less distorted octahedral 0×0- (O^{2-}) or hydroxo- (OH^-) coordination. The $Mn(O,OH)_6$ octahedra can be connected to each other by sharing common corners or edges. Face-sharing (as it is known from Nb cluster compounds containing close Nb–Nb bonds) does not usually occur, since in this case the central atoms of the polyhedra would come into too close contact. So far, a metal cluster is unknown in the structural chemistry of manganese dioxide and related compounds. The closest Mn–Mn distance in the various modifications of MnO_2 usually occurs along the shortest crystallographic axis. Many compounds contain a short translation period ranging approximately from 280 to 290 pm, which represents the distance between the central atoms of two edge-sharing octahedra. The octahedral unit is the fundamental building element for manganese oxides. How the octahedra are connected together can be used to classify the crystal structures. Similarly to silicate chemistry, the large family of manganates (II, III, or IV) can be divided into subgroups which contain characteristic building blocks of edge/corner-sharing $Mn(O,OH)_6$ octahedra [9].

A common structural feature is the formation of one-dimensionally infinite strings of edge-sharing octahedra, which extend along the shortest translation period. Two or three of these strings can be connected to one another by further edge-sharing, thus forming double or triple chains. Four such MnO_6 strings, connected by corner-sharing, enclose, a one-dimensionally infinite tunnel of various dimensions. This category of compounds is generally described as *chain* or *tunnel structure*. The other frequently occurring structural element is formed from two-dimensionally infinite layers of edge-sharing $Mn(O,OH)_6$ octahedra. The stacking sequence of the octahedral layers and the kind/number of the interlayer atoms or molecules (metal cations, water, hydroxide anions) are further criteria for a structural classification of these *layer* or *sheet* structures ('phyllomanganates').

3.2
Tunnel Structures

3.2.1
β-MnO₂

The crystal structure of pyrolusite, or β-MnO_2, is the simplest one within the family of compounds with tunnel structures. The manganese atoms occupy half of the octahedral voids in the hexagonal close packing of oxygen atoms in an ordered manner, thus forming a rutile-type structure. The distorted MnO_6 units build up strings of edge-sharing octahedra extending along the crystallographic c-axis. These chains are crosslinked with neighboring chains by sharing common corners, resulting in the formation of narrow [1 × 1] channels in the structure. The voids within the channels are too small for larger cations, but there is enough space for an intercalation of hydrogen or lithium ions. The crystallographic data for β-MnO$_2$ and other tunnel structures are summarized in Table 3.1. The crystal structure of β-MnO$_2$ is shown in Figure 3.1a.

The β-modification is the thermodynamically stable form of MnO$_2$. Hence, pyrolusite frequently occurs in natural ores and it is easily prepared in a high-purity form by thermal decomposition of manganese nitrate, Mn(NO$_3$)$_2 \cdot n$H$_2$O [29, 30]. Another method for the synthesis of β-MnO$_2$ is by heating γ-MnO$_2$ in closed reaction vessels in the presence of strong acids (H$_2$SO$_4$ or HNO$_3$) at 130–150 °C [11, 31] or under hydrothermal conditions within a wide range of temperature and pressure [11, 32]. Naturally occurring β-MnO$_2$ as well as the synthetic materials with a rutile-type structure usually have a stoichiometry very close to the ideal ratio of Mn:O = 1 : 2. The stability region of MnO$_{2-x}$ is assumed to be in the range of $x = 0$–0.1 [33–35]. Pyrolusite is the only modification in which the ideal composition MnO$_2$ can be reached. Hence, the β-modification can be regarded as a *true* MnO$_2$ compound.

3.2.2
Ramsdellite

The manganese and oxygen atoms in the ramsdellite modification of MnO$_2$ occupy the same crystallographic sites as the aluminum and oxygen atoms in the diaspore (AlOOH) structure and may thus be considered to be *isopointal* (according to Parthé and Gelato the term '*isotypic*' should be reserved for compounds in which the occupancy of the same sites results in identical coordination polyhedra and the same stoichiometry [36]) with this aluminum oxide hydroxide. The crystal structure of ramsdellite is very similar to that of pyrolusite except that the single chains of octahedra in β-MnO$_2$ are replaced by double chains in ramsdellite. Consequently, the tunnels extending along the short c-axis of the orthorhombic structure ($a = 446$ pm, $b = 932$ pm, $c = 285$ pm; see also Table 3.1) have a larger dimension [1 × 21] compared with those of β-MnO$_2$. The cell volume of ramsdellite is approximately double the cell volume of β-MnO$_2$. Similarly to pyrolusite, the [1 × 2] channels are too small to allow the presence of cations other than protons or Li$^+$. However, the transport properties of the ramsdellite crystal structure for protons or lithium ions as well as the structural integrity of the protonated or lithiated compounds are extremely important for the performance of electrochemical cells. As has already been pointed out for pyrolusite, the oxygen atoms in ramsdellite occupy the positions of a hexagonally close-packed lattice. The manganese atoms are located in every second pair of neighboring octahedral voids, which share a common edge. One-half of the oxygen atoms has more or less ideal trigonal planar coordination of manganese; the other half is located at the apex of a flat trigonal

3 Structural Chemistry of Manganese Dioxide and Related Compounds

Table 3.1 Crystallographic data for manganese oxides with tunnel structures.

Compound or mineral	Approximate formula	Symmetry	Space group	Lattice constants						References	Tunnel size
				a (pm)	b (pm)	c (pm)	α (°)	β (°)	γ (°)		
Pyrolusite	MnO_2	Tetragonal	$P4_2/mnm$	440.4	440.4	287.6	90	90	90	[10]	[1 × 1]
β-MnO_2	MnO_2	Orthorhombic	Pnma	446	932	285	90	90	90	[11]	[1 × 2]
Ramsdellite	$MnO_{2-x}OH_x$	Orthorhombic		446.2	934.2	285.8	90	90	90	[12]	[1 × 1]/[1 × 2]
γ-MnO_2	$MnO_{2-x}OH_x$	Hexagonal		278.3	278.3	443.7	90	90	120	[12]	[1 × 1]/[1 × 2]
α-MnO_2 group											
Kryptomelane	$K_{2-x}Mn_8O_{16}$	Tetragonal	I4/m	986.6	986.6	287.2	90	90	90	[13]	[2 × 2]
	$(K, Ba)_{2-x}Mn_8O_{16}$	Monoclinic	C2/m	979	288	994	90	90.37	90	[14]	[2 × 2]
Hollandite	$(Ba, K, Pb)_{2-x}(Mn, Fe, Al)_8O_{16}$	Tetragonal	I4/m	980.0	980.0	286.0	90	90	90	[15]	[2 × 2]
	$(Ba, K)_{2-x}Mn_8O_{16}$	Monoclinic	C2/m	1002.6	287.8	972.9	90	91	90	[16]	[2 × 2]
	$Ba_{2-x}(Mn, Fe)_8O_{16}$	Monoclinic	$P2_1/c$	1003	576	990	90	90.42	90	[17]	[2 × 2]
Coronadite	$Pb_{2-x}Mn_8O_{16}$	Tetragonal	I4/m	988.7	988.7	284.2	90	90	90	[18]	[2 × 2]
	$Pb_{1.34}Mn_8O_{16}$	Monoclinic	C2/m	991.3	286.5	984.3	90	90.2	90	[19]	[2 × 2]
Manjiroite	$(Na, K)_{2-x}Mn_8O_{16}$	Tetragonal	I4/m	991.6	991.6	286.4	90	90	90	[20]	[2 × 2]
	$Na_{2-x}Mn_8O_{16}$	Monoclinic	C2/m	991	286	962	90	90.93	90	[21]	[2 × 2]
Larger tunnel structures											
Romanèchite	$BaMn_9O_{10}\cdot 2H_2O$	Orthorhombic		1340	286.4	925.4	90	90	90	[22]	[2 × 3]
	$(Ba, H_2O)_xMn_5O_{10}$	Orthorhombic		1390	572	945	90	90	90	[23]	[2 × 3]
	$Ba_{1.6}Mn_5O_{10}\cdot 1.5H_2O$	Monoclinic	C2/m	1393.1	284.6	967.6	90	90.24	90	[24]	[2 × 3]
Rb manganate	$Rb_{16.64}Mn_{24}O_{48}$	Monoclinic	C2/m	1419.1	285.1	2432.3	90	91.3	90	[25]	[2 × 4]
Rb manganate	$Rb_{0.27}MnO_2$	Monoclinic	C2/m	1464.0	288.6	1504.0	90	92.4	90	[26, 27]	[2 × 5]
Todorokite	$(Mg, Ca, K)_x(Mn, Mg)_6O_{12}\cdot xH_2O$	Orthorhombic		951.6	1031	297	90	90	90	[6]	[3 × 3]
	$(Na, Ca, K)_{0.59}(Mn, Mg)_6O_{12}\cdot 3.1H_2O$	Monoclinic	P2/m	978.9	283.4	955.1	90	93.7	90	[28]	[3 × 3]

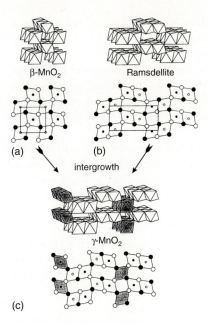

Figure 3.1 Crystal structure of (a) β-MnO$_2$, (b) ramsdellite, and (c) the intergrowth structure of these two compounds, γ-MnO$_2$. The structures are shown as three-dimensional arrangements of the MnO$_6$ octahedra and as projections along the short crystallographic c-axis, respectively. (Small circles: manganese atoms; large circles: oxygen atoms; open circles: height $z = 0$; filled circles: height $z = 1/2$.) The shaded octahedra in (c) represent the β-MnO$_2$ parts of the intergrowth structure of γ-MnO$_2$.

pyramid formed by one oxygen and three manganese atoms. The crystal structure is shown in Figure 3.1b.

Ramsdellite is thermodynamically unstable toward a transformation into the stable β-modification. Hence, it is rarely found in natural deposits. Natural ramsdellite has a stoichiometry close to the composition of MnO$_2$ and can be considered another true modification of manganese dioxide. Attempts to synthesize ramsdellite in the laboratory usually lead to materials of questionable composition and structural classification. It is very likely that synthetic 'ramsdellite' materials are more or less well-crystallized samples of the γ-modification that will be described in more detail below.

3.2.3
γ-MnO$_2$ and ε-MnO$_2$

For a long time there was uncertainty about the crystal structure of γ-MnO$_2$ or the naturally occurring species nsutite. Single-crystal material could be taken neither from natural deposits nor from synthetic manganese oxides prepared by various methods in the laboratory. Powder diffraction patterns of only a very poor quality with diffuse peaks, a high background and a selective peak broadening made the structure determination very complicated. Additionally, a large number of significantly different patterns could be observed, depending strongly on the preparation conditions. Some XRD patterns resembled the diffractograms of pyrolusite, others were similar to the line-rich patterns of ramsdellite samples, and many showed only a few broad peaks, that could be indexed on the basis of hexagonal close packing. A typical XRD pattern of an electrolytically prepared manganese dioxide (EMD sample from 'Chemetals') is shown in Figure 3.2.

In the early literature this wide variety of different powder patterns led to the distinction between $\gamma, \gamma', \gamma''$, and ε-MnO_2 [37]. De Wolff [38, 39] was the first to propose a plausible structural model for these phases. The De Wolff model is based on the assumption that the oxygen atoms in γ-MnO_2 are hexagonally close packed. The manganese atoms occupy half of the octahedral voids in this matrix, as in pyrolusite or ramsdellite. The only difference between these two modification is the crystallographic b-axis of the $Pnma$ setting of ramsdellite (see Table 3.1). The atomic arrangements in the a- and c-directions are very similar. β-MnO_2 and ramsdellite differ only in the arrangement of the manganese atoms, which form single chains of edge-sharing octahedra in the β-modification and double chains in ramsdellite.

Therefore De Wolff proposed in his model that the crystal structure of γ-MnO_2 intergrowth of pyrolusite and ramsdellite domains. An idealized section of the γ-MnO_2 intergrowth structure is shown in Figure 3.1c together with the parent

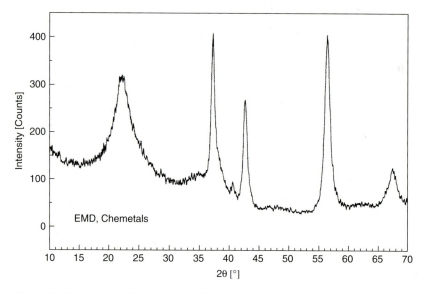

Figure 3.2 XRD pattern of an EMD sample (Chemetals). The diffractogram is taken with a Bruker AXS D5005 diffractometer using CuKα radiation and a scintillation counter. The step width is 0.02° with a constant counting time of 10 s/step.

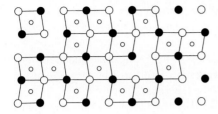

Figure 3.3 Schematic drawing of the crystal structure of ε-MnO$_2$. The manganese atoms arc randomly distributed in the octahedral voids of the hexagonal close packing of oxygen atoms. (Adapted from Ref. [41].)

structure of β-MnO$_2$ and ramsdellite. Depending on the relative fraction of the two parent components contributing to the crystal structure, the XRD patterns may resemble either one or the other component. The De Wolff model of γ-MnO$_2$ was confirmed by HRTEM investigations [4, 5, 40]. In this study the [1 × 1] and [1 × 2] tunnel domains in the structure could be visualized. Additionally, a large number of discontinuities and structural faults were observed. Even larger tunnels (e.g., [2 × 2]) exist in the real lattice of γ-MnO$_2$. These findings explain the relatively large amount of water in the compounds and the presence of foreign cations and anions (e.g., sulfate), incorporated during the chemical or electrochemical synthesis.

As described above, the XRD patterns of most γ-MnO$_2$ samples differ significantly. Some patterns can be indexed on the basis of the orthorhombic ramsdellite lattice according to the De Wolff model. In these samples the intergrown β-MnO$_2$ slices in the structure clearly do not destroy the orthorhombicity of the lattice. With an increasing number of defects and a decreasing order within the domains, the distribution of manganese in the octahedral voids of the hexagonally close packed oxygen atoms becomes more and more statistical. In this case the contribution of the manganese atoms to the coherent scattering of X-rays becomes very small and is sometimes only indicated by the presence of the very broad reflection around 21° (in 2θ, or at about 3.9–4.2 Å). Ignoring this peak, all other reflections of such a diffraction pattern can be indexed with a small hexagonal cell (e.g., $a = 278.3$, 443.7 pm [12]), describing the close packing of the oxygen matrix. These samples with a high degree of disorder at the manganese sites are called ε-MnO$_2$ (see Figure 3.3).

This modification contains tunnels of irregular shape and a statistical distribution in the structure. The only limitation for the distribution of the manganese atoms, and thus for the shape and size of the tunnels, is that no face-sharing voids can be occupied by manganese atoms, since in this case the interatomic distance of the Mn^{4+} ions would become too small. Usually, it is quite difficult to distinguish between the XRD patterns of γ- and ε-MnO$_2$. This can only be done by careful analysis of the diffractograms using effective profile-fitting routines and an accurate determination of the orthorhombic or hexagonal lattice parameters [12].

The De Wolff disorder model has been extended to the cation vacancy model for γ-MnO$_2$ and ε-MnO$_2$ by Ruetschi [42]. In this model the occurrence of manganese cation vacancies and the non stoichiometry of electrochemical MnO$_2$ have been taken into account. Furthermore, the vacancy model deals with the explanation of the different water contents of manganese dioxide. Ruetschi makes some simple assumptions:

1) The manganese atoms are distributed in a more or less ordered manner at the slightly distorted octahedral voids in the hexagonally close-packed oxygen atoms (as described above).
2) A fraction x of the Mn^{4+} ions is missing in the manganese sublattice. For charge compensation, each Mn^{4+} vacancy is coordinated by four protons in the form of OH^- anions at the sites of the O^{2-} ions.
3) A fraction y of the Mn^{4+} ions are replaced by Mn^{4+}. This fraction determines the average valence of the manganese atoms. For each Mn^{3+} there is a further OH^- ion in the lattice, replacing an O^{2-} anion in the coordination sphere of the Mn^{3+} cation. A schematic drawing of the Ruetschi model is shown in Figure 3.4.
4) Therefore the crystal structure is composed of Mn^{4+}, Mn^{3+}, O^{2-}, OH^-, and vacant sites. The water in the structure is present in the form of OH^- anions and Mn^{4+} vacancies.
5) Chemisorbed water is considered to be present in the form of surface OH^- groups.
6) Electronic conductivity arises from delocalized electrons, tunneling or hopping processes.

Therefore, Ruetschi proposes a general formula for γ-MnO_2:

$$Mn^{4+}_{1-x-y}Mn^{3+}_y O^{2-}_{2-4x-y} OH_{4x+y}$$

which replaces the formerly widely accepted, but unsatisfactory, formula:

$$MnO_n \cdot (2 - n + m)H_2O$$

The terms x, y, n, and m can easily be transformed into one another by applying Equations 3.1–3.4:

$$x = \frac{m}{2+m} \qquad (3.1)$$

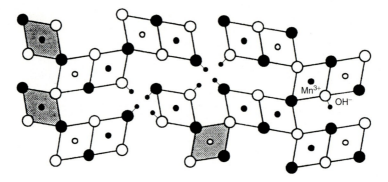

Figure 3.4 Vacancy (Ruetschi) model for the crystal structure of γ-MnO_2. The shaded octahedra represent the β-MnO_2 parts of the lattice. The small gray circles represent protons attached to oxygen atoms.

$$y = \frac{4(2-n)}{2+m} \tag{3.2}$$

$$n = \frac{4(1-x) - x}{2(1-x)} \tag{3.3}$$

$$m = \frac{2x}{1-x} \tag{3.4}$$

which allowed Ruetschi to predict the density, proton-transfer rates, electronic behavior, theoretical (maximum) electrochemical capacities, and electrode potentials for a wide range of x and y.

Further improvements on the previously discussed models were proposed in the latest model for γ- and ε-MnO$_2$ by Chabre and Pannetier [12, 43, 44]. Starting from De Wolff's model they developed a structural description of manganese dioxides that accounts for the scattering function of all γ- and ε-MnO$_2$ materials and provides a method of characterizing them quantitatively in terms of structural defects. All γ- and ε- MnO$_2$ samples can be described on the basis of an ideal ramsdellite lattice affected by two kinds of defects:

1) A stacking disorder *(De Wolff disorder*: intergrowth of ramsdellite- and pyrolusite- type units, as already described above). This kind of disorder can be quantified by two parameters:
   a. the probability P_r of occurrence of rutile-like slabs in the crystal structure: the probability of the presence of ramsdellite building blocks is $P_R = 1 - P_r$.
   b. the junction probability, which describes, for example, the probability $P_{r,r}$ that a rutile-like layer r is followed by a similar layer. Analogous $P_{R,R}$, $P_{c,R}$, and $P_{R,r}$ parameters can be defined for the three other possibilities of conjunctions.

Starting from the four general possibilities that can occur (completely ordered, partly ordered, segregated, and completely random), Chabre and Pannetier found that the commercially available samples are best described by a truly random sequence of rutile and ramsdellite slabs. Furthermore, an extended simulation study of different rutile and ramsdellite fractions in the structure led to the findings that, for example, even a small amount P_r of rutile changes the diffraction pattern of ramsdellite significantly. The reflections with an odd value of k are shifted and broadened significantly, while the reflections with $k/2 + 1$ (e.g., (0 2 1), (1 2 1), (2 4 0), (0 6 1)) are not affected by De Wolff disorder. Starting with $P_r = 1$ (from a pure rutile-type structure) the typical (1 1 0) peak of pyrolusite disappears even at low values of P_r. When a value of 0.5 was reached, a broad peak at about 20° (in 2θ) appeared in the simulated patterns. On the basis of these simulations Pannetier developed a method to estimate the pyrolusite concentration in orthorhombically indexable lattices. After a careful refinement of the lattice constants by profile refinements, the expected (theoretical) d value for the (110) reflection of ramsdellite is calculated from the lattice constants:

$$d_{(110),\text{expected}} = [1/a^2 + 1/b^2]^{1/2} \tag{3.5}$$

Subsequently, the difference $\Delta d_{(110)}$ (in angstrom) is calculated from the theoretical value and the measured d value of the (1 1 0) reflection. Using the empirical calibration curve (a second-order polynomial, Equation 3.5), the pyrolusite concentration can be calculated or it may be taken from a diagram, as shown in Figure 3.5.

2) The model of De Wolff disorder gives no explanation for the line broadening of reflections which are not affected by this type of lattice disorder. Chabre and Pannetier ascribed this effect to a micro twinning of the ramsdellite/rutile lattice on the planes [0 2 1] and [0 6 1]. These faces are believed to be growth planes of EMD [45, 46]

It is well known that in rutile-like structures the planes [0 1 1] and [0 3 1] are twinning planes. Hence, Chabre and Pannetier concluded that twinning faults in the planes [0 2 1] and [0 6 1] (the equivalent planes in the ramsdellite doubled unit cell) are the explanation for some features in the diffraction patterns of γ-MnO$_2$: for example, the lineshift of the (1 1 0) reflection toward lower angles or the merging of the reflection groups $(h\,2\,1)/(h\,4\,0)$ and $(h\,6\,1)/(h\,0\,2)$.

In Figure 3.6 the arrangement of the manganese atoms is shown in a projection along the a-axis. The unit cells are marked by the shaded regions. It can easily be seen, that no lattice distortion is necessary to form the [0 2 1] and [0 6 1] twins. The very low activation energy for the twinning usually results in a very high number of micro twin boundaries in the crystal structure. The exact number of micro twin domains is difficult to estimate from the diffraction patterns and according to Chabre and Pannetier it is difficult to distinguish between the effects of the two

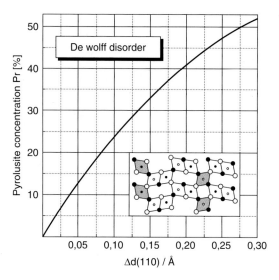

Figure 3.5 Calibration curve for the determination or the pyrolusite concentration of orthorhombically indexed γ-MnO$_2$ samples by comparison of the calculated with the observed $d_{(110)}$ value.

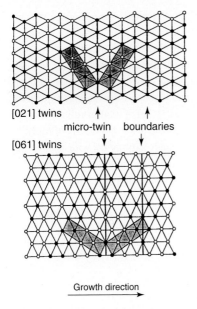

Figure 3.6 Projection of the manganese atoms in the ramsdellite lattice onto the *bc*-plane. The oxygen atoms are not shown. The twinning planes [021] (above) and [061] (below) are marked with arrows. The twins at these planes are generated by rotating the shaded ramsdellite cells by either 60° or 120° around the *a*-axis. (Adapted from Ref. [41].)

kinds of disorder in the lattice, particularly when there is a large amount of micro twinning. Although the model of Chabre and Pannetier seems to be very close to reality, there might be some features in the structural chemistry of γ-MnO$_2$ which still have to be integrated into an optimized model such as the vacancies at the Mn^{4+} sites, the presence of Mn^{3+}, and the presence of larger tunnels (e.g., [1 × 3] or [2 × 2]) in the real lattice.

As already mentioned above the γ-MnO$_2$/ε-MnO$_2$ modification is usually obtained by electrochemical deposition at graphite, lead, or titanium anodes from a solution of Mn^{2+} salts in strong acids (usually H$_2$SO$_4$) (for references see Ref. [47]). Electrochemically prepared manganese dioxide (EMD) usually grows with a fibrous texture along the [0 2 1] and [0 6 1] faces, as described above. Nsutite, the naturally occurring modification of γ-MnO$_2$, shows a similar, needle-shaped, crystallite morphology. The fibrous habit can best be seen in chemically prepared samples of γ-MnO$_2$. These materials typically consist of bundles and small balls of tiny needles with an average length of about 1–4 pm an average width of about 100–200 nm and an average thickness of 20–100 nm. In contrast to EMD, which usually has a compact but porous consistency, the chemically prepared manganese dioxides (CMDs) have a very low volumetric density and are poorly compressible. The degree of oxidation in EMD and in CMD is lower than in pure pyrolusite or ramsdellite. Depending on the (chemical) preparation method, the value of

x in γ-$MnO_2 \cdot nH_2O$ ranges between 1.7 and 1.98; the ideal value of $x = 2$ cannot be reached. Many methods are known for the preparation of γ-MnO_2 samples. In general, most syntheses can be modified by the way in which γ-MnO_2 is obtained. Samples of a good quality can be synthesized by oxidation of Mn (II) salts with peroxodisulfates, permanganates, halogens, chlorates, bromates, and hypohalogenates. Another preparation method is the reaction of permanganates with organic or inorganic reducing agents (for a good collection of further references, see Ref. [46]). More than 100 years of research on MnO_2 led to a huge variety of reaction types, suitable for the production of manganese dioxide samples of widely differing properties. Some CMD materials can be used as active battery components for Le-clanché cells or as catalysts; other samples may have ideal properties for organic synthesis. However, the γ-modification of MnO_2, regardless of the preparation method (EMD or CMD), is one of most important basic, inorganic chemicals.

3.2.4
α-MnO_2

The crystal structure of α-MnO_2 consists of a series of $[2 \times 2]$ and $[1 \times 1]$ tunnels extending along the short crystallographic c-axis of the tetragonal unit cell. These tunnels are formed by double chains of edge-sharing Mn_6 octahedra cross linked by sharing corners. A three-dimensional view of the linked octahedra in the α-MnO_2 structure as well as a projection of the structure onto the ab plane is shown in Figure 3.7. In contrast to the β-MnO_2, ramsdellite, and γ-MnO_2 chain structures

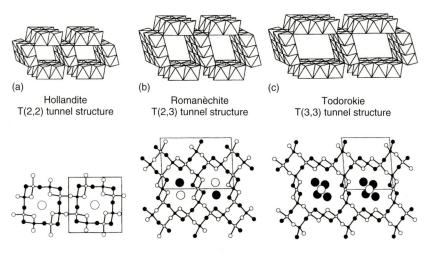

Figure 3.7 Crystal structures of (a) hollandite, (b) romanechite (psilomelane), and (c) todorokite. The structures are shown as three-dimensional arrangements of the MnO_6 octahedra (the tunnel-filling cations and water molecules, respectively, are not shown in these plots) and as projections along the short axis. Small, medium, and large circles represent the manganese atoms, oxygen atoms, and the foreign cations or water molecules, respectively. Open circles, height $z = 0$; filled circles, height $z = \frac{1}{2}$.

discussed above, the larger [2 × 2] tunnel allow various cations to be located in the middle of the cavity.

Hence, it is not surprising that a large number of different minerals with more or less ideal α-MnO$_2$ types of structure are known. Minerals containing sodium (manjiroite), potassium (cryptomelane), barium (hollandite), and lead (coronadite) have been structurally characterized in detail. The crystallographic data for some α-MnO$_2$ compounds are summarized in Table 3.1, from which it can be seen that the members of the structural family with [2 × 2] tunnels can be described either by an ideal tetragonal or by a monoclinic cell with very similar dimensions, but an angle slightly differing from $\beta = 90°$. Generally, α-MnO$_2$ type compounds have a stoichiometry of $A_{2-y}B_{8-x}O_{16}$ (A = large cations, e.g., K$^+$, NH$_4^+$, Ba^{2+}, or water; B = small cations, Mn^{4+}, Mn^{3+}, V^{4+}, Fe^{3+}, Al^{3+}). Each large cation is surrounded by eight oxygen atoms forming a slightly cubic environment and additionally by four oxygen atoms outside the lateral faces of the cube. The octahedrally coordinated manganese atoms can be replaced by other small transition-metal cations with a similar ionic radius. Natural α-MnO$_2$ samples usually contain one large cationic species as the major component (e.g., Ba^{2+} in hollandite) and the other possible elements (e.g., K$^+$) in minor amounts. Water molecules have similar dimensions to the large ions mentioned above and therefore they can replace these cations in the tunnels. The crystallographic c-axis of the tetragonal description of α-MnO$_2$ or the b-axis of the monoclinic setting, respectively, has a dimension of about 280–290 pm. Hence, the shortest distances between the large cations A would be in the same range if their respective sites were completely filled. Since such a short distance does not usually occur in oxides (in contrast, in intermetallic compounds or the elemental structures such distances *can* be observed), the A site has an occupancy factor of about 50% or lower. This does not mean that a superstructure due to cation ordering will necessarily be observed, although some examples are known in the literature (see Table 3.2, Ba$_{2-x}$(Mn,Fe)$_8$O$_{16}$, with a doubled monoclinic b-axis). Usually the ordered domains of the cations in the α-MnO$_2$ are too small to be detected by the occurrence of superstructure reflections. Thus, the α-MnO$_2$ structure is mostly described by the tetragonal, pseudotetragonal (all angles 90°,

Table 3.2 $T(m,n)$ nomenclature scheme for manganese oxides, according to Turner and Buseck [4].

Common dimension	Variable dimension					
	$n = 1$	$n = 2$	$n = 3$	$n = 4$	$n = 5$	$n = \infty$
$m = 1$	$T(1,1)$	$T(1,2)$	$T(1,3)$	$T(1,4)$	$T(1,5)$	$T(1,\infty)$
Examples	β-MnO$_2$	Ramsdellite				Birnessite
$m = 2$	$T(2,1)$	$T(2,2)$	$T(2,3)$	$T(2,4)$	$T(2,5)$	$T(2,\infty)$
Examples		α-MnO$_2$	Romanèchite	Rb$_{16.64}$Mn$_{24}$O$_{48}$	Rb$_{0.27}$MnO$_2$	Buserite
$m = 3$	$T(3,1)$	$T(3,2)$	$T(3,3)$	$T(3,4)$	$T(3,5)$	$T(3,\infty)$
Examples			Todorokite			

true symmetry respective to the atomic arrangement is monoclinic), or monoclinic cell.

The presence of the foreign cations stabilizes the crystal structure of α-MnO_2 compounds. This manganese dioxide modification (more exactly it is *not a real MnO_2 modification*, since the structure contains a considerable proportion of foreign atoms) can be heated to relatively high temperatures (300–400 °C) without destruction of the lattice. Although Thackeray *et al.* reported the synthesis of cation-and water- free α-MnO_2 [48, 49], which is reported to be stable up to 300 °C without destruction of the [2 × 2] tunnel structure, it is commonly believed that a small, but significant, amount of water or foreign cations is necessary to prevent the collapse of the lattice. Otherwise submicro heterogeneities are formed, in which pyrolusite, ramsdellite, and intergrowth domain of these two structural elements coexist with the [2 × 2] tunnel structure of α-MnO_2 within the real lattice.

An interesting property of the sieve-like structure of α-MnO_2 is that it shows pronounced cation exchange. For example, an α-MnO_2 with a high barium content is easily prepared by stirring a sample of $(NH_4)_{2-x}$ Mn_8O_{16} (obtained from oxidation of $MnSO_4$ with a concentrated solution of $(NH_4)_2S_2O_8$ in the presence of additional NH_4 ions) in an acidified solution of barium nitrate at elevated temperatures. Consequently, the crystal structure of α-MnO_2 must contain OH^- groups, cation vacancies, and/or a proportion of manganese with an oxidation state below Mn^{4+} in order to maintain the charge balance. This is reflected in the significantly longer Mn–O distances in the MnO_6 ocathedra in α-MnO_2 (198 pm) compared with those in β-MnO_2 (188 pm).

The wide variety of natural minerals with [2 × 2] tunnels already indicates that a huge number of different compounds of the α-MnO_2 type can be obtained by a laboratory synthesis. Most chemical synthesis can be modified by working with concentrated solutions of the chosen foreign cations or by adding larger quantities, for example, of potassium salts or ammonium salts during the reaction in order to produce the α-modification as the major product (for references, see Ref. [47]). Similarly, it is possible to produce samples with a controlled α-MnO_2/γ-MnO_2 ratio by electrolysis in the presence of various amounts of K_2SO_4 or $CaSO_4$ [50].

3.2.5
Romanèchite, Todorokite, and Related Compounds

The crystal structure of romanèchite (or psilomelane) is closely related to that of α-MnO_2. Whereas the α-modification of manganese dioxide consists of corner-sharing double chains of MnO_6 octahedra connected through common edges, the romanèchite structure is build up by crosslinking of chains of double and triple octahedra, as shown in Figure 3.7b. The resulting [2 × 3] tunnels, extending in the b direction of the monoclinic cell, are partially filled by potassium cations or barium cations and by water molecules. Both an orthorhombic setting [22] and a monoclinically distorted structure [24] have been described in the literature (Table 3.1). Mukherjee found a doubling of the short *b*-axis of an orthorhombic subcell [23] due to cation/water ordering within the [2 × 3] tunnels. The most reliable and realistic structure determination seems to be the monoclinic structure refinement from single crystal X-ray data collected from a natural romanèchite crystal [24]. The presence of water, Ba^{2+}, and K^+ ions in the structure was found to be essential for the stability of romanèchite. After being heated to higher temperatures the crystal structure collapses and forms the much more stable hollandite-type compound $(Ba, K)_{2-x} Mn_8O_{16}$. Furthermore, in order to account for the charge balance of a romanèchite of the typical composition $Ba_{0.66}Mn_5O_{10} \cdot 1.5\ H_2O$, a certain fraction of the manganese atoms in the structure must have a lower oxidation state than Mn(IV), or manganese vacancies must occur. In contrast to the other MnO_2 modification described above, romanèchite contains three nonequivalent manganese sites. Two of these sites have average Mn–O distances of about 191 pm, where as the third Mn octahedral position has a significantly larger Mn–O distance of 199 pm. This indicates that manganese species with a lower valence accumulate at this crystallographic site.

Investigation of the rubidium–manganese–oxygen ternary systems revealed the existence of two manganates (III, IV) with tunnel structures comparable with the mineral romanèchite. $Rb_{16.64}Mn_{24}O_{48}$ [25] contains [2 × 4] tunnels formed by cross linking (through common corners) of double chains with a building element consisting of four edge-sharing MnO_6 octahedra chains. A similar compound, $Rb_{0.27}MnO_2$ [26, 27], consists of MnO_6 octahedra chains, which are connected in a way that [2 × 5] tunnels are formed. The rubidium atoms occupy ordered positions within the tunnels. Although it may be considered speculative, it seems likely that compounds with larger tunnels (e.g., [2 × 6] or [2 × 7]) might exist.

For a long time the structural classification of the mineral todorokite was uncertain, until Turner and Buseck [4] could demonstrate by HRTEM investigations that the crystal structure of that mineral consists of triple chains of edge-sharing octahedra, which form [3 × 3] tunnels by further corner-sharing. These tunnels are partially filled by Mg^{2+}, Ca^{2+}, Na^+, K^+, and water (according to the chemical analysis of natural todorokites). In 1988 Post and Bish could perform a Rietveld structure determination from XRD data taken for a sample of natural todorokite [28]. This diffraction study confirmed the results of Turner and Buseck. The cations and water molecules in the [3 × 3] tunnels show a high degree of disorder, as could be expected. The mineral itself is rarely found in natural deposits [51]. A reliable laboratory synthesis was not reported until Shen *et al.* [6] demonstrated by HRTEM and XRD investigations that synthetic todorokite can be obtained by mild hydrothermal synthesis in alkaline solutions in the presence of Mg^{2+} ions. These samples show pronounced ion exchange and have been tested as absorbing agents for organic molecules (i.e., cyclo-C_6H_{12}, CCl_4). The authors found that relatively large amounts (18–20 g/100 g todorokite) of the organic compounds can be absorbed. The tunnels in todorokite have a typical diameter of ~690 pm, which seems – similarly to zeolites – to be well suited for the incorporation of small organic molecules.

The intensive investigation of manganese oxides during recent decades led to the discovery of a large number of closely related tunnel structures. The various ways in which the one dimensionally infinite $Mn(O,OH)_6$ octahedra strings in these compounds can be linked to each other resulted in the occurrence of manganese oxides with tunnel sizes ranging from [1 × 1] (pyrolusite) to the very large [3 × 3] channels present in todorokite. The latest investigations of the rubidium manganese oxides $Rb_{0.27}MnO_2$ and $Rb_{16.64}Mn_{24}O_{48}$ [19–21] demonstrated that even [2 × 4] and [2 × 5] channel structures do exist. Furthermore, the HRTEM investigations of Turner and Buseck [4] confirmed the intergrowth model of De Wolff for γ-MnO_2 and it could also be shown that an intergrowth of larger tunnel structures may occur. The authors reported the occurrence of intergrown todorokite [3 × 3] with [3 × 7] structures as well as random arrangements of hollandite ([2 × 2], α-MnO_2) with romanèchite units ([2 × 3] tunnels). Therefore they proposed a classification scheme (see Table 3.2) which describes the crystal structures as a system of tunnels $T(m, n)$ with a common dimension m and a variable dimension n. For example, pyrolusite, which contains only –[1 × 1] tunnels, is denoted as $T(1, 1)$, and a compound that contains [2 × 2] (hollandite-like) and [2 × 3] (romanèchite-type) tunnels can be considered as an inter-growth of $T(2, 2)$ and $T(2, 3)$ types. With increasing n the structures approach the layered compounds (e.g., $T(2, 4)$ and $T(2, 5)$ structures with broad channels). Finally, structural features denoted as $T(1, \infty)$ and $T(2, \infty)$ can be regarded as representatives of phyllomanganates.

3.3
Layer Structures

Similarly to the tunnel structures of β-MnO_2, ramsdellite, γ-MnO_2, the layered manganese dioxides, and many related compounds are based on a more or less distorted hexagonal close packing of oxygen atoms. In layer manganates or phyllomanganates, the manganese atoms occupy the octahedral voids in such a way that two-dimensionally infinite sheets of edge-sharing MnO_6 groups are formed. In the direction perpendicular to the layer plane, the empty and filled layers alternate. In general, the structure of layered manganese oxide is similar to the C6 type (CdI_2 or $Mg(OH)_2$ (brucite) structure). A schematic drawing of this lattice type, consisting of stacked layers formed by edge-sharing octahedra, is shown in Figure 3.8.

The enormous variety of manganese oxides with structures similar to the one shown in Figure 3.8 arises from the different cation and water contents of the

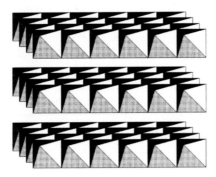

Figure 3.8 Schematic drawing of the layered manganese oxides. The structure consists of a stacking of empty and Mn(III, 1V)-tilled layers of edge-sharing octahedra.

space between the MnO_6 octahedra sheets, from the various ways the layers may be stacked, and from a large number of possible defects and superstructures in this family of crystal structures. Table 3.3 gives an overview of the crystallographic properties of some manganese oxides with a layer structure.

3.3.1
Mn_5O_8 and Similar Compounds

The compound Mn_5O_8 was first described in 1934 by Le Blanc and Wehner [67]. At that time the compound was believed to be a modification of Mn_2O_3. About 30 years later Oswald and Wampetich correctly determined the crystal structure of Mn_5O_8 and the isotypic compound $Cd_2Mn_3O_8$ [68] from single-crystal data. These two manganese oxides, as well as the isotypic copper- and zinc- containing phases $Cu_2Mn_3O_8$ [53] and $Zn_2Mn_3O_8$ [69], crystallize monoclinically. Mn_5O_8 represents a mixed-valence compound containing manganese in the oxidation states Mn^{2+} and Mn^{4+}. Hence, the formula can written as $(Mn^{2+})_2(Mn^{4+})_3O_8$, suggesting that in the isotypic compounds Zn^{2+}, Cu^{2+}, and Cd^{2+} replace the Mn(II) atoms at their respective sites. The crystal structure (see Figure 3.9) is best described as a strongly distorted pseudohexagonal layer structure, derived from the CdI_2-type structure. The lattice is build up of wave-like sheets of heavily distorted MnO_6 octahedra, in which every fourth manganese atom is missing. The Mn–O distances in the octahedral units range from 185 to 192 pm. The Mn^{2+} cations are placed below and above the Mn^{4+} vacancies, occupying distorted trigonal-prismatic voids. Riou and Lecerf [54] found that the cobalt compound $Co_2Mn_3O_8$ crystallizes with the higher-symmetry space group $Pmn2$, compared with Mn_5O_8 (space group $C2/m$). The authors suggested that this might be due to the finding that the cobalt atoms occupy two kinds of sites with differing coordination polyhedra, where as in structures of the Mn_5O_8 type only one coordination occurs for the Mn^{2+} site. The relatively short interlayer distances of 472 pm ($Cu_2Mn_3O_8$), 488 pm (Mn_5O_8), and 510 pm ($Cd_2Mn_3O_8$) indicate a strong interaction between the layers and the interlayer atoms. None of the compounds occurs in natural deposits; they can only be prepared in the laboratory, either by classical solid state chemical methods (e.g., as in Refs [53, 54]) or by soft-chemical reaction paths under mild conditions. Mn_5O_8 can be obtained by hydrothermal oxidation of MnO in the temperature range 120–910 °C at a water pressure of up to 1 kbar and at oxygen partial pressures ranging from 1 to 100 bar [70]. Another method is the oxidation of manganese oxide hydroxides in air or in oxygen at moderate temperatures [71–73].

A manganate (III, IV) similar to Mn_5O_8 is known in the literature: ternary lead manganese oxide $Pb_3Mn_7O_{15}$. In this compound the sheets of MnO_6 octahedra contain defects at the manganese sites. Four out of 14 manganese atoms are missing within the layer and are positioned in an octahedral environment above and below the vacancy in the sheets. Additionally, the structure contains Pb–O layers that separate the Mn–O sheets. Thus the stacking sequence in $Pb_3Mn_7O_{15}$ is: MnO_6 (main layer) – MnO_6 (interlayer) – PbO–MnO_6 (interlayer) – MnO_6 (main layer). The distance between the main MnO_6 sheets is larger (678 pm) than

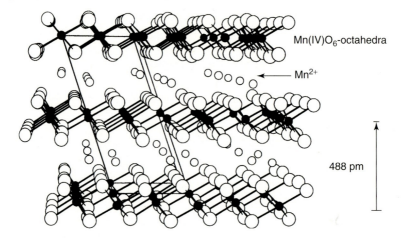

Figure 3.9 Perspective drawing of the crystal structure of Mn$_5$O$_8$. Small, filled balls represent the Mn^{4+} ions; small, open circles mark the positions of the Mn^{2+} ions. The oxygen atoms are shown as large, open circles.

in Mn$_5$O$_8$ (488 pm) due to the insertion of a PbO layer. The crystal structure is described in the hexagonal space group $P6_3/mcm$ [60] and in its orthorhombic subgroup $Cmcm$ [59] (Table 3.3).

A third compound of comparable stoichiometry and atomic arrangement has been described by Chang and Jansen [58]. The author prepared the sodium manganate (IV) Na$_2$Mn$_3$O$_7$ and determined the crystal structure from single-crystal data. The lattice is built up by two-dimensionally infinite sheets of [Mn$_3$O$_7$]$^{2-}$. Similarly to Mn$_5$O$_8$ and chalcophanite (see Section 3.3.3) a certain fraction of manganese atoms (one in seven) is missing in the layers. The sodium atoms are located above these vacancies in a distorted octahedral coordination. The distance between the MnO$_6$ main layers in Na$_2$Mn$_3$O$_7$ (490 pm) is comparable with the interlayer spacing in Mn$_5$O$_8$ and in isotypic compounds (472–510 pm).

3.3.2
Lithiophorite

The mineral lithiophorite, (Al,Li)MnO$_2$(OH)$_2$, can be found in natural deposits together with other manganese oxides. It is somewhat unusual within the family of manganese oxides containing foreign cations or anions, since it has been found that only small amounts of other transition metals and *no* larger alkaline metals (e.g., Na$^+$ or K$^+$) can replace the manganese and lithium atoms, respectively, in the crystal structure [74–77]. Lithiophorite has a layer structure (see Figure 3.10), in which sheets of edge-sharing MnO$_6$ octahedra alternate with very similar layers of (Al,Li)(OH)$_6$ octahedra. The two layer types in the crystal structure are connected by O–H bridging bonds between the OH$^-$ groups of the (Al,Li)–OH sheet and the oxygen atoms in the Mn–O layers. The distance between two Mn–O layers is

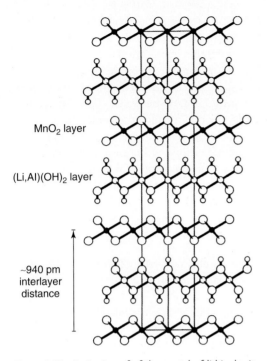

Figure 3.10 Projection of of the crystal of lithiophorite, (Li,Al)MnO$_2$(OH)$_2$, along the [110] direction of the hexagonal cell [57]. The connections within the MnO$_6$ and (Li,Al)(OH)$_6$ octahedra layers are emphasized. For a better understanding the O–H bridging bounds between the two layer types are not shown.

in the region of 940 pm. The sheets are stacked in such a way that each OH group is located directly above or below an oxygen atom of a neighboring MnO$_6$ layer.

Since the first structure determination by Wadsley [55] in 1952 there has been confusion about the correct cell dimensions and symmetry of natural as well as synthetic lithiophorite. Wadsley determined a monoclinic cell (for details see Table 3.3) with a disordered distribution of the lithium and aluminum atoms at their respective sites. Giovanoli *et al.* [74] found, in a sample of synthetic lithiophorite, that the unique monoclinic *b*-axis of Wadsley's cell setting has to be tripled for correct indexing of the electron diffraction patterns. Additionally, they concluded that the lithium and aluminum atoms occupy different sites and show an ordered arrangement within the layers. Thus, the resulting formula given by Giovanelli *et al.* is LiAl$_2$Mn$_3$O$_6$(OH)$_6$. Another structure determination was performed by Pauling and Kamb [56]; the large superstructure unit cell that they deduced has a trigonal symmetry. A model for the ordering of lithium and aluminum atoms has been proposed and the authors suggested that one of 21 octahedra in the (A1,Li)–OH layer remain vacant. The latest structure model was suggested by Post and Appleman [57] in 1994. They refined the crystal structure with a trigonal unit, which has a similar c-axis ($c = 2816.9$ pm) to the cell proposed by Pauling et al. (2820 pm), but a very short a-axis of only 282.5 pm. In the structural model of Post and Appleman the lithium and aluminum atoms are statistically distributed at their respective crystallographic sites (this model is shown in Figure 3.10).

All the structure models discussed above describe generally the same structural feature: the alternation of Mn–O and (Al,Li)–OH sheets. The only difference is that each model deals with a slightly different atomic arrangement in terms of ordering at the Al/Li site and the commensurability of the layer stacking in the various superstructures of the simple monoclinic model proposed by Wadsley. Since all of these authors investigated different samples with possibly slightly differing structural properties, none of the models is necessarily to be preferred.

3.3.3
Chalcophanite

The crystal structure of the mineral chalcophanite, $ZnMn_3O_7 \cdot 3H_2O$ (see Figure 3.11), was one of the first layer structures of manganese oxides that has been determined.

Wadsley [78] described the crystal structure with a triclinic unit cell ($a = h = 754$ pm. $c = 822$ pm, $\alpha = 90°, \beta = 117.2°, \gamma = 120°$), while Post and Appleman [61] found a trigonal symmetry ($a = 753.3$ pm, $c = 2079.4$ pm). However, the general features of the crystal structure of chalcophanite are described by both symmetries, although the latter might considered the more reliable one. The crystal lattice consists of single sheets of edge-sharing MnO_6 octahedra, which are separated by layers of water molecules. The zinc atoms are located in octahedral coordination spheres between the water and the MnO_6 layer. The distance between two Mn–O sheets was found to be 717 pm. One in seven of the manganese sites in each layer is unoccupied, so the composition of the layers is $[Mn_3O_7]^{2-}$ and not MnO_2.

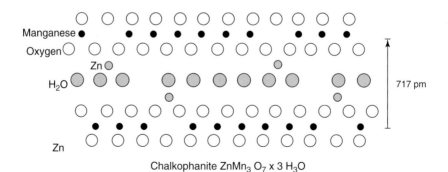

Chalkophanite $ZnMn_3O_7 \times 3 H_3O$

Figure 3.11 Projection of the chalcophanite ($ZnMn_3O_7 \cdot 3H_2O$) structure along the crystallographic b-axis. The structure consists of layers or edge-sharing MnO_6 octahedra, a separating water layer and single zinc ions which are octahedrally coordinated by three oxygen atoms from the Mn–O sheets and three oxygen atoms from the water layer.

Associated with each Mn–O layer and directly above and below the unoccupied Mn position, the Zn^{2+} ions are located in a somewhat distorted octahedral void, formed by three oxygen atoms of the Mn–O layer and three oxygen atoms from the water layer. The two kinds of sheets in the structure are held together by H–O bridging bonds between the water molecules and the oxygen atoms of the Mn–O layer. Natural chalcophanites differ quite significantly from the ideal composition. Not only can the water content be variable, but there are some Mn^{4+} defects in the lattice and the sum of the cations usually exceeds four per formula unit. This indicates that some additional interlayer atoms are present and that some Mn^{4+} ions must be replaced by manganese atoms with a lower oxidation state, for example, Mn^{2+} and Mn^{3+}.

Similarly to Mn_5O_8 and related compounds, the chalcophanite structure can be interpreted as a 'filled' CdI_2-type structure. The space in the octahedral layer is filled by an additional layer of water molecules and some foreign cations. A comparable situation is found in several hydroxozincates, for example, $Zn_5(OH)_8Cl_2 \cdot H_2O$ or $Zn_5(OH)_6(CO)_3$. In these compounds the layers are formed by edge-sharing zinc hydroxide octahedra, $Zn(OH)_6$, and the space between the layers is filled with chloride and carbonate anions and some Zn^{2+} cations, which are located above and below vacancies in the Zn–OH layers.

3.3.4
δ-MnO_2 Materials

A large number of natural mineral and synthetic materials with a layered structure and strongly varying water and foreign-cation content have been collected in the δ-MnO_2 group. Most of these materials have a very poor crystallinity and a wide range of existence. Therefore many authors have described 'subgroups' of the layered (III, IV) manganates and named them in a mostly confusing way [79–85], for example, manganous manganates, δ-manganese dioxide, 7 Å phyllomanganates. Giovanoli et al. [86] suggested that all these compounds belong to only one group, so only the name δ-MnO_2 should be used to describe these layered manganates. The differences in the XRD patterns arise from the strongly differing composition and crystallinity, but the general arrangements of the structural units for the δ-MnO_2 compounds are the same. The crystal structure is built up from layers of edge-sharing MnO_6 octahedra with a certain number of water molecules and foreign cations between the layers. Hence, the chalcophanite type structure might be considered as a well-crystallized prototype for the structural chemistry of δ-MnO_2 materials. In 1956 Jones and Milne [87] described a mineral of composition $(Na_{0.7}Ca_{0.3})Mn_7O_{14} \cdot 2.8H_2O$ that was found in a deposit near Birness in Scotland. Since no mineralogical name had been given to the ore, they called this mineral '*birnessite*'; this name is now used synonymously for the designation δ-MnO_2. A number of additional studies have revealed that a relatively large number of natural manganese oxide deposits contain materials of the birnessite type [88–91]. Additionally, it has been shown that layered manganese oxides of birnessite-type compounds are the major components in the manganese nodules found on the sea floor [91–94].

Because of the low crystallinity of δ-MnO_2 and birnessite samples, Giovanoli et al. used high-resolution diffraction techniques, XRD powder methods, and chemical analyses in order to study a number of synthetic layer manganates [1, 62, 63]. The crystal structures of the sodium-manganese-(II, III) manganate-(IV) hydrate ($Na_4Mn_7O_{27}\cdot 9H_2O$) and the manganese-(III) manganate-(IV) hydrate $Mn_7O_{13}\cdot 5H_2O$ have been modeled on the basis of the structure of chalcophanite. It could be shown that the lattice of the sodium-manganese manganate (IV) is built up by a stacking of alternating sheets of water and hydroxide ions and layers of edge-sharing MnO_6 octahedra with a distance of about 713 pm between two Mn–O layers. One of every six manganese sites in these layers is unoccupied, and Mn^{2+} and Mn^{3+} are considered to lie above and below these vacancies in a distorted octahedral arrangement formed by three oxygen atoms of the Mn–O layer and three hydroxide ions or water molecules of the intermediate layer. The position of the sodium atoms in this layer remained uncertain. The orthorhombic lattice constants for $Na_4Mn_7O_{27}\cdot 9H_2O$ determined by Giovanoli et al. [62] can be taken from Table 3.3. The sodium-free compound $Mn_7O_{13}\cdot 5H_2O$ is obtained by leaching of $Na_4Mn_7O_{27}\cdot 9H_2O$ with HNO_3. The structure of the main Mn–O layer (including the Mn^{3+} ions above and below the Mn^{4+} vacancies) is the same as in the sodium-sheets of containing sample. The sodium ions are missing in the water/hydroxide layer separating the MnO_6 octahedra. The distance between the main Mn–O layers is slightly larger (727 pm) than that in the sodium-containing compound (713 pm). For $Mn_7O_{13}\cdot 5H_2O$ a small hexagonal cell of $a = 284$ pm and $c = 727$ pm was found. Recently Chen et al. [64] described a potassium-containing birnessite, $K_{0.27}MnO_2\cdot 0.54H_2O$, with a similar rhombohedral symmetry ($R\ 3m$), but a tripled c-axis with $c = 2153.6$ pm as compared with 727 pm in $Mn_7O_{13}\cdot 5H_2O$, while the a lattice parameters are quite similar (see Table 3.3). The distance between two Mn–O layers in $K_{0.27}MnO_2\cdot 0.54H_2O$ (718 pm) is comparable with that in $Na_4Mn_7O_{27}\cdot 9H_2O$ (713 pm). Giovanoli found that the 'pure' $Mn_7O_{13}\cdot 5H_2O$ compounds are less stable than the samples containing foreign ions. Hence, one can conclude that foreign ions (e.g., sodium or potassium) have a stabilizing effect on the layer structure [63]. Additionally, the presence of water in layered manganates might assist in stabilizing the layered arrangement. Figure 3.12a shows a schematic drawing of the crystal structures of a birnessite-type material. Generally, according to Stouff and Boulègue [8], the foreign cations can occupy two different positions: between the water and the Mn–O layer (above an Mn^{4+} vacancy, position B1) or directly between the two main Mn–O layers (at the same height as the water molecules or hydroxide ions, position B2).

Post and Veblen [65] investigated the crystal structures of several synthetic sodium, potassium, and magnesium birnessites by electron diffraction methods and Rietveld refinements of XRD patterns. Starting from the crystal structure of chalcophanite they determined the atomic arrangement in these compounds. The materials of the composition $A_xMnO_2\cdot yH_2O$ (A = Na, K, Mg) have monoclinic symmetry (see Table 3.3). Post and Veblen reported a number of detailed structural

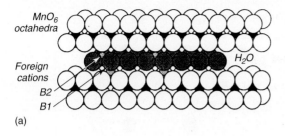

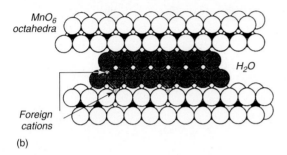

Figure 3.12 Schematic drawing of the theoretical positions of the foreign metal ions in (a) 7 Å phyllomanganates and (b) 10 Å phyllomanganates. The foreign ions can be located above a manganese vacancy (between the Mn–O layer and the sheet of water molecules) and within the layer, respectively (adapted from Ref. [41]).

features for these layered manganates. The average Mn–O bond length of 194 pm is significantly greater than that in chalcophanite with a Mn–O distance 190.6 pm. This might be due to a substitution of Mn^{4+} by Mn^{3+} in the MnO_6 octahedra. In contrast to the findings of many other authors, they detected only a few manganese vacancies in the Mn–O layer. According to chemical analyses, Post and Veblen found that water was the predominant species in the separating layer between the MnO_6 octahedra. The positions of the foreign metal ions depend strongly on the chemical nature of the ions themselves. Figure 3.13 is a schematic drawing of the atomic arrangement in $Na_{0.58}MnO_2 \cdot 1.5H_2O$ and $Mg_{0.29}MnO_2 \cdot 1.7H_2O$. In the sodium compound the alkaline metal ions occupy sites within the layer, whereas the magnesium atoms are located in an octahedral environment between the Mn–O layer and the sheet of water molecules. These two positions correspond to the sites B1 and B2 for foreign metal ions, as proposed by Stouff and Boulègue in Figure 3.12a.

It has been mentioned above that birnessite-type samples can show a wide variety of different XRD patterns. Mostly, the samples show only XRD peaks around 240 pm ($2\theta \approx 37°$ for CuKα radiation) and 142 pm ($2\theta \approx 66°$). These peaks correspond to the (1 0 0) and (1 1 0) reflections of the simple hexagonal setting of the δ-MnO_2 unit cell. Additionally, in some natural as well synthetic materials the basal plane

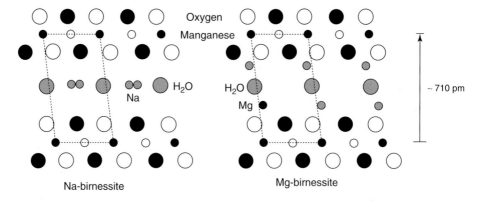

Figure 3.13 Schematic drawing of the crystal structures of ($Na_{0.58}MnO_2 \cdot 1.5H_2O$) and ($Mg_{0.29}MnO_2 \cdot 1.7H_2O$).

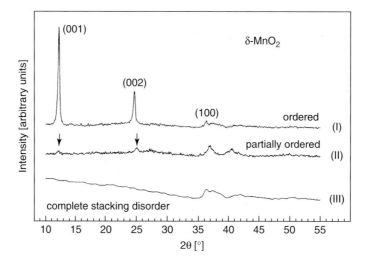

Figure 3.14 XRD patterns of various δ-MnO_2 samples.

(0 0 l) reflections (0 0 1) and (0 0 2) do occur at about 700 pm ($2\theta \approx 12°$) and 350 pm ($2\theta \approx 25°$), respectively.

Figure 3.14 shows the XRD patterns of three different δ-MnO_2-type samples. Sample I was prepared by thermal decomposition of $KMnO_4$ at 450 °C for 24 h. Subsequently the resulting dark brown powder was leached with water and dried at 80 °C. The birnessite-type compound II was obtained by synproportionation of $Mn(OH)_2$ with $BaMnO_4$ in an alkaline solution, and sample III is the reaction product of the disproportionation of K_2MnO_4 into $KMnO_4$ and an Mn(IV) compound, which occurs during dilution of a slightly alkaline solution of this manganate–(VI). Sample I shows (0 0 l) reflections as well as the (1 0 0) peak, which indicates a high degree of order in the crystal lattice, whereas sample II only shows a very weak (0 0 l) reflection. In the XRD pattern of sample III the basal (0 0 l) peaks are completely missing. Giovanoli [3] has developed a model that explains this somewhat extraordinary behavior of different δ-MnO_2 materials. In the model of birnessite-type materials proposed by Giovanoli, several arrangements of the Mn–O layer and the interlayer cations or water molecules are possible. The ideal is the perfect ordering of both types of sheets in the structure (see Figure 3.15a), thus building up a fully commensurate crystal lattice such as is realized in chalcophanite. The XRD pattern of sample I in Figure 3.14 may be of a structure type with a high degree of order. In the second case a certain kind of disorder appears within the cation or water layer in terms of an inhomogeneous distribution of the interlayer species, while the distance between the main layers is constant (Figure 3.15b). A very similar situation is demonstrated in Figure 3.15c: a disorder of the interlayer species is combined with a shift of the main Mn–O layer within its own plane. This results in an incommensurate arrangement of the layers in the direction perpendicular to the layers. As a result the XRD patterns show only weak

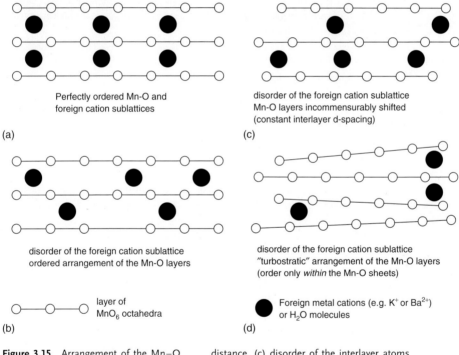

Figure 3.15 Arrangement of the Mn–O layers and separating sheets according to Giovanoli [3]. The layer structure can be (a) completely ordered or (d) completely disordered (turbostratic disorder). The cases (b) and (c) represent situations between the two extremes: (b) disorder of the interlayer atoms or molecules but an ordered stacking of the Mn–O layers with constant layer distance, (c) disorder of the interlayer atoms and an incommensurate shift of the complete Mn–O sheet within the layer plane, resulting in an incommensurate superstructure along the c-direction (perpendicular to the layer) and in a diffuse distribution of the electron density in this layer, resulting in a lower contribution of this layer to the 0 0 l reflections. (Adapted from Ref. [41].)

and broadened (0 0 l) peaks because of the imperfection of the crystal lattice or (in other words) because of the very small homogeneously scattering domains in the crystallites. Sample II in Figure 3.14 may be of this kind. The highest degree of disorder is described by the model in Figure 3.15d.

In this case the foreign cations and water molecules are inhomogeneously distributed and the Mn–O layers are stacked with no periodicity. In some regions of the lattice, where no foreign cations or water molecules are present, the distance between the layers becomes quite small (comparable with Mn_5O_8), while in regions with separating cations or water molecules the Mn–O sheets are at a regular distance from one another. The XRD spectra of such compounds show only the (1 0 0) and (1 1 0) peaks because of a relatively high degree of order within the Mn–O layers, as is also the case for Figure 3.15a–c. The situation for birnessites is similar to that found in another layer compound with different degrees of crystallinity, namely *graphite* in comparison with its disordered species *carbon black*. Graphite has a high crystallinity and a more or less perfect and commensurate order of the carbon layers, while the hexagonal nets of carbon atoms in the structure of carbon black exhibit a large number of different interlayer distances and stacking faults, also called a *turbostratic disorder*.

3.3.5
10 Å Phyllomanganates of the Buserite Type

Buserite is structurally closely related to the 7 Å manganates discussed above. The crystal structure is built up by slabs of edge-sharing MnO_6 octahedra, which are separated by two layers of water molecules or hydroxide anions. The latter layer contains various amounts of foreign cations (e.g., Na^+, or Mn^{2+}, Mn^{3+}). The main Mn–O layers are at a distance of about 1000 pm. Buserite-type materials were found to be one of the major components of marine manganese deposits. Synthetic buserites of the composition $(Na,Mn) Mn_3O_7 \cdot xH_2O$ have been studied by Wadsley [66]. The symmetry of this hygroscopic material was found to be hexagonal with the lattice constants $a = 841$ pm (see Table 3.3). Additionally, Wadsley found that this 10 Å manganate contains not only various amounts of water, part of which can be reversibly extracted and re-introduced into the crystal lattice without destruction of the structure, but also of a significant amount of sodium atoms, which can be easily exchanged within the water and hydroxide layer. In further experiments Giovanoli et al. [95] observed that even bivalent metals (Ca^{2+} or Mg^{2+}) can be incorporated into the buserite structure. In acidic solution the stability of the 10 Å manganates decreases with increasing valence of the interlayer ions and with decreasing number of foreign cations in the structure. On complete dehydration the 10 Å manganates decompose irreversibly to the 7 Å manganates. Hence, the 10 Å phase can be interpreted as a hydrated form of the 7 Å phase, as shown in the schematic drawing of Figure 3.12b. Due to the ion-exchange properties of sodium-rich buserites it is possible to replace sodium by large organic cations (e.g., *n*-dodecyl ammonium cations) as Paterson did [96]. This ion exchange increases the layer distance from 1000 pm to about 2600 pm.

Table 3.3 Crystallographic data of layered manganese(IV, III) oxides.

Compound or mineral	Approximate formula	Symmetry	Space group	Lattice constants						Interlayer distance (pm)	References
				a (pm)	b (pm)	c (pm)	α (°)	β (°)	γ (°)		
Mn_5O_8 type	Mn_5O_8	Monoclinic	$C2/m$	1034.7	572.4	485.2	90	109.4	90	488	[52]
	$Cd_2Mn_3O_8$	Monoclinic	$C2/m$	1080.6	580.8	493.2	90	109.4	90	510	[52]
	$Cu_2Mn_3O_8$	Monoclinic	$C2/m$	969.5	563.5	491.2	90	103.3	90	472	[53]
	$Co_2Mn_3O_8$	Orthorhombic	$Pmn2_1$	574.3	491.5	936.1	90	90	90	468	[54]
Lithiophorite	$Al_{0.58}Li_{0.32}MnO_2(OH)_2$	Monoclinic	$C2/m$	506.0	291.6	955.0	90	100.5	90	939	[55]
	$Al_{14}Li_6Mn_{21}O_{42}(OH)_{42}$	Rhombohedral	$P3_1$	1337.0	1337.0	2820.0	90	90	120	940	[56]
	$Al_{0.65}Li_{0.33}MnO_2(OH)_2$	Rhombohedral	$R\bar{3}m$	292.5	292.5	2816.9	90	90	120	939	[57]
$Na_2Mn_3O_7$	$Na_2Mn_3O_7$	Triclinic	$P1$	663.6	685.4	754.8	105.8	106.9	111.6	490	[58]
$Pb_3Mn_7O_{15}$	$Pb_3Mn_7O_{15}$	Orthorhombic	$Cmcm$	1728	998	1355	90	90	90	678	[59]
	$Pb_3Mn_7O_{15}$	Hexagonal	$P6_3/mcm$	998	998	1355	90	90	120	678	[60]
Chalcophanite	$ZnMn_3O_7 \cdot 3H_2O$	Rhombohedral	$R\bar{3}$	753.3	753.3	2079.4	90	90	120	693	[61]
$\delta\text{-}MnO_2$ group	$Na_4Mn_{14}O_{27} \cdot 9H_2O$	Orthorhombic	–	854	1539	1426	90	90	90	713	[62]
	$Mn_{14}O_{27} \cdot 5H_2O$	Hexagonal	–	284	284	727	90	90	120	272	[63]
	$K_{0.27}MnO_2 \cdot 0.54H_2O$	Rhombohedral	$R\bar{3}m$	284.9	284.9	2153.6	90	90	120	718	[64]
Synth. birnessite	$Na_{0.58}Mn_2O_4 \cdot 1.5H_2O$	Monoclinic	$C2/m$	517.4	285.0	733.6	90	103.2	90	714	[65]
	$K_{0.46}Mn_2O_4 \cdot 1.4H_2O$	Monoclinic	$C2/m$	514.9	284.3	717.6	90	100.8	90	705	[65]
Buserite	$Mg_{0.29}Mn_2O_4 \cdot 1.7H_2O$	Monoclinic	$C2/m$	505.6	284.6	705.4	90	96.6	90	700	[65]
	$(Na,Mn)Mn_3O_7 \cdot nH_2O$	Hexagonal	–	841	841	1010	90	90	120	1010	[66]

3.4
Reduced Manganese Oxides

The reduced manganese oxides and oxide – hydroxides will also be considered briefly in this section, since many of them are of interest for the performance of aqueous manganese dioxide electrodes (e.g., in primary cells), but also in rechargeable alkaline manganese dioxide/zinc cells (RAM cells; e.g., Ref. [97]). Compounds of the composition MnOOH are the reaction product of the electrochemical reduction of manganese dioxides, the electrochemically inactive spinel-type compounds (e.g., Mn_3O_4 or Mn_2O_3) are formed during cycling of a MnO_2 cathode, and the manganese hydroxide $Mn(OH)_2$ is the manganese compound with the lowest oxidation state occurring in aqueous systems. The lattice constants and symmetry data of several reduced manganese oxides are summarized in Table 3.4.

3.4.1
Compounds of Composition MnOOH

3.4.1.1 Manganite (γ-MnOOH)

The crystal structure of manganite is closely related to that of pyrolusite in that it consists of single chains of edge- and corner-sharing $Mn(O,OH)_6$ octahedra. The coordination polyhedra of the Mn^{3+} ions in the structure are strongly distorted due to the Jahn–Teller effect of the trivalent manganese ions and the substitution of O^{2-} counter-ions by OH^- ions. This results in the formation of four short Mn–O bonds (with Mn–O distances ranging from 188 to 198 pm) and two longer apical Mn–OH bonds (220–233 pm). The formation of hydrogen bridging bonds leads to a pseudo-orthorhombic (or monoclinic) superlattice of the tetragonal β-MnO_2 parent structure. Manganite is the thermodynamically stable modification of MnOOH; therefore it can found in natural deposits as well as being easily prepared in the laboratory. It also occurs as the reaction product during electrochemical reduction of β-MnO_2 in batteries.

3.4.1.2 Groutite (α-MnOOH)

In the same way as manganite may be regarded as the structurally closely related reduction product of β-MnO_2, groutite or α-MnOOH was found to be isostructural with ramsdellite. The arrangement of the $Mn(O,OH)_6$ octahedra in α-MnOOH is very similar to that of the ramsdellite modification of MnO_2. The structure consists of double chains of octahedra. As has already been described for manganite, the protons occupy positions in the crystal structure which allow them to build up a hydrogen-bound network within the [2 × 1] tunnels. The situation for MnOOH is comparable with that of the compounds Li_xMnO_2, in which the lithium ions occupy sites in the tunnel of a ramsdellite host lattice. In both compounds the Mn–O octahedra are strongly distorted because of the insertion of foreign cations into the lattice and the resulting reduction from Mn^{4+} to Mn^{3+}. A schematic drawing of the protonated ramsdellite or α-MnOOH structure (diaspore type)

Table 3.4 Crystallographic data for reduced manganese oxides and manganese oxide–hydroxides.

Compound or mineral	Approximate formula	Symmetry	Space group	Lattice constants						References
				a (pm)	b (pm)	c (pm)	α(°)	β(°)	γ(°)	
Groutite	α-MnOOH	Orthorhombic	$Pnma$	1076.0	289.0	458.0	90	90	90	[98]
Feitknechtite	β-MnOOH	Trigonal	$P3m1$	332	332	471	90	90	120	[99]
Manganite	γ-MnOOH	Orthorhombic	$B2_1/d$[a]	880.0	525.0	571.0	90	90	90	[100]
Hausmannite	Mn_3O_4	Tetragonal	$I4_1/amd$	814.0	814.0	942.0	90	90	90	[101]
	α-Mn_2O_3	Cubic	$Ia3$	943.0	943.0	943.0	90	90	90	[102]
	γ-Mn_2O_3	Tetragonal	$I4_1/amd$ P	815	815	944	90	90	90	[103]
Pyrochroite	$Mn(OH)_2$	Trigonal	$P3m1$	332.2	332.2	473.4	90	90	120	[104]

[a] Nonstandard setting of this space group.

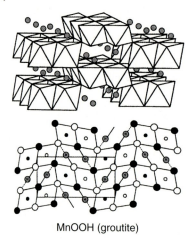

MnOOH (groutite)

Figure 3.16 Crystal structure of α-MnOOH. The structure is shown as a three-dimensional arrangement of the Mn(O,OH)$_6$ octahedra with the protons filling the [2 × 1] tunnels, and as a projection along the short crystallographic c-axis. Small circles, manganese atoms; large circles, oxygen atoms; open circles, height $z = 0$; filled circles, height $z = \frac{1}{2}$. The shaded circles represent the hydrogen ions.

is shown in Figure 3.16. The lattice parameters for α-MnOOH are significantly larger than those of ramsdellite, but the symmetry of the parent structure is maintained.

3.4.1.3 δ-MnOOH

It can be easily realized that an intergrowth structure of β-MnO$_2$ and ramsdellite (i.e., γ-MnO$_2$) is protonated (reduced) in a very similar way. By analogy with the De Wolff model for γ-MnO$_2$, the crystal structure of δ-MnOOH can be interpreted as an intergrowth of manganite and groutite [105] domains. δ-MnO$_2$ is believed to be the reaction product of γ-MnO$_2$ during single-electron discharge in alkaline solutions. The unit cell of δ-MnOOH can be described in terms of an orthorhombic cell, as is the case for many γ-MnO$_2$ samples.

3.4.1.4 Feitknechtite β-MnOOH

The reduced manganese oxide–hydroxides described above are based on tunnel structures. The layered manganates-(III, IV) can be reduced as well. The respective product is best described by a stacking of two-dimensionally infinite sheets of edge-sharing Mn(O,OH)$_6$ octahedra, which are held together by hydrogen bridging bonds between the layers. The crystal structure is very closely related to that of Mn(OH)$_2$ (see Section 3.4.1.1 in this chapter). The symmetry is hexagonal with unit cell parameters ($a = 332$ pm, $c = 471$ pm) which are very close to these of the brucite-type Mn(OH)$_2$ ($a = 332.2$ pm, $c = 473.4$ pm). δ-MnO$_2$ samples are topotactically reduced via β-MnOOH to Mn(OH)$_2$ and reoxidized without the need for significant change in the crystal structure.

3.4.2
Spinel-Type Compounds Mn_3O_4 and $\gamma\text{-}Mn_2O_3$

Both compounds Mn_3O_4 (i.e., hausmannite) and $\gamma\text{-}Mn_2O_3$ crystallize with a tetragonally distorted spinel-type structure (see Figure 3.17).

Hausmannite has the composition $(Mn^{2+})(Mn^{3+})_2O_3$. In this tetragonal spinel the bivalent cations are coordinated tetrahedrally, while the Mn^{3+} ions have an octahedral environment. The Mn^{2+} ion can be replaced by other divalent ions with nearly the same radius (e.g., Zn^{2+}; in zinc-containing manganese dioxide batteries heterolyte $ZnMn_2O_4$, which is isotypic with Mn_3O_4 and may form a solid solution with it, can be observed). If the synthesis of Mn_3O_4 is performed under oxidizing conditions, materials with XRD patterns very similar to that of Mn_3O_4 can be found, but the oxidation state is significantly higher than that of hausmannite. Verwey and De Boer [106] proposed the name of $\gamma\text{-}Mn_2O_3$ for samples of the composition $MnO_{1.39}$–$MnO_{1.5}$. Goodenough and Loeb [107] found that $\gamma\text{-}Mn_2O_3$ crystallizes with the tetragonally distorted spinel-type structure of Mn_3O_4 but with significant defects (one-third) at the tetrahedrally coordinated Mn^{3+} site.

3.4.3
Pyrochroite, $Mn(OH)_2$

Pyrochroite, $Mn(OH)_2$, is isotypic with brucite, $Mg(OH)_2$. The crystal structure consists of layers of edge-sharing $Mn(OH)_6$ octahedra. The layers are held together by hydrogen bonding, as shown in Figure 3.18. $Mn(OH)_2$ represents the manganese compounds with the lowest valence occurring in aqueous systems. Pyrochroite is easily oxidized in the presence of oxygen, even at ambient condition, and therefore

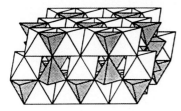

Mn_3O_4 / $\gamma\text{-}Mn_2O_3$ / $ZnMn_2O_4$

Figure 3.17 Schematic drawing of the spinel-type structure of Mn_3O_4 and $\gamma\text{-}Mn_2O_3$. The structure is built up of MnO_6 octahedra (white) and MnO_4 tetrahedra (shaded).

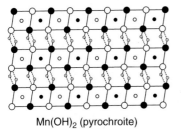

$Mn(OH)_2$ (pyrochroite)

Figure 3.18 Crystal structure of $Mn(OH)_2$, shown as a projection along the crystallographic a-axis. Small circles, manganese atoms; large circles, oxygen atoms; open circles, height $z = 0$; filled circles, height $z = \frac{1}{2}$. The shaded circles represent the hydrogen ions.

it does not occur in natural ores. The instability toward oxidation might be due to the easy topotactical transformation to the thermodynamically more stable MnOOH and birnessite-type materials containing manganese in the oxidation states III and IV, respectively.

3.5
Conclusion

More than 100 years of research on manganese dioxides has resulted in an enormous body of knowledge on the subtle structural details of a huge variety of crystalline materials. This detailed information about the structural chemistry of manganese dioxides and their related physical properties is important for further development of electrochemical systems based on manganese oxides, such as RAM cells or lithium-ion cells, and to enable us to elaborate systematic reaction paths toward the synthesis of new materials with improved properties. The crystal structure, morphology, composition, and physical properties of various manganese oxides still have to be optimized for a large number of different applications.

References

1. Giovanoli, R. and Stähli, E. (1970) *Chimia (Aarau)*, **24**, 49–61.
2. Ramdohr, P. and Frenzel, G. (1958) *Congr. Geol. Int. Mex. City*, **1**, 19–73.
3. Giovanoli, R. (1980) Natural and synthetic manganese nodules, in *Geology and Geochemist of Manganese*, vol. 1 (eds I.M. Varentsov and G. Grassely), Akadémiai Kiadó, Budapest, pp. 159–202.
4. Turner, S. and Buseck, P.R. (1981) *Science*, **212**, 1024.
5. Turner, S. and Buseck, P.R. (1983) *Science*, **304**, 134.
6. Shen, Y.F., Zerger, R.P., DeGuzman, R.N., Suib, S.L., McCurdy, L., Potter, D.I., and O'Young, C.L. (1993) *Science*, **260**, 511–515.
7. Potter, R.M. and Possman, G.R. (1979) *Am. Mineral.*, **64**, 1199.
8. Stouff, P. and Boulègue, J. (1988) *Am. Mineral.*, **73**, 1162.
9. Burns, R.G. and Burns, V.M. (1975) in *Proceedings MnO$_2$ Symposium*, vol. 1 (eds A. Kozawa and R.J. Brodd), Cleveland, p. 306.
10. Bolzan, A.A., Fong, C., Kennedy, B.J., and Howad, C.J. (1993) *Aust. J. Chem.*, **46**, 939–944.
11. Byström, A.M. (1949) *Acta Chem. Scand.*, **3**, 163–173.
12. Pannetier, J. (1992) *Prog. Bart. Batt. Mater*, **11**, 51–55.
13. Vicat, J., Fanchon, E., Strobel, P., and Tran Qui, P. (1986) *Acta Ctystallogr. Sect. B*, **42**, 162–167.
14. Mathieson, A.M. and Waldsley, A.D. (1950) *Am. Mineral.*, **35**, 485–499.
15. Byström, A. and Byström, A.M. (1950) *Acta Crystallogr.*, **3**, 146–154.
16. Post, J.E., von Dreele, R.B., and Buseck, P.R. (1982) *Acta Crystallogr. Sect. B*, **38**, 1056–1065.
17. Mukherjee, B. (1960) *Acta Crystallogr*, **13**, 164–165.
18. Faller, M. (1989) JCPDS Database Coll. No. 42 −1347, University of Berne.
19. Post, J.E. and Bish, D.L. (1989) *Am. Mineral.*, **74**, 913–917.
20. Nambu, Y. and Tanida, T. (1967) *J. Jpn. Assoc. Mineral. Petrol. Econ. Geol.*, **58**, 39.

21. Faller M. (1989) JCPDS Database Col. No. 42 −1349, University of Berne.
22. Mukherjee, B. (1959/60) *Mineral. Mag.*, **32**, 166–171.
23. Mukherjee, B. (1965) *Mineral. Mag.*, **35**, 643.
24. Turner, S. and Post, J.E. (1988) *Am. Mineral.*, **73**, 1055–1061.
25. Rziha, T., Gies, H., and Rius, J. (1996) *Eur. I. Mineral.*, **8**, 675–686.
26. Yamamoto, N. and Tamada, O. (1985) *J. Cryst. Growth*, **73**, 199.
27. Tamada, O. and Yamamoto, N. (1986) *Mineral. J.*, **13**, 130–140.
28. Post, J.E. and Bish, D.L. (1988) *Am. Mineral.*, **73**, 861–869.
29. Berzelius, J.J. (1817) *Ann. Chem. Phys.*, **6**, 204–205.
30. Nossen, E.S. (1951) *Ind. Eng. Chem.*, **43**, 1695–1700.
31. Kondrashev, Y.D. (1957) *Russ. J. Inorg. Chem.*, **2**, 9–18.
32. Klingsberg, C. and Roy, R. (1960) *J. Am. Ceram. Soc.*, **43**, 620–626.
33. Bode, H. (1961) *Angew. Chem.*, **73**, 553–560.
34. Wadsley, A.D. (1964) in *Non Stoichiometric Compounds* (ed. L. Mandelkorn), Academic Press, New York, pp. 98–209.
35. Pons, L. and Brenet, J. (1965) *Compt. Rend*, **260**, 2483–2486.
36. Parthé, E. and Gelato, L.M. (1984) *Acta Crystallogr. Sect. A*, **40**, 169.
37. Glemser, O., Gattow, G., and Meisiek, H. (1961) *Z. Anorg. Allg. Chem.*, **309**, 1.
38. De Wolff, P.M. (1959) *Acta Crystallogr.*, **12**, 341.
39. De Wolff, P.M., Visser, J.W., Giovanoli, R., and Brutsch, R. (1978) *Chimia*, **32**, 257.
40. Turner, S. and Buseck, P.R. (1979) *Science*, **203**, 4.56.
41. Donne, S.W. (1996) Doctoral Thesis, University of Newcastle, Australia.
42. Ruetschi, P. (1984) *J. Electrochem. Soc.*, **131**, 2737–2744.
43. Ripert, M., Pannetier, J., Chabre, Y., and Poinsignon, C. (1991) *Mater. Res Soc. Symp. Proc.*, **210**, 359–365.
44. Chabre, Y. and Pannetier, J. (1995) *Prog. Solid State Chem.*, **23**, 1.
45. Preisler, E. (1976) *J. Appl. Electrochem.*, **6**, 301.
46. Preisler, E. (1976) *J. Appl. Electrochem.*, **6**, 311.
47. (1973) *Gmelins Handbook of Inorganic Chemistry Manganese*, Part 56C1, pp. 157–166.
48. Roussow, M.H., Liles, D.C., Thackeray, M.M., David, W.I.F., and Hull, S. (1992) *Mater Res. Bull.*, **27**, 221.
49. Johnson, C.S., Dees, D.W., Mansuetto, M.F., Thackeray, M.M., Vissers, D.R., Argyriou, D., Loong, C.K., and Christensen, L. (1997) *J. Power Sources*, **68**, 570.
50. Hoechst, A.G., Preisler, E., Harnisch, H., and Mietens, G. (197l) German Patent 2 026 597.
51. Anthony, K.E. and Oswald, J. (1984) Proceedings of the 14th International Power Sources Symposium, pp. 199–215.
52. Oswald, H.R. and Wampetich, M.J. (1967) *Helv. Chim. Acta*, **50**, 2023–2034.
53. Riou, A. and Lecerf, A. (1977) *Acta Crystallogr. Sect. B*, **33**, 1896–1900.
54. Riou, A. and Lecerf, A. (1977) *Acta Crystallogr. Sect. B*, **31**, 2487–2490.
55. Wadsley, A.D. (1952) *Acta Crystallogr.*, **5**, 676–680.
56. Pauling, L. and Kamb, B. (1982) *Am. Mineral.*, **67**, 817–821.
57. Post, J.E. and Appleman, D.E. (1994) *Am. Mineral.*, **79**, 370–374.
58. Chang, F.M. and Simon, A. (1985) *Z. Anorg. Allg. Chem*, **531**, 177–182.
59. Marsh, R.E. and Herbstein, F.H. (1983) *Acta Crystallogr. Sect. B*, **39**, 280–287.
60. Le Page, Y. and Calvert, L.D. (l983) *Acta Crystallogr. Sect. B*, **40**, 1787–1789.
61. Post, J.E. and Appleman, D.E. (1988) *Am. Mineral.*, **73**, 1401–1404.
62. Giovanoli, R., Stähli, E., and Feitknecht, W. (1970) *Helv. Chim. Acta*, **53**, 209.
63. Giovanoli, R., Stähli, E., and Feitknecht, W. (1970) *Helv. Chim. Acta*, **53**, 453.
64. Chen, R., Zavalij, P., and Wittingham, M.S. (1996) *Chem. Mater*, **8**, 1275–1280.

65. Post, J.E. and Veblen, D.R. (1990) *Am. Mineral.*, **75**, 477–489.
66. Wadsley, A.D. (1950) *J. Am. Chem. Soc.*, **72**, 1781.
67. Le Blanc, M. and Wehner, G. (1934) *Z. Phys. Chem. A*, **168**, 59–78.
68. Oswald, H.R. and Wampetich, M.J. (1967) *Helv. Chim. Acta*, **50**, 2023–2034.
69. Lecerf, A. (1974) *C.R. Seances Acad. Sci. Paris, Sér. C*, **279**, 879.
70. Klingsberg, C. and Roy, R. (1960) *J. Am. Ceram. Soc.*, **43**, 620–626.
71. Giovanoli, R. and Leuenberger, U. (1969) *Helv. Chim. Acta*, **52**, 2333–2347.
72. Davis, R.J. (1967) *Mineral. Mag.*, **36**, 274–279.
73. Dent Glasser, L.S. and Smith, I.B. (1967) *Mineral. Mag.*, **36**, 976–987.
74. Giovanoli, R., Bühler, H., and Sokolowska, K. (1973) *J. Microsc.*, **18**, 97.
75. Oswald, J. (1984) *Mineral. Mag.*, **48**, 383.
76. Manceau, A., Llorca, S., and Calas, G. (1987) *Geochim. Cosmochim. Acta*, **51**, 105.
77. Manceau, A., Buseck, P.R., Miser, D., Rask, J. and Nahon, D. (1990) *Am. Mineral.*, **75**, 90.
78. Wadsley, A.D. (1955) *Acta Crystallogr.*, **8**, 165.
79. McMurdie, H.F. (1944) *Trans. Electrochem. Soc.*, **86**, 313–326.
80. Cole, W.F., Wadsley, A.D., and Walkley, A. (1947) *Trans. Electrochem. Soc.*, **92**, 133–158.
81. Gorgeu, A. (1862) *Ann. Chim. Phys.*, **66**, 153–161.
82. Dubois, P. (1936) *Ann. Chim. Phys.*, **5**, 411–482.
83. Feitknecht, W. and Marti, W. (1945) *Helv. Chim. Acta*, **28**, 149–156.
84. Glemser, O., Gattow, G., and Meisick, H. (1961) *Z. Anorg. Allg. Chem.*, **309**, 1–19.
85. Buser, W., Graf, P., and Feitknecht, W. (1954) *Helv. Chim. Acta*, **37**, 2322–2333.
86. Giovanoli, R., Stähli, E., and Feitknecht, W. (1969) *Chimia*, **23**, 264–266.
87. Jones, L.H.P. and Milne, A.A. (1956) *Mineral. Mag.*, **31**, 283.
88. Fleischer, M. and Richmond, W.E. (1943) *Econ. Geol*, **381**, 269.
89. Finkelman, R.B., Evans, H.T., and Matzko, J.J. (1974) *Mineral. Mag.*, **39**, 549.
90. Park, S., Nahon, D., Tardy, Y., and Vieillard, P. (1989) *Am. Mineral.*, **74**, 466.
91. Burns, R.G. (1976) *Geochim. Cosmochim. Acta*, **40**, 95.
92. Crerar, D.A. and Barnes, H.L. (1974) *Geochim. Cosmochim. Acta*, **38**, 279.
93. Glover, E.D. (1977) *Am. Mineral.*, **62**, 278.
94. Chukhrov, F.V., Gorshkov, A.I., Beresovskaya, V.V., and Sivtsov, A.V. (1979) *Mineral. Deposita*, **14**, 249.
95. Giovanoli, R., Bürki, P., Guiffredi, M., and Stumm, W. (1975) *Chimia*, **29**, 517.
96. Paterson, E. (1981) *Am. Mineral.*, **66**, 424.
97. Kordesch, K. and Weissenbacher, M. (1994) *J. Power Sources*, **51**, 61–78.
98. Callin, R.L. and Libscomb, W.N. (1949) *Acta Crystallogr.*, **2**, 104–106.
99. Meldau, R., Newelesy, H., and Strunz, H. (1973) *Naturwissenschaften*, **60**, 387.
100. Dachs, H. (1963) *Z. Kristallogr.*, **118**, 303–326.
101. Aminoff, G. (1926) *Z. Kristallogr.*, **64**, 475–490.
102. Geller, S. (1971) *Acta Crystallogr. Sect. B*, **27**, 821–828.
103. Sinha, K.P. and Sinha, A.P.B. (1957) *J. Phys. Chem.*, **61**, 758–761.
104. Christensen, A.N. and Ollivier, G. (1972) *Solid State Commun.*, **10**, 609–614.
105. Maskell, W.C., Shaw, J.E.A., and Tye, F.L. (1981) *Electrochim. Acta*, **26**, 1403.
106. Verwey, E.J.W. and De Boer, J.H. (1936) *Recl. Trav. Chim.*, **55**, 531.
107. Goodenough, J.B. and Loeb, A.L. (1955) *Phys. Rev.*, **98**, 391.

Further Reading

Barber, J. (2004) *Biochim. Biophys.Acta-Bioenerg.*, **1655**, 123–132.

Chen, J. and Cheng, F.Y. (2009) *Acc. Chem. Res.*, **42**, 713–723.

Ducrocq, C., Blanchard, B., Pignatelli, B., and Ohshima, H. (1999) *Cell. Mol. Life Sci.*, **55**, 1068–1077.

Peters, H., Radeke, K.H., and Till, L. (1966) *Z. Anorg. Allg. Chem.*, **346**, 1–11.

4
Electrochemistry of Manganese Oxides

Akiya Kozawa, Kohei Yamamoto, and Masaki Yoshio

4.1
Introduction

Most primary dry cells nowadays use MnO_2 as a cathode. The total annual world consumption of MnO_2 for dry cells exceeds 600 000 tons, about 50% being electrolytic manganese dioxide (EMD).

The total world population was about 5.5 billion (5.5×10^9) in 1997, of whom 640 million (about 12%) were in the developed areas (USA, Japan, Western Europe). They consume 10–15 dry cells per person in a year. In the developing areas (China, South-East Asia, India, Africa, South America), which contain is 88% of the world population, they use only 2–3 cells per person per year. In the USA, Japan, and Western Europe, 60–80% of the dry cells are now alkaline MnO_2 (EMD) cells and $ZnCl_2$ cells using EMD. Therefore, the use of EMD has been increasing, as shown in Figure 4.1, and in the future, for dry cells it will grow to at least two to three times the current production, since the use of alkaline MnO_2 cells will increase steadily in the developing countries. The most popular dry cell sizes are AA and AAA, since portable electronic devices are becoming smaller. Table 4.1 compares the capacity of the 1.5 V AA-size cells on the market today. In the developed countries, a rapid growth in rechargeable cells is taking place for various devices such as portable telephones, camcorders, and portable computers. One big future application of rechargeable batteries is for electric vehicles (EVs). The most promising rechargeable cells and batteries for small portable devices are Ni–MH cells, alkaline MnO_2 cells, and Li-ion batteries using $LiCoO_2$, $LiMn_2O_4$, or $Li_{0.33}MnO_2$. In this chapter, we describe the important properties of EMD and new methods of determination of these properties.

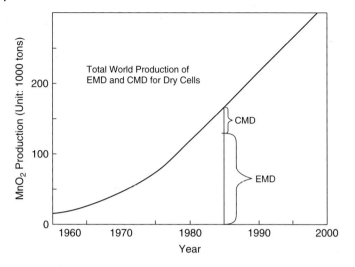

Figure 4.1 Total world production of EMD and chemical manganese dioxide (CMD).

Table 4.1 Capacity of primary AA-size cells.

	Capacity (mAh)
Regular carbon–zinc cells using natural MnO_2 one	500 ± 50
Zinc chloride-type cells using EMD	850 ± 85
Alkaline MnO_2 cells using EMD	2200 ± 220
Li–FeS_2 cells	3000 ± 300

4.2
Electrochemical Properties of EMD

4.2.1
Discharge Curves and Electrochemical Reactions

The basic electrochemical properties of EMD are summarized schematically in Figure 4.2 [1] based on the original work by Kozawa et al. [2–5]. Electrolytic MnO_2 discharges in two steps in $9\,mol\,L^{-1}$ KOH. The reaction during each step is shown below.

- First step (curve abc in Figure 4.2a) in $9\,mol\,L^{-1}$ KOH:

$$MnO_2 + H_2O \rightarrow MnOOH + OH^- \tag{4.1}$$

- Second step (curve cde in Figure 4.2a) in $9\,mol\,L^{-1}$ KOH:

$$MnOOH + H_2O \rightarrow Mn(OH)_2 + OH^- \tag{4.2}$$

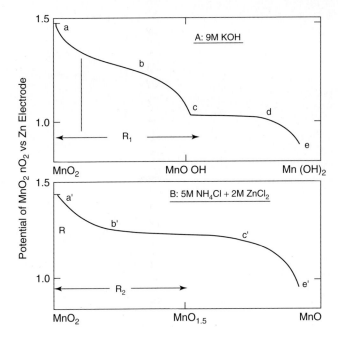

Figure 4.2 Schematic semi-ideal discharge curves of MnO_2 in 9 and 5 mol L^{-1} $NH_4Cl_2 + 2$ mol L^{-1} $ZnCl_2$ solutions. R_1, range of discharge capacity of commercial alkaline MnO_2–Zn; R_2, range of discharge capacity of commercial Leclanché or zinc chloride cells.

In $ZnCl_2$ solutions (with or without NH_4Cl), MnO_2 discharges as shown in Figure 4.2b. The first 25% (from a' to b') is essentially the same Equation 4. , and the essential part of curve b'c' is Equation 4.3:

$$MnO_2 + 4H^+ \rightarrow Mn^{2+} + 2H_2O \tag{4.3}$$

R_1 and R_2 indicate the actual discharge ranges in practical cells.

Under certain conditions, MnO_2 shows a four-step discharge curve [6], as shown in Figure 4.3, because MnO_2 can recrystallize during the discharge. Such a recrystallization takes place easily when the discharge is stopped at a deeper stage and at a high temperature (45 °C), as seen in Figure 4.3b–d. Even when MnO_2 is continuously discharged, if the current is low (10 mA), MnO_2 recrystallizes during the discharge showing a four-step curve (see Figure 4.3a), since there is enough time for recrystallization. This structural change in MnO_2 is very critical (unfavorable) for using MnO_2 in rechargeable batteries, since the recrystallized oxide is not rechargeable. In practical primary dry cells (D, C, and AA sizes) we do not see these four-step curves even when the cell is left at a deeply discharged stage. This is probably because of the very limited amount of KOH electrolyte in the cell. This structure change probably occurs through the dissolved Mn(III) ions, as Mn(III) oxide and Mn(II) oxide significantly dissolve when the KOH concentration

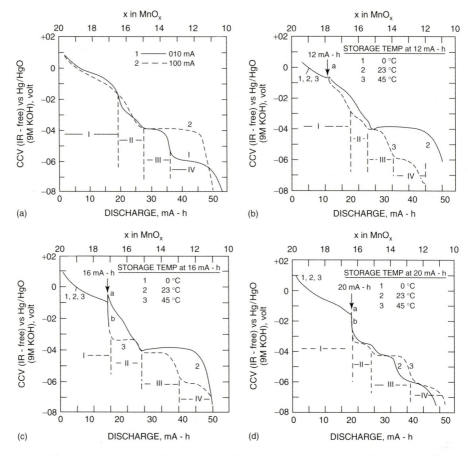

Figure 4.3 Four-step discharge curves of MnO$_2$ [2]: 100 mg of EMD was discharged continuously in 9 mol L^{-1} KOH (a) or stored at a defined discharge capacity of (b) 12 mAh, (c) 16 mAh, and (d) 20 mAh for 96 h before continuing. Three cells containing 100 mg of γ-MnO$_2$ were discharged at 23 °C at 1 mA to 12, 16, and 20 mAh. The cells were kept on open circuit at the temperature shown for 96 h, then the discharge was continued at 1 mA at 23 °C. The cathode potential for the cells stored at 0 and 23 °C recovered to point 'a' but that of the cell stored at 45 °C decreased to point 'b'.

is high (see Figure 4.4). To minimize this structure change, 2–4 mol L^{-1} KOH solution would be a better choice.

4.2.2
Modification of Discharge Behavior of EMD with Bi(OH)$_3$

Both Swinkels *et al.* [7] and Chabre and Pannetier [8] described the process of EMD reduction as three overlapping processes. Recently Donne *et al.* reported [9] that the presence of Bi(OH)$_3$ on the EMD surface modified the discharge curve considerably and the rechargeability was increased. Formation of the birnessite

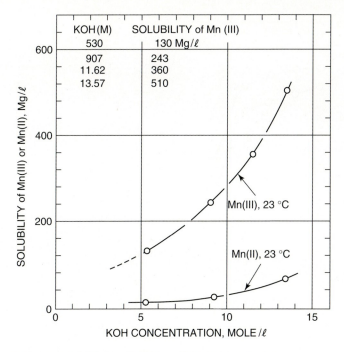

Figure 4.4 Solubilities of Mn(III) and Mn(II) in KOH [12].

structure from EMD and $Bi(OH)_3$ or Bi_2O_3 (mechanically mixed with EMD) [10] is the cause of the increase in rechargeability.

4.2.3
Factors which Influence MnO_2 Potential

4.2.3.1 Surface Condition of MnO_2

The electrode potential should be a reflection of the ΔF value of the oxide, representing its total energy. However, Kozawa and Sasaki reported [11] that surface conditions of the battery active MnO_2 have some effect on the measured potential. A stable β-MnO_2 was prepared by heating EMD at 400 °C for 10 days. Using the structurally stable oxide, a surface equilibrium with O_2 gas in air was obtained at various temperatures between 100 and 400 °C, based on the weight change (Figure 4.5). As seen in Figure 4.5, the surface oxide layer decomposes (with release of O_2) and reversibly absorbs O_2 between 300 and 400 °C. When the heated oxide at 400 °C is quickly cooled, the surface condition of 400 °C is maintained (no weight change). When the heated oxide is cooled slowly for 2–3 h at 300 °C, oxygen is adsorbed.

The potentials of β-MnO_2 samples which were heated at various temperatures and quickly cooled were determined in 1 mol L^{-1} KOH (Figure 4.6). The potential of the β-MnO_2, which was heated at 400 °C and cooled slowly (exposing the oxide surface at 300 °C for sufficient time), is higher (note the potential of points a and b).

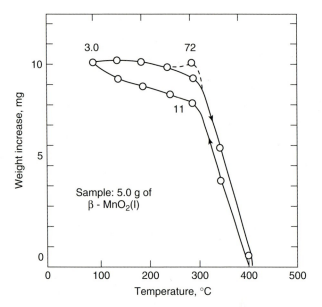

Figure 4.5 Thermogravimetric equilibrium study of β-MnO$_2$. The weight change of the sample was measured in air between 100 and 400 °C by changing from one temperature to another after attaining a constant weight at each temperature. The reference point for the thermogravimetric study was the initial steady-state weight at 400 °C in air. In a few cases, the sample was kept for 3–72 h at the same temperature to see whether the weight changed further or not. It was found that the weight essentially reached in equilibrium at each temperature with 3 h. The numbers by the points (3, 11, 72) indicate the extra hours waited for this purpose.

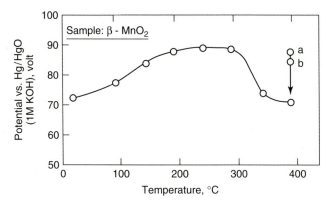

Figure 4.6 Static potential of β-MnO$_2$ after heating in air between 100 and 400 °C. All the samples were heated at each temperature for 2 h and cooled quickly to 25 °C. The potential was measured in 1 mol L^{-1} KOH solution. Point 'a' and 'b' are the values obtained with the β-MnO$_2$ samples which were cooled slowly (allowing samples to reach room temperature overnight).

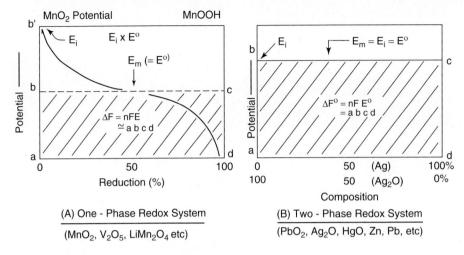

Figure 4.7 Two types of redox systems [13]: (a) one-phase (MnO$_2$, V$_2$O$_5$, LiMn$_2$O$_4$, etc.) and (b) two-phase (PbO$_2$, Ag$_2$O, HgO, Zn, Pb, etc.).

These results indicate that the potential of MnO$_2$ can change, depending on the surface condition, by as much as 18 mV. This is important for obtaining the ΔF value.

4.2.3.2 Standard Potential of MnO$_2$ in 1 mol L^{-1} KOH

Two types of redox systems (Figure 4.7) are used for batteries [13]. The standard potential ($E°$) of MnO$_2$ should be a good representation of the total energy of the oxide. For two-phase systems such as PbO$_2$, Ag$_2$O, HgO, and so on, the initial potential (E_i) and middle potential (E_m) are equal to $E°$, from which we can calculate $\Delta F(=nFE°)$. For MnO$_2$, a one-phase system, as shown in Figure 4.7a, the E_i (initial potential) cannot be used as $E°$. Kozawa proposed the middle potential (E_m) of the S-shaped curve to be used as the $E°$ value since it can be a fairly good representation of the system to calculate the ΔF value. We need to obtain $E°(=E_m)$ of MnO$_2$ in 1 mol L^{-1} KOH. In our attempt to obtain a repeatable static potential of MnO$_2$ in 0.1–1 mol L^{-1} KOH solutions, we found that the potential of MnO$_2$ changes slowly, and it is difficult to reach a steady value. This is probably due to the high porosity and presence of a large amount of fine cavities in the EMD; therefore we still do not have a good $E°$ value for EMD and other MnO$_2$ samples.

4.2.4
Three Types of Polarization for MnO$_2$

As shown in Figure 4.8, three types of polarization exist during the discharge of porous MnO$_2$. The battery active EMD or CMD (chemical manganese dioxide) is highly porous and the concentration polarization due to the pH change, $\eta(\Delta pH)$, is very important. Kozawa studied the three types of polarization for 10 (IC) MnO$_2$

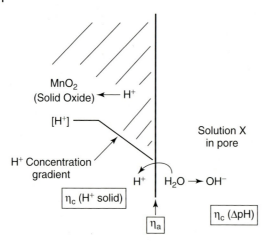

Figure 4.8 Three types of polarization of MnO$_2$: (i) η_c (H$^+$ solid), due to proton diffusion in solid; (ii) η_a, due to the solution–solid interface; and (iii) η_c, (ΔpH), due to a pH change of the electrolyte in the pores.

samples in 9 mol L^{-1} KOH and 25% ZnCl$_2$ (+5% NH$_4$Cl) solutions [14] and found $\eta(\Delta$pH) is much smaller for CMDs. Tables 4.2 and 4.3 show the polarization values at 7, 14, and 21 mAh for a 1.0 mA continuous discharge of 100 mg MnO$_2$ sample having 27 mAh for one-electron reaction. The total polarization, η_τ, is the sum of the three polarizations.

$$\eta_\tau = \eta_a + \eta_c \text{ (solid)} + \eta(\Delta\text{pH}) \tag{4.4}$$

Table 4.2 Open-circuit voltage (OCV) and polarization (P) at 1.00 mA per 100 mg sample in 9 mol L^{-1} KOH solution[a].

IC	MnO$_2$ sample[b]	7 mAh		14 mAh		21 mAh	
		OCV (V)	P (V)	OCV (V)	P (V)	OCV (V)	P (V)
1	EMD (Ti anode)	0.022	0.052	−0.050	0.060	−0.261	0.021
2	EMD (Pb anode)	0.023	0.053	−0.048	0.061	−0.250	0.028
3	EMD (C anode)	0.021	0.048	−0.060	0.049	−0.264	0.046
4	EMD	0.013	0.054	−0.080	0.043	−0.272	0.059
5	CMD	−0.047	0.101	−0.170	0.042	−0.249	0.100
7	Natural ore	−0.077	0.188	−0.183	0.168	−0.292	0.119
8	CMD	−0.033	0.090	−0.149	0.047	−0.282	0.068
9	EMD	0.014	0.054	0.065	0.075	−0.241	0.100
10	EMD	0.013	0.052	−0.083	0.054	−0.314	0.016
11	CMD (chlorate process)	0.025	0.085	−0.026	0.119	−0.158	0.157

[a] Polarization: voltage different between OCV and IR-free CCV (closed-circuit voltage) vs. Hg/HgO (9 mol L KOH).
[b] EMD, electrolytic MnO$_2$; CMD, MnO$_2$ made by chemical process.

Table 4.3 OCV and polarization (P) at 1.00 mA per 100 mg sample in 25% $ZnCl_2$ + 5% NH_4Cl solution[a].

IC	MnO$_2$ sample	7 mAh		14 mAh		21 mAh	
		OCV (V)	P (V)	OCV (V)	P (V)	OCV (V)	P (V)
1	EMD (Ti anode)	0.458	0.127	0.411	0.122	0.431	0.182
2	EMD (Pb anode)	0.443	0.093	0.411	0.108	0.424	0.177
3	EMD (C anode)	0.452	0.086	0.412	0.102	0.422	0.162
4	EMD	0.462	0.112	0.428	0.126	0.441	0.178
5	CMD	0.408	0.097	0.397	0.110	0.432	0.155
7	Natural ore	0.389	0.188	0.383	0.175	0.420	0.231
8	CMD	0.415	0.095	0.393	0.092	0.415	0.134
9	EMD (coarse particles)	0.460	0.096	0.414	0.094	0.442	0.172
10	EMD	0.453	0.0093	0.407	0.095	0.431	0.163
11	CMD (chlorate process)	0.476	0.266	0.463	0.204	0.488	0.248

[a] Polarization voltage different between OCV and IR-free CCV. OCV vs. SCE.

Since the pH change $\eta(\Delta pH)$ is practically zero for the discharge in 9 mol L^{-1} KOH solution, we can assume that $\eta_a + \eta_c$ (solid) is the same for the two solutions (9 mol L^{-1} KOH and 25% $ZnCl_2$). Therefore, the difference in the polarization values (in Tables 4.1 and 4.2) is $\eta(\Delta pH)$, where

$$\eta(\Delta pH) = \eta_\tau - (\eta_a + \eta_c(\text{solid})) \tag{4.5}$$

Table 4.4 shows $\eta(\Delta pH)$ at the 7 mAh stage. We can see that $\eta(\Delta pH)$ is very small for CMD and nonporous manganese dioxide (NMD). This is very reasonable since the pore diameter of CMD and NMD is much bigger than that of EMD. This is why EMD for $ZnCl_2$ cells is usually produced at high current, in order to increase its pore dimensions.

Table 4.4 Polarization and $\eta_c(\Delta pH)$ (in millivolts).

IC	MnO$_2$ sample	(a) 9 mol L^{-1} KOH	(b) η_T ZnCl$_2$[a]	(b–a) $\eta_c(\Delta pH)$
1	EMD (Ti)	52	127	75
2	EMD (Pb)	53	93	40
3	EMD (C)	48	86	38
4	EMD	54	112	58
5	CMD	101	97	−4
7	NMD	188	188	0
8	CMD	90	95	5
9	EMD	54	96	42
10	EMD	52	93	41

[a] Measured in 25% $ZnCl_2$ + 5% NH_4Cl.

4.2.5
Discharge Tests for Battery Materials

Testing various battery materials (MnO_2, MH, $LiCoO_2$, etc.) requires a simple and repeatable electrochemical test using 0.1–0.5 g of sample. Kozawa proposed a plastic cell test [4] using a 100 mg MnO_2 sample mixed with a large amount of graphite and discharging it at 1.0 mA. This method provides reproducible results for open-circuit voltage (OCV), closed-circuit voltage (CCV), polarization, and capacity

Table 4.5 Composition and application of TABS available.

	Teflon (%)	Acetylene black (%)	Surfactant (%)	Applications[a]
TAB-1	34	66	0	Gas-diffusion electrodes
TAB-2	33	66	1	Lithium battery electrodes
TAB-3	32	64	4	Aqueous battery electrodes

[a] All three are available for research, in 100 g bag, from ITE, Aichi. Japan (Fax: (81) 586-81-1988).

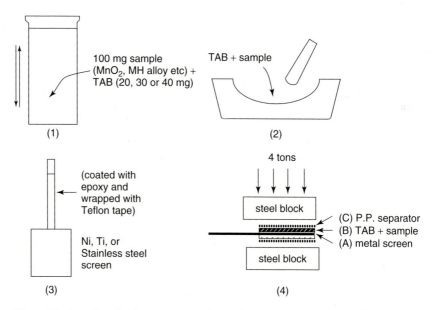

Figure 4.9 Procedure for the preparation of the test electrode for aqueous electrolytes (9 mol L^{-1} KOH or $ZnCl_2$ solution). (1) The sample is mixed by shaking in a plastic container 20 mm (diam.) × 40 mm (height); (2) the mixture is made into a thin film by grinding with a pestle in a ceramic mortar; (3) the metal screen is prepared; and (4) the three layers (A, B, C) are pressed between the steel blocks.

(milliamperes hour) if the mixing and packing technique is practiced thoroughly and the same graphite is used. However, the measured voltage has a considerable effect on the IR drop [15]. In 1992, Kozawa *et al.* [16] proposed an easy, repeatable test method using Teflonized acetylene black (TAB) developed by the Bulgarian laboratory [17]. The three useful TABs available today are shown in Table 4.5.

TABS are a mixture of a Teflon emulsion and acetylene black, which is prepared by a special vigorous mixing technique.

They are very useful not only for testing. They are being used in practical commercial cells. Figure 4.9 shows electrode preparation with TAB. Figures 4.10–4.12 show the test results for a MnO_2 sample (IC No. 17) in $9\,mol\,L^{-1}$ KOH and 25% $ZnCl_2$ (+5% NH_4Cl) solutions.

Figure 4.12 shows discharge tests of MnO_2 in 3, 6, and $9\,mol\,L^{-1}$ KOH solutions with and without O_2. With the $9\,mol\,L^{-1}$ KOH solution, the effect of dissolved O_2

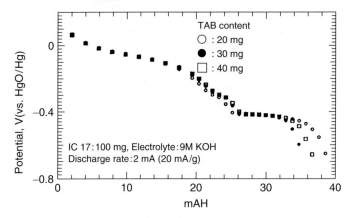

Figure 4.10 Effect of the amount of TAB on the discharge behavior of IC No. 17 (EMD) in a moles per liter KOH.

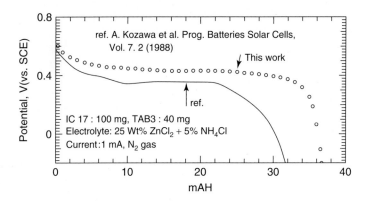

Figure 4.11 Comparison of discharge behavior of IC No. 17 (EMD) for currently proposed method and a previously published method.

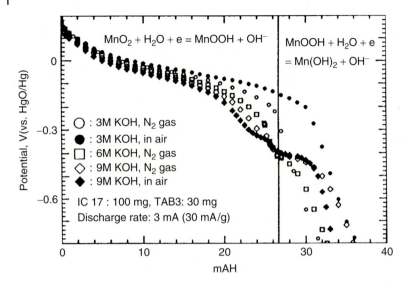

Figure 4.12 Effect of KOH concentration and dissolved oxygen on the discharge behavior of IC No. 17 (EMD) in 3–9 mol L^{-1} KOH.

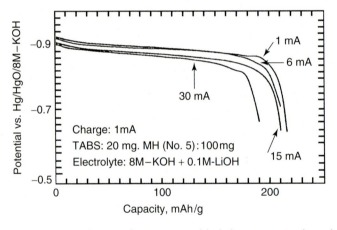

Figure 4.13 Application of TAB-3 to metal hydride (IBA No. 5) electrode.

gas is very small, but the effect is significant with 3 and 6 mol L^{-1} KOH solutions since the O$_2$ solubility increases. Figure 4.13 shows that this test method is useful for MH alloys. LiCoO$_2$ and other materials have also been successfully tested with TAB [16].

Recently, Tachibana et al. [18] used a nickel mesh electrode containing a mixture of Teflon emulsion, graphite (or acetylene black), and oxides (MnO$_2$, LiCoO$_2$, etc.). In this method the electrode is very thin, and there is no *IR* drop within the electrode. Therefore, measurements can be made by a simple, direct method (no potentiostat is needed).

4.3
Physical Properties and Chemical Composition of EMD

Table 4.6 shows the physical properties and chemical composition of typical battery active EMDs, and Table 4.7 shows a typical chemical analysis of EMD. Figure 4.14 shows a schematic model of an EMD particle showing three types of pores. Figure 4.15 shows the calculated surface area of nonporous solid cubes having a specific gravity (SG) of 1.0, 4.0, and 5.0 g mL^{-1}. As the measured SG of EMD is 4.3 g mL^{-1} and the surface area is 25–35 m^2 g^{-1}, the superficial surface area is less than 0.1 m^2 g^{-1}. Therefore, most of the electrochemical reaction takes place on the pore wall of the fine pores and cavities since the pore diameter is 50–100 Å.

The ideal battery material should be highly porous, but should have a high density in order to pack as much as possible in to the limited space of the cell. EMD is almost the ideal MnO$_2$ since it is dense and has fine pores (actually cavities).

4.3.1
Cross-Section of the Pores

Based on the gas adsorption behavior, Kozawa and Yamashita proposed a hypothesis [19, 20] that the cross-section of the fine pores of EMD is a cavity shape as shown in Figure 4.16. The cross-section of the pore by computer calculation is a circle (Figure 4.16a). Nobody knows the real shape of the cross-section as yet. Kozawa's belief in the cavity shape (Figure 4.16b) is based on the results of experiments involving the oxygen adsorbed and desorbed from the pore walls [19].

Table 4.6 Physical properties and chemical composition of EMD.

Physical properties	
Electrical resistivity[a]	50–100 Ω cm
BET surface area	40–50 m^2 g^{-1}
Pore diameter	40–60 Å
Pore volume (<150 Å pores)	0.032–0.035 mL g^{-1}
Density	
True density	4.0–4.3 g mL^{-1}
Apparent density	2.2–2.3 g mL^{-1}
Average particle size	10–45 in.
Chemical composition	
MnO$_2$	92%
H$_2$O	2–4%
Mn$_2$O$_4$ and MnO	3–4%
SO$_4^{2-}$	~1%
Metallic impurity	Very low

[a]Measured with powder sample in tablet from under high pressure.

Table 4.7 Specification and typical analysis.

Component	Specification	Typical analysis
MnO_2	91.0% min.	92.0%
Mn	60.0% min.	60.4%
H_2O	2.0% min.	1.8%
HCl insoluble	0.2% min.	0.02%
SO_4	1.3% min.	1.20%
Fe	150 ppm max.	80 ppm
Pb	5 ppm max.	<1 ppm
Cu	5 ppm max.	1 ppm
Ni	10 ppm max.	3 ppm
Co	15 ppm max.	3 ppm
Sb	15 ppm max.	<1 ppm
As	1 ppm max.	<1 ppm
Mo	5 ppm max.	1 ppm
Cr	15 ppm max.	5 ppm
V	2 ppm max.	<1 ppm
K	0.15% max.	0.08%
PH		
JIS method[a]	5.0–5.6	5.2
USA method	8.0–9.0	8.5

[a] Neutralization with NH_4OH or NaOH. JEC Sample, 1997. available from ITE Japanese Office.

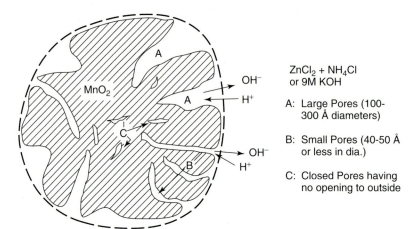

Figure 4.14 Porous MnO_2 particle with three of pores: (A) large pores, 100–300 Å diameter; (B) small pores, 40–50 Å or less diameter; and (C) closed pores with no outside opening; ------ superficial surface; ——— true surface including the pore walls.

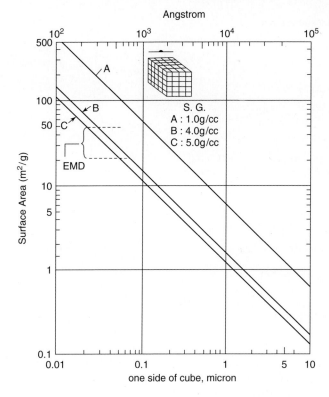

Figure 4.15 Surface area of solid cubes of various specific gravities.

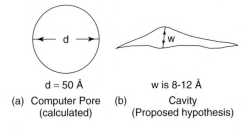

Figure 4.16 (a,b) Cross-section of pore and cavity in EMD.

4.3.2
Closed Pores

The presence of closed pores was demonstrated by Kozawa [21] by measuring the BET surface area of EMD samples of various particle sizes. Kozawa's new method for the determination of the closed pore is based on the relationship of the BET surface area and the particle size, by extrapolating the surface area value to zero particle size (Figure 4.17). Table 4.8 shows the percentage of closed pores of various EMD samples.

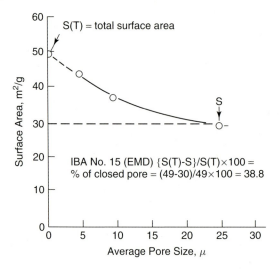

Figure 4.17 Example of BET surface area vs average particle size (APS).

4.3.3
Effective Volume Measurement

A new, simple, and practical method for pore volume measurement was proposed by Kozawa [22]. The method requires only a 100 mL graduated cylinder as explained in Figure 4.18. The principle is illustrated in Figure 4.19. In the method, water is added to an MnO_2 sample (50 g) in a 100 mL cylinder. For each water addition (0.5 mL), the MnO_2 sample and water are mixed as shown in Figure 4.18 step (3). Water vapor is adsorbed very quickly on the pore walls or condensed in the cavities. Therefore, shaking the mixture 10 times is sufficient. The volume of the sample is measured after tapping 5 or 10 times. As soon as the pores are completely filled with water, the water level in the cylinder begins to rise, as shown in Figure 4.19.

From the variation of the volume of the EMD vs the amount of added water, the effective pore volume (EPV) is obtained as shown in Figure 4.19.

This method is practical and useful for battery engineers. As seen in Figure 4.20, when the MnO_2 (EMD) sample (IC No. 9) was heated to 120 and 230 °C, the EPV increased from 1.55 mL (for 25 °C) to 3.2 mL (for 120 °C) and to 3.5 mL (for 230 °C), confirming our previous results. Table 4.9 shows pore volume values measured by this method.

4.4
Conversion of EMD to LiMnO$_2$ or LiMn$_2$O$_4$ for Rechargeable Li Batteries

An industrial quantity of high-purity EMD is now being produced for alkaline MnO_2–Zn cells. Therefore, EMD is an excellent source of future cathode materials ($Li_{0.3}MnO_2$ and $LiMn_2O_4$) for rechargeable lithium batteries. EVs need large

Table 4.8 Percentage of closed pores (CPs) in MnO_2 samples[a].

MnO_2 sample	Description	APS (50%) (50%) (μM)	S ($m^2\,g^{-1}$)	S(T) ($m^2\,g^{-1}$)	CP (%)
IC No. 1 (EMD)	Regular EMD made by the electrodeposition on a Ti anode from $MnSO_4 + H_2SO_4$ solution (pore diameter is 50 Å)	45.0	48.2	58.0	16.9
IC No. 8 (CMD)	Chemical MnO_2 prepared by the thermal decomposition of $MnCO_3$ with additional process to deposit γ–MnO_2 (pore diameter is 70 Å)	45.0	92.8	94.3	1.6
IC No. 9 (EMD)	Coarse EMD sample	77.0	66.0	72.0	8.3
IC No. 13 (NMD)	A typical natural MnO_2 for dry cells	25.0	24.0	30.0	20.0
IBA No. 15 (EMD)	Dense EMD deposited from a suspension bath of $MnSO_4$ solution; the pore size is considerably finer than regular EMD, such as IC No. 1	25.0	30.0	48.0	37.5
IBA No. 17 (EMD)	Regular EMD deposited on Ti anode	25.7	48.0	57.0	15.8
IBA No. 20 (EMD)	EMD deposited on Ti anode with a medium particle size	24.0	42.5	50.0	15.0
IBA No. 22 (CMD)	A CMD made by thermal decomposition of $MnCO_3$ around 320 °C in oxygen with steam	77.0	52.0	59.0	13.4
IBA No. 27 (NMD)	Natural ore from Mexico; excellent for carbon-zinc cells	17.5	27.0	30.0	10.0

[a] Summary of closed pore percentages: EMD, 16.9–37.5%; CMD, l.6–13.4%; NMD, 10–20%. APS, average particle size of the original sample at 50 wt%; S, BET surface area of original sample; S(T), total BET surface area obtained by extrapolation.

batteries using inexpensive electrode materials. $LiMn_2O_4$ or $Li_{0.3}MnO_2$ would be a potential cathode material for EV batteries. Today's rechargeable Li batteries mostly use $LiCoO_2$, since these small batteries are for portable telephones, camcorders, and other portable electronic devices. $Li_{0.3}MnO_2$ will be for 3 V systems and $LiMn_2O_4$ for 4 V systems. Both oxides can be easily produced from EMD.

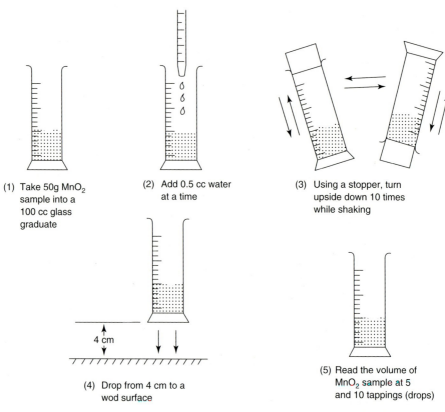

Figure 4.18 Procedure for EPV (effective pore volume) measurement: (1) a 50 g MnO$_2$ sample is placed in a 100 mL graduated cylinder; (2) water is added gradually in 0.5 mL portions; (3) with a stopper in place, the cylinder is turned upside down 10 times while being shaken; (4) the cylinder is dropped 4 cm onto a wooden surface; and (5) the MnO$_2$ sample volume is read after 5 and 10 taps (i.e., drops).

4.4.1
Melt-Impregnation (M–I) Method for EMD

Since EMD is highly porous, having 50–100 Å pores, we can use the M–I method developed by Yoshio [23]. This process consists of a two-step heating. In the first step, a thorough mixture of a Li salt (LiNO$_3$ or LiOH) and EMD powder is heated at a temperature which is slightly above the melting point of the salt in order to allow the molten salt to penetrate into the pores of the EMD. The mixture is then heated at 350 or 650–800 °C, depending on the intended material, LiMnO$_4$ or LiMn$_2$O$_4$ respectively. The melting point of LiNO$_3$ is 260 °C and that of LiOH is 420 °C. When a nonporous MnO$_2$ is used to produce LiMn$_2$O$_4$, the MnO$_2$ must be ground to fine particles with the Li salt before heating. The heated product must be ground 5–10 times after each repeated treatment. The entire process is very time consuming.

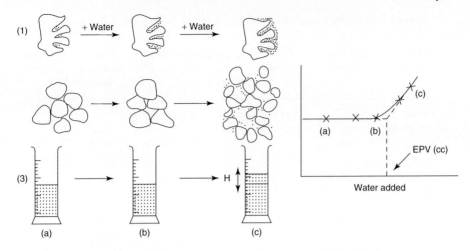

Figure 4.19 Model of the sudden volume increase of MnO_2 powder sample at EPV (effective pore volume) point. (a) Water fills 50% of the pores; (b) water fills almost 100% of the pores; and (c) when excess water (1–2 mL more than the pore volume) is added, the MnO_2 volume suddenly increases by 5–10 mL since the particles stick to each other. The sudden increase (far more than the amount of water added) is shown as H in (3), stage (c), above.

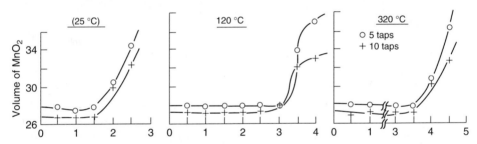

Figure 4.20 Volume of 50 g MnO_2 (IC No. 9) upon addition of water (the volume was measured after 5 and 10 taps).

4.4.2
Preparation of $Li_{0.3}MnO_2$ from EMD

An EMD and $LiNO_3$ mixture (molar ratio 3 : 1) is heated at 260 °C (the melting point of $LiNO_3$) and then further heated to 350 °C for 5 h [24].

The MnO_2 for this purpose can be CMD such as IC No. 12 or Cellmax, but an EMD which was prepared at a high current density of 1.5–5 A dm^{-2} during the electrolysis of a $MnSO_4$–H_2SO_4 bath (95 °C) is very suitable since the high-current-density EMD has much larger pores and high surface area (60 m^2 g^{-1} or higher). The $Li_{0.3}MnO_2$ produces 150–180 mAh g^{-1}. An AA-size 3 V Li cell using this oxide has 300 Wh l^{-1} and 140 Wh kg^{-1}, which is larger than the 4 V cells of 225 Wh l^{-1} and 95 Wh kg^{-1}. The discharge curves are shown in Figure 4.21.

4 Electrochemistry of Manganese Oxides

Table 4.9 EPVs (effective pore volumes) IC and IBA MnO$_2$ samples.

MnO$_2$ sample	Heat treatment temperature (°C)	Initial volume of 50 g (mL)	Pore volume by N$_2$ desorption (mL g^{-1})	EPV (mL g^{-1})	% Based on N$_2$ pore volume	Remarks[a]
IC No. 1 (EMD, Ti)	(25)	24.0	–	0.024	–	–
	120	23.5	0.0416	0.044	105.8	0
	230	23.0	0.0539	0.050	92.8	0
IC No. 2 (EMD, Pb)	(25)	25.5	–	0.026	–	–
	120	24.0	0.0332	0.040	120.5	0
	230	26.0	0.0449	0.044	98.0	0
IC No. 3 (EMD, C)	(25)	26.0	–	0.032	–	–
	120	27.0	0.0407	0.044	108.1	0
	230	26.5	0.0524	0.064	122.1	0
IC No. 4 (EMD)	(25)	32.0	–	–	(No clear step)	–
	120	31.0	0.0367	–	–	–
	230	30.5	0.0479	–	–	–
IC No. 5 (CMD)	(25)	38.0	–	0.208	–	–
	120	38.0	0.408	0.240	170.5	x
	230	40.0	0.1515	0.244	161.1	x
IC No. 7 (NMD)	(25)	27.0	–	0.010	–	–
	120	27.5	0.0242	0.014	58.3	x
	230	27.0	0.0284	0.014	50.0	x
IC No. 8 (CMU)	(25)	33.0	–	0.164	–	–
	120	33.5	0.177	0.170	96.1	0
	230	34.0	0.183	0.170	93.0	0
IC: No. 9 (EMD)	(25)	27.0	–	0.032	–	–
	120	27.5	0.0545	0.062	113.7	0
	230	27.5	0.0664	0.072	108.4	0
IC No. 10 (EMD)	(25)	27.0	–	0.046	–	–
	120	27.0	0.0490	0.060	122.4	0
	230	27.0	0.0571	0.062	108.8	0
IC No. 11 (CMD)	(25)	20.0	–	0.010	–	–
	120	20.5	0.0062	0.026	419.4	x
	230	20.5	0.0089	0.016	179.7	x
IBA No. 14 (EMD)	(25)	22.5	–	0.020	–	–
	120	22.0	0.036	0.030	83.3	0
IBA No. 15 (EMD)	(25)	26.0	–	0.020	–	–
	120	26.0	0.038	0.046	121.0	0
IBA No. 16 (EMD)	(25)	24.5	–	0.024	–	–
	120	25.0	0.037	0.034	91.9	0

Table 4.9 (continued)

MnO_2 sample	Heat treatment temperature (°C)	Initial volume of 50 g (mL)	Pore volume by N_2 desorption (mL g^{-1})	EPV (mL g^{-1})	% Based on N_2 pore volume	Remarks[a]
IBA No. 17 (EMD)	(25) 120	26.0 26.0	– 0.043	0.040 0.034	– 79.1	– 0
IBA No. 18 (EMD)	(25) 120	23.0 24.0	– 0.040	0.024 0.040	– 100.0	– 0
IBA No. 19 (EMD)	(25) 120	22.5 22.0	– 0.041	0.024 0.038	– 92.7	– 0
IBA No. 20 (EMD)	(25) 120	25.0 26.0	– 0.055	0.038 0.054	– 98.2	– 0
IBA No. 21 (EMD)	(25) 120	42.5 42.5	– 0.099	– –	(No clear step) –	– –
IBA No. 22 (CMD)	(25) 120	26.0 26.5	– 0.181	0.174 0.182	– 100.6	– 0
IBA No. 23 (EMD)	(25) 120	23.0 23.5	– 0.042	0.036 0.056	– 133.3	– Δ
IBA No. 24 (EMD)	(25) 120	24.0 23.0	– 0.058	0.040 0.052	– 123.8	– 0
IBA No. 25 (EMD)	(25) 120	25.0 25.5	– 0.084	0.062 0.078	– 92.8	– 0
IBA No. 26 (EMD)	(25) 120	24.5 25.0	– 0.058	0.040 0.062	– 106.9	– 0

[a]Remarks: 0, difference is within 25% (75–125%); Δ, difference is within 50% (50–150%); x, difference is over 50%.

4.4.3
Preparation of LiMn$_2$O$_4$ from EMD

A mixture of LiOH and EMD is heated at 420 °C for 2–3 h in order to allow molten LiOH to penetrate into the pores of the EMD [24, 26]. The mixture is then heated from 650 to 800 °C to produce LiMn$_2$O$_4$. The amount of LiOH and EMD in the mixture must be stoichiometric (LiOH : MnO$_2$ = 1 : 2). The product, LiMn$_2$O$_4$, is usually tested by cyclic voltammetry (Figure 4.22): a good LiMn$_2$O$_4$ does not have peaks at a and b (peak a (3.3 V) would be due to the oxygen deficiency and peak b (4.5 V) to replacement of the Li ion sites by Mn^{4+} ion). Since EMD has many fine pores and the Li salt and MnO$_2$ are mixed intimately in the M–I method, an excellent LiMn$_2$O$_4$ is produced which does not show peaks a and b.

146 | 4 Electrochemistry of Manganese Oxides

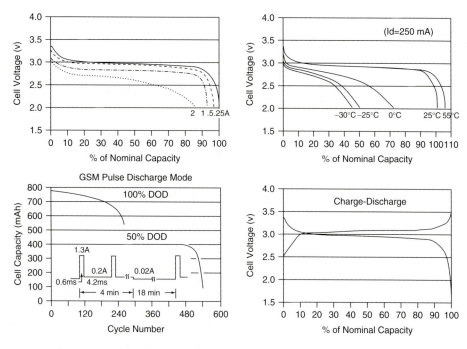

Figure 4.21 Cell performance of LiMnO$_2$ oxide. The anode is metallic Li [25].

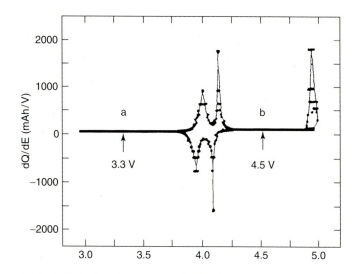

Figure 4.22 Cyclic voltammogram of LiMn$_2$O$_4$.

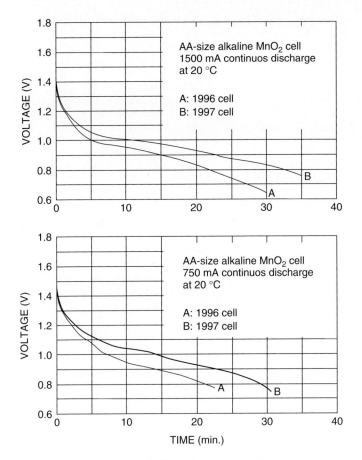

Figure 4.23 New Japanese alkaline cells using expanded graphite and other recent improvement.

4.5
Discharge Curves of EMD Alkaline Cells (AA and AAA Cells)

Figure 4.23 shows discharge curves of a new type of alkaline cells, which is able to discharge at a very high rate.

References

1. Kozawa, A. (1979) *Power Sources*, 7, 485.
2. Kozawa, A. and Yeager, J.F. (1965) *J. Electrochem. Soc.*, 112, 959.
3. Kozawa, A. and Powers, R.A. (1966) *J. Electrochem. Soc.*, 113, 870.
4. Kozawa, A. and Powers, R.A. (1967) *J. Electrochem. Tech.* 5 (1), 535.
5. Kozawa, A. and Powers, R.A. (1968) *J. Electrochem. Soc.*, 115, 122, 1003.
6. Kozawa, A. and Powers, R.A. (1969) *J. Electrochem. Soc. Jpn.*, 37, 31.
7. Swinkels, D.A.J., Anthony, K.E., Fredericks, P.M., and Osborn, P.R. (1984) *J. Electroanal. Chem.*, 168, 433.

8. Chabre, Y. and Pannetier, J. (1995) *Prog. Solid State Chem.*, **23**, 1.
9. Donne, S.W., Laurance, G.A., and Swinkels, D.A.J. (1997) *Prog. Batteries Batt. Mater.*, **16**, 79.
10. Tschikawa, K., Matsuki, K., and Kozawa, A. (1997) Paper presented at IBA Singapore Meeting, September 5–12.
11. Kozawa, A. and Sasaki, K. (1954) *Denki Kagaku*, **22**, 569, 571.
12. Kozawa, A., Kalnoki-Kis, T., and Yeager, J.F. (1966) *J. Electrochem. Soc.*, **113**, 405.
13. Kozawa, A. and Power, R.A. (1972) *J. Chem. Educ*, **49**, 587.
14. Kozawa, A. (1982) *Denki Kangaku*, **50**, 763.
15. Kozawa, A., Kana, G., Horiba, K., and Takeuchi, Y. (1988) *Denki Kagaku*, **7**, 2.
16. Kozawa, A., Yoshio, M., Noguchi, H., Piao, G., and Uchiyama, A. (1992) *Prog. Batteries Batt. Mater.*, **11**, 235.
17. Manev, V., Momchilov, A., Tagawa, K., and Kozawa, A. (1993) *Prog. Batteries Batt. Mater*, **12**, 157.
18. Tachibana, K., Matsuki, K., and Kozawa, A. (1997) *Prog. Batteries Batt. Mater.*, **16**, 322.
19. Kozawa, A. (1989) in *Handbook of MnO$_2$* (eds D. Glover, B. Schumm Jr., and A. Kozawa), IBA Inc., p. 295; *Idem* (1988) *Prog. Batteries Batt. Mater.*, **7**, 59.
20. Yamashita, M., Ida, M., Takemura, H., and Kozawa, A. (1993) *Prog. Batteries Batt. Mater.*, **12**, 19.
21. Kozawa, A. (1989) in *Handbook of MnO$_2$* (eds D. Glover, B. Schumm Jr., and A. Kozawa), IBA Inc., p. 287; *Idem* (1988) *Prog. Batteries Batt. Mater.*, **7**, 320.
22. Kozawa, A. (1988) *Prog. Batteries Batt. Mater.*, **7**, 327.
23. Yoshio, M., Inoue, S., Hyakutake, M., Piao, G., and Nakamura, H. (1991) *J. Power Sources*, **34**, 147.
24. Yoshio, M. and Kozawa, A. (1995) *Prog. Battery Batt. Mater.*, **14**, 87.
25. Tadiran (1997) *ITE Battery Newsletter*, **2**, 45.
26. Xia, Y., Zhou, H., and Yoshio, M. (1997) *J. Electrochem. Soc.*, **144**, 2593.

Further Reading

Koksbang, R., Barker, J., Shi, H., and Saidi, M.Y. (1996) *Solid State Ionics*, **84**, 1–21.

Kozawa, A. (1985) *Proceedings of Symposium on MnO$_2$ Electrode*, vol. 85-4, US Electrochemical Society, Pennington, PA, 19xx.

Rabaey, K., Rodriguez, J., Blackall, L.L., Keller, J., Gross, P., Batstone, D., Verstraete, W., and Nealson, K.H. (2007) *ISME J.*, **1**, 9–18.

Rolison, D.R. and Dunn, B. (2001) *J. Mater. Chem.*, **11**, 963–980.

5
Nickel Hydroxides
James McBreen

5.1
Introduction

Nickel hydroxides have been used as the active material in the positive electrodes of several alkaline batteries for over a century [1]. These materials continue to attract much attention because of the commercial importance of nickel–cadmium and nickel–metal hydride batteries. In addition to being the active material of the cathode in nickel–metal hydride batteries, $Ni(OH)_2$ is an important corrosion product of the anode during cycling. There are several reviews of work in this field [2–10].

Progress in understanding the reactions of nickel hydroxide electrodes has been very slow because of the complex nature of these reactions. Exercises which are normally trivial for most battery electrodes, such as the determination of the open-circuit potential, the overall reaction, and the oxidation state of the charged material, have required much effort and ingenuity. The materials have been studied with the aid of an enormous array of spectroscopic, structural, and electrochemical techniques. The most significant advance in the understanding of the overall reaction was made by Bode and his co-workers [11]. They established that both the discharged material ($Ni(OH)_2$) and the charged material (NiOOH) could exits in two forms. One form of $Ni(OH)_2$, which was designated as β-$Ni(OH)_2$, is anhydrous and has a layered brucite ($Mg(OH)_2$) structure. The other form, α-$Ni(OH)_2$, is hydrated and has intercalated water between brucite-like layers. Oxidation of β-$Ni(OH)_2$ on charge produces γ-NiOOH, and oxidation of α-$Ni(OH)_2$ also produces γ-NiOOH. Discharge of β-NiOOH yields β-$Ni(OH)_2$, and discharge of γ-NiOOH yields α-$Ni(OH)_2$. During voltage hold in discharge mode, the α-$Ni(OH)_2$ can dehydrate and recrystallize in the concentrated alkaline electrolyte to form β-$Ni(OH)_2$. Bode et al. also found that β-NiOOH could be converted to γ-NiOOH when the electrode is overcharged. Their overall reaction scheme is shown schematically in Figure 5.1.

All subsequent work has in general validated these conclusions. The two reaction schemes are often referred to as the β/β and the α/γ cycles.

Handbook of Battery Materials, Second Edition. Edited by Claus Daniel and Jürgen O. Besenhard.
© 2011 Wiley-VCH Verlag GmbH & Co. KGaA. Published 2011 by Wiley-VCH Verlag GmbH & Co. KGaA.

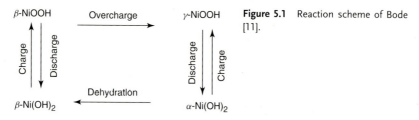

Figure 5.1 Reaction scheme of Bode [11].

This section gives a brief overview of the structure of nickel hydroxide battery electrodes and a more detailed review of the solid-state chemistry and electrochemistry of the electrode materials. Emphasis is on work done since 1989.

5.2
Nickel Hydroxide Battery Electrodes

Conventional nickel hydroxide battery electrodes are designed to operate on the β/β cycle, to accommodate the volume changes that occur during cycling, and to have adequate electronic conductivity to yield high utilization of the active material on discharge. The β/β cycle is preferred because there is less swelling of the active material on cycling. The conductivity of β-NiOOH is more than 5 orders of magnitude higher than that of Ni(OH)$_2$ [12]. As a result, there is usually no problem in charging the electrode because the NiOOH that forms increases the conductivity of the active material. However, on discharge the charged material can become isolated in a resistive matrix of the discharged product and cannot be discharged at useful rates [13]. Operation on the β/β cycle is ensured by control of the electrolyte composition and the use of a combination of additives such as Co and Zn. Provisions have to be made for electronic conduction to the active material and confinement of the active material on cycling. Over the years, several electrode designs have been used. These include incorporation of the active material in pocket plates, perforated metal tubes, sintered nickel plaques, plastic-bonded electrodes with graphite as the conductive diluent, nickel foams, and fibrous nickel mats.

Pocket and tubular electrodes have been described in detail by Falk and Salkind [1]. McBreen has reviewed work on both sintered-plate and plastic-bonded electrode technology [9]. More recent work is on the use of nickel foams and nickel mats.

Early work on the use of foams and mats has been reviewed [9]. Nickel fiber, nickel-plated steel fiber, or nickel-plated graphite fiber mats are preferred because they have smaller pores ($\sim$50 μm) [14]. The most recently developed mats can have porosities as high as 95% [13] and are much lighter than the sintered nickel plaques, which typically have porosities between 80 and 90%. Initially, standard cathodic impregnation methods were used to load the active material into the foam [9]. More recently, the preferred method is to incorporate the Ni(OH)$_2$ in the form of a slurry into the mat [13, 14]. This has been called the '*suspension impregnation method*' [14]. Considerable improvement in the Ni(OH)$_2$ has been achieved by the addition of divalent Co compounds to the slurry. The best results were achieved with the addition of 10% CoO [13, 15]. The following mechanism has been proposed for this improvement [13]. In the alkaline electrolyte, CoO dissolves to form the blue cobaltite ion. The ion precipitates on the Ni(OH)$_2$ particles to form insoluble β-Co(OH)$_2$. On charge, the β-Co(OH)$_2$ is oxidized to a highly conductive β-CoOOH, which is not reduced on subsequent discharges. The β-CoOOH provides interparticle contact and access of electrons to the active material.

Considerable increases in the capacity density of the electrodes have been achieved through the use of high-density β-Ni(OH)$_2$ with a uniform particle size, a narrow range of pore sizes, and a high tapped density (1.9–2.0 g cm^{-3}) [15, 16]. Conventional β-Ni(OH)$_2$ consists of irregular particles with 30% inner pore volume, a large range of pore sizes, and a tapped density of ~1.6 g cm^{-3}. With the new material, it is possible to increase the active material filling by 20%. Using this material, it was possible to make electrodes with capacity densities exceeding 550 mAh cm^{-3}. Conventional sintered plates have capacity densities of ~400 mAh cm^{-3}.

A major problem with pasted-plastic-bonded and fiber mat electrodes is swelling of the electrode on cycling. This is due to the formation of γ-NiOOH. This causes comminution of the active material and an increase in the pore volume. This problem can be largely avoided by the use of β-Ni(OH)$_2$ containing ~7% of coprecipitated Cd or Zn. The coprecipitation of one of these additives along with 7% Co also greatly improves the charge acceptance of the electrode at elevated temperatures (up to 45 °C) [15]. This additive combination also greatly improves the charge retention at elevated temperatures [15]. Zinc is preferred over Cd as an additive because of its lower toxicity and the detrimental effect of Cd on metal hydride electrodes. These advances in the nickel hydroxide electrode represent considerable progress, and have increased the capacity of sealed nickel–cadmium AA cells, at the C rate, from 500 to 800 mAh [15].

5.3
Solid-State Chemistry of Nickel Hydroxides

5.3.1
Hydrous Nickel Oxides

β-Ni(OH)$_2$, α-Ni(OH)$_2$, β-NiOOH, and γ-NiOOH are considered to be the model divalent and trivalent materials for the nickel hydroxide electrode.

5.3.1.1 β-Ni(OH)$_2$
β-Ni(OH)$_2$ can be made with a well-defined crystalline structure and is in many ways similar to the active material in chemically prepared battery electrodes that are made by the method described by Fleischer [17]. Several methods of preparation have been reported. One is to precipitate the hydroxide at 100 °C from a nickel nitrate solution by addition of a KOH solution. Further enhancement

in the crystallinity of this material has been obtained by hydrothermal treatment in an aqueous slurry containing NH$_4$OH, KOH [18], or NaOH [19]. A method which produces good crystals has been described by Fievet and Figlarz [20]. The hydroxide is prepared in two steps. First, an ammonia solution is added to a nickel nitrate solution at room temperature. The precipitate is washed, and then hydrothermally treated at 200 °C. Another method is to precipitate the hydroxide by dropwise addition of 3 mol L^{-1} Ni(NO$_3$)$_2$ to hot (90 °C) 7 mol L^{-1} KOH with constant stirring. The precipitate is washed and dried. The Ni(OH)$_2$ is then dissolved in 8 mol L^{-1} NH$_4$OH, and the resulting blue solution of Ni(NH$_3$)$_6$(OH)$_2$ is transferred to a desiccator containing concentrated H$_2$SO$_4$ and kept there for several days. Slow removal of the NH$_3$ by H$_2$SO$_4$ yields well-formed glassy flakes of β-Ni(OH)$_2$ [21]. α-Ni(OH)$_2$ can be prepared electrochemically, and can be converted to β-Ni(OH)$_2$ by heating in 6–9 mol L^{-1} KOH at 90 °C for 2–3 h [11].

The definitive structural determination of the β-hydroxide is the powder neutron diffraction work of Greaves and Thomas on β-Ni(OD)$_2$ [22]. They did neutron diffraction studies on well-crystallized deuterated Ni(OD)$_2$ that had been prepared by a hydrothermal method. They also investigated a high-surface-area Ni(OH)$_2$ that was prepared by precipitation by addition of KOH to an NiSO$_4$ solution. The X-ray and neutron diffraction results indicate that β-Ni(OH)$_2$ has a brucite C6-type structure that is isomorphous with the divalent hydroxides of Ca, Mg, Fe, Co, and Cd. The structure is shown in Figure 5.2.

The crystal consists of stacked layers of nickel–oxygen octahedra. The nickels are all in the (0 0 0 1) plane and are surrounded by six hydroxyl groups, each of which is alternately above and below the (0 0 0 1) plane. The fractional coordinates

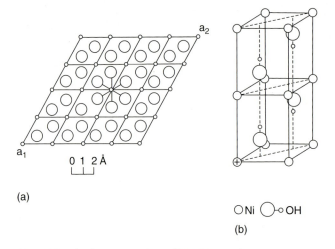

Figure 5.2 The brucite structure of Ni(OH)$_2$: (a) hexagonal brucite layer, in which the small circles are the Ni atoms and the large circles the O atoms and alternate O atoms are below and above the plane of the Ni atoms and (b) stacking of the planes showing the orientation of the O–H bonds.

Table 5.1 Crystallographic parameters for β-Ni(OD)$_2$ [22].

Parameter	(Å)
a_0	3.126
c_0	4.593
Ni–O bond length	2.073
O–H bond length	0.973
Ni–Ni bond length	3.126

are 0, 0, 0 for nickel and 1/3, 2/3, z and 2/3, 1/3, z for oxygen. Values for the crystallographic parameters of well-crystallized β-Ni(OD)$_2$ are given in Table 5.1.

Because of anomalous scattering by H, the results for the as-precipitated Ni(OH)$_2$ could not be refined. Nevertheless, cell constants and the O–H bond distance could be determined. The results showed that the as-precipitated material was different from the well-crystallized material. The unit cell dimensions were $a_0 = 3.119$ Å and $c_0 = 4.686$ Å. Also the O–H bond length was 1.08 Å, a value similar to that previously reported by Szytula et al. in a neutron diffraction study of Ni(OH)$_2$ [23]. The O–H bond in both well-crystallized and as-precipitated materials is parallel to the c-axis. The difference between well-crystallized and as-precipitated material is important since the well-crystallized material is not electrochemically active. The differences between the materials are attributed to a defective structure that arises from the large concentration of surface OH$^-$ ion groups in the high-surface-area material [22]. These are associated with absorbed water. This is a consistent with an absorption band in the infrared at 1630 cm^{-1}. This is not seen in the well-crystallized material.

Infrared spectroscopy has also confirmed the octahedral coordination of nickel by hydroxyl groups [24, 25]. No evidence for hydrogen bonding has been found. In battery materials, evidence was also found for a small amount of absorbed water [24, 26]. Even though these materials contain small amounts of water they are still classified as β-Ni(OH)$_2$ because of an (0 0 1) X-ray reflection corresponding to a d spacing of 4.65 Å. Thermogravimetric analysis (TGA) indicates that the water is removed at higher temperatures [26–28]. Kober [24, 26] has proposed that this water is associated with nickel ions in the lattice and suggested the formula [Ni(H$_2$O)$_{0.326}$](OH)$_2$ for the chemically prepared battery material. A similar formula was proposed by Dennsted and Loser [27]. However, this has been disputed [29]. The evidence is that well-crystallized β-Ni(OH)$_2$ does not contain absorbed water [22, 29]. However, the high-surface-area material that is used in batteries does. This is consistent with the expansion in the c-axis of the crystal from 4.593 to 4.686 Å, the increase in the average O–H bond distance from 0.973 to 1.08 Å [22], and the presence of broad absorption bond in the infrared spectrum at 1630 cm^{-1} [22, 24, 26]. TGA results indicate that this water is removed in a single process over a temperature range of 50–150 °C [30].

The Raman spectroscopic work of Jackovitz [31], Cornilsen et al. [32, 33], and Audemer et al. [34] is the most direct spectroscopic evidence that the discharge product in battery electrodes operating with the β/β cycle is different from well-crystallized β-Ni(OH)$_2$. The O–H stretching modes and the lattice modes in the Raman spectra are different from those found for well-crystallized Ni(OH)$_2$ prepared by recrystallization from the ammonia complex, and are more similar to those found for the initial material prepared by Barnard et al. [21] by precipitation of the Ni(OH)$_2$ by adding 3 mol L^{-1} Ni(NO$_3$)$_2$ to hot 7 mol L^{-1} KOH. In discharged electrodes, a Raman band at 3605 cm^{-1} is observed. This has been ascribed to absorbed water molecules on the surface of the Ni(OH)$_2$ [34]. There have been discrepant results in the Raman evidence for adsorbed water. However, some water cannot be ruled out since the O–H modes are very poor Raman scatterers. Infrared spectroscopy is much better at detecting water, and Jackovitz has seen water-stretching modes in both the nondeuterated and deuterated material after discharge [31]. Audemer et al. have also seen this band at 1630 cm^{-1}. Furthermore, they have confirmed that both the Raman band at 3605 cm^{-1} and the IR band at 1630 cm^{-1} decrease at temperatures above 100 °C and completely vanish at 150 °C. Neutron diffraction work on Ni(OH)$_2$ and Raman and IR spectroscopy clearly show that discharge product in battery electrodes is closely related, but not identical, to well-crystallized. β-Ni(OH)$_2$. It probably has a defect structure, which facilitates water adsorption and the electrochemical reactions.

5.3.1.2 α-Ni(OH)$_2$

α-Ni(OH)$_2$, which has a highly hydrated structure, was first identified by Lotmar and Fectknecht [35]. α-Ni(OH)$_2$ is a major component of the active material in battery electrodes when nickel battery plaques are cathodically impregnated from an aqueous Ni(NO$_3$)$_2$ solution at temperatures below 60 °C [36]. α-Ni(OH)$_2$ can be prepared chemically by precipitation from dilute solutions at room temperature. One method is simply to add an ammonia solution to a nickel nitrate solution [20]. Another method is to add 0.5 or 1 mol L^{-1} KOH to 1 mol L^{-1} Ni(NO$_3$)$_2$ [21]. In both cases, the precipitate is filtered and washed. Methods for electrochemical preparation of α-Ni(OH)$_2$ films on nickel substrates have been described [37, 38]. One method consists of cathodically polarizing a cleaned nickel sheet in a quiescent 0.1 mol L^{-1} Ni(NO$_3$)$_2$ solution at 8 mA cm^{-1}. There is reduction of nitrate and a concomitant increase in pH at the electrode surface. This causes precipitation of an adherent coating of α-Ni(OH)$_2$ on the nickel. A 100 s period of deposition will produce 0.5 mg cm^{-2} of α-Ni(OH)$_2$.

Determination of the structure of α-Ni(OH)$_2$ has been difficult, since sometimes it exhibits no diffraction pattern [39]. After washing with water, a diffuse pattern develops. Hydrothermal treatment eventually leads to well-crystallized β-Ni(OH)$_2$ [39, 40]. The evolution of the X-ray diffraction patterns is shown in Figure 5.3. Bode proposed a layered structure for α-Ni(OH)$_2$ similar to that for β-Ni(OH)$_2$ [11]. His suggested structure was essentially identical to that shown for β-Ni(OH)$_2$ in Figure 5.2, except that between the (0 0 0 1) planes there are water molecules that result in an expansion of the c-axis spacing to about 8 Å. Bode proposed a

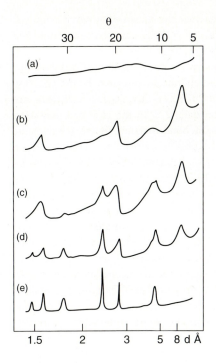

Figure 5.3 X-ray diffraction patterns (Co K_α) for α-Ni(OH)$_2$: (a) as-precipitated and (b–d) increasing in crystallinity with time when aged in water. The pattern (e) for β-Ni(OH)$_2$ eventually develops [42].

3Ni(OH)$_2$·2H$_2$O unit cell and assigned definite positions to the intercalated water molecules in which two-thirds of the available nickel sites were occupied with water molecules [39]. The model gives unit cell dimensions of $a_0 = 5.42$ Å and $c_0 = 8.05$ Å. In addition to the increase in c-axis spacing, Bode reported a small contraction in the lattice parameters within the layer planes of α-Ni(OH)$_2$. Later work by Figlarz and Le Bihan, using the X-ray diffraction line profiles, showed that α-Ni(OH)$_2$ was turbostratic and that it consisted of brucite-like layers randomly oriented along the c-axis [41]. Subsequently, McEwen [19], too, used line profile analysis to arrive at the same conclusion. He also concluded that the intercalated water layer was not ordered. He disputed the contraction in the basal plane that was proposed by Bode, and ascribed the diffraction peak shifts to disorder and particle size effects. Le Bihan and Figlarz use a combination of X-ray diffraction, electron microscopy, and infrared spectroscopy to study α-Ni(OH)$_2$ as-prepared and after repeated washing in water [42]. They confirmed that in the α-structure the Ni(OH)$_2$ planes are essentially identical to those shown for β-Ni(OH)$_2$ in Figure 5.2. The layers are stacked with random orientation. The c-axis spacing is constant, but the layers are randomly oriented. The layers are separated by water molecules that are hydrogen-bonded to the Ni–OH groups in the basal planes. In electron micrographs, turbostratic nickel hydroxide appears as thin crumpled sheets. The crystallites have a mean size of 30 Å along [0 0 1 1], which corresponds to a stacking of five layers. The basal plane dimensions are about 80 Å [42]. Because of the high degree of division, α-Ni(OH)$_2$ retains adsorbed surface water and a small amount

(<3%) of nitrate ions [43]. TGA indicates that the adsorbed water is removed between 50 and 90 °C, whereas the intercalated water is removed between 90 and 180 °C [30].

Pandya et al. have used extended X-ray ascription fine structure (EXAFS) to study both cathodically deposited α-Ni(OH)$_2$ and chemically prepared β-Ni(OH)$_2$ [44]. Measurements were done at both 77 and 297 K. The results for β-Ni(OH)$_2$ are in agreement with the neutron diffraction data [22]. In the case of α-Ni(OH)$_2$, they found a contraction in the first Ni–Ni bond distance in the basal plane. The value was 3.13 Å for β-Ni(OH)$_2$ and 3.08 Å for α-Ni(OH)$_2$. The fact that a similar significant contraction of 0.05 Å was seen at both 77 and 297 K when using two reference compounds (NiO and β-Ni(OH)$_2$) led them to conclude that the contraction was a real effect and not an artifact due to structural disorder. They speculate that the contraction may be due to hydrogen bonding of OH groups in the brucite planes with intercalated water molecules. These *ex situ* results on α-Ni(OH)$_2$ were compared with *in situ* results in 1 mol L^{-1} KOH. In the *ex situ* experiments, the α-Ni(OH)$_2$ was prepared electrochemically, washed with water, and dried in vacuum. In the *in situ* experiment, the hydroxide film, after preparation, was simply rinsed without drying and immediately immersed in the cell containing 1 mol L^{-1} KOH. The coordination numbers for Ni–O for the *in situ* samples were consistently higher. The significance of this is not clear: it might suggest some water association with nickel, as was postulated by Kober [24, 27].

Raman spectroscopy results indicate that the structure of α-Ni(OH)$_2$ is very dependent on how it is prepared [32, 33, 45]. Data on chemically prepared [32, 33], cathodically deposited [32, 33, 45], and electrochemically reduced γ-NiOOH have been obtained [32, 33, 45]. There are differences in all the spectra, both in the lattice modes and the O–H stretching modes. The shifts in the lattice modes for the reduced γ-NiOOH may be due to this material having an oxidation state higher than two. The changes in the O–H stretching modes may be due to changes in water content and hydrogen bonding.

Figlarz and his co-workers have suggested that the formula Ni(OH)$_2 \cdot n$H$_2$O is not the correct one for α-Ni(OH)$_2$ [46, 47]. They studied α-Ni(OH)$_2$ materials made by precipitation of the hydroxide by the addition of NH$_4$OH to solutions of various nickel salts. In addition to Ni(NO$_3$)$_2$ and NiCO$_3$, they used nickel salts with carboxylic anions of various sizes. They found that the interlaminar distance in the α-Ni(OH)$_2$ depended on the nickel salt anion size. For instance, when nickel adipate was used, the interlaminar distance was 13.2 Å. Infrared studies of α-Ni(OH)$_2$ precipitated from Ni(NO$_3$)$_2$ indicated that NO$_3^-$ was incorporated into the hydroxide and was bonded to Ni. They suggested a model based on hydroxide vacancies and proposed a formula Ni(OH)$_{2-x}$ A$_y$ B$_z \cdot n$H$_2$O, where A and B are mono- or divalent anions and $x = y + 2z$. Chemical analysis of α-Ni(OH)$_2$ precipitated from Ni(NO$_3$)$_2$ indicates OH vacancies in the range of 20–30%.

α-Ni(OH)$_2$ is unstable in water and is slowly converted to β-Ni(OH)$_2$. Transmission electron micrographs of the reactants and products indicate that the reaction proceeds via the solution [42, 45]. In concentrated KOH, the reaction is much more rapid and the product has a smaller particle size. For instance, the α-Ni(OH)$_2$

in an electrochemically impregnated battery electrode is completely converted to β-Ni(OH)$_2$ 30 min after immersion in 4.5 mol L^{-1} KOH [36].

Infrared studies of the reaction product in water indicate that the β-Ni(OH)$_2$ that is initially formed also contains anions and adsorbed water. As the particle size of the product increases, the amount of anions and adsorbed water decreases [45].

Delmas and co-workers have proposed the existence of an intermediate phase between α- and β-Ni(OH)$_2$ [48]. This phase consist of interleaved α and β material and can be formed on aging of α-Ni(OH)$_2$. Recent Raman results confirm the existence of such a phase [49].

5.3.1.3 β-NiOOH

β-Ni(OH)$_2$ has been identified as the primary oxidation product of electrodes containing β-Ni(OH)$_2$ [11, 50, 51]. Glemser describes a method for preparation of β-NiOOH [52, 53]. A solution of 100 g of Ni(NO$_3$)$_2 \cdot$6H$_2$O in 1.5 L H$_2$O was added dropwise to a solution of 55 g KOH and 12 mL Br$_2$ in 300 mL H$_2$O, while keeping the temperature at 25 °C. The black precipitate was washed with CO$_2$-free water until both K$^+$ and NO$_3^-$ were removed and then dried over H$_2$SO$_4$. Structural determinations of the higher oxides of nickel are complicated because of their amorphous nature [53]. However, it appears that β-Ni(OH)$_2$ is oxidized to the trivalent state without major modifications to the brucite structure. The unit cell dimensions change from $a_0 = 3.126$ Å and $c_0 = 4.605$ Å for β-Ni(OH)$_2$ to $a_0 = 2.82$ Å and $c_0 = 4.85$ Å for β-NiOOH. X-ray diffraction clearly indicates expansion along the c-axis. Asymmetry in the h k lines indicates the turbostratic nature of β-NiOOH. Even after correcting the a_0, values for this, McEwen found that there was a real contraction in the basal plane [19]. Transmission EXAFS has been used to investigate the oxidation products of β-Ni(OH)$_2$ [54, 55]. *In situ* measurements on plastic-bonded electrodes showed that in the charged state a two-shell fit was necessary for the first Ni–O coordination shell. This suggests that the oxygen coordination in β-NiOOH is a distorted octahedral coordination with four oxygens at a distance of 1.88 Å and two oxygens at a distance of 2.07 Å. The distorted coordination is consistent with the edge features of the X-ray absorption spectra [55, 56]. The overall reaction for the electrochemical formation of β-NiOOH is usually given as

$$\beta\text{-Ni(OH)}_2 + \beta\text{-NiOOH} + H^+ + e^- \tag{5.1}$$

During the reaction, protons are extracted from the brucite lattice. Infrared spectra [24, 25, 31] show that during charge the sharp hydroxyl band at 3644 cm^{-1} disappears. This absorption is replaced by a diffuse band at 3450 cm^{-1}. The spectra indicate a hydrogen-bonded structure for β-NiOOH with no free hydroxyl groups. β-NiOOH probably has some adsorbed and absorbed water. However, TGA data on charged materials are very limited [57, 58], and it is not always clear that the material is pure β-NiOOH. Unless electrochemical experiments are done very carefully there is always the possibility of the presence of γ-NiOOH [13].

5.3.1.4 γ-NiOOH

γ-NiOOH is the oxidation product of α-Ni(OH)$_2$. It is also produced on overcharge of β-Ni(OH)$_2$, particularly when the charge is carried out at high rates in high concentrations of alkali [11, 59]. Use of the lighter alkalis (LiOH and NaOH) favor the formation of γ-NiOOH, whereas use of RbOH inhibits its formation [60]. The material was first prepared by Glemser and Einerhand [52] by fusing one part Na$_2$O$_2$ with three parts NaOH in a nickel crucible at 600 °C. Hydrolysis of the product yields γ-NiOOH. They gave cell dimensions for a rhombohedral system with $a = 2.8$ Å and $c = 20.65$ Å. The material has a layer structure with a spacing of 7.2 Å between layers. γ-NiOOH always contains small quantities of alkali metal ions and water between the layers, whereas β-NiOOH does not. The X-ray diffraction patterns have more, and much sharper, lines than those of either α-Ni(OH)$_2$ or β-NiOOH [19, 53]. γ-NiOOH prepared by the method of Glemser and Einerhand has the formula NiOOH·0.51H$_2$O. TGA analysis shows that this water is lost between 50 and 180 °C [58].

5.3.1.5 Relevance of Model Compounds to Electrode Materials

The reaction scheme of Bode [11] was derived by comparison of the X-ray diffraction patterns of the active materials with those for the model compounds. How the β-Ni(OH)$_2$ in battery electrodes differs from the model compound is discussed in Section 5.3.1.3. In recent years, the arsenal of *in situ* techniques for electrode characterization has greatly increased. Most of the results confirm Bode's reaction scheme and essentially all the features of the proposed α/γ cycle. For instance, recent atomic force microscopy (AFM) of α-Ni(OH)$_2$ shows results consistent with a contraction of the interlayer distance from 8.05 to 7.2 Å on charge [61–63]. These are the respective interlayer dimensions for the model α-Ni(OH)$_2$ and γ-NiOOH compounds. Electrochemical quartz crystal microbalance (ECQM) measurements also confirm the ingress of alkali metal cations into the lattice upon the conversion of α-Ni(OH)$_2$ to γ-NiOOH [45, 64, 65]. However, *in situ* Raman and surface-enhanced Raman spectroscopy (SERS) results on electro-chemically prepared α-Ni(OH)$_2$ in 1 mol L^{-1} NaOH show changes in the O–H stretching modes that are consistent with a weakening of the O–H bond when compared with results for the model α- and β-Ni(OH)$_2$ compounds [66]. This has been ascribed to the delocalization of protons by intercalated water and Na$^+$ ions. Similar effects have been seen in passive films on nickel in borate buffer electrolytes [67].

Recent ECQM work and X-ray diffraction have confirmed the conversion of the α/γ cycle to the β/β cycle upon electrochemical cycling in concentrated alkali. Earlier ECQM studies of α-Ni(OH)$_2$ films had shown a mass inversion in the microgravimetric curve after prolonged cycling [64]: there is a mass decrease in charge instead of a mass increase. More recent work has confirmed that this mass inversion is due to conversion of the α/γ cycle to the β/β cycle [65].

5.3.2
Pyroaurite-Type Nickel Hydroxides

Allmann found that when suitable trivalent ions were introduced during the precipitation of the hydroxides of Mg, Zn, Mn, Fe, Co, and Ni, these were incorporated in the lattice, and the structure changed from the brucite ($Mg(OH)_2$) to the pyroaurite ($Mg_6Fe_2(OH)_{16}(CO_3) \cdot 4H_2O$) type of structure [68]. One of the nickel materials he prepared was an Ni/Al hydroxide. Axmann et al. [69–71] have given the nickel compounds the general formula

$$[\underbrace{Ni^{II}_{1-x} M^{III}_x (OH_2)]^{x+}}_{K^{x+}} \underbrace{[(X^{n-})_{x/n}(H_2O)_z]^{x-}}_{A^{x-}}$$

where $0.2 \leq x \leq 0.4$ and X^{n-} are the anions of the precursor salts. The resultant structure consists of brucite cationic layers intercalated with anions and water molecules. The cationic nature of the brucite layers is due to the higher valence of the substituent cations, and the anions in the intercalated anion layers provide electro-neutrality. As a result, the interlayer distance increases from 4.68 to 7.80 Å, when compared with β-$Ni(OH)_2$. The structure is shown schematically in Figure 5.4. Electrochemical and chemical oxidation transforms the pyroaurite structure to a product that is isostructural with γ-NiOOH [72]. In the early work, the pyroaurite compounds were prepared by precipitation. Buss et al. report on the preparation of $Ni_4Al(OH)_{10}NO_3$ prepared in a computer-controlled apparatus wherein a solution of 0.4 mol L^{-1} $Ni(NO_3)_2 6H_2O + 1 \text{ mol L}^{-1} Al(NO_3)_3 \cdot 9H_2O$ in doubly distilled water with the pH adjusted to 2 with 1 mol L^{-1} HNO_3 was simultaneously sprayed with a carbonate-free 1 mol L^{-1} KOH solution into a receptor solution maintained at pH 11.5 and a temperature of 32 °C [70, 71]. The precipitate was filtered, washed, and dried at 50 °C for three days at 0.01 bar and carbon dioxide was excluded at all stages of the preparation.

Delmas and his co-workers have done extensive work on pyroaurite-type materials, which has recently been reviewed [74]. In addition to precipitation methods, they have prepared the materials by mild oxidative hydrolysis of nickelates that were prepared by thermal methods similar to those used for the preparation of $LiNiO_2$ [75]. A cobalt-substituted material $NaCo_x(Ni_{1-x}O_2)$ was prepared by the reaction of Na_2O, Co_3O_4, and NiO at 800 °C under a stream of oxygen. The material was then treated with a 10 mol L^{-1} NaClO + 4 mol L^{-1} KOH solution for 15 h to form the oxidized γ-oxyhydroxide. The pyroaurite phase was prepared by subsequent reduction in a solution of 0.1 mol L^{-1} H_2O_2 in 4 mol L^{-1} KOH [76]. These mild chemical treatments are referred to as 'chemie douce' reactions [69]. The thermally prepared nickelates have a layered R3m structure. The 'chemie douce' treatments essentially leave the covalently bonded $Co_x Ni_{1-x}O_2$ layers intact, as a result of which more crystalline materials with larger particle sizes can be made by this method.

Most of the work on pyroaurite materials has been done on materials with Fe [68–73, 77], Co [68, 76, 78], Mn [72, 79], or Al [68, 70, 72] substitutions. When at least 20% of the Ni atoms are replaced by the trivalent substituent, the materials are stable in concentrated KOH. In many ways, the pyroaurite phase is similar to

α-Ni(OH)$_2$. Thus, substitution of 20% of the Ni with these trivalent ions stabilizes the operation of the electrode in the α/γ cycle in concentrated KOH.

Because of the possibility of applying Mössbauer spectroscopy, the solid-state chemistry of the Fe-substituted material is the best understood [69, 72, 73]. Mössbauer spectroscopy confirms that the Fe in the pyroaurite type material is Fe(III). Glemser and co-workers have found that electrochemical oxidation of the material converts about 30% of the Fe(III) to Fe(IV) [69, 72]. The results were consistent with a high-spin configuration with the Fe(IV) in FeO$_6$ octahedra with O$_h$ symmetry. The O$_h$ symmetry can only occur if the surrounding NiO$_6$ octahedra also have an O$_h$ symmetry. Hence, the Fe(IV) ions in the layer must be surrounded by six NiO$_6$ octahedra with the Ni in the Ni(IV) state. Delmas and coworkers found evidence for Fe(IV) in both high- and low-spin states for oxidized materials prepared by the 'chemie douce' method [73]. The difference in the results may be due to the effect of the platelet size on the pyroaurite structure.

In the pyroaurite structure, the brucite layers are cationic. However, on oxidation the resultant brucite layers in γ-NiOOH are anionic. To preserve electroneutrality, cations and anions are exchanged in the intercalated layer during the oxidation–reduction process. This is illustrated in Figure 5.4. In the case of Mn-substituted materials, some Mn can be reduced to Mn(II). This neutralizes

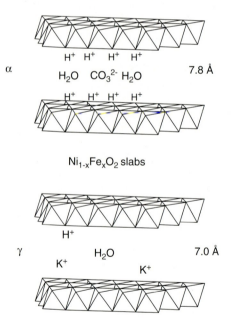

Figure 5.4 Structure of the Fe(III)-substituted pyroaurite phase in the discharged (α) and charged (γ) state. The edge-shared NiO$_6$ and FeO$_6$ octahedra are shown. Also shown is the incorporation of anions and water in the galleries of the discharged material. On charging, the anions are replaced by cations [73].

the charge in the brucite layer; this part of the structure reverts to the β-Ni(OH)$_2$ structure and the intercalated water and anions are expelled from the lattice. With this, there is a concomitant irreversible contraction of the interlayer spacing from 7.80 to 4.65 Å [72].

5.4
Electrochemical Reactions

5.4.1
Overall Reaction and Thermodynamics of the Ni(OH)$_2$/NiOOH Couple

In normal battery operation, several electrochemical reactions occur on the nickel hydroxide electrode. These are the redox reactions of the active material, oxygen evolution, and, in the case of nickel–hydrogen and nickel–metal hydride batteries, hydrogen oxidation. In addition, there are parasitic reactions such as the corrosion of nickel current collector materials and the oxidation of organic materials from separators. The initial reaction in the corrosion process is the conversion of Ni to Ni(OH)$_2$.

Because of the complexity of the redox reactions, they cannot be conveniently presented in a Pourbaix pH-potential diagram. For battery applications, the revised diagram given by Silverman [80] is more correct than that found in the Pourbaix *Atlas* [81]. The diagram is shown in Figure 5.5.

The respective literature values for the free energy of formation of Ni(OH)$_2$, NiOOH, H$_2$O, and HgO are -78.71, -109.58, -56.69, and -13.98 kcal mol^{-1} [80]. The calculated Ni(OH)$_2$/NiOOH reversible potential is 0.41 V vs Hg/HgO, and the reversible oxygen potential is 0.30 V vs Hg/HgO. Unlike other battery positive

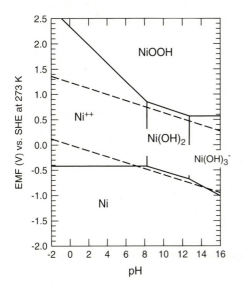

Figure 5.5 The modified Pourbaix diagram for Ni [80].

electrode materials, such as AgO or PbO, the nickel hydroxide electrode is a good catalyst for oxygen evolution. Toward the end of charge, oxygen evolution occurs in all nickel batteries, and during constant voltage hold in charge mode self-discharge occurs via a couple involving the reduction of NiOOH and the oxidation of water to oxygen.

The self-discharge process has made experimental determination of the reversible potential of the $Ni(OH)_2$/NiOOH couple very difficult. A major advance was the realization by Bourgault and Conway that the open-circuit potential of a charged nickel oxide electrode was a mixed potential, not a true equilibrium potential [82], and was the result of two processes: the discharge of NiOOH and oxygen evolution. They devised an extrapolation technique for the determination of the open-circuit potential of $Ni(OH)_2$ as a function of charge state. Later, the work was expanded by Barnard and his co-workers to include oxidation of both the α/γ and the β/β couples [83]. The open-circuit potentials depended on pretreatment, such as formation cycles and aging in concentrated KOH electrolytes. The β/β couples had open-circuit potentials in the range of 0.44–0.47 V vs Hg/HgO, whereas the α/γ couples had values in the range 0.39–0.44 V. In cyclic voltammetry experiments, the respective anodic and cathodic peaks for the α/γ couple occur at 0.43 and 0.34 V. For the β/β couple, the peaks are at 0.50 and 0.37 V. The reversible potentials of the β/β couple are essentially invariant with KOH^- concentration, whereas those of the α/γ couple vary with OH^- concentration, and aging of α-$NiOH)_2$ reduces the OH^- dependence of the reversible potential [84]. This is due to the conversion to the β/β couple. The reactions for both the β/β and the α/γ couples are highly reversible. Barnard and Randell, in a simple experiment, showed that β-NiOOH could oxidize α-Ni(OH) to γ-NiOOH [85]. This reaction is possible during cyclic voltammetry on α-$Ni(OH)_2$ thin film electrodes in KOH electrolyte, and some of the α material gets transformed to the β form. This could account for the negative drift that is seen in the anodic peaks in the early stages of cycling [66]. Reactions of this type can introduce distortions and features in cyclic volammograms that are difficult to interpret.

5.4.2
Nature of the $Ni(OH)_2$/NiOOH Reaction

The $Ni(OH)_2$/NiOOH reaction is a topochemical type of reaction that does not involve soluble intermediates. Many aspects of the reaction are controlled by the electrochemical conductivity of the reactants and products. Photoelectrochemical measurements [86, 87] indicate that the discharged material is a p-type semiconductor with a bandgap of about 3.7 eV. The charged material is an n-type semiconductor with a bandgap of about 1.75 eV. The bandgaps are estimates from absorption spectra [87].

The simple experiments of Kuchinskii and Ershler have provided great insights into the nature of the $Ni(OH)_2$/NiOOH reaction [88, 89]. They investigated oxidation and reduction of a single grain of $Ni(OH)_2$ with a platinum point contact. On charge, the $Ni(OH)_2$ turned black and oxygen was evolved preferentially on the black material and not on the platinum. This implies that NiOOH is conductive and has a lower oxygen overvoltage than platinum oxide. They found that discharge started at the point contact and that formation of resistive Ni(OH)$_2$ at this interface could stop current flow and result in an incomplete discharge. These results provide a good macroscopic picture of how the electrode works.

This type of mechanism has been considered by Barnard et al. [83]. They postulate the initiation of the charging reaction at the Ni(OH)$_2$/current collector interface with the formation of a solid solution of Ni^{3+} ions in Ni(OH)$_2$. With further charging when a fixed nickel ion composition (Ni^{2+})$_x$·(Ni^{3+})$_{1-x}$ is reached, phase separation occurs with the formation of two phases, one with the composition (Ni^{2+})$_{1-x}$·(Ni^{3+})$_x$ in contact with the current collector and the other with the composition (Ni^{2+})$_x$·(Ni^{3+})$_{1-x}$ further out into the active mass. This scheme is consistent with the observations of Briggs and Fleischman on thick α-Ni(OH)$_2$ films [90]. In microscopic observations of cross-sections of partially charged electrodes, they observed a green layer of uncharged Ni(OH)$_2$ in front of the electrode. The central part of the electrode was coated with a black material, and a thin layer in contact with the current collector had a yellowish metallic luster. On discharge, the reverse process occurred. It is possible for some of the NiOOH to be isolated in the poorly conducting matrix of Ni(OH)$_2$ and not to be discharged. This has been confirmed in recent in Raman spectroscopy studies *in situ* [66].

Sometimes two discharge voltage plateaus are seen on nickel oxide electrodes. Early observations are documented in previous reviews [2, 9]. Normally, nickel oxide electrodes have a voltage plateau on discharge in the potential range 0.25–0.35 V vs Hg/HgO. The second plateau, which in some cases can account for up to 50% of the capacity, occurs at -0.1 to -0.6 V. At present, there is a general consensus that this second plateau is not due to the presence of a new, less-active, compound [91–94]. Five interfaces have been identified for a discharging NiOOH electrode [93]. These are

1) the Schottky junction between the current collector and the n-type NiOOH, polarized in the forward direction,
2) the p–n junction between Ni(OH)$_2$ and NiOOH, polarized in the forward direction,
3) the NiOOH/electrolyte interface,
4) the Ni(OH)$_2$/electrolyte interface, and
5) the Schottky junction between the current collector and Ni(OH)$_2$, polarized in the forward direction.

At the beginning of discharge only junctions (1) and (3) are present. As discharge progresses, junction (5) develops. The passage of current shifts the electrode potential to more negative values. The hole conductivity of the Ni(OH)$_2$ increases and a second discharge plateau appears. A quantitative modeling effort by Zimmerman [94] supports this hypothesis.

5.4.3
Nickel Oxidation State

Like all other facets of the electrode, determination of the overall redox process has been difficult and many aspects are still disputed. The presence of Ni(IV) species in charged materials has been proposed by many authors. The early work has been reviewed [9]. The evidence for Ni(IV) is based mostly on coulometric data [95] or determinations of active oxygen by titration with iodide or arsenious oxide. Active oxygen contents corresponding to a nickel valence of 3.67 have been reported for α-Ni(OH)$_2$ films charged in 1 mol L^{-1} KOH [95], and values of 3.48 were found for overcharged β-Ni(OH)$_2$ battery electrodes in 11 mol L^{-1} KOH [96]. When these electrodes of high active oxygen content are discharged, or even over-discharged, an appreciable amount of active oxygen remains [57]. Cycling between a nickel valence of 2.5 and 3.5 has been proposed [97]. X-ray absorption has been used to study the problem [98, 99]. In one case, results consistent with an Ni oxidation state of 3.5 were found for a charged electrode [99]. In the case of the α/γ couple, indications are that a nickel oxidation state of at least 3.5 can be reached on charge. It is not clear that this is the case with the β/β couple. *In situ* experiments with simultaneous X-ray diffraction and X-ray absorption measurements should be done on the β/β couple to check for the presence of γ-NiOOH. Experiments on materials stabilized with both Co and Zn additives are also necessary.

The existence of these high nickel oxidation states offers the possibility of a 'two-electron' electrode. This is one of the incentives for stabilizing the α/γ cycle through the use of the pyroaurite structures [74]. With this approach, it has been possible to achieve a 1.2 electron exchange for the overall reaction. However, none of the pyroaurite structures is satisfactory for battery electrodes. The Co- and Mn-substituted materials are unstable with cycling [72, 74]. The end-of-charge voltages for both the Fe- and Al-substituted materials are high and the charging efficiencies are low [72, 74]. However, the use of mixed substitution, such as combinations of Co and Al, can lower the charging voltage [74].

5.4.4
Oxygen Evolution

Oxygen evolution occurs on nickel oxide electrodes throughout charge, on overcharge, and on standby. It is the anodic process in the self-discharge reaction of the positive electrode in nickel–cadmium cells. Early work in the field has been reviewed [9]. No significant new work has been reported in recent years.

5.4.5
Hydrogen Oxidation

The reaction of hydrogen at the nickel electrode determines the rate of self-discharge in nickel–hydrogen batteries.

Under typical operating pressures (30–50 atm), a nickel–hydrogen battery will lose 50% of its capacity in a week. The self-discharge rate is about five times that encountered in sealed nickel–cadmium batteries, where the rate-determining step is oxygen evolution [100]. Tsenter and Sluzhevskii [101] developed a set of kinetic equations to describe the self-discharge process; in their model its rate depends on the hydrogen pressure and the amount of undischarged NiOOH in the cell. Experimental results of Srinivasan and co-workers confirm many aspects of this model [102]. They used a combination of microcalorimetry, open-circuit voltage measurements, and capacity measurements to study the problem. By doing measurements on the active material and substrate, separately and combined, they were able to establish that hydrogen oxidation occurs predominantly on the charged active material with simultaneous reduction of the oxide.

References

1. Falk, S.U. and Salkind, A.J. (1969) *Alkaline Storage Batteries*, John Wiley & Sons, Inc., New York.
2. Milner, P.C. and Thomas, U.B. (1967) in *Advances in Electrochemistry and Electrochemical Engineering* (ed. C.W. Tobias), John Wiley & Sons, Inc., New York, p. 1.
3. Briggs, G.W.D. (1974) in *Electrochemistry*, Specialist Periodical Reports, Vol. 4, The Chemical Society, London, p. 33.
4. Arvia, A.J. and Posadas, D. (1975) in *Encyclopedia of Electrochemistry of the Elements*, Vol. III (ed. A.J. Bard), Marcel Dekker, New York, p. 212.
5. Weininger, J.L. (1982) in *Proceedings of the Symposium on the Nickel Electrode* (eds R.G. Gunther and S. Gross), The Electrochemical Society, Pennington, NJ, p. 1.
6. Oliva, P., Leonardi, J., Laurent, J.F., Delmas, C., Bracconier, J.J., Figlarz, M., Fievet, F., and de Guibert, A. (1982) *J. Power Sources*, **8**, 229.
7. Halpert, G. (1984) *J. Power Sources*, **12**, 177.
8. Halpert, G. (1990) in *Proceedings of the Symposium on Nickel Hydroxide Electrodes* (eds D.A. Corrigan and A.H. Zimmerman), The Electrochemical Society, Pennington, NJ, pp. 3–17.
9. McBreen, J. (1990) in *Modern Aspects of Electrochemistry*, Vol. 21 (eds R.E. White, J.O'M. Bockris, and B.E. Conway), Plenum Press, New York, pp. 29–64.
10. Zimmerman, A.H. (1994) in *Proceedings of the Symposium on Hydrogen and Metal Hydride Batteries* (eds P.D. Bennett and T. Sakai), The Electrochemical Society, Pennington, NJ, pp. 268–283.
11. Bode, H., Dehmelt, K., and Witte, J. (1966) *Electrochim. Acta*, **11**, 1079.
12. Zimmerman, A.H. and Phan, A.H. (1994) in *Proceedings of the Symposium on Hydrogen and Metal Hydride Batteries* (eds P.D. Bennett and T. Sakai), The Electrochemical Society, Pennington, NJ, pp. 341–352.
13. Oshitani, M., Yufu, H., Takashima, K., Tsuji, S., and Matsumaru, Y. (1989) *J. Electrochem. Soc.*, **136**, 1590.
14. Ferrando, W.A. (1985) *J. Electrochem. Soc.*, **132**, 2417.
15. Watada, H., Ohnishi, M., Harada, Y., and Oshitani, M. (1990) in *Proceedings of the 25th IECEC Meeting*, IEEE, Piscataway, NJ, pp. 299–304.
16. Oshitani, M. and Yufu, H. (1989) US Patent 4 844 999.
17. Fleischer, A. (1948) *Trans. Electrochem. Soc.*, **94**, 289.
18. Aia, M. (1966) *J. Electrochem. Soc.*, **113**, 1045.
19. McEwen, R.S. (1971) *J. Phys. Chem.*, **75**, 1782.
20. Fievet, F. and Figlarz, M. (1975) *J. Catal.*, **39**, 350.

21. Barnard, R., Randell, C.F., and Tye, F.L. (1981) in *Power Sources 8* (ed. J. Thompson), Academic Press, London, p. 401.
22. Greaves, C. and Thomas, M.A. (1986) *Acta Crystallogr., Sect. B*, **42**, 51.
23. Szytula, A., Murasik, A., and Balanda, M. (1971) *Phys. Stat. Sol.*, **43**, 125.
24. Kober, F.P. (1965) *J. Electrochem. Soc.*, **112**, 1064.
25. Kober, F.P. (1967) *J. Electrochem. Soc.*, **114**, 215.
26. Kober, F.P. (1967) in *Power Sources 1966* (ed. D.H. Collins), Pergamon, Oxford, p. 257.
27. Dennstedt, W. and Loser, W. (1971) *Electrochim. Acta*, **16**, 429.
28. Mani, J.P. and de Neufville, J.P. (1984) *Mater. Res. Bull.*, **19**, 377.
29. Le Bihan, S. and Figlarz, M. (1973) *Electrochim. Acta*, **18**, 123.
30. Mani, B. and de Neufville, J.P. (1988) *J. Electrochem. Soc.*, **135**, 801.
31. Jackovitz, J.F. (1982) in *Proceedings of the Symposium on the Nickel Electrode* (eds R.G. Gunther and S. Gross), The Electrochemical Society, Pennington, NJ, p. 48.
32. Cornilsen, B.C., Karjala, P.J., and Loyselle, P.L. (1988) *J. Power Sources*, **22**, 351.
33. Cornilsen, B.C., Shan, X.-Y., and Loyselle, P.L. (1990) *J. Power Sources*, **29**, 453.
34. Audemer, H., Delahaye, A., Farhi, R., Sac-Epée, N., and Tarascon, J.-M. (1997) *J. Electrochem. Soc.*, **144**, 2614.
35. Lotmar, W. and Fectknecht, W. (1936) *Kristallogr. Mineral. Petrogas Abt. Z*, **A93**, 368.
36. Portemer, F., Delahaye-Vidal, A., and Figlarz, M. (1992) *J. Electrochem. Soc.*, **139**, 671.
37. MacArthur, D.M. (1970) *J. Electrochem. Soc.*, **117**, 422.
38. Corrigan, D.A. (1987) *J. Electrochem. Soc.*, **134**, 377.
39. Bode, H. (1961) *Angew. Chem.*, **73**, 553.
40. Le Bihan, S., Guenot, J., and Fialarz, M. (1970) *C.R. Acad. Sci. Ser. C*, **270**, 2131.
41. Figlarz, M. and Le Bihan, S. (1971) *C.R. Acad. Sci. Ser. C*, **272**, 580.
42. Le Bihan, S. and Figlarz, M. (1972) *J. Cryst. Growth*, **13/14**, 458.
43. Delahaye-Vidal, A. and Figlarz, M. (1987) *J. Appl. Electrochem.*, **17**, 589.
44. Pandya, K.I., O'Grady, W.E., Corrigan, D.A., McBreen, J., and Hoffman, R.W. (1990) *J. Phys. Chem.*, **94**, 21.
45. Cordoba-Torresi, S.I., Gabrielli, C., Hugot-Le Goff, A., and Torresi, R. (1991) *J. Electrochem. Soc.*, **138**, 1548.
46. Genin, P., Delahaye-Vidal, A., Portemer, F., Tekkia-Elhsissen, K., and Figlarz, M. (1991) *Eur. J. Solid State Inorg. Chem.*, **28**, 505.
47. Delahaye-Vidal, A., Beaudoin, B., Sac-Epee, N., Tekkia-Elhsissen, K., Auderner, A., and Figlarz, M. (1996) *Solid State Ionics*, **84**, 239.
48. Faure, C., Delmas, C., and Fouassier, M. (1991) *J. Power Sources*, **35**, 279.
49. Bernard, M.C., Bernard, P., Keddam, M., Senyarich, S., and Takenouti, H. (1996) *Electrochim. Acta*, **41**, 91.
50. Briggs, G.W.D. and Wynne-Jones, W.F.K. (1956) *Trans. Faraday Soc.*, **52**, 1273.
51. Falk, S.U. (1960) *J. Electrochem. Soc.*, **107**, 661.
52. Glemser, O. and Einerhand, J. (1950) *Z. Anorg. Chem.*, **261**, 26.
53. Melandres, C.A., Paden, W., Tani, B., and Walczak, W. (1987) *J. Electrochem. Soc.*, **134**, 762.
54. McBreen, J., O'Grady, W.E., Pandya, K.I., Hoffman, R.W., and Sayers, D.E. (1987) *Langmuir*, **3**, 428.
55. Pandya, K.I., Hoffman, R.W., McBreen, J., and O'Grady, W.E. (1990) *J. Electrochem. Soc.*, **137**, 383.
56. McBreen, J., O'Grady, W.E., Tourillon, G., Dartyge, E., Fontaine, A., and Pandya, K.I. (1989) *J. Phys. Chem.*, **93**, 6308.
57. Aia, M. (1965) *J. Electrochem. Soc.*, **112**, 418.
58. Greaves, C., Thomas, M.A., and Turner, M. (1983) in *Power Sources 9* (ed. J. Thompson), Academic Press, London, p. 163.

59. Harivel, J.P., Morignat, B., Labat, J., and Laurent, J.F. (1967) in *Power Sources 1966* (ed. D.H. Collins), Pergamon Oxford, p. 239.
60. Lim, H.S., Verzwyvelt, S.A., and Clement, S.K. (1988) *Proceedings of the 23rd IECEC Meeting*, Vol. 2, American Society of Mechanical Engineers, New York, pp. 457–463.
61. Chen, R.-R., Mo, Y., and Scherson, D.A. (1994) *Langmuir*, **10**, 3933.
62. Häring, P. and Kötz, R. (1995) *J. Electroanal. Chem.*, **385**, 273.
63. Kowal, A., Niewiara, R., Peroczyk, B., and Haber, J. (1996) *Langmuir*, **12**, 2332.
64. Mo, Y., Hwang, E., and Scherson, D.A. (1996) *J. Electrochem. Soc.*, **143**, 37.
65. Kim, M.S., Hwang, T.S., and Kim, K.B. (1997) *J. Electrochem. Soc.*, **144**, 1537.
66. Kostecki, R. and McLarnon, F. (1997) *J. Electrochem. Soc.*, **144**, 485.
67. Oblonsky, L.J. and Devine, T.M. (1995) *J. Electrochem. Soc.*, **142**, 3677.
68. Allmann, R. (1970) *Chimia*, **24**, 99.
69. Axmann, P., Erdbrügger, C.F., Buss, D.H., and Glemser, O. (1996) *Angew. Chem. Int. Ed. Engl.*, **35**, 1115.
70. Buss, D.H., Diembeck, W., and Glemser, O. (1985) *J. Chem. Soc., Chem. Commun.*, 81.
71. Glemser, O., Buss, D.H., and Bauer, J. (1988) US Patent 4 735 629.
72. Axmann, P. and Glemser, O. (1997) *J. Alloys Compd.*, **246**, 232.
73. Demourges-Guerlou, L., Fournès, L., and Delmas, C. (1995) *J. Solid State Chem.*, **114**, 6.
74. Delmas, C., Faure, C., Gautier, L., Guerlou-Demourgues, L., and Rougier, A. (1996) *Phil. Trans. R. Soc. Lond.*, **354A**, 1545.
75. Ohzuku, T., Ueda, A., and Nagayama, M. (1993) *J. Electrochem. Soc.*, **140**, 1862.
76. Delmas, C., Braconnier, J.J., Borthomeiu, Y., and Hagenmuller, P. (1982) *Mater. Res. Bull.*, **17**, 117.
77. Demourges-Guerlou, L. and Delmas, C. (1994) *J. Electrochem. Soc.*, **141**, 713.
78. Delmas, C., Faure, C., and Borthomeiu, Y. (1992) *Mater. Sci. Eng.*, **B13**, 89.
79. Cuerlou-Demourgues, L. and Delmas, C. (1996) *J. Electrochem. Soc.*, **143**, 561.
80. Silverman, D.C. (1981) *Corrosion*, **37**, 546.
81. Deltombe, E., de Zoubov, N., and Pourbaix, M. (1974) in *Atlas of Electrochemical Equilibria in Aqueous Solutions* (ed. M. Pourbaix), NACE, Houston, TX, p. 330.
82. Bourgault, P.L. and Conway, B.E. (1960) *Can. J. Chem.*, **38**, 1557.
83. Barnard, R., Randell, C.F., and Tye, F.L. (1980) *J. Appl. Electrochem.*, **10**, 109.
84. Barnard, R., Randell, C.F., and Tye, F.L. (1981) *J. Electroanal. Chem.*, **119**, 17.
85. Barnard, R. and Randell, C.F. (1983) *J. Appl. Electrochem.*, **13**, 97.
86. Madou, M.J. and McKubre, M.C.H. (1983) *J. Electrochem. Soc.*, **130**, 1056.
87. Carpenter, M.K. and Corrigan, D.A. (1988) in Papers presented at the Atlanta Meeting of the Electrochemical Society, May 15–20, Abstract No. 490, p. 700.
88. Kuchinskii, E.M. and Erschler, B.V. (1940) *J. Phys. Chem. (USSR)*, **14**, 985.
89. Kuchinskii, E.M. and Erschler, B.V. (1946) *Zh. Fiz. Khim.*, **20**, 539.
90. Briggs, G.W.D. and Fleischman, M. (1971) *Trans. Faraday Soc.*, **67**, 2397.
91. Klapste, B., Mrha, J., Micka, K., Jindra, J., and Maracek, V. (1979) *J. Power Sources*, **4**, 349.
92. Barnard, R., Crickmore, G.T., Lee, J.A., and Tye, F.L. (1980) *J. Appl. Electrochem.*, **10**, 61.
93. Klapste, B., Micka, K., Mrha, J., and Vondrak, J. (1982) *J. Power Sources*, **8**, 351.
94. Zimmerman, A.H. (1994) *Proceedings of the 29th IECEC Meeting*, American Institute of Aeronautics and Astronautics, pp. 63–68.
95. Desilvestro, J., Corrigan, D.A., and Weaver, M.G. (1988) *J. Electrochem. Soc.*, **135**, 885.
96. Tuomi, D. (1965) *J. Electrochem. Soc.*, **112**, 1.
97. Corrigan, D.A. and Knight, S.L. (1989) *J. Electrochem. Soc.*, **136**, 613.

98. Mansour, A.N., Melandres, C.A., Pankuch, M., and Brizzolara, R.A. (1994) *J. Electrochem. Soc.*, **141**, L69.
99. O'Grady, W.E., Pandya, K.I., Swider, K.E., and Corrigan, D.A. (1996) *J. Electrochem. Soc.*, **143**, 1613.
100. Font, S. and Goulard, J. (1975) in *Power Sources 5* (ed. D.H. Collins), Academic Press, London, p. 331.
101. Tsenter, B.I. and Sluzhevskii, A.I. (1980) *Elektrokhimiya*, **54**, 2545.
102. Kim, Y.J., Visintin, A., Srinivasan, S., and Appleby, A.J. (1992) *J. Electrochem. Soc.*, **139**, 351.

6
Lead Oxides
Dietrich Berndt

6.1
Introduction

Lead oxides play an important role in lead–acid batteries.

- Lead dioxide (PbO_2) forms the charged state of the active material in the positive electrode.
- Lead oxide (PbO) (also called *litharge*) is formed when the lead surface is exposed to oxygen. Furthermore, it is important as a primary product in the manufacturing process of the active material for the positive and negative electrodes. It is not stable in acidic solution but it is formed as an intermediate layer between lead and lead dioxide at the surface of the corroding grid in the positive electrode. It is also observed underneath lead sulfate layers at the surface of the positive active material.
- Minium (Pb_3O_4) represents a more highly oxidized form of lead oxide that enhances the electrochemical oxidation of lead oxide to lead dioxide.

The history of the lead–acid battery goes back to 1854 when Sinsteden published performance data on this battery system for the first time (cf. Ref. [1]). The practical importance of the lead–acid battery system was detected, in 1859, independently of Sinsteden's work, by Planté, who produced a rechargeable battery by alternately charging and discharging lead sheets immersed in sulfuric acid [2]. Lead dioxide (PbO_2) as the 'active material' is thereby directly generated from lead that is used as the conducting substrate. Planté plates are still in use, and are in principle produced by this method [3]. Separate production of the active material was introduced by Fauré in 1881. It is the basis of the pasted-plate design, mainly used today.

Since those early days, the lead–acid battery has always been the most important rechargeable electrochemical storage system, maintaining its prime position unchallenged now for more than a century. This seems surprising, because the fundamental data concerning the amount of energy that can be stored are rather modest on account of the high weight of the reacting substances. One main reason is the comparatively moderate price of lead–acid batteries. But besides that, there are many features that favor this system, which to a large extent are due to the substances that form the reacting components in the electrodes, for example:

- The same chemical element forms the active material in both electrodes: lead as metal (Pb) in the negative and as lead dioxide (PbO_2) in the positive electrode.
- The reactants are solids of low solubility, and the reactions are highly reversible.
- The reactants compounds lead (Pb), lead sulfate ($PbSO_4$), and lead dioxide (PbO_2) are well-defined chemical compounds, and there are no intermediate states of oxidation. As a consequence, any voltage above the open circuit voltage results in complete charge, and equalizing charges are not required when the battery is used under continual charging in standby applications.
- The electronic conductivity of lead dioxide is comparatively high; thus there is no need for conducting additives.
- Due to the high potential of the $PbO_2/PbSO_4$ electrode, the cell voltage of 2 V is fairly high, and a comparatively small number of cells is sufficient for a certain battery voltage.
- The high potential of the positive electrode, on the other hand, does not allow the use of conducting metals like copper within the positive electrode. Lead can be used instead due to its passive properties caused by a (PbO_2) layer that largely protects the underlying material but conducts the electronic current and so allows electrochemical reactions at its surface.

This list could be extended (cf. Ref. [4]). It simultaneously indicates the large number of parameters that influence the properties of a battery.

6.2
Lead/Oxygen Compounds

Lead forms two types of chemical compounds: lead (II) and lead (IV) compounds based on Pb^{2+} and Pb^{4+} ions, where those based on Pb^{2+} ions are the more stable. The metal is oxidized even at room temperature to lead oxide (PbO) and also by water that contains oxygen and forms lead hydroxide ($Pb(OH)_2$). In the lead–acid battery, the (less stable) lead (IV) oxide (lead dioxide, PbO_2), is of greatest importance. Besides these two, a number of oxides that are mostly mixtures are observed in the battery. A brief survey will now be given of those compounds that are of interest for lead–acid batteries.

6.2.1
Lead Oxide (PbO)

Lead oxide is formed by oxidation of a lead surface according to Equation 6.1.

$$2Pb + O_2 \rightarrow 2PbO \tag{6.1}$$

One technical process involves blowing air above the surface of molten lead (cf. the Barton process in Section 6.4.2.1), but at room temperature reaction 6.1 also soon covers any piece of lead exposed to air with a dull gray layer of lead oxide (cf. the milling process in Section 6.4.2.1).

Two modifications are known:

1) red PbO, the tetragonal modification (litharge);
2) yellow PbO, the rhombic modification (massicot).

The crystallographic structure is described in Ref. [5], p. 11.

Red PbO is stable at low temperatures, whereas yellow PbO is the high-temperature modification. The temperature of conversion is 488 °C:

$$\text{Red PbO} \rightarrow 488\,°C \rightarrow \text{Yellow PbO} \tag{6.2}$$

At low temperatures the conversion is a slow reaction, and yellow PbO also exists at room temperatures as a metastable modification.

6.2.2
Minium (Pb_3O_4)

Minium, also called *red lead*, is formed when lead oxide is exposed to air at about 500 °C according to Equation 6.3 Roasting ovens are used for the technical process.

$$3PbO + 1/2 O_2 \rightarrow Pb_3O_4 \tag{6.3}$$

Minium contains Pb^{2+} and Pb^{4+} ions, but it is not simply a mixture of PbO and PbO_2. According to its composition and structure it can be regarded as a lead(II) salt of the lead(IV) acid ($Pb_3O_4 = 2PbO \cdot PbO_2$) (cf. [5], Figure 2.4, p. 17). In the technical process, the stoichiometric composition Pb_3O_4 is usually not attained and the (fictive) percentage of PbO_2 is often used to specify the grade of oxidation. Stoichiometric Pb_3O_4 contains 34% PbO_2 (239.2 g of PbO_2 per 685 g Pb_3O_4, cf. Table 6.2), and the technical products contain PbO_2 in the range between 25 and 30%.

When minium comes in to contact with sulfuric acid it is converted into lead sulfate and lead dioxide according to Equation 6.4.

$$Pb_3O_4 + 2H_2SO_4 \rightarrow 2PbSO_4 + PbO_2 + 2H_2O \tag{6.4}$$

6.2.3
Lead Dioxide (PbO_2)

Lead dioxide exists in two modifications:

1) a rhombic modifications called α-PbO_2;
2) a tetragonal modification called β-PbO_2.

Besides the crystalline material, a certain portion of amorphous lead dioxide is always observed. In the working electrode such amorphous material is apparently hydrated and forms a gel structure at the phase boundary between the solid material and the electrolyte (cf. Ref. [6]).

In battery electrodes, the stoichiometric composition is usually not accomplished, and oxidation ends at a composition of about $PbO_{1.98}$ [7].

α-PbO_2, is formed in an alkaline environment, whereas β-PbO_2 is produced in an acidic medium. Both modifications can be prepared by chemical and electrochemical methods (for detailed descriptions of preparations methods and structure see Ref. [5], p. 19, and Ref. [8]). In the positive electrodes of lead–acid batteries, a certain proportion of α-PbO_2 is formed during the electrochemical conversion of the electrode (cf. Sections 6.4.2.2 and 6.4.2.3). When the battery has been discharged and is charged again, only β-PbO_2 is formed on account of the acidic environment. For this reason, the content of α-PbO_2 decreases with the number of discharge–charge cycles, especially for a high initial α-PbO_2 content (cf. Ref. [5], p. 267, especially Figures 3.27 and 3.28). Plates containing much α-PbO_2 show a reduced initial capacity, which increases gradually on account of the conversion of α-PbO_2 to β-PbO_2. There was some indication that plates with high α-PbO_2, content would outperform other plates in standby applications. Therefore, the α/β-PbO_2 ratio has occasionally been specified for stationary lead–acid batteries, but these observations have not been confirmed in general [6]. Other experiments indicate that the structure of lead dioxide agglomerates, which can be influenced by the formation process, is important for cycle stability [9] (cf. also the description of 'paste mixing' in Section 6.4.2.1).

Under normal conditions β-PbO_2, is the more stable modification. Under a high pressure (>8500 bar) the β modification can be transformed to α-PbO_2.

6.2.4
Nonstoichiometric PbO_x Phases

PbO_x phases with x between 1.42 and 1.58 can be formed by oxidation of PbO (cf. Ref. [5], p. 18). It is assumed that such compounds are formed underneath the protecting PbO_2 layer at a corroding lead surface (cf. Figure 6.8).

6.2.5
Basic Sulfates

Basic sulfates are intermediate compounds that contain lead oxide and lead sulfate and to some extent also water (Table 6.1). They are stable only in alkaline environment.

Basic sulfates are important intermediates during the manufacturing process, since they determine the structure of the active material in the positive electrode, which again is decisive with respect to the performance data and service life of the battery (cf. Ref. [9]).

6.2.6
Physical and Chemical Properties

Some physical and chemical properties of the lead oxides are compiled in Table 6.2.

Table 6.1 Basic sulfates that are formed as intermediate compounds when lead oxide is mixed with sulfuric acid.

Compound	Formula	pH range	Remark
Monobasic sulfate	$PbO \cdot PbSO_4$	6.28–7.31	–
Dibasic sulfate	$2PbO \cdot PbSO_4$	–	–
Tribasic sulfate	$3PbO \cdot H_2O \cdot PbSO_4$	7.31–8.99	–
Tetrabasic sulfate	$4PbO \cdot PbSO_4$	–	$T > 50\,°C$

Table 6.2 Molecular weight, density (cf. Ref. [5], Table VII), and electrical resistance of the chemical compounds used in lead–acid batteries.

Substance	Molecular weight ($g\,mol^{-1}$)	Specific weight ($g\,cm^{-3}$)	Electrical resistance ($\Omega\,m^{-1}$)
Pb	207.2	11.34	2×10^{-7}
PbO (red)	232.2	9.35	$10^{13}–10^{14}$
PbO (yellow)	223.2	9.64	10^{12}
Pb_3O_4	685.6	9.1	9.6×10^9
α-PbO_2	239.2	9.1–9.4	$10^5–10^{-6}$
β-PbO_2	239.2	9.1–9.4	$10^{-5}–10^{-6}$
$PbSO_4$	303.25	6.1–6.4	–
H_2O	18.02	0.997	$\approx 10^{4a}$

[a] Distilled water.
The specific resistance of the oxides depends on pressure (cf. Ref. [5], Table 2.3).

6.3
The Thermodynamic Situation

The exchange of energy connected with a chemical or electrochemical reaction is described by thermodynamic laws and data, as shown in Chapter 1 of this book. Since these laws apply only to the state of equilibrium, all reactions are balanced. In an electrochemical cell, these data can only be measured when no current flows through the cell or its electrodes. On account of this balance, the thermodynamic parameters do not depend on the reaction path; they only depend on different energy levels between the final and initial components (the 'products' and the 'reactants' of the chemical or electrochemical reaction). For the same reason, the laws of thermodynamics describe the possible upper limit of energy that can be delivered by a reaction, or the minimum of energy that is required for its reversal. Thermodynamic data only indicate whether a reaction is possible at a given electrode potential or not; the actual rate of a reaction is largely determined by the laws of electrode kinetics.

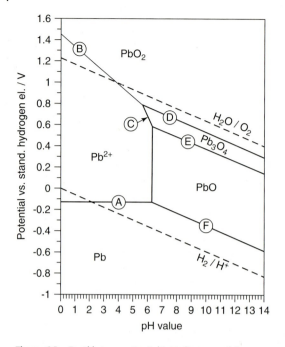

Figure 6.1 Equilibrium potential/pH diagram of the Pb/H$_2$O system at 25 °C, according to Pourbaix [10], but simplified for $a = 1$ mol L^{-1}. The pH value is used to express the acidity of the solution. Its definition is pH $= -\log(a_{H+})$; pH stands for the negative logarithm of the activity of the H$^+$ ions.

A survey of the thermodynamic situation is provided by so-called Pourbaix diagrams [10], which show equilibrium potentials versus the pH value. Figure 6.1 shows such a diagram for lead and its oxides in a very simplified form that considers only the standard concentrations of the dissolved components. The complete diagram contains a great number of parallel lines that express the various concentrations.

6.3.1
Water Decomposition

The electrolyte in lead–acid batteries is dilute sulfuric acid that contains the component 'water.' Its stability is an important factor since it can be decomposed into hydrogen and oxygen, and the two broken lines in Figure 6.1 represent the borderlines of this stability. They show the equilibrium potentials of hydrogen and oxygen evolution and their dependence on the pH value.

The H$_2$/H$^+$ line in Figure 6.1 represents the equilibrium potential E^0 of the hydrogen evolution according to Equation 6.5

$$H_2 \leftrightarrow 2H^+ + 2e^- \tag{6.5}$$

and its dependence on the pH is described by the Nernst equation (Equation 6.6),

$$E^0 = E^{0,S}_{H^+/H_2} + \frac{RT}{2F} \ln \frac{a^2_{H^+}}{a_{H_2}} \tag{6.6}$$

with E^0 = equilibrium potential, $E^{0,S}$ = standard value of the equilibrium potential (when a_{H^+} and a_{H_2} are 1 mol L^{-1}, and a_{H^+}, a_{H_2} = activities (i.e., effective concentrations) of H$^+$ and H$_2$, respectively.

Using the definition of the pH value, pH = $-\log(a_{H^+})$, and the relation between the gas pressure (p_{H_2}) and the concentration of dissolved gas (a_{H_2}) (Henry's law), Equation 6.6 can be written:

$$E^0 = E^{0,S}_{H^+/pH_2} - 2.303 \frac{RT}{F} \left\{ pH + \frac{1}{2}\log(p_{H_2}) \right\} \tag{6.7}$$

with $-2.303\frac{RT}{F} = -0.0592 V$.

This is the curve H$_2$/H$^+$ in Figure 6.1 for p_{H_2} = 1 atm. The standard value of this equation (proton activity a_{H^+} = 1 mol^{-1}; p_{H_2} = 1 atm)

$$E^{0,S}_{H^+/pH_2} = 0 \tag{6.8}$$

Equation 6.8 represents by definition the zero point of the electrochemical potential scale (standard hydrogen electrode, often denoted SHE).

The corresponding relation for oxygen evolution:

$$H_2O \rightarrow \frac{1}{2}O_2 + 2H^+ + 2e^- \tag{6.9}$$

has the equilibrium potential

$$E^0_{O_2} = E^{0,S}_{O_2} + \frac{RT}{2F} \ln \frac{a^{1/2}_{O_2} a^2_{H^+}}{a_{H_2O}} \tag{6.10}$$

with the standard value

$$E^{0,S}_{O_2} = 1.229\ V \tag{6.11}$$

In Figure 6.1, this is represented by the H$_2$O/O$_2$, curve.

$E^0_{O_2}$ is the decomposition voltage of water. Above this value, water is not stable but decomposes with formation of oxygen (O$_2$).

In Figure 6.1, H$_2$O/O$_2$ and H$_2$/H$^+$ curves are straight lines which have the same slope of -0.059 V per pH unit.

6.3.2
Oxidation of Lead

Curve A in Figure 6.1 corresponds to the oxidation of lead to its divalent ion, described by the reaction

$$Pb \leftrightarrow Pb^{2+} + 2e^- \tag{6.12}$$

with the dependence on activities

$$E^0_{Pb/Pb^{2+}} = E^{0,S}_{Pb/Pb^{2+}} + \frac{RT}{2F} \ln(a_{Pb^{2+}}) \tag{6.13}$$

and the standard value

$$E^0_{Pb/PbSO_4} = -0.126 \text{ V} \tag{6.14}$$

$E^0_{Pb/Pb^{2+}}$ does not depend on the pH value. In Figure 6.1, the activity of SO_4^{2-} equals 1 mol L^{-1}.

At small pH values, the curve A is below the H_2/H^+ curve, which means that lead is not stable in an aqueous solution under these conditions, but is converted into Pb^{2+} ions, and simultaneously water is decomposed with formation of hydrogen H_2.

Curve B describes the corresponding relation for the oxidation of divalent lead ions $Pb^{2+} \rightarrow Pb^{4+}$, which implies the formation of lead dioxide since Pb^{4+} does not exist alone:

$$Pb^{2+} + 2H_2O \rightarrow PbO_2 + 4H^+ + 4e^- \tag{6.15}$$

The dependence on H^+ concentration (activity) is given by

$$E^0_{Pb^{2+}/PbO_2} = E^{0,S}_{Pb^{2+}/PbO_2} + \frac{RT}{2F} \ln \frac{a_H^4}{a_{H_2O}^2} \tag{6.16}$$

and leads to a slope of $2 \times 0.0592 = 0.1184$ V per pH unit.

The standard value is:

$$E^0_{Pb^{2+}/PbO_2} = 1.459 \text{ V} \tag{6.17}$$

At small pH values, the curve B is above the H_2O/O_2 curve, which means that lead dioxide (Pb^{4+} ions) is not stable in such an acidic aqueous solution, but is reduced to Pb^{2+} ions, and simultaneously water is decomposed with formation of oxygen (O_2).

PbO and Pb_3O_4 are oxides of lead that are important primary products. Under certain circumstances, PbO is also observed in lead–acid batteries (cf. Section 6.3.3). Figure 6.1 shows that these oxides are only stable in a neutral or alkaline environment. Their equilibrium potentials are represented by curves C–F and their standard values are compiled in Table 6.3.

Table 6.3 Equilibrium potentials shown by the curves C–F in Figure 6.1, their standard values and dependence on pH (cf. also Ref. [5], p.36).

Curve	Equilibrium	Standard value ($E^{0,S}$) (V)	pH dependence (V/pH unit)
C	Pb^{2+}/Pb_3O_4	2.094	−0.2368
D	Pb_3O_4/PbO_2	1.124	−0.0592
E	PbO/Pb_3O_4	0.972	−0.059
F	Pb/PbO	0.248	−0.059

6.3.3
The Thermodynamic Situation in Lead–Acid Batteries

In the lead–acid battery, sulfuric acid has to be considered as an additional component of the charge–discharge reactions. Its equilibrium constant influences the solubility of Pb^{2+} and so the potential of the positive and negative electrodes. Furthermore, basic sulfates exist as intermediate products in the pH range where Figure 6.1 shows only PbO (cf. corresponding Pourbaix diagrams in Ref. [5], p. 37, or in Ref. [11]; the latter is cited in Ref. [8]). Table 6.2 shows the various compounds.

The charge–discharge reaction of the negative electrode corresponds to curve A in Figure 6.1, but the Pb^{2+} ion activity is now determined by the solubility of lead sulfate ($PbSO_4$). Thus Equation 6.12 has to be modified into

$$Pb + H_2SO_4 \rightarrow PbSO_4 + 2H^+ + 2e^- \quad (6.18)$$

The dependence of the equilibrium potential on the activities of hydrogen and sulfuric acid is given by the corresponding Nernst equation:

$$E^0_{Pb/PbSO_4} = E^{0,S}_{Pb/PbSO_4} + \frac{RT}{2F} \ln \frac{a^2_{H^+}}{a_{H^+} a_{HSO_4^-}} \quad (6.19)$$

with the standard value

$$E^{0,S}_{Pb/PbSO_4} = -0.295 \text{ V} \quad (6.20)$$

The charge–discharge equation of the positive electrode corresponds to curve B in Figure 6.1 and the corresponding Equation 6.15. But here also the Pb^{2+} activity is now determined by the solubility of lead sulfate, and Equation 6.15 has to be modified into:

$$PbSO_4 + 2H_2O \rightarrow PbO_2 + H_2SO_4 + 2H^+ + 2e^- \quad (6.21)$$

with the corresponding Nernst equation

$$E^0_{PbO_2/PbSO_4} = E^{0,S}_{PbO_2/PbSO_4} + \frac{RT}{2F} \ln \frac{a^2_{H^+} a_{HSO_4^-}}{a^2_{H_2O}} \quad (6.22)$$

and the standard value

$$E^{0,S}_{PbO_2/PbSO_4} = 1.636 \text{ V} \quad (6.23)$$

Note: Here the calculations of the standard potentials are referred to the dissociation of H_2SO_4, into H^+ and HSO_4 which fairly, closely corresponds to the actual situation. Sometimes such calculations are based on the assumption of completely dissociated sulfuric acid,

$$H_2SO_4 \rightarrow 2H^+ + SO_4^{2-} \quad (6.24)$$

This causes different standard values, but the results are identical (cf., e.g., Ref. [12]).

Figure 6.2 illustrates the resulting situation. Due to the strong acidic solution in the battery, it corresponds to Figure 6.1 for small pH values, but here the electrode potential is drawn on the vertical axis. The values are referred to the above-mentioned SHE. To enlarge the scale, the range between 0 and 1.2 V is omitted.

The columns just above the potential scale represent the equilibrium potential of the charge–discharge reaction, given by Equations 6.19 and 6.22 for the negative and positive electrode, respectively. The width of each column expresses the dependence on acid concentration. The column that represents the $PbSO_4/PbO_2$ equilibrium potential corresponds to line B in Figure 6.1. Its standard value is shifted to the more positive potential $E^0 = 1.636\,V$, since the Pb^{2+} concentration amounts to only about 10^{-6} mol L^{-1} due to the low solubility of lead sulfate ($PbSO_4$). Above this equilibrium potential, lead dioxide (PbO_2) is formed.

Water decomposition is not directly influenced by sulfate formation and occurs as described in Figure 6.1. Thus, in the strong acid hydrogen and oxygen evolution occurs below 0 V and above 1.23 V respectively, as shown in Figure 6.2. It means that for thermodynamic reasons, oxygen evolution and hydrogen oxidation (the reversal of hydrogen evolution) are possible at an electrode potential far below that of the positive electrode. The same applies to the corrosion of lead, also indicated by the corresponding arrow in Figure 6.2. Thus couples of reactions are to be expected that reduce lead dioxide and so cause self-discharge of the active material:

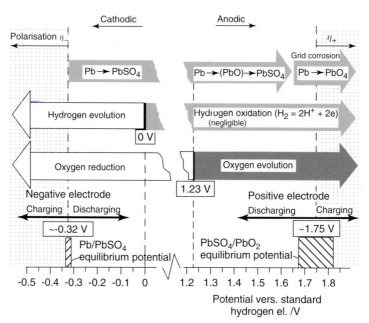

Figure 6.2 Reactions that occur in lead–acid batteries versus electrode potential (thermodynamic situation). Their equilibrium potentials are inserted as boxed numbers. Equilibrium potentials of the charge–discharge reactions ($Pb/PbSO_4$ and $PbSO_4/PbO_2$) are represented by hatched columns, to indicate their dependence on acid concentration. The inserted equilibrium potentials (−0.32 and +1.75 V) of the charge–discharge reactions correspond to an acid density of 1.23 g cm^{-3}.

1) **Oxygen evolution** according to Equation 6.5 is possible above 1.23 V and forms the couple with the discharge reaction (the reversal of Equation 6.21):
   Discharge of the positive electrode

   $$PbO_2 + H_2SO_4 + 2H^+ + 2e^- \rightarrow PbSO_4 + 2H_2O$$

   and oxygen evolution

   $$H_2O \rightarrow 1/2 O_2 + 2H^+ + 2e^-$$

   with the result

   $$PbO_2 + H_2SO_4 \rightarrow 1/2 O_2 + PbSO_4 + H_2O \tag{6.25}$$

2) **Hydrogen oxidation** according to Equation 6.5 is possible above 0 V. If hydrogen evolution occurs at the negative electrode and the H_2 evolved reaches the positive electrode, from the thermodynamic situation the reaction, that is, to be expected is:
   Discharge of the positive electrode

   $$PbO_2 + H_2SO_4 + 2H^+ + 2e^- \rightarrow PbSO_4 + 2H_2O$$

   and hydrogen oxidation

   $$H_2 \rightarrow 2H^+ + 2e^-$$

   with the result

   $$PbO_2 + H_2SO_4 + 2H^+ \rightarrow PbSO_4 + 2H_2O \tag{6.26}$$

   In practice, however, this reaction can be neglected since its rate is extremely low at the PbO_2 surface.

3) **Corrosion of lead** starts at the equilibrium potential of the negative electrode. It induces self-discharge of the positive electrode on account of the following couple of reactions:
   Discharge of the positive electrode

   $$PbO_2 + H_2SO_4 + 2H^+ + 2e^- \rightarrow PbSO_4 + 2H_2O$$

   and grid corrosion

   $$Pb + H_2O \rightarrow PbO + 2H^+ + 2e^-$$

   with the result

   $$PbO_2 + H_2SO_4 + 2H^+ + 2e^- \rightarrow PbSO_4 + 2H_2O \tag{6.27}$$

Above the potential given by line F in Figure 6.1, the formation of PbO is possible, but this is stable only in an alkaline environment. As soon as it comes to contact with sulfuric acid, it is converted into lead sulfate. For this reason PbO is in parentheses in Figure 6.2. The final result of these reactions is:

$$PbO_2 + Pb + 2H_2SO_4 \rightarrow 2PbSO_4 + 2H_2O \tag{6.28}$$

Above the equilibrium potential of the positive electrode, lead is oxidized to PbO_2.

Fortunately, the kinetic parameters reduce the rates of these reactions so far that the gradual self-discharge of the PbO_2 is such a slow reaction that it usually does not affect the performance of the battery.

- Self-discharge by oxygen generation (Equation 6.25) occurs equivalent to a current in the range of 0.5 mA/100 Ah, which means ~0.4% of nominal capacity per month (starting at higher values).
- Corrosion of the positive grid (Equation 6.28) occurs equivalent to about 1 mA/100 Ah at open-circuit voltage and intact passivation layer. It depends on electrode potential, and is at minimum about 40–80 mV above the $PbSO_4/PbO_2$ equilibrium potential. The corrosion rate depends furthermore to some extent on alloy composition and is increased with high-antimony alloys.
- As already mentioned, hydrogen oxidation can be neglected.

Although the rate of these reactions is slow, according to its thermodynamic situation the lead dioxide electrode is not stable. Since a similar situation applies to the negative electrode, the lead–acid battery system as a whole is unstable and a certain rate of water decomposition cannot be avoided.

6.3.4
Thermodynamic Data

The thermodynamic data for the substances employed in lead–acid batteries are compiled in Table 6.4.

Table 6.4 Standard values ($T = 25\ °C$) of the thermodynamic data for the chemical compounds in the active material of lead–acid batteries (cf. Ref. [5], p. 366). In older tables, the energy often is given in calories: $1\ cal = 4.187\ J$.

Substance	Enthalpy of formation, $H^{0,S}$ (kJ mol^{-1})	Free enthalpy of formation, $G^{0,S}$ (kJ mol^{-1})	Entropy, $S^{0,S}$ (J K^{-1} mol^{-1})
Pb	0	0	64.8
Pb^{2+}	1.67	−24.39	10.5
PbO (red)	−219.0	−199.0	66.5
PbO (yellow)	−217.3	−187.9	68.7
Pb_3O_4	−718.4	−601.2	211
α-PbO_2	−265.8	−217.3	92.5
β-PbO_2	−276.7	−219.3	76.4
$PbSO_4$	−919.9	−813.2	149
H^+	0	0	0
H_2O	−219.0	−237.2	189

6.4
PbO₂ as Active Material in Lead–Acid Batteries

The active material comprises the substances that constitute the charge–discharge reaction. In the positive electrode of lead–acid batteries, the active material in the charged state is lead dioxide (PbO_2), which is converted into lead sulfate ($PbSO_4$) when the electrode is discharged. The active material is the most essential part of a battery, and battery technology has to aim at optimum constitution and performance for the expected application. This concerns not only the chemical composition but also the physical structure and its stability. Specialized methods have been developed to fulfill these requirements, and the primary products as well as the manufacturing process are usually specified by the individual battery manufacturer.

It is characteristic for battery manufacture that lead dioxide (PbO_2) as the charged state of the active material is always generated by electrochemical oxidation. Thus, electron-conducting bridges are established between the fine particles, and a matrix is formed of comparatively low electronic resistance. Three general types of positive electrodes are mainly used today: Planté, pasted, and tubular plates, which vary not only in their design but also in the way they are manufactured.

The charge–discharge reactions occur at the phase boundary between the active material and the electrolyte. To make sure that a sufficient rate of reaction is achieved, the surface of the reacting materials has to be large. Otherwise, the kinetic parameters would reduce the reaction rate too much. Table 6.5 shows the surface areas of the active materials in the positive and the negative electrode.

Figure 6.3 shows the typical microscopic appearance in the charged and discharged states. Although certain features are characteristic, microscopic pictures of this kind vary considerably, because of the different parameters that influence the formation of the crystals when a substance is precipitated. Furthermore, the charge–discharge conditions and the age of the battery influence the morphology of the active material (cf., e.g., Refs [8, 13]). The 'lump' structure is typical of the charged active material of the positive electrode, at least for a fairly new electrode. These 'lumps' are porous agglomerates. About 50% of their volume is occupied by lead dioxide; the other 50% is pores. A large share of micropores produces the high surface area shown in Table 6.5.

In the lead–acid battery, the reactions at both electrodes include the dissolved state, which means that the reacting species are dissolved in the course of the

Table 6.5 Surface areas of the active materials in lead–acid batteries.

Substance	BET surface ($m^2 \, g^{-1}$)
Lead dioxide, PbO_2	4–6
Lead, Pb	0.3–0.6

 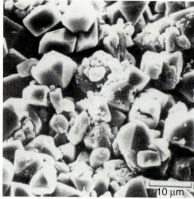

Figure 6.3 Active material of a lead dioxide electrode, charged (a) and discharged (b). The charged state shows the typical 'lump' structure; in the discharged state lead sulfate crystals predominate [14].

reaction. The new chemical compounds formed during the reaction are precipitated again as solid matter. This explains the completely different appearance of the material in the charged and discharged states.

The discharge reaction is the reverse of Equation 6.21:

$$PbO_2 + H_2SO_4 + 2H^+ + 2e^- \rightarrow PbSO_4 + 2H_2O \qquad (6.29)$$

In this reaction at the positive electrode bivalent lead ions are formed by tetravalent lead ions acquiring two electrons according to $Pb^{4+} + 2e^- \rightarrow Pb^{2+}$. The Pb^{2+} ions dissolve but immediately form lead sulfate ($PbSO_4$) on account of its low solubility. Equation 6.28 shows that water is formed in addition, since oxygen ions are released from the lead dioxide (PbO_2) and combine with the protons (H^+) to form H_2O molecules. When the battery is being charged, these reactions occur in the opposite direction. Lead dioxide (PbO_2) is formed from lead sulfate ($PbSO_4$) and water, while electrons are released.

The charge–discharge process can be repeated quite often, since the decisive parameters, solubility and dissolution rate of the various compounds, are well matched in the lead–acid battery system. The chemical conversions occur close to each other, and most of the material transport takes place in the micrometer range. Nevertheless, a gradual disintegration of the active material is observed.

The required amount of lead dioxide can be calculated with the aid of Equation 6.29, as shown in Table 6.6. The amount of electricity required per multiple of this reaction is $2F = 192\,970$ As $= 53.61$ Ah.

6.4.1
Planté Plates

In positive Planté plates the lead dioxide is generated by direct oxidation of lead that forms the conducting substrate. Figure 6.4 shows a cross-section of such a

Table 6.6 Molar weight of lead dioxide, required quantity, and utilization.

Molar weight (g)	Required quantity (g Ah^{-1})	Utilization^a		
		Theory		Practice
		(Ah kg^{-1})	(Ah lb^{-1})	(Ah kg^{-1})
239.2	4.462	224.1	101.7	60–130

^{a}Utilization in theory derived from Equation 6.29; utilization in practice mainly depends on electrode thickness (cf. Ref. [12]).

Figure 6.4 Cross-section of a Planté plate. The fine lamellas of the casting enlarge its surface area by a factor of 8–12. On the top, the current collector, the plate lug, is to be seen.

plate. The plate is cast from pure lead. Its surface is enlarged by lamellas that can be seen in Figure 6.4. Planté formed the dioxide layer by a large number of charge–discharge cycles. Nowadays, the electrochemical oxidation is enhanced by the addition of perchloric acid (HClO$_4$) during the 'formation process.' It increases the solubility of lead and increases the corrosion rate so far that a single charging cycle is sufficient. The precisely controlled process results in a layer of active material, that is, about 220 μm thick.

An advantage of this plate design is the short distance that has to be passed by electrons to reach the current conducting core; the disadvantage is the weight of this thick core made of pure lead.

Negative electrodes in Planté batteries are of the pasted type.

6.4.2
Pasted Plates

The plate support in lead–acid batteries is usually called the '*grid*.' In most batteries the grid has to provide both mechanical support for the active material and electronic conductivity for the collected current.

Figure 6.5 shows the design of a grid and its horizontal and vertical cross-sections. This figure indicates the origin of term '*grid*' for this kind of plate support.

The network of lead wires must provide optimum mechanical support to the pellets of active material that fill the void space. Sufficient conductivity has also to be provided by the grid. Grids for positive and negative electrodes are usually similar. In batteries designed for extended service life, the positive grid is made heavier to provide a corrosion reserve. For very thin electrodes, a lead foil is used as the substrate and current conductor.

The mechanical support of the active material is of minor importance when additional support is provided, for example, by envelopes or tubes (cf. Section 6.4.3).

6.4.2.1 Manufacture of the Active Material

The production of the active material for positive and negative electrodes starts with the same substance: a mixture of lead oxide (PbO) and metallic lead called *gray oxide* or *lead dust*. It is a fine powder that contains 20–30 wt% of lead (Pb). The size of the primary particles is in the range of 1–10 µm. Larger agglomerates are usually formed.

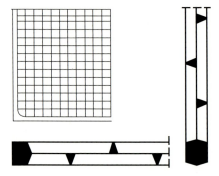

Figure 6.5 Grid of a lead–acid battery with thicker walls and a rectangular mesh. One corner and cross-sections in vertical and horizontal directions (from Ref. [14]).

Gray oxide can be produced by a **milling process**, which, strictly speaking, does not mill the material. A rotating drum is filled with solid balls or ingots of lead.

Flakes are shared, crushed, and at the same time partly oxidized by an airstream that flows through the drum. Temperature and airflow rate are used to control the process to achieve the desired powder. At the end the oxidized material is carried away by an airstream and classified. Particles that are too coarse are fed back into mill.

Another process, the **Barton process**, is based on molten lead. The core of such a device is the 'Barton reactor,' a heated pot that is partly filled with molten lead. It is continuously refilled by a fine stream of molten lead. Fine droplets of lead are produced by a fast rotating paddle that is partly immersed under the surface of the molten lead within the 'Barton reactor.' The surface of each droplet is transformed by oxidation into a shell of PbO by an airstream that simultaneously carries away the oxidized particles if they are small enough; otherwise, they fall back into the melt and the process is repeated. Thus the airstream acts as a classifier for particle size.

A short description of both processes is given in Ref. [15]. Nowadays, the Barton process is preferred for a number of reasons: it can more easily be installed in small units and it can be controlled faster [16]. (In a mill it takes a number of hours for the material to run through the drum, so controlling actions are slow.)

Paste mixing means the addition of sulfuric acid and water. The result is a fairly stiff paste with a density between 1.1 and 1.4 g cm^{-3} containing 8–12 wt% of lead sulfate. The water content of this mix determines the porosity of the active material achievable later (cf. 'curing' below). In the paste, a mixture of lead sulfate and basic lead sulfate is formed (cf. Table 6.1). In the usual mixing process between room temperature and 50 °C, tribasic lead sulfate is formed. The generation of the tetrabasic modification (4PbO·PbSO$_4$) is favored at temperatures above 70 °C [17].

To a certain extent, the formation of the tetrabasic variant is desired, because 4PbO·PbSO$_4$ forms fairly large crystals when transformed into lead dioxide (PbO$_2$). This results in a mechanically stable active material, but there are disadvantages because it is more difficult to transform this material into lead dioxide, that is, the formation process (see below) is more expensive (and takes longer) and the initial capacity is slightly reduced (cf., e.g., Ref. [18]. For 'long-life batteries' (Bell systems cell), a special process has been developed to produce pure tetrabasic material [19].

Sometimes red lead or minium (Pb$_3$O$_4$) is added to the paste. The addition is usually 5–10 wt%, and is mainly made for easier formation of the final compound lead dioxide (PbO$_2$) (cf. 'curing' and 'formation,' below).

A general problem in open mixing machines is that the exothermic process

$$PbO + H_2SO_4 \rightarrow PbSO_4 + H_2O \tag{6.30}$$

causes a temperature increase of the mix ($\Delta H = -173$ kJ mol^{-1}). In mixing machines the evaporation of water is often used for cooling, but then water is lost and the composition of the mix changes. For this reason a certain surplus of water is scheduled in the paste recipe. However, evaporation rate depends

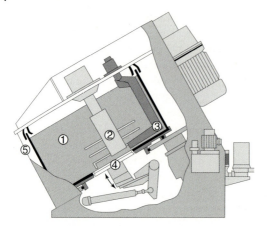

Figure 6.6 Vacuum mixer for lead paste: l, mixing compartment; 2, fast-rotating mixing tools; 3, material-deflecting plate; 4, discharge opening; and 5, static, vacuum-sealed enclosure [20].

on environmental parameters like temperature and humidity, and this limits the possibility of keeping the specified water content of the paste, which is important in regard to the porosity of the active material (as mentioned below).

For accurate mixing, independently of environmental conditions, vacuum mixers are used as shown in Figure 6.6.

In the vacuum mixer, water evaporation is also used for the temperature control, since the evaporation rate can be influenced by the grade of the vacuum. The water vapor, however, does not escape from the mixer, but is condensed and returned into the mix, the composition of which is thus not changed.

At the end of the mixing process, the paste contains about 10 wt% of metallic lead and about 50 vol% of water. The water is evaporated during the subsequent production steps, and the resulting void space represents the pore volume of the dried active material.

Pasting means that the paste and the grids are supplied to a machine that smears the mix into the grid. Single plates are superficially dried after pasting, to prevent sticking when they are stacked afterwards. Continuously cast grids leave the pasting machine as an endless ribbon, usually enveloped by paper. Pasting of flat elements or foils is achieved with a slurry instead of a stiff paste (cf., e.g., Ref. [21]).

The subsequent production step, the 'curing,' is especially important for the positive plate, because the structure of the active material can be influenced by environmental conditions such as temperature and humidity [22]. During the curing step, the lead content in the active material is reduced by gradual oxidation from about 10 to less than 3 wt%. Furthermore, the water (about 50 vol%) is evaporated. This evaporation must be done quite carefully, to ensure that the volume occupied by the water actually gives rise to porosity and is not lost by shrinkage, which again might lead to the formation of cracks. As in the paste-mixing process, the transformation into tetrabasic lead oxide ($PbO \cdot PbSO_4$) is favored at curing temperatures above 70 °C [23], which is of prime interest for the later mechanical stability of the positive active material.

Single plates are usually cured in special devices (curing ovens or curing chambers, cf., e.g., Ref. [24]) that control humidity as well as temperature. In continuous plate production, the drying of the pasted ribbon is correspondingly controlled. Furthermore, in continuous manufacture final curing can occur after the plates are separated and inserted into the containers.

6.4.2.2 Tank Formation

Tank formation means that the cured positive and negative 'raw plates' are inserted alternately in special tanks filled with fairly dilute sulfuric acid (generally in the range 1.1–1.15 g cm^{-3}) and positive and negative plates are connected, a number of each, in parallel with a rectifier. The formation process means that the active material of the plates is electrochemically transformed into the final stage, namely:

- lead dioxide (PbO_2) in the positive electrode and
- spongy metallic lead (Pb) in the negative electrode.

A survey of formation techniques is given in Ref. [25].

Because of the porous material in the raw plate, both substances are produced in a spongy state with a porosity of about 50 vol%. Tank formation takes between 8 and 48 hours, depending on the plate thickness and formation schedule. When the formation process is finished, the plates are washed and dried. They can be stored and later assembled in batteries.

6.4.2.3 Container Formation

The fundamental difference from tank formation is that the battery is assembled first, then filled with electrolyte, and finally the formation process is carried out with the complete battery.

During the **formation process**, the battery is considerably overcharged and generates both hydrogen and oxygen, with resulting water loss. The concentration of the filling acid is adjusted in such a way that the desired final acid concentration is approximated at the end of the formation, and only minor corrections are required; another method includes dumping of the acid and refilling of the battery during the formation step. These methods are known as 'one-shot formation' and 'two-shot formation,' respectively (cf., e.g., Ref. [25], p. 17).

6.4.3
Tubular Plates

The tubular-plate design for the positive electrodes, shown in Figure 6.7, is common mainly in European countries for batteries with larger capacities. In this plate design, the conducting elements are separated from the components that contribute mechanical support. The grid consists of vertical lead rods in the centers of tubes that are formed by woven, braided, or nonwoven fabrics.

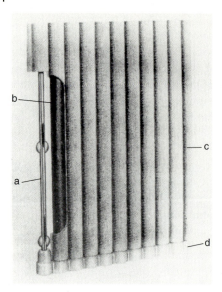

Figure 6.7 Section of a tubular plate: a, lead-alloy spine (grid); b, active material (PbO_2); c, tube, (in this example, fabric of polyester fibers); and d, bottom seal of plastic caps.

The advantage of tubular plates is the comparatively high utilization of the active material, which results in a rather low weight in relation to capacity. These features have two causes, namely:

- The central current-collecting spine produces uniform current flow across the active material.
- The mechanical support by the tube allows the use of fairly light active material. This means high porosity and a high utilization factor.

A disadvantage for tubular plates is the fact that a minimum tube diameter between 6 and 8 mm is required for economic production, but the tube diameter corresponds with the plate thickness, and lead–acid batteries with such thick plates are inferior for high-rate discharge.

The production of tubular positive plates is in principle similar to that of pasted plates. A number of manufacturers use the same gray oxide as the basic filling substance. Sometimes the portion of red lead or minium (Pb_3O_4) is increased above 25 or even to 100 wt%. The latter is more economic when the manufacturer runs his own minium plant; then the expense of the chemical oxidation of lead oxide (PbO) to minium (Pb_3O_4) may be compensated by reduced formation cost. Furthermore, curing is not required, because of the high oxidation state, and the battery starts with full capacity when formed.

Different methods are in use for plate filling. The material can be filled as a powder with the aid of vibrators. Other techniques use a slurry of lead oxide or even a paste, as described above [26].

When dry material or a slurry has been filled, 'pickling' is required, which means that the plate is stored in sulfuric acid for a short time. The material is soaked by the acid and transformed, at least partly, into lead sulfate ($PbSO_4$), as in the paste-mixing process (Section 6.4.2.1). When minium is used, during the 'pickling' process lead dioxide is also formed according to Equation 6.4

The subsequent procedures, formation, washing, drying, and battery assembly, are similar to those described above.

6.5
Passivation of Lead by Its Oxides

Corrosion of the current-conducting elements in the positive electrode, as of the plate support (grid), bus bars, and terminals, is a side-effect of the high cell voltage of this battery system, which implies a high potential of the positive electrode. Metals that are usually applied as current conductors, and even noble metals like gold, would be dissolved by oxidation when connected to the positive electrode of the lead–acid battery.

Lead can be used, because the corrosion itself forms a rather dense passivating layer of lead dioxide that protects the underlying material against fast corrosion [27]. If foreign metals like copper are used they have to be covered thoroughly by a dense layer of lead.

However, the protecting PbO_2 layer does not establish a stable situation at the phase boundary between metal and oxide layer. Rather, the corrosion process gradually penetrates into the bulk material, and the corrosion of the positive grid represents a restriction of the lead–acid battery that finally limits the useful life, if no other reasons cause earlier failure. Figure 6.8 illustrates the situation: the

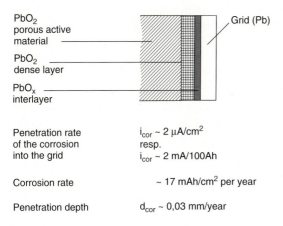

PbO_2 porous active material	
PbO_2 dense layer	
PbO_x interlayer	
	Grid (Pb)
Penetration rate of the corrosion into the grid	$i_{cor} \sim 2\ \mu A/cm^2$ resp. $i_{cor} \sim 2\ mA/100Ah$
Corrosion rate	$\sim 17\ mAh/cm^2$ per year
Penetration depth	$d_{cor} \sim 0{,}03\ mm/year$

Figure 6.8 Structure of the corrosion layer at the grid surface. Penetration and corrosion rates are approximated for room temperature and normal float voltage (2.23–2.25 V/cell) (see text).

Table 6.7 Density and volume ratio of corrosion products related to lead.

	Density (g cm^{-3})	Volume ratio relative to Pb
Pb	11.34	1
PbO$_{(red)}$	9.64	1.26
α-PbO$_2$	9.87	1.32
β-PbO$_2$	9.3	1.40
PbSO$_4$	6.29	2.64

active material is represented by the area on the left; the grid is shown on the right. Underneath the porous lead dioxide that constitutes the active material, a dense layer, also of lead dioxide, covers the grid surface. This layer is formed by corrosion and protects the grid. On account of acid depletion a rather stable oxide layer (mainly of α-PbO$_2$) is formed [28]. However, lead dioxide and lead cannot exist beside each other for thermodynamic reasons, and a thin layer of less-oxidized material is always formed between the grid and the lead dioxide (PbO$_x$ in Figure 6.8) [29]. The existence of lead oxide (PbO) in this layer has been determined; the existence of higher oxidized species is assumed (PbO$_x$ phases: cf. Ref. [5], p. 18), but their structure is not yet known exactly [30].

The PbO$_2$/PbO$_x$ border slowly penetrates into the metal, but only at a very slow rate as a solid-state reaction. Cracks are formed when the oxide layer exceeds a given thickness, on account of the growth in volume when lead becomes converted into lead dioxide (Table 6.7). Underneath the cracks the corrosion process starts again and again. As a whole, the corrosion proceeds at a fairly constant rate. It never comes to a standstill, and a continually flowing anodic current, the corrosion current, is required to re-establish the corrosion layer.

When the grid material (Pb) is converted into lead dioxide (PbO$_2$), the basic electrochemical reaction is

$$Pb \rightarrow Pb^{4+} + 4e^- \tag{6.31}$$

Twice the amount of electricity is required compared with the discharge reaction at the negative electrode according to Equation 6.18, since corrosion involves four valences, which means $4F = 107.21$ Ah per multiple of Equation 6.31. Consequently, for the corrosion reaction according to Equation 6.31 the equivalent values are:

$$\frac{207.19}{107.21} = 1.9326 \text{ g Pb/Ah or } 517.4 \text{ Ah/kgPb} \tag{6.32}$$

This value means that 1 cm^3 of lead (11.34 g; cf. Table 6.2) is equivalent to 5.89 Ah.

A current of 1 μA cm^{-2} means 8760 μAh cm^{-2} per year. Referred to 5.89 Ah cm^{-3}, a penetration rate of 1.49×10^{-3} cm/year results, assuming that the corrosion attack occurs uniformly and progresses at a constant rate. Thus the value in Figure 6.8 means that a corrosion current of 2μA cm^{-2} implies a penetration rate of about 0.03 mm/year.

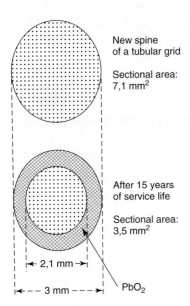

Figure 6.9 Conversion of grid material into lead dioxide (PbO_2) by corrosion: spine of a positive tubular plate. New plate: 3 mm diameter means 7.1 mm² cross-section (πr^2 with $r = 1.5$ mm). Aged plate: reduction of r by $0.03 \times 15 = 0.45$ mm means $\pi r^2 = 3.5$ mm².

Figure 6.9 illustrates the consequences for battery practice. The above penetration rate would reduce the cross-section of a grid spine in a tubular electrode by about 50% within the usual service life of 15 years. This result is confirmed by field experience and shows that long-life batteries must have a corresponding 'corrosion reserve' in their positive grids.

Since grid material is converted into lead dioxide, a slight increase in the actual capacity is often observed with lead–acid batteries. The reduced cross-section in Figure 6.9 does not affect the performance of batteries that are used for discharge durations in the order of 1 h or more. Attention must, however, be paid to batteries that are loaded with high currents, because the conductivity of the grid gains importance with increased current flow.

The 'corrosion capacity of the positive grid' can be estimated from the above figures. The positive grid in lead–acid batteries for stationary and traction applications contains about 10 g of lead/Ah^{-1} (usually slightly more). This means a positive-grid weight of about 1 kg/100 Ah. With the values of Equation 6.32, the 'corrosion capacity' is 500/100 Ah of battery capacity. The corrosion rate of 1 or 2 mA/100 Ah means 8.76 and 17.52/100 Ah per year respectively. Related to the 500 Ah of the total 'corrosion capacity,' 2–4% of the grid material would be converted into lead dioxide per year under these assumptions.

6.5.1
Disintegration of the Oxide Layer at Open-Circuit Voltage

In Section 6.3.3 it has been shown that corrosion is one of the reactions that cause self-discharge of the positive electrode. In connection with Figure 6.8 it has been mentioned that an anodic current, the **corrosion current**, must flow continuously to stabilize the lead dioxide layer at the grid surface. Then the **PbO$_x$** layer remains thin because **PbO$_x$** is always converted into **PbO$_2$** by further oxidation.

At open-circuit voltage, no anodic current flow through the positive electrode occurs that can oxidize the PbO (or **PbO$_x$**) layer, but the corrosion reaction

$$Pb + PbO_2 \rightarrow 2PbO \tag{6.33}$$

continues between grid and passivating layer. Consequently the PbO (or **PbO$_x$**) layer grows between the grid and **PbO$_2$**.

However, as mentioned in Section 6.3.3, PbO and **PbO$_x$** are not stable against sulfuric acid, and react very fast according to

$$PbO + H_2SO_4 \rightarrow PbSO_4 + H_2O \tag{6.34}$$

as soon as these substances come in contact with it. The protecting lead dioxide layer, shown in Figure 6.8, would be destroyed by this reaction, and severe grid corrosion is one of the problems that occur when the battery stands for prolonged periods without any charging. Battery manufacturers therefore recommend recharging a lead–acid battery filled with electrolyte within regular periods, which must be shortened when the battery is stored at elevated temperatures.

6.5.2
Charge Preservation in Negative Electrodes by a PbO Layer

The drying of negative plates is not possible without precautions, because of the tendency to spontaneous oxidation. This oxidation reaction is much accelerated by water, and the active material of a moist negative electrode is spontaneously converted into lead oxide when exposed to air. When, on the other hand, the charged plate is dry, a thin layer of oxide covers the surface of the active material, and prevents further oxidation. So, prevention of access of oxygen as long as the plates are wet is a common feature of various methods to achieve **dry charged negative plates**. As a result of the superficial oxidation, a loss of about 10% of capacity is always incurred with the dry charge process, regardless of the method applied.

The dried plates can be stored for a practically unlimited time without losing capacity or ageing. This is true also for complete batteries that are assembled but not yet filled with electrolyte.

6.6
Ageing Effects

The active material of the positive electrode is prone to lose its mechanical strength when repeated discharge/charge cycles occur, because the alternating dissolution and precipitation processes convert the agglomerate structure into an accumulation of fine crystals [31]. So, the active material suffers degradation, and part of it may fall off the plate as fine particles. This process is called '*shedding*.' Shedding of the positive active material is a characteristic feature of ageing conventional lead–acid batteries when they are charged and discharged frequently. It is likewise also described as '*soft positives.*' Shedding is only the outward appearance of a more general ageing process which means that the active material is prone to disintegration of its electronic conductivity and mechanical strength. This causes the so-called '**premature capacity-loss**' [32] (a survey with references is given in Ref. [33]). It becomes evident as a decreasing utilization factor with increasing cycles.

In a model that considers the active material as an aggregate of spheres ('*Kugelhaufen*') it is explained by a gradual increase in the ohmic resistance, mainly in the connecting region of the individual particles of the active material. The connecting regions establish the electrical contact between the individual particles of the active material. They are decisive for the ohmic resistance because of the minimized cross-sectional areas in these bridging zones. The structure of the connecting regions between the particles of active material is largely influenced by the conditions when these regions are reestablished during the charging process. For this reason, it is understandable that the charging conditions are important for the stability of the active material and that in many cases, after a premature decay, full capacity can be regained with suitable charge/discharge procedures [34]. For this reason, the premature capacity loss sometimes is called 'reversible capacity decay.'

Another model assumes that gel zones are formed by hydrated lead dioxide $(PbO(OH)_2)$ and act as bridging elements between the crystallite particles. Electrons can move along the polymer chains of this gel and so cause electronic conductivity between the crystalline zones [35].

Quite often, simultaneously with the capacity decay, the formation of a barrier layer of lead sulfate $(PbSO_4)$ is observed between the grid and the active material [36]. In view of the explanation given above, this layer may be the final stage of the process. When the ohmic resistance of the active material is increased, the charge/discharge reaction is restricted to the area close to the grid surface. Then, deep discharge must happen to this part of the active material, causing a high concentration of sulfate.

6.6.1
The Influence of Antimony, Tin, and Phosphoric Acid

Antimony (Sb) and tin (Sn) are usually not added to the active material, but both are alloying components of the grid. They are gradually released from the grid by corrosion, and permeate the active material by dissolution and diffusion.

The 'premature capacity loss' described above, that is, a decay of the utilization factor, became especially evident when antimony-free alloys were introduced and such batteries were operated in charge/discharge cycle regimes. For this reason, this effect is likewise called the '*antimony-free effect*,' although it is also observed with grids containing antimony. The mechanism of this effect has not yet been explained in detail, but antimony has a strong influence on the stability of the active material that cannot be compensated by special pretreatment or design of the electrodes [37].

When specimens of pure lead and a 5% antimony alloy were periodically oxidized and reduced, lead oxide layers were observed with different structures:

- coarse and insulating with antimony-free electrode,
- fine and low resistance with the antimony alloy [38].

The origin of such insulating layers may explain the high resistance also established within the active material when antimony is not present.

The '*Kugelhaufen*' model mentioned in the preceding section explains the beneficial influence of antimony by improved conductivity of the zones that connect the spheres. According to the gel model, antimony decreases the crystallinity of PbO_2 and so increases the conductivity by the gel zones [39], and especially influences the structure of the corrosion layer intermediate between the grid and active material [40].

Addition of tin to the positive-grid alloy also has a capacity-stabilizing effect, but this apparently concerns only the boundary between the grid and active material. **Phosphoric acid** (H_3PO_4) is added in small amounts to the electrolyte. A beneficial effect on cycle stability has long been known for this acid, which has been used as an additive in conventional lead–acid batteries for many years to improve cycle stability, although the disadvantage of a slightly reduced capacity had to be accepted [41]. Addition of phosphoric acid to the electrolyte improves long-term capacity and reduces the formation of sulfate layers around the grid [42]. The addition of 20–35 g dm^{-3} phosphoric acid was protected by patent for valve-regulated lead–acid batteries with gelled electrolyte [43]. Extensive experiments [44] showed that, at low H_3PO_4 concentrations, $Pb_3(PO_4)_2$ acts as an intermediary in the corrosion of Pb to PbO_2. Clearly, the phosphoric acid influences the formation of lead dioxide (PbO_2) on account of its strong adsorption and leads to a fine grain structure of the positive active material [45]. However, in spite of the repeated use of phosphoric acid in lead–acid batteries, some questions on its interaction are still to be elucidated [46].

References

1. (a) Bullock, K.R. (1987) in *Proceedings of the Symposium History of Buttery Technology*, Electrochemical Society Proceedings, Vol. 87-14 (ed. A.J. Salkind), The Electrochemical Society, Pennington, NJ, p. 106; (b) Garche, J. (1990) *J. Power Sources*, **31**, 401.
2. Vinal, G.W. (1955) *Storage Batteries*, John Wiley & Sons, Inc. New York, p. 2ff.
3. Kordesch, K.V. (1955) *Batteries*, Marcel Dekker, New York, p. 419.
4. Bullock, K.R. (1984) *Proceedings of the Symposium Advances in Lead-Acid Batteries*, The Electrochemical Society, Pennington, NJ, p. l.
5. Bode, H. (1977) *Lead Acid Batteries-ohn*, John Wiley & Sons, Inc., New York, p. 366.
6. Burbank, J., Simon, A.C., and Willihnganz, E. (1971) *Advances in Electrochemistry and Electrochemiral Engineering*, Vol. 8, John Wiley & Sons, Inc., New York, p. 170.
7. Pohl, J. and Rickert, H. (1975) in *Proceedings of the 9th International Symposium Brighton 1974* (ed. D.H. Collins), Academic Press, p. 15.
8. Carr, J.P. and Hampson, N.A. (1972) The lead dioxide electrode. *Chem Rev.*, **72**, 679–703.

9. Pavlov, D., Bashtavelova, E., and Iliev, V. (1984) in *Advances in Lead-Acid Batteries*, Electrochemical Society Proceedings, Vol. 84-14 (eds K.R. Bullock and D. Pavlov), The Electrochemical Society, Pennington, NJ, p. 16.
10. Pourbaix, M. (1966) *Atlas of Electrochemical Equilibria in Aqueous Solution*, Pergamon, New York, p. 485.
11. Barnes, S.C. and Mathieson, R.T. (1963) in *Batteries* (ed. D.H. Collins), Pergamon, New York, p. 41.
12. Berndt, D. (1993) *Maintenance-Free Batteries*, 1st edn, Research Studies Press, Taunton, UK, John Wiley & Sons, Inc., New York, p. 40; (1997) 2nd edn, p. 102.
13. Yamashita, J., Yufu, H., and Matsumaru, Y. (1990) *J. Power Sources*, **30**, 13.
14. VARTA Battery AG (ed.) (1986) *Bleiakkumulatoren*, 11th edn, VDI-Verlag, Düsseldorf, p. 79 (in German).
15. (1994) *Batteries Int.*, **21**, 84.
16. Busdieker, E.A. and Maurer, K.W. (1993) *Batteries Int.*, **16**, 56.
17. Pavlov, D. and Kapazov, G. (1976) *J. Appl. Electrochem.*, **6**, 339.
18. Pavlov, D. and Kapkov, N. (1970) *J. Electrochem. Soc*, **137**, 1305.
19. Biagetti, R.V. and Weeks, M.C. (1970) *Bell Syst. Tech. J.*, **49**, 1305.
20. (a) *Processing Technology for Lead Acid Battery Paste*, Maschinenfabrik Gustav Eirich, Hardheim, Germany; (b) Vogel, H.J. (1994) *Power Sources*, **48**, 71.
21. Tamura, H. (1988) *Prog. Batteries Solar Cells*, **7**, 205.
22. Lam, L.T. and Rand, D.A.J. (1990) *Batteries Int.*, **137**, 21.
23. Pavlov, D. and Kapkov, N. (1990) *J. Electrochem. Soc.*, **137**, 21.
24. Clerici, G. (1991) *J. Power Sources*, **33**, 67, Fig. 11.
25. Kiessling, R. (1992) *Lead Acid Battery Formation Techniques*, Digatron/Firing circuits, Digatron, Industrie Elektronik GmbH, Aachen, Germany; Firing Circuits Inc., Norwlk, CT.
26. (a) Lüdecke, W. (1990) *Batteries Int.*, **22**, 60; (b) Mittermaier, F.X. (1996) *Batteries Int.*, **28**, 43.
27. (a) Pavlov, D. (1967) *Ber. Bunsenges Phys. Chem.*, **71**, 398; (b) Pavlov, D. (1984) *Advances in Lead-Acid Batteries*, Electrochemical Society Proceedings, Vol. 84-14, The Electrochemical Society, Pennington, NJ, p. 110.
28. Ruetschi, P. and Angstadt, R.T. (1964) *J. Electrochem. Soc.*, **111**, 1323.
29. Pavlov, D. and Dinev, Z. (1980) *J. Electrochem. Soc.*, **127**, 855.
30. Bullock, K.R., Trischan, G.M., and Burrow, R.G. (1980) *J. Electrochem. Soc.*, **130**, 1283.
31. Atlung, S. and Zachau-Christiansen, B. (1990) *J. Power Sources*, **30**, I31.
32. Hollenkamp, A.F., Constati, K.K., Huey, A.M., Koop, M.J., and Aputeanu, L. (1992) *J. Power Sources*, **40**, 125.
33. Meissner, E. and Rabenstein, H. (1992) *J. Power Sources*, **40**, 157.
34. Hullmeine, U., Voss, E., and Winsel, A. (1989) *J. Power Sources*, **25**, 27.
35. (a) Pavlov, D. (1992) *J. Electrochem. Soc.*, **139**, 3075; (b) Pavlov, D. and Monahov, B. (1996) *J. Electrochem. Soc.*, **143**, 3616.
36. Tudor, S., Wesstuch, A., and Davang, S.H. (1965) *Electrochem. Technol.*, **4**, 406 [esp. 408, Fig. 2].
37. Chang, T.G. (1984) *Advances in Lead-Acid Batteries*, Electrochemical Society Proceedings, Vol. 84-14, The Electrochemical Society, Pennington, NJ, p. 86.
38. Burbank, J. (1971) *Power Sources*, Vol. 3, Oriel Press, Newcastle upon Tyne, p. 13.
39. Pavlov, D., Dakhouche, A., and Rogachev, T. (1993) *Power Sources*, **42**, 71.
40. (a) Monahov, B. and Pavlov, D. (1994) *J. Electrochem. Soc.*, **141**, 2316; (b) Pavlov, D. (1994) *Power Sources*, **48**, 179.
41. Drotschmann, C. (1951) *Bleiakkumulatoren*, Verlag Chemie, Weinheim, p. 161.
42. Tudor, S., Weisstuch, A., and Davang, S.H. (1967) *Electrochem. Technol.*, **5**, 21. [esp. 23, Fig. 3].
43. (1967) Accumulatorenfabrik Sonnenschein GmbH, Büdingen, German Patent 1671693, January 12, 1967.

44. Bullock, K.R. and McClelland, D.H. (1977) *J. Electrochem. Soc.*, **124**, 1478.
45. Garche, J., Döring, H., and Wiesener, K. (1991) *J. Power Sources*, **33**, 213.
46. Voss, E. (1988) *J. Power Sources*, **24**, 171.

Further Reading

Brunauer, S., Emmett, P.H., and Teller, E. (1938) *J. Am. Chem. Soc.*, **60**, 309.

Chen, H.Y., Li, A.J., and Finlow, D.E. (2009) *J. Power Sources*, **191** (1), 22–27.

Ellis, T.W. and Mirza, A.H. (2010) *J. Power Sources*, **195** (14), 4525–4529.

Isastia, V. and Meo, S. (2009) *Int. Rev. Electrical Eng.-IREE*, **4** (6), 1122–1144.

Karden, E. (2008) *Energy Sav. Veh. Electron.*, **2033**, 95–116.

Pickard, W.F., Shen, A.Q., and Hansing, N.J. (2009) *Renew. Sustain. Energy Rev.*, **13** (8), 1934–1945.

Soria, M.L., Trinidad, F., Lacadena, J.M., Valenciano, J., and Arce, G. (2007) *J. Power Sources*, **174** (1), 41–48.

Winsel, A., Voss, E., and Hullmeine, U. (1990) *J. Power Sources*, **30**, 209. [esp. p. 220, Fig. 12].

7
Bromine-Storage Materials
Christoph Fabjan and Josef Drobits

7.1
Introduction

As a positive active material for rechargeable cells, bromine offers various attractive properties such as high voltage and specific energy and power. The main difficulty encountered with the operation of a battery using bromine/bromide electrode is the necessity to develop a suitable for the storage of this aggressive and toxic elemental halogen.

Several systems using different negative materials such as Zn [1–3], Li [4], and Al [5] were investigated, but at present practical success has been achieved only with the zinc–bromine battery. A satisfactory approach for bromine storage was the formation of organic nonaqueous polybromine complex phases by the reaction of quaternary ammonium salts, dissolved in the aqueous electrolyte solution, with elemental bromine generated in the charge process, reducing energy losses due to self-discharge to extraordinarily low values.

A variety of complexes exist in a solid or liquid state at ambient temperatures in the range required for battery operation. Liquid polybromine phases are preferred since they enable storage of the active material externally to the electrochemical cell stack in a tank, hence enhancing the storage capacity of the system and reducing energy losses in standby periods to very low values. This design requires electrolyte circulation and application of the flow battery concept, utilizing its characteristic advantages and accepting the drawbacks associated with this type of storage system.

The operating principle and the conceptual design of a multicell bipolar zinc-flow battery are presented in Figure 7.1.

From an aqueous $ZnBr_2$ solution zinc is plated at the cathode during charge while bromine is generated at the anode, forming the water-immiscible polybromine phase with the complexing agents. In the discharge process zinc metal is dissolved anodically, and the active bromine is consumed from an emulsion of the complex phase and aqueous solution pumped over the electrode surface in the cathodic reaction. Hence the active materials are reconverted to form aqueous $ZnBr_2$ solution. The net reaction, $Zn^s + Br_2^{naq} \leftrightarrow ZnBr_2^{aq}$, provides a cell voltage of approximately 1.82 V and a theoretical specific energy of 430 Wh kg^{-1}.

Handbook of Battery Materials, Second Edition. Edited by Claus Daniel and Jürgen O. Besenhard.
© 2011 Wiley-VCH Verlag GmbH & Co. KGaA. Published 2011 by Wiley-VCH Verlag GmbH & Co. KGaA.

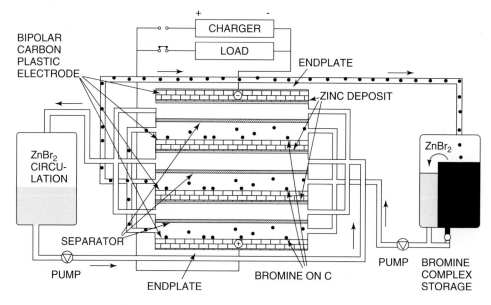

Figure 7.1 Operating principle of a zinc-flow battery.

The system consists of three essential components.

1) The cell stack with the bipolar electrodes.
2) The electrolyte flowing in two independent streams through the electrochemical module and the individual cell compartments, which are divided by microporous separators. In the discharged state the electrolyte is a homogeneous aqueous solution, whereas during charge a two-phase electrolyte is produced in the bromine loop.
3) Auxiliary equipment, including two pumps, tubing valves, and electrolyte reservoirs is also necessary.

The bromine-storing equipment, as an essential component of the system, has to achieve various objectives to ensure the practical success of the battery:

1) a decrease in the concentration of active bromine in the aqueous solution, which is in equilibrium with the organic bromine-rich phase, to very low values (0.01 mol L^{-1}) to minimize self-discharge;
2) a decisive reduction in bromine vapor pressure to a small percentage of the values obtained with elemental bromine over the full range of operating temperatures (from approximately -10 to $50\,°C$);
3) provision of the liquid state under the conditions described above;
4) satisfactory electrochemical reactivity in the emulsion with the aqueous electrolyte phase, to minimize polarization and voltage losses;
5) higher specific weight than that of the aqueous solution to enable fast separation of the two phases;

6) long-term stability against oxidation or bromination of organic substituents by bromine;
7) minimization of health and safety risks and environmental impact in the case of battery failure (electrolyte leakage) or damage.

The ionic conductivities of the polybromide complexes are considerable, approaching the values of ordinary aqueous salt solutions, due to their 'fused-salt' nature. This property means an additional advantage since the electrolyte resistivity, the ohmic voltage drop, and polarization effects remain low.

The properties and behavior of the bromine-storing polybromide complexes will now be treated in detail from the fundamental and technological viewpoints, including economic and ecological (safety risks, recyclability, disposal, etc.) aspects.

7.2
Possibilities for Bromine Storage

7.2.1
General Aspects

Previously studied possibilities for bromine storage systems are listed in Table 7.1. The widely known reduction of the Br_2 vapor pressure by formation of adducts with various carbon materials results from strong chemisorption interactions and has been investigated in the case of activated carbon [4]. Adducts containing up to ~85 wt% Br_2 were reported to be stable at ambient temperature; however, considerable equilibrium bromine vapor pressures were found.

Intercalation of molecular Br_2 into graphite yields compounds with higher thermal stabilities [6, 7] forming so-called residual compounds which do not quantitatively release all the included Br_2 even at temperatures of ~2000 °C [26, 31]. Bromine is fixed in the graphitic lattice, bound in two different ways: (i) chemisorption on structural crystalline defects and layer edges and (ii) intercalation, thus producing separated islands without contacting the crystal edges. Repeated intercalation and extraction of intercalant species leads to exfoliation and destruction of the graphite material. Formation of bromine adducts with zeolites has been reported [8]. A considerable number of crystalline organic compounds containing polyiodide and polybromide anions such as TMA·0.7 $H_2O \cdot x Hal_5 H$ (TMA = trimesic acid, $x = 0.09$ for Hal = I, and $x = 0.103$ for Hal = Br) [32] or $M(dpq)_2 Hal$ (M = Ni; Pd; dpg = diphenylglyoxime) [33] have been tested with the intention of modeling new conducting materials. Electrical conductance was reported to depend only negligibly on the nature of the anions. Infinite linear chains of essentially Hal_5^- units were found in XRD analyses [32].

As early as 1888 Horton reported the formation of stable urotropin-bromide adducts [23]. The storage of bromine in organic solids and particularly as a liquid complex phase stabilized by quaternary ammonium salts is of outstanding importance for modern battery applications. In conjunction with polymeric and porous carbon matrices, these species were used in investigations aiming at the

Table 7.1 General possibilities for bromine storage.

Chemical stabilization	Examples and references
Storage in inorganic solid matrix	Intercalation into graphite [6, 7]
	Carbon-bromine adduct [4]
	In zeolites [8]
Polymetric matrix	Polydiallyldimethylammonium bromide [9]
	Polypyrrole [10]
	Poly (N,N-dimethyl)-3,4-pyrrolidinium bromide [11]
	Styrene-divinyl benzene copolymers [4]
	Polyacrylamide [12]
Ion-exchanger resins	De-acidite FF anion exchange resin [13]
Charge-transfer complexes	Dioxane, pyridine, polyvinyl, pyrrolidone, poly-2-vinyl pyridine, polyethyleneoxide [4]
Further organic storing materials	Phenyl bromide [14], pyridine, 1-picoline, 2,6-lutidine [15–17]
	Arsonium salts [18, 19]
	Phosphonium salts [20]
	Pyridinium bromides [21]
	Aromatic amines [22]
	Urotropin-bromine adduct [23]
	Pyridinium and sulfonium salts [24]
	Propionitrile [25]

development of solid-state batteries [12]. However, high self-discharge rates have remained an unsolved problem for this cell type until today. The zinc-flow battery using liquid polybromide phase from the reaction with quaternary ammonium salts is currently of practical importance in the field of alternative systems of energy storage.

7.2.2
Quaternary Ammonium-Polybromide Complexes

The tendency of the halogens to form chain-like polyanions that are stabilized by delocalization of the negative charge [15, 34] is a basic chemical principle. Donor–acceptor interactions between Lewis-acidic Br_2 and halide anions, but also with polyhalides acting as Lewis bases, give rise to the formation of a variety of homo- and heteroatomic adducts. The maximum number of atoms in these chains increases with the atomic weights of the halogens (i.e., $Cl < Br < I$). Stabilities of the predominant species, Br_3^- and Br_5^-, in aqueous [35, 36] and organic solvents [36, 37] were studied by means of IR, Raman, and UV spectroscopy. In Br_3^- a substantial elongation of bond distances occurs compared with the Br_2 molecule (2.55 vs 2.28 Å). The asymmetric and symmetric Br–Br stretching modes at $\sim$201–210 and 160–170 cm^{-1} of dissolved tribromide ions in various solvents were found in Raman spectroscopic studies [36] to be approximately 100 cm^{-1} lower than the characteristic value for elemental Br_2 [38]. While theoretical studies predict that the free Br_3^- ion has a linear symmetric structure [39–43], a broadly scattered variety of linear and nonlinear symmetric and asymmetric Br_3^- ions were reported to exist in a crystalline environment [44, 45] and in solution [35, 36], according to the size and shape of the corresponding countercations.

Br_5^- forms V-shaped planar species [42, 43] with relatively strong terminal and weak central bonds. Raman spectra in aqueous and acetonitrile solutions [36] exhibit terminal stretching modes of 250 and 257 cm^{-1} which are essentially higher than those in Br_3^-. Structural data from experimental studies exist only for crystalline compounds in which the Br_5^- ions adopt linear geometry. In a theoretical study using density functional theory [43], a central bond angle of 114.6° and a central terminal bond ratio of ∼1.111 were obtained; the latter is somewhat smaller than the 1.181 estimated from XRD experiments on linear Br_5^- units in TMA·0.7H_2O·xHal$_5$H [32].

It is evident that the shapes and relative stabilities of the polybromide anions depend to a large extent on the nature of the countercations. The tendency to form a nonaqueous phase is determined by the chemical environment, and depends particularly on the degree of hydration of the ions. Hence, important properties of bromine-storing phases can be varied over a wide range by adjusting the type and composition of the quaternary ammonium salts used. Additional parameters, such as pH, concentration of supporting salts (i.e., Cl$^-$) also producing heteroatomic polyhalide ions [46], and temperature effects, enhance the difficulties of a systematic investigation of the polybromide structure in practical battery electrolytes.

An overview of the most important quaternary ammonium salts tested for possible applicability in zinc–bromine batteries is presented in Table 7.2. A rough classification has been applied according to the substance classes of the substituents attached to the nitrogen.

The two basic requirements for efficient bromine storage in zinc–bromine batteries, which need to be met in order to ensure low self-discharge are a substantial reduction of equilibrium vapor pressure of Br_2 of the polybromide phase in association with low solubility of active bromine in the aqueous phase. As mentioned by Schnittke [4] the use of aromatic N-substituents for battery applications is highly problematic due to their tendency to undergo bromination. Based on Bajpai's pioneering work on vapor pressures [56] (see Figure 7.2) and Eustace's [73] study of Br_2 distribution between the complex phase and overstanding aqueous solution in combination with stability considerations N-methyl N-ethylmorpholinium bromide (MEM), and N-chloromethyl N-methylpyrrolidinium bromide (C-MEP) and analogous aliphatic heterocyclic compounds have confirmed their advantages for possible application in zinc-flow batteries.

Eustace noticed the correlation between asymmetric N-substitution and low melting points (at temperatures ≥ 15 °C) of the substances [73] observed in this study. The importance of sufficiently high densities of the 'fused salts' for an efficient separation of the complex phase and the aqueous solution was emphasized. Mixtures of various quaternary ammonium bromides were tested subsequently [48, 49, 66, 75], aiming particularly at the avoidance of crystallization in the

Table 7.2 Complexation of Br_2 with quaternary ammonium salts.

Substance class	Examples and references
Aliphatic	$Me_2Et_2N^+Br^-$ [47, 48], $MeEt_3N^+Br^-$ [47, 48]
	$Me_2EtPrN^+Br^-$ [47, 48], $Me_2Et_2PrN^+Br^-$ [47–49]
	$Et_3PrN^+Br^-$ [48]
	Bu_4N^+Br [14, 33, 45, 49–51]
	$Et_4N^+Br^-$ [14, 45, 49, 52, 53], Et_4N^+Cl [14, 53]
	$Me_4N^+Br^-$ [45, 49, 51, 54–56]
	$Oct_4N^+Br^-$ [45]
	Me_3EtN^+Br [48, 57], Me_3PrN^+Br [48]
	$MeEt_2CMN^+Br^-$ [47]
	$Me_3CMN^+Br^-$ [48], $Et_3CMN^+Br^-$ [48]
	$Me_2EtCMN^+Br^-$ [48], $MeEt_2CMN^+Br^-$ [48]
	$Oct_3MeN^+Cl^-$ [56] $(isoamyl)_3NHCl$ [50]
	Me_3NHBr [50, 58], Me_3NHCl [50]
	$NH_4^+Br^-$ [59]
	$Et_3hexadecN^+Br^-$ [60], $Me_2hexadec_2N^+Br^-$ [60]
	Me_2dodec-2-hydoxy-EtN^+Br^- [60]
Aromatic	Et_3PhN^+Br [53, 60]
	$Me_3PhN^+Br^-$ [51, 61, 62]
Heterocyclic	Benzyldimethyl-(3-isobutoxy-2-hydroxy propyl)-N^+Br^- [60]
	N-Methyl-N-ethyl pyrrolidinium bromide (MEP) [3, 14, 47–49, 63–72]
	N-Methyl-N-ethyl morpholinium bromide (MEM) [3, 14, 47–49, 53, 56, 63–73]
	N-Chloromethyl-N-methyl pyrrolidinium bromide (C-MEP) [47, 48, 73]
	N-Chloromethyl-N-methyl morpholidinium bromide (C-MEM) [48]
	2-Bromo-cyclohexylpyridinium bromide [53, 60, 63, 74]
	2-Chloro-cyclohexylpyridinium bromide [53, 60]
	N-Bromo-t-butyl pyridinium bromide [60]
	N-Methoxymethyl-N-methyl piperidinium bromide [73]
Tenside-like	$[PhCH_2NMe_2(CH_2)_6NMe_2CH_2Ph]^{2+}2Br^-$ [53, 60]
	$[C_{16}H_{33}NMe_2(CH_2)_4NMe_2C_{16}H_{33}]^{2+}2Br^-$ [60]
	$[C_{12}H_{25}NMe_2(C_2H_4O)_3H]\,Cl^-$ [53]

temperature range (0–50 °C), which is required for successful operation of flow batteries. No single compound was able to fulfill this demand. From tests of various ratios of MEP, MEM, and aliphatic ammonium bromides at 0, 50, and 100% state of charge (SOC) and different temperatures, mixtures of MEP and MEM were found to be particularly useful for bromine storage. The highly efficient potential of MEP for reducing the concentration of free Br_2 in the aqueous phase was pointed out. This fact was also mentioned by Bellows *et al.* [76] in their study of possibilities for improving the battery electrolyte. Table 7.3 [48] illustrates the results of that work.

Kawahara [64] studied the influence of the molar ratio MEP:MEM on the rate of self-discharge in model batteries. This effect is strongly enhanced by high

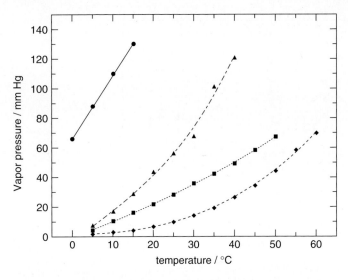

Figure 7.2 Vapor pressures of bromine/quaternary ammonium salt complexes: elemental Br_2, $Me_4N^+Br^-$, MEMBr, $Oct_3MeN^+Cl^-$. From Ref. [56].

Table 7.3 Equilibrium bromine concentrations in the aqueous electrolyte phase.

SOC	QBr	Br_2 Concentraion (mol L^{-1})		
		0 °C	25 °C	50 °C
0% SOC	MEM	0.095	0.187	0.270
	MEP	0.043	0.065	0.143
50% SOC	MEM	0.049	0.080	0.174
	MEP	0.033	0.037	0.123
100% SOC	MEM	0.115	0.141	0.174
	MEP	0.077	0.054	0.123

Taken from Ref. [48].

concentrations of bromine in the aqueous electrolyte phase. As suggested by the curves in Figure 7.3, the ratio 3 : 1 seems to be the most effective among the mixtures under consideration over one complete charge process, which is represented here by the percentage of electrochemical deposition of the available Zn^{2+}. An evident decrease in the aqueous bromine concentration is a direct consequence of the high storage capacity of the complex phase. A small series of zinc-flow batteries using this MEP:MEM ratio are produced at present by the Powercell Co. (Austria).

It should be kept in mind that not only are quaternary ammonium salts useful as complexing agents for bromine storage, but they provide further advantages, in particular concerning the electrochemical deposition of zinc. Their behavior as leveling agents [77] and their dendritic growth-inhibiting properties [78, 79] are

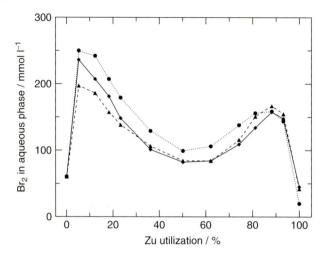

Figure 7.3 Bromine concentration in the aqueous phase in equilibrium with complexes of different MEP:MEM ratios: 1:1, 3:1, 9:1. Taken from Ref. [64].

well known. Positive effects on the elimination of hydrogen evolution during the charge process by adding $NH_4^+Cl^-$ (~1:3 $NH_4Cl:ZnBr_2$) to the electrolyte have been reported [80].

7.3
Physical Properties of the Bromine Storage Phase

7.3.1
Conductivity

The conductivity of a number of bromine containing complexes with different quaternary ammonium cations was studied by Gerold (see Ref. [53]) with respect to the dependence on temperature and bromine content. Most of the substances yielded phases that were crystalline at ambient temperature but that could liquefied by addition of large amounts of Br_2.

Electrical conductivities of polybromide complexes containing MEP and MEM were studied by Arbes [71]. Pure MEPBr complexes always show higher conductivity than those containing only MEM, as long as equal amounts of Br_2 are added. According to these investigations the conductivities of the fused polybromide salts increase exponentially with the concentration of Br_2, reaching values typical for the aqueous electrolyte phases (11–20 Ω/cm) at very high bromine contents such as 3 mol Br_2/mol complexing agent.

The dependence on the temperature of the specific resistance (Ω/cm) of the pure MEPBr and MEMBr complexes, and a 1:1 mixture thereof, as obtained in Ref. [71], is listed in Table 7.4. It is remarkable that within the complex phases consisting of

Table 7.4 Specific resistance and states of aggregation of pure MEMBr and MEPBr complexes and a 1:1 mixture.

Temperature (°C)	MEMBr		MEPBr		MEMBr/MEPBr (1:1)	
	Specific resistance (Ω cm^{-1})	State	Specific resistance (Ω cm^{-1})	State	Specific resistance (Ω cm^{-1})	State
0	287	s	194	s	180	s
10	173	s	113	s	93.4	1
20	124	s	28.7	1	43.1	1
30	43.1	1	24.2	1	33.3	1
40	32.3	1	19.3	1	25.6	1
50	24.3	1	16.2	1	21.1	1
60	20.1	1	14.1	1	17.5	1

Taken from Ref. [71]

Br$_2$ and either pure MEP or MEM the change of specific resistance at the liquid → solid phase transition amounts to about 1 order of magnitude, whereas the value is only doubled in the 1:1 mixture. The table also indicates that MEMBr complexes possess higher melting temperatures.

Cathro et al. determined conductivity data of aqueous electrolyte phases containing MEP and MEM, varying the concentrations of ZnBr$_2$ and of the complexing agents as well as the temperature conditions [66]. Tables 7.5 and 7.6 contain a compilation of the results obtained at concentrations close to those occurring during operation of zinc-flow batteries. Aqueous phases containing MEM were found to provide better conductivity than those containing MEP.

Eustace [73] studied the specific resistance of samples of bromine-fused salt phase produced by electrolysis of 3.0 mol L^{-1} ZnBr$_2$ and 1.0 mol L^{-1} MEM at 23 °C. As is shown in Figure 7.4, a considerable specific resistance is observed in the initial phase of the charge process, dropping to approximately one-third at

Table 7.5 Specific resistance (Ω^{-1} cm) of aqueous electrolyte containing MEM and 3 mol L^{-1} ZnBr$_2$.

MEM (mol L^{-1})	3 mol L^{-1} ZnBr$_2$		
	0 °C	25 °C	50 °C
0.0	17.2	9.9	7.4
0.5	22.5	12.7	9.0
0.7	24.5	14.0	11.4
1.0	30.8	15.9	11.6

Taken from Ref. [66].

Table 7.6 Specific resistance of aqueous electrolyte containing MEM or MEP at 25 °C.

Concentration (mol L^{-1})		Specific resistance (Ω cm^{-1})	
ZnBr$_2$	QBr	QBr = MEP	QBr = MEM
1	0.3	11.4	9.8
2	0.65	12.1	11.2
3	1.0	16.5	15.9

Taken from Ref. [66].

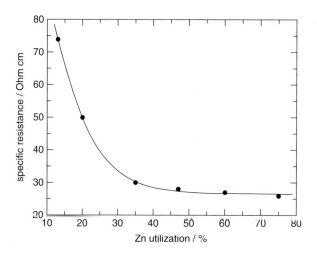

Figure 7.4 Specific resistance of a pure MEM–polybromide complex phase at 23 °C at various states of charge (represented by zinc utilization). Taken from Ref. [73].

30% Zn utilization. At higher states of charge the increase in the conductivity is significantly slower.

Corresponding to the SOC, variable concentrations of Zn-containing complex anions ($[ZnX_n(OH)_m]^{q-}$ with X being Cl or Br, $q = 1.2$, and integers n, m ranging from 0 to 4) were detected in the complex polybromide phase. As is evident from Table 7.7 compiled by Hauser [70], the electrical conductivity decreases with increasing amounts of zinc. This effect is associated with a rise in the concentration of Br$_2$ in the equilibrated aqueous phases. The increase in specific conductivity of the complex phase with temperature shown in Table 7.4 was confirmed by Hauser [70] and Niepraschk [63]; however, the latter reported essentially higher values for the polybromide phase than for the aqueous phase of the electrolyte. He estimated the maximum conductivity to be achieved at ~9 mol L^{-1} Br$_2$ complex phase. Figure 7.5 shows the conductivity of MEM and MEP at different temperatures and Br$_2$ concentrations. From the temperature dependence of the conductivity the activation

Table 7.7 Conductivity λ of the polybromide phase at 50% SOC and various contents of Zn^{2+}.

$[Zn^{2+}]$ (mol L^{-1})	λ (S cm^{-1})	Temperature (°C)
0.00	3.42×10^{-2}	20
0.07	3.53×10^{-2}	20
0.60	2.76×10^{-2}	20
0.95	2.07×10^{-2}	20
2.05	0.58×10^{-2}	21

Taken from Ref. [70].

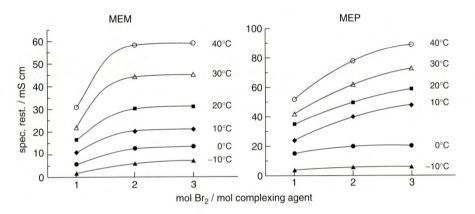

Figure 7.5 Isotherms for the conductivity of pure MEM and MEP-complex phases (mS cm^{-1}) versus degree of added Br_2. Taken from Ref. [63].

energy for the transport of charge can be obtained applying Arrhenius' equation,

$$k = k_0 e^{-\frac{E_A}{RT}} \qquad (7.1)$$

Niepraschk [63] found 23.4 kJ mol^{-1} for pure MEMBr complex and 16.5 kJ mol^{-1} for pure MEPBr complexes at 3 mol Br_2/mol complexing agent. These values increase slightly with decreasing concentration of Br_2. A value of E_A of 11.5 kJ mol^{-1} for a complex phase containing MEP:MEM in the ratio 3:1 at comparable contents of Br_2 was reported by Hauser [70].

7.3.2
Viscosity and Specific Weight

The kinematic viscosity of MEM containing aqueous electrolytes at different concentrations of MEM and $ZnBr_2$ and at different temperatures has been studied 66 (see Table 7.8).

Table 7.8 Kinematic viscosity ($m^2 \, s^{-1}$) of aqueous electrolyte containing MEM and 3 mol L^{-1} $ZnBr_2$.

MEM (mol L^{-1})	3 mol L^{-1} $ZnBr_2$		
	0 °C	25 °C	50 °C
0.0	2.786	1.371	0.786
0.5	3.722	1.665	0.960
0.7	4.153	1.809	1.048
1.0	4.996	2.175	1.215

Taken from Ref. [66].

Kinematic viscosities of aqueous electrolyte phases containing $Et_4N^+Br^-$ and $Bu_4N^+Br^-$ and various concentrations of $ZnBr_2$ were studied by Cedzynska [75]. Ionic conductivity of bromine-storing phases was estimated [53] by applying the Pisarzhevski–Walden equation to measured values of the dynamic viscosity. However, use of this relation is only correct for solutions in the limit of zero concentration and no change in solution mechanism.

Eustace [73] reported a dynamic viscosity of 25 cP for a pure MEMBr complex phase at 23 °C. (The specific weight was 2.3 g cm^{-3}).

Typical specific weights of bromine-storing complex phases between 10 and 70 °C at bromine concentrations in the range of 1–4 mol/mol complexing agent lie around 2.45 ± 0.1 g cm^{-3}. Densities as a function of temperature at various bromine contents of a number of quaternary ammonium salts were given by Gerold [53].

Viscosities and specific weights of complexes and the corresponding aqueous phases, with the aim of simulating realistic battery conditions with MEP:MEM ratio of 1 : 1, 3 : 1, and 6 : 1 in the electrolyte at 50, 75, and 100% SOC, were studied in a temperature range between ~10 and 50 °C [81]. Kinematic viscosities between 5×10^{-6} and 30×10^{-6} $m^2 \, s^{-1}$ of the complex phases were found.

7.3.3
Diffusion Coefficients

From the value of the diffusion coefficient of Br_2 in electrolyte solutions, conclusions can be drawn concerning mass transport in the diffusion layer on the anode surface. Cathro et al. [66] used the rotating disc electrode in order to determine the diffusion coefficients of bromine in aqueous electrolyte phases containing various concentrations of MEM, MEP, and $ZnBr_2$. The results are listed in Table 7.9. It was found that aqueous electrolytes containing MEP exhibit diffusion coefficients higher by ~6% compared with those with MEM. A compilation of diffusion coefficients of Br_2 in various electrolytes is given in Table 7.10.

Table 7.9 Diffusion coefficients of Br_2 in the aqueous electrolyte phase at 25 °C.

Concentration (mol L^{-1})		Specific resistance (Ω cm^{-1})	
ZnBr$_2$	QBr	QBr = MEP	QBr = MEM
1	0.3	1.00	0.95
2	0.65	0.61	0.59
3	1.0	0.38	0.35

Taken from Ref. [66].

Table 7.10 Diffusion coefficients of Br_2 in aqueous electrolytes at 25 °C.

Diffusion coefficient (10^{-9} m^2 s^{-1})	Electrolyte	References
0.99	1.0 mol L^{-1} ZnBr$_2$	[82]
1.23	1.0 mol L^{-1} ZnBr$_2$	[66]
1.21	0.1 mol L^{-1} KBr	[83]
1.44	2.2 mol L^{-1} KCl	[84]
2.00	Not reported	[85]

Karigl [69] defined a format diffusion coefficient for bromine transport through a polyethylene separator of a zinc-flow battery by considering the separator as a diffusion layer. A value of $D_{sep}(Br_3^-) = 2.77 \times 10^{-10}$ m^2 s^{-1} was obtained.

Diffusion coefficients of Br_2 in aqueous electrolyte phases containing $Et_4N^+Br^-$ and $Bu_4N^+Br^-$ were studied by Cedzynska [75] at various concentrations of $ZnBr_2$. Attempts at finding an ideal modified (MOD) electrolyte consisting of MEM, MEP, $MeEt_2PrN^+Br^-$, and $Bu_4N^+Br^-$ were made [49].

7.3.4
State of Aggregation

An indispensable requirement for an efficient complexing agent usable in a zinc-flow battery is a sufficiently low melting point of the corresponding complex phase. By this criterion the application of a vast number of quaternary ammonium cations is ruled out. Low melting points of the bromine-storing phase can commonly be achieved by combinations of two or more different complexing agents. As was anticipated by Gibbard [86], asymmetric substitution reduces the tendency of the complex to crystallize.

Cathro et al. [48] found that the use of mixtures of QBr-compounds (Q = quaternary ammonium) may give a liquid polybromide phase, even though the individual components form solid or highly viscous polybromides. With the two chloro compounds *N*-chloromethyl-*N*-methyl morpholidinium bromide (C-MEM) and C-MEP, white precipitates were formed at 0 °C upon mixing with an aqueous solution of 1 mol L^{-1} ZnBr$_2$. Symmetrically substituted QBr compounds such as Me$_4$N$^+$Br$^-$ and Et$_4$N$^+$Br$^-$ yielded solid polybromide phases over wide ranges of temperature and composition [48]. A pronounced influence not only of the temperature but also of the bromine contents of the complex phases is reviewed. Results for the corresponding aqueous phases of the electrolyte can be found in the original paper, which also contains studies of a number of electrolyte phases made up of MEM:MEP mixtures.

7.4
Analytical Study of a Battery Charge Cycle

In order to improve the bromine-storing capacity and hence the battery efficiency of a zinc-flow cell, knowledge of the structure and consistency of the complex phase during the entire charge–discharge cycle is an essential requirement.

Until today the only available data, obtained by direct sampling of a prototype battery system concerning mass flow of the complexing agents as well as the Br$_2$ produced in both the aqueous and nonaqueous electrolyte phases, have been gained by application of Raman spectroscopy [87, 88].

The most interesting results of a study of a real 12 V/1 kWh zinc-flow battery (Powercell Lda.) with a charge capacity of 92 Ah are reviewed in Figures 7.6 and 7.7.

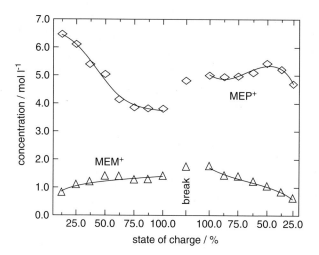

Figure 7.6 Concentration of the complexing cations MEP^1 (△) and MEM^1 (◇) in the complex electrolyte phase during one total charge–discharge cycle of a model zinc-flow battery. Taken from Ref. [88].

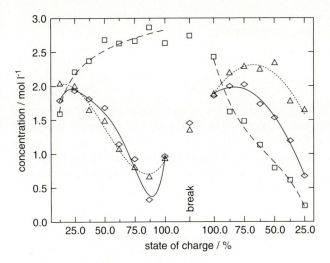

Figure 7.7 Concentration of Br_2 (◇), Br_3 (△), and Br_5 (□) in the complex electrolyte phase during one total charge–discharge cycle of a model zinc-flow battery. Taken from Ref. [88].

phase of the charging procedure shows a very high MEP:MEM ratio (~8:1), reaching the value 3:1 only after 50% SOC. This effect is determined by reaction kinetics and availability of the complexing An electrolyte composition of 3:1 MEP:MEM was used from Figure 7.6 it is evident that the complex formed in the initial agents in the cell compartment. During the discharge process, when complex of essentially similar composition was delivered to the anode, MEM was found to be released from the complex more easily then MEP. From these results higher stability of MEP-rich complex phases under the battery operation conditions can be deduced.

An estimate of the concentrations of the individual polybromide anions and of Br_2 in the complex phase during the same charge–discharge cycle, derived from spectral data, is plotted in Figure 7.7. The curves indicate a pronounced tendency toward Br_5^- formation during the charge period, but domination of Br_3^- and Br_2 species while the battery is discharging. Highest concentrations of pentabromide seem to be correlated with low contents of complexing agents. Further conclusions from these data must be drawn with extreme caution due to the sensitivity of the equilibrium between the bromine species to thermal effects, influences of the electric fields, and the uncertainties inherent in the underlying mathematical model used for data analysis.

The authors suggested testing of the application zinc-flow batteries with higher MEP:MEM ratios. It should be noted that the results from a study of the mass flow in the aqueous parts of the system [87] correlated favorably with the findings described above.

7.5
Safety, Physiological Aspects, and Recycling

7.5.1
Safety

Safety risks and environmental impact are of major importance for the practical success of bromine storage system. The nonaqueous polybromide complexes in general show excellent physical properties, such as good ionic conductivity (0.1–0.05 Ω cm^{-1}), oxidation stability (depending on the nature of the ammonium ion), and a low bromine vapor pressure. The concentration of active bromine in the aqueous solution is reduced by formation of the complex phase up to 0.01–0.05 mol L^{-1}, hence ensuring a decisive decrease of self-discharge.

Figure 7.2 demonstrates that the bromine vapor pressure over a complex phase remains remarkably low with increasing temperature and is not a critical factor restricting battery operation. Even at ~60 °C, vapor pressures of Br$_2$ and elemental bromine reaching only a small percentage of the atmospheric pressure are obtained.

Moreover, calculations on the evaporation rate of bromine from the complex phase were carried out assuming a worst-case scenario, namely a complete spillage of the total bromine inventory (as polybromide complex) of a fully charged (100% SOC) 15 kWh module, which means ~32.5 kg of available Br$_2$, forming a 10 m^2 pool on the ground as a consequence of battery damage. Rates of bromine evaporation from the complex phase in air were measured in laboratory tests [89] under various conditions (flow, temperature). The results are presented in Figure 7.8.

Figure 7.9 shows the distribution of bromine emissions (concentrations) as a function of distance from the source of emission, assuming various atmospheric conditions (air flow) at 20 °C. The maximum admissible concentration (MAK value) of 0.7 mg m^{-3} (0.01 ppm) is reached within about 50 m under worst-case atmospheric condition, whereas higher and dangerous values are observed close to the place of complex spillage. However, the assumptions serving as the basis for these estimates and calculations appear rather unrealistic, considering the low probability of a complete release of the complex phase at 100% SOC from the reservoir and of the distribution on the ground, and moreover neglecting the role of the aqueous electrolyte solution which is also present and tends to spread on the complex surface due to its lower density.

The most important safety-relevant measures involving automatic control during battery operation are:

- thermal management, at a maximum of 42 °C;
- automatic pumps and flow control, leakage sensor and pumps that stop automatically in the case of bromine escaping from the stack;
- overcharge control.

As shown by several investigations [89], the bromine-rich polybromide phase by itself is hardly flammable and fire-extinguishing properties have been reported

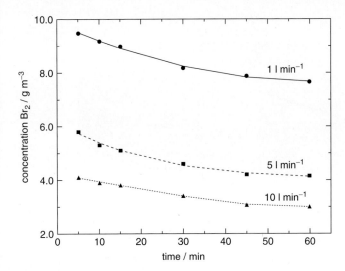

Figure 7.8 Evaporation rate of bromine from the complex phase at 100% SOC in air is shown on each curve. Taken from Ref. [67].

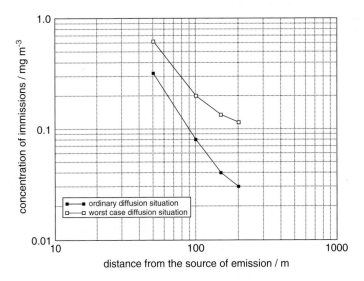

Figure 7.9 Diffusion of bromine after evaporation from a pool of complex phase at 20 °C. Taken from Ref. [67].

occasionally. The formation of polybrominated dibenzo-dioxins (PBrDDs) and polybrominated dibenzo-furans (PBrDFs) due to the plastic-containing housing of a zinc-flow battery cannot be totally neglected in the case of a fire, but their concentrations are far away from the tetrachloro-dibenzodioxin (TCDD) toxic equivalents even in a worst-case scenario.

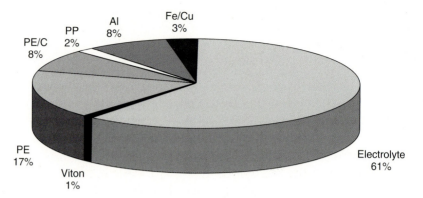

Figure 7.10 Materials used in a battery, and their recyclability. Taken from Ref. [67].

7.5.2
Physiological Aspects

Elemental bromine is a readily evaporating liquid (p_{Br} at $1\,°C = 0.23$ bar) with high reactivity. Because of the good solubility of Br_2 in lipids, its aggressive and toxic properties affect skin and mucous membranes (bronchi). The MAK value of elemental Br is defined as 0.1 ppm (0.7 mg m^{-3}) while the sense of smell is affected at a value of 0.01 ppm. [89, 90].

Storage of bromine by formation of a polybromide phase with a lowering of the vapor pressure by more than one order of magnitude, to at least 10% of the value of Br_2 at maximum, is the basic requirement for safe application in zinc-flow batteries [89, 90]. No information is available concerning negative health effects of the complexing agents MEM and MEP.

7.5.3
Recycling

Recycling of the major valuable battery components is an important factor influencing the introduction into the market and the economic development of the system. Figure 7.10 shows a breakdown of the materials and components, their weight fractions and recyclability.

Recycling of the electrolyte from used, damaged, or faulty batteries and reuse in new stacks have gained considerable attention. The electrolyte is an essential constituent from the technical and economic viewpoint, showing extraordinary stability and no ageing effects.

References

1. Zito, R. (1968–1973) US Patent 3 382 105, 3 640 770, 3 640 771, 3 642 738, 3 719 526.

2. Eskra, M., Eidler, P., and Miles, R. (1991) Proceedings of the 24th International Symposium Automotive

technology and Automation, Florence.
3. EXXON Research Cooperation (1980) Monthly Progress Report, April, 1980.
4. Schnittke, A. (1992) PhD Thesis. Ernst-Moritz-Arndt-University Greifswald (in German).
5. Movahedi, R. (1977) PhD Thesis. TU Vienna (in German).
6. Hrold, A. (1979) in *Intercalated Materials* (ed. F. Levy), D. Reidel, Dordrecht, pp. 323–421.
7. Rubim, J.C. and Sala, O. (1980) *J. Raman Spectrosc.*, **9**, 155.
8. Rubim, J.C. and Sala, O. (1981) *J. Raman Spectrosc.*, **11**, 320.
9. Mastragostino, M. and Valcher, S. (1983) *Electrochem. Acta*, **28** (4), 501–505.
10. Mengoli, G., Musiani, M.M., Tomat, R., Valcher, S., and Pletcher, D. (1985) *J. Appl. Electrochem.*, **15**, 697–704.
11. Ottenbrite, R.M. (1981) *Polym. Bull.*, **6**, 225–228.
12. Manassen, J. and Cabasso, I. (1989) *J. Electrochem. Soc.*, **136**, 578.
13. Irving, H. and Wilson, P.D. (1964) *J. Inorg. Nucl. Chem.*, **26**, 2235.
14. Vogel, I. (1990) PhD Thesis. TU Dresden, Sektion Chemie (in German).
15. Popov, A.I. (1967) in *Halogen Chemistry* (ed. V. Gutmann), Academic Press, New York, pp. 225–264.
16. Surles, T. and Popov, A.I. (1969) *Inorg. Chem.*, **8** (10), 2049.
17. Scaife, D.B. and Tyrell, H.J.V. (1958) *Chem. Soc.*, **386**, as reported by Gerold A. (1991) Master's Thesis. TU Dresden, Sektion Chemie (in German).
18. Bogaard, M.P., Peterson, J., and Rae, A.D. (1979) *Cryst. Struct. Commun.*, **8**, 347.
19. Ollis, J., James, V.J., Ollis, D., and Bogaard, M.P. (1976) *Cryst. Struct. Commun.*, **5**, 39.
20. Bogaard, M.P. and Rae, A.D. (1982) *Cryst. Struct. Commun.*, **11**, 175.
21. Gabes, W., Stufkens, D.J., and Gerding, H. (1973) *J. Mol. Struct.*, **17**, 329.
22. Fries, K. (1906) *Ann. Chem.*, **346**, 217.
23. Horton, H.E.L. (2000) *Chem. Ber.*, **1888**, 21.
24. Ajami, A.M. *et al.* (1977) US Patent.
25. Singh, P., White, K., and Parker, A.J. (1983) *J. Power Sources*, **10**, 309.
26. Henning, G. and McClelland, J.D. (1975) *J. Chem. Phys.*, **23**, 1431.
27. Rouillon, J.C. and Marchand, A. (1972) *C.R. Acad. Sci. Paris*, **112**, 274.
28. Aoki, K. (1971) *J. Mater. Sci.*, **6**, 140.
29. Mazieres, C., Colin, G., Jegoudez, J., and Setton, R. (1975) *Carbon*, **13**, 289.
30. Miyauchi, K. and Kakahashi, Y. (1976) *Carbon*, **14**, 35.
31. Bloc, F. (1964) PhD Thesis. Nancy, 1964.
32. Herbstein, F.H., Kapon, M., and Reisner, G.M. (1981) *Proc. R. Soc Lond., Ser. A*, **376**, 301.
33. Kalina, D.W., Lyding, J.W., Ratajack, M.T., Kannewurf, C.R., and Marks, T.J. (1980) *J. Am. Chem. Soc.*, **102**, 7854.
34. Tebbe, K.F. (1977) in *Homoatomic Rings, Chains and Macromolecules of Main-Group Elements* (ed. A.L. Rheingold), Elsevier, Amsterdam, pp. 551–606.
35. Person, W.B., Anderson, G.R., Fordemwalt, J.N., Stammreich, H., and Forneris, R. (1961) *J. Chem. Phys.*, **35**, 908.
36. Evans, J.C. and Lo, G.Y.-S. (1967) *Inorg. Chem.*, **6**, 1483.
37. Bellucci, G., Bianchini, R., Chiappe, C., and Ambrosetti, R. (1989) *J. Am. Chem. Soc.*, **111**, 199.
38. Greenwood, N.N. and Earnshaw, A. (1990) *Chemie der Elemente*, Wiley-VCH Verlag GmbH, Weinheim (in German).
39. Bowmaker, G.A., Boyd, P.D., and Sorrenson, R.J. (1984) *J. Chem. Soc., Faraday Trans. 2*, **80**, 1125.
40. Bertelot, J., Guette, C., Desbène, P.-L., Basselier, J.-J., Chaquin, P., and Masure, D. (1990) *Can. J. Chem.*, **68**, 464.
41. Gutsev, G.L. (1992) *Russ. J. Phys. Chem.*, **66**, 1596.
42. Lin, Z. and Hal1, M.B. (1993) *Polyhedron*, **12**, 1499.
43. Schuster, P., Bauer, G., Mikosch, H., Fabjan, C., and Drobits, J. to be published.
44. Breneman, G.L. and Willet, R.D. (1986) *Acta Crystallogr. Sect. C*, **42**, 1614.
45. Burns, G.R. and Renner, R.M. (1991) *Spectrochim. Acta Sect. A*, **47**, 991.
46. Wang, T.X., Kelley, M.D., Cooper, I.N., Beckwith, R.C., and Margerum, D.W. (1994) *Inorg. Chem.*, **33**, 5872.

47. Hoobin, P.M., Cathro, K.J., and Niere, J.O. (1989) *J. Appl. Electrochem.*, **19**, 943–945.
48. Cathro, K.J., Cedzynska, K., Constable, D.C., and Hoobin, P.M. (1986) *J. Power Sources*, **18**, 349.
49. Cedzynska, K. (1995) *Electrochim. Acta*, **40** (8), 971.
50. Mercier, P.L. and Kraus, C.A. (1956) *Proc. Nat. Acad. Sci. U.S.A.*, **42**, 487.
51. (a) Rallo, F. and Silvestroni, P. (1973) *Gazz. Chem. Ital.*, **103**, 1011; (b) Rallo, F. and Silvestroni, P. (1972) *J. Electrochem. Soc.*, **119** (11), 1471.
52. Rubinstein, I., Bixon, M., and Gileadi, E. (1980) *J. Phys. Chem.*, **84**, 715.
53. Gerold, A. (1991) Master's Thesis. TU Dresden, Sektion Chemie (in German).
54. Clerici, G., Rossi, M., Marcheto, M., and Collins, D.H. (1974) *Power Sources 5*, Academic Press, London, p. 167.
55. Bloch, R. (1951) US Patent 2 566 114.
56. Bajpai, S.N. (1981) *J. Chem. Eng. Data*, **26** (1), 2.
57. Kinoshita, K., Leach, S.C., and Ablow, C.M. (1982) *J. Electrochem. Soc.*, **129** (11), 2397.
58. Mesen, G.W. and Kraus, C.A. (1952) *Proc. Nat. Acad. Sci. U.S.A.*, **38**, 1023.
59. Vogel, I. and Moebius, A. (1991) *Electrochim. Acta*, **36** (9), 1403.
60. Jacob, G. (1986) Master's Thesis. TU Dresden Sektion Chemie (in German).
61. Chattaway, F.D. and Hoyle, G. (1923) *J. Chem. Soc.*, 654.
62. Kume, Y. and Nakamura, D.J. (1976) *J. Magn. Reson.*, **21**, 235.
63. Niepraschk, H. (1988) Master's Thesis. TU Dresden (in German).
64. Kawahara, K. (1987) *4th International Zinc-bromine Battery Symposium Perth, Australia*, Toyota Central R&D Inc., Handout.
65. Fabjan, C. and Hirss, G. (1986) *Dechema Monographie*, vol. 102, Wiley-VCH Verlag GmbH, Weinheim, p. 149 (in German).
66. Cathro, K.J., Cedzynska, K., and Constable, D.C. (1985) *J. Power Sources*, **16**, 53.
67. Fabjan, C. and Kronberger, H. (1993) *Sterreich. Z. Energiewirt. (ZE)*, **46** (9), 451 (in German).
68. Stckl, W. (1988) PhD Thesis. TU Vienna (in German).
69. Karigl, B. (1993) PhD Thesis. TU Vienna (in German).
70. Hauser, R. (1993) PhD Thesis. TU Vienna (in German).
71. Arbes, G. (1984) PhD Thesis. TU Vienna (in German).
72. Fabjan, C. and Kordesch, K. (1982) *Studie und Vorprojekt: Zink-Halogen Batterie/Elektrofahrzeug*, vol. 54, VEW AG, Wien, p. 1 (in German).
73. Eustace, D.J. (1980) *J. Electrochem. Soc.*, **127**, 528.
74. Moebius, A., Vogel, J., Jakob, G., Grosser, K., Beger, J., Jacobi, R., and P schmann, C. (1987) Word Patent 308 178.
75. Cedzynska, K. (1989) *Electrochim. Acta*, **34** (10), 1439.
76. Bellows, R.J. (1985) First International Zinc-Bromine Battery Symposium, Jan 1985, Final Progress Report, at S.E.A. Mürzzuschlag, Austria.
77. Bauer, G., Drobits, J., Fabjan, C., Kronberger, H., Mikosch, H., and Schuster, P. (1996) Elektrochemie der Elektronenleiter, GDCh-Monographie 3, pp. 121–135 (in German).
78. Meidensha Electric Mfg. Co. Ltd., Japan (1982) Japanese Patent 82-129626 820727 (in Japanese).
79. Meidensha Electric Mfg. Co. Ltd. Japan (1982) Japanese Patent 82-88911 820527 (in Japanese).
80. Ando, Y. (1987) Meidensha Electric Mfg. Co. Ltd., Japan, Japanese Patent 87-209151 870825 (in Japanese).
81. Drobits, J., Schuster, P., and Fabjan, C. in preparation.
82. Lee, J. and Selman, R. (1983) *J. Electrochem. Soc.*, **130**, 1237.
83. Osipov, O.R., Novitskii, M.A., Povarov, Y.M., and Lukovtsev, P.D. (1972) *Sov. Electrochem.*, **8**, 317.
84. Voloshina, G.N., Ksenzenko, V.I., Abdyev, S., and Chemleva, T. (1978) *Isv. Akad. Nauk Turkm. SSR, Ser. Fiz-Tekh. Khim. Geol. Nauk*, 64, as reported in Ref. 67.
85. Bellows, R.J., Eustace, D.J., Grimes, P., Shropshire, J.A., Tsien, H.S., and Venero, A.F. (1979) in *Power Sources*,

vol. 7 (ed. J. Thompson), Academic Press, London, p. 301.
86. Gibbard, H.F. (1979) UK Patent Application Great Britain.
87. Bauer, G., Drobits, J., Fabjan, C., Mikosch, H., and Schuster, P. (1996) *Chem. Ing. Tech.*, **68**, 100 (in German).
88. Bauer, G., Drobits, J., Fabjan, C., Mikonch, H., and Schuster, P. (1997) *J. Electroanal. Chem.*, **427**, 123.
89. Steininger, T. (1992) V-Bayer, G3-UTM50, November 25 1991 and 1 October 1992 (in German).
90. Schallaböck, K.O. and Lichtl, E. (1984) Vorausschauende Umweltverträglichkeitsprüfung für Zink-Brom-Batterien, inst. für Umweltforschung, Forschungsgesellschaft Johanneum, Graz, Austria (in German).

8
Metallic Negatives
Leo Binder

8.1
Introduction

Although many different methods of representing primary and secondary (or storage) batteries are used, the correct form of displaying a battery system is the following:

(−) Anode material/Electrolyte/Cathode material (+).

Therefore, being the negative electrode in the system, the anodes are frequently called '*negatives.*' When the battery is ready for use (discharge) the typical battery anode consists of a metal in the form of either sheet, powder, or an electrolytic deposit. The last version is usually found in some of the secondary (storage) batteries. During discharge, the metal atoms are oxidized to metal ions, delivering a number of electrons corresponding to the positive charge of the cation:

$$M \rightarrow M^{n+} + ne^- \qquad (8.1)$$

The electrons sustain the current via external load and are used to reduce the active material of the cathode (positive). In the case of storage batteries, the ideal anodic and cathodic reactions are completely reversible.

8.2
Overview

Since there are numerous metals used as anodes (negatives) in a variety of battery systems with aqueous electrolytes, there are different ways of arranging them in groups to enable easier access to the required information. Useful selection categories may be:

1) the chemical elements (iron, zinc, lead, etc., . . .)
2) battery types (primary, storage, etc., . . .)
3) methods of anode preparation (metal, metal oxide, or hydroxide + subsequent reduction, other metal compounds + reduction, etc., . . .).

Handbook of Battery Materials, Second Edition. Edited by Claus Daniel and Jürgen O. Besenhard.
© 2011 Wiley-VCH Verlag GmbH & Co. KGaA. Published 2011 by Wiley-VCH Verlag GmbH & Co. KGaA.

The most suitable way to organize the different metallic negatives in groups seems to be a combination of these three classifications, using one as the main criterion and the others to create subdivisions. Such a system is demonstrated in the following example:

- **First parameter**: (A) Element name (in alphabetical order)
- **Second parameter**: (B) Battery type;
  - (AB1) primary;
  - (AB11 ... AB1n) cathodes;
  - (AB2) storage;
- **Third parameter**: (C) Method of preparation;
  - (AB2C1) introduced in metallic form;
  - (AB2C2) introduced as metal oxide or hydroxide;
  - (AB2C3) introduced as other metal compound.

The paragraphs AB2C1–AB2C3 may be similarly subdivided by considering the different cathode materials and thus defining a particular battery system.

8.3
Battery Anodes ('Negatives')

8.3.1
Aluminum (Al)

Aluminum is directly applied in its metallic form when it serves as battery anode, and it contributes significantly to the success of these battery systems by its outstanding electrochemical equivalent of 2980 Ah kg^{-1}.

Traditional battery concepts are in general single-use types (primary batteries). The most developed systems belong to the 'metal–air'-batteries using the reduction of atmospheric oxygen as the cathode reaction, for example, (−) Al/KOH/O$_2$ (+) or (−) Al/seawater/O$_2$ (+).

In some cases oxygen may be generated by decomposition of hydrogen peroxide [1].

Aluminum/air batteries with their high (theoretical) specific energy of 8.1 kWh kg^{-1} of Al, energy density of 219 kWh L^{-1}, and nearly unlimited access to the cathode active material (sometimes these batteries are called '*semi-fuel cells*'), found particular applications in the past and will be considered for electric vehicle propulsion in the near future [2]. The theoretical open-circuit voltage (OCV) of 2.73 V is another argument for the choice of this battery system. A review of the long history of aluminum as battery anode was given by Li [3].

The main discharge reactions are:

$$\text{Anode: } 4\text{Al} \rightarrow 4\text{Al}^{3+} + 12\text{e}^- \tag{8.2}$$

$$\text{Cathode: } 3\text{O}_2 + 6\text{H}_2\text{O} + 12\text{e}^- \rightarrow 12\text{OH}^- \tag{8.3}$$

$$\text{Overall cell reaction: } 4\text{Al} + 3\text{O}_2 + 6\text{H}_2\text{O} \rightarrow 4\text{Al}(\text{OH})_3 \tag{8.4}$$

In principle, all these reactions may be run in pure water, but in order to obtain sufficiently high currents, electrolytes with better conductivity (aqueous solutions of KOH or NaOH or seawater [4]) are necessary. The latter is found only in naval applications.

While in alkaline solutions, the problem of aluminum corrosion has to be overcome by protecting the metal by adding inhibitors like zinc oxide [5], zincate ions [6], a combination of zinc species with organic additives [7], or by replacing water in the electrolyte with methanol [8]; the aluminum anode in seawater has to be activated by breaking the passive surface layer with suitable etchants [9–12].

Although high-purity aluminum (up to 99.999%) is used in science and research dealing with the main discharge reactions and battery design concepts, most prototypes and batteries are made with aluminum alloys. Such materials are Alcan AB, ERC 3–4, Alcan BDW [13, 14], or S1–S6 (binary aluminum–tin alloys) [15]. Continuous efforts are made to replace expensive aluminum alloys by cheaper unalloyed Al qualities [16, 17] or to replace the normal sheet electrode by a polymer/aluminum composite electrode [18]. A survey of widely used alloys has been carried out by Zein el Abedin [19].

8.3.2
Cadmium (Cd)

Although one of the most common storage batteries is called the 'Nickel/Cadmium'-system ('NiCad'), correctly written (−) Cd/KOH/NiO(OH) (+), cadmium is not usually used to form a battery anode. The same can be said with regard to the silver/cadmium [(−) Cd/KOH/AgO (+)] and the 'MerCad'-battery [(−) Cd/KOH/HgO (+)]. The 'metallic negative' in these cases may be formed starting with cadmium hydroxide incorporated in the pore system of a sintered nickel plate or pressed upon a nickel-plated steel current collector (pocket plates), which is subsequently converted to cadmium metal by electrochemical reduction inside the cell (type AB2C2). This operation is performed by the purchaser on first use of these (storage) batteries in accordance with the users manual by inserting them into the appropriate charger for a certain time (e.g., overnight). Cadmium hydroxide for anode formation usually contains some additives (e.g., iron, nickel, graphite) and – in some cases – organic inhibitors and/or polymer binding agents [20–25].

The other method is cadmium electrodeposition on a nickel-plated steel foil (serving as current collector) using a plating bath containing acidified cadmium sulfate (type AB2C3). In this case the user is supplied with a battery in charged ('ready for use') condition.

The reversible anodic charge/discharge reaction is:

$$Cd(OH)_2 + 2e^- \longleftrightarrow Cd + 2OH^- \tag{8.5}$$

Because of the high hydrogen overvoltage of cadmium in the caustic electrolyte, no amalgamation is needed.

The electrochemical equivalent of about 480 Ah kg^{-1} is one of the lowest for all metallic anodes, and the OCV of 1.35 V for the 'Nicad' is not favorable for many applications. Studies of failure mechanisms [26] revealed that the cadmium electrode is responsible for capacity loss and memory effect of the nickel/cadmium battery. Additionally, it is desirable to restrict the use of cadmium for environmental reasons. The consequence is a continuous retreat of this system from many applications, and battery packs for electric tools may eventually be the only remaining use.

The replacement of nickel/cadmium batteries by nickel/metal hydride cells may be seen in this light. The better performance of the latter (about 30%) is a further strong argument to pay the only slightly higher price. Additional advantages are the flat discharge curve and the extremely good cycle life.

Recycling of valuable materials from used 'Nicad' batteries is an issue of growing importance [27].

8.3.3
Iron (Fe)

Probably the best-known battery system using an iron anode is the *nickel/iron battery*. It should be written: (−) Fe/KOH/NiO(OH) (+), and has its merits as a heavy duty accumulator [28]. By far less famous and much more recent are the applications of iron anodes in (rechargeable) iron/air cells [(−) Fe/KOH/O$_2$ (+)] [29, 30] and in iron/silver oxide batteries [(−) Fe/KOH(+LiOH)/AgO (+)] [31, 32].

The composition of an iron anode includes Fe$_3$O$_4$ (produced by partial reduction of Fe$_2$O$_3$ with hydrogen), iron powder, and additives (e.g., sulfur, FeS, HgO). One group of inventors claims 2000 cycles for an iron electrode containing ZnS as a main additive [33], others describe additive systems containing FeS (to retard passivation of the iron electrode) and ammonium sulfate (to provide porosity) [34], FeS and PbS to retard self-discharge [35], FeS and potassium sulfide to suppress hydrogen evolution and to improve cyclability [36], or potassium sulfide together with bismuth sulfide for the same reasons [37]. Relatively new is the addition of carbon nanoparticles to increase conductivity of the discharged electrode [38, 39]. Another method of electrode precursor preparation is filling carbon nanotubes with iron nitrate followed by decomposition of this compound in argon atmosphere yielding pure Fe$_2$O$_3$ [40].

The precursor mixture is converted to the active iron anode either by internal reduction (AB2C2) or by high-temperature external reduction (AB2C1) [41, 42].

The discharge/charge of this electrode is done in two steps, but only the first step (Fe $\leftrightarrow$ Fe^{2+} + 2e$^-$) is of practical use. For the iron/nickel oxide-hydroxide system these steps (or voltage plateaus) may be written as:

$$Fe + 2NiO(OH) + 2H_2O \longleftrightarrow 2Ni(OH)_2 + Fe(OH)_2 \tag{8.6}$$

$$3Fe(OH)_2 + 2NiO(OH) \longleftrightarrow 2Ni(OH)_2 + Fe_3O_4 + 2H_2O \tag{8.7}$$

The overall reaction is:

$$3Fe + 8NiO(OH) + 4H_2O \longleftrightarrow 8Ni(OH)_2 + Fe_3O_4 \tag{8.8}$$

The electrochemical equivalent of iron (if only the first reduction step is taken into account) is 960 Ah kg^{-1}, and the OCV of the 'nickel/iron' cell is 1.4 V.

8.3.4
Lead (Pb)

The lead electrode used as the anode in the well-known lead–acid battery is a rather complex structure consisting of a metallic grid (lead–antimony, lead–calcium, or other alloys [43–53]) filled with a paste made from particles of the active mass, sulfuric acid, and various additives (e.g., expanders, gelling agents, coatings [54–67], inert materials [68–75], or conductive particles [76–79]). The active mass originally consists of oxidized lead grains with a residual metal content of about 30%. These grains come either from a Barton reactor or from a ball mill. Their particular properties are described in another chapter (Part II, Chapter 9).

The pasted and cured anode plates have to pass through the formation process, which is nothing else but an external electrochemical reduction done in sulfuric acid (AB2C1) [80].

The regular discharge reaction follows a dissolution/precipitation mechanism simplified as follows:

$$Pb \longleftrightarrow Pb^{2+} + 2e^- \tag{8.9}$$
$$Pb^{2+} + SO_4^{2-} \longleftrightarrow PbSO_4 \tag{8.10}$$

Reaction mechanisms have been described in general [81–83] or in detail with regard to passivation [84], oxygen recombination [85], or the applied mode of charging [86], particularly pulse charging [87]. Current and potential distributions were studied by Li and co-authors [88].

Many concepts were developed to overcome one of the main drawbacks of the lead–acid system: the heavy supporting lead structures (grids, connectors, etc.). Lead foam [89], lead-plated carbon rods [90], electroplated vitreous carbon [91], flexible-graphite grids [92], or graphite foams [93] were tested, also lead-plated materials like titanium [94], Ebonex [95], copper mesh [96], polymeric structures [97], polymer foam [98], or glass fiber mesh [99]. Warlimont and Hofmann [100] describe the development of multilayer composite grids.

Some attempts have been made to transform the conventional lead accumulator into a 'dissolution' accumulator by replacing sulfuric acid of any particular concentration [101] with tetrafluoroboric acid (HBF$_4$), but this highly corrosive and toxic acid was not finally accepted [102]. More recently, methanesulfonic acid was tried [103]. The electrochemical equivalent of lead is the lowest of all metallic anodes (260 Ah kg^{-1}), but in many applications the high OCV of 2.1 V per cell compensates for this disadvantage.

8.3.5
Lithium (Li)

Most battery systems in which lithium is applied as anode material belong to the group using nonaqueous electrolytes, but there is one system that works with water serving as solvent and reactant as well. This is only possible because lithium forms a passive layer in solutions containing higher amounts of caustic [104–107].

Nevertheless, the water is decomposed with evolution of hydrogen but under controlled conditions and with a reasonable reaction rate. With water as the active cathode material, the battery system – used in military underwater applications – can be designed as: (−) Li/KOH/H$_2$O (+) [108].

The regular discharge reactions are:

$$\text{Anode: Li} \rightarrow \text{Li}^+ + \text{e}^- \qquad (8.11)$$

$$\text{Cathode: H}_2\text{O} + \text{e}^- \rightarrow \text{OH}^- + {}^1\!/{}_2\,\text{H}_2 \uparrow \qquad (8.12)$$

$$\text{Overall reaction: Li} + \text{H}_2\text{O} \rightarrow \text{LiOH} + {}^1\!/{}_2\text{H}_2 \uparrow \qquad (8.13)$$

The electrochemical equivalent of 3860 Ah kg^{-1} is the highest of all metal anodes and the OCV of 2.7–2.8 V (depending on electrolyte concentration) is also rather high.

8.3.6
Magnesium (Mg)

Magnesium alloys as anode materials are found in (at least) four types of primary cells: the 'dry cell' type, (−) Mg/KOH/MnO$_2$(+) [109], magnesium/air batteries (−) Mg/KOH/O$_2$ (+) including a particular version operating on oxygen dissolved in sea water [110–112], water-activated reserve batteries, (−) Mg/KOH/K(+), where K = AgCl [113], CuCl, PbCl$_2$, Cu$_2$I$_2$ [114], CuSCN, CuO [115], and again MnO$_2$ [116], and finally cells equipped with organic cathodes on copper current collectors (−) Mg(Zn)/chlorides, perchlorates/organic active material (e.g., 2-nitrophenylpyruvic acid) (+) [117]. All of these cells are primaries (AB1) using the advantage of a high electrochemical equivalent (2200 Ah kg^{-1}) and relatively high OCVs (e.g., 2.8 V for the Mg/MnO$_2$-system).

The dominating discharge reactions (for the Mg/air system as an example) are:

$$\text{Anode: Mg} \rightarrow \text{Mg}^{2+} + 2\text{e}^- \qquad (8.14)$$

$$\text{Cathode: } 1/2\,\text{O}_2 + \text{H}_2\text{O} + 2\text{e}^- \rightarrow 2\text{OH}^- \qquad (8.15)$$

$$\text{Overall reaction: Mg} + 1/2\,\text{O}_2 + \text{H}_2\text{O} \rightarrow \text{Mg(OH)}_2 \qquad (8.16)$$

$$\text{Side reaction: Mg} + 2\text{H}_2\text{O} \rightarrow \text{Mg(OH)}_2 + \text{H}_2 \uparrow \qquad (8.17)$$

The most important anodes for battery use are ternary Mg/Al/Zn-alloys (AZ 61, AZ 63 – Norsk Hydro), an alloy containing 1% of lithium (AZ 21), and the rather complex alloy AZ 31 (Mg-3 Al-1 Zn-0.2 Mn-0.15 Ca) [118].

Table 8.1 Open circuit voltages of battery systems with zinc anodes.

Open circuit voltage (V)	Cathode material
1.6	MnO_2, acidic
1.5	MnO_2, alkaline
1.73	NiOOH
1.85	AgO
1.65	O_2/air
2.12	Cl_2
1.85	Br_2
1.84	$K_3[Fe(CN)_6]$

8.3.7
Zinc (Zn)

The zinc electrode is probably the most widely used metallic negative. The material is relatively cheap, has a good electrochemical equivalent (820 Ah kg^{-1}), and shows high OCVs in most systems (Table 8.1).

It is so universally applied that it may be found in combination with metal oxide cathodes (e.g., HgO, AgO, NiOOH, MnO_2), with catalytically active oxygen electrodes, with inert cathodes using aqueous halide or ferricyanide solutions as active materials ('zinc-flow' or 'redox' batteries), and with reducible organic materials on metal carriers [119–121].

The cell (battery) sizes vary from small button cells for hearing aids or watches up to kilowatt-hour modules for electric vehicles (electrotraction). Primary and storage batteries exist in all categories except for flow batteries where only storage types are found. Acidic, neutral, and alkaline electrolytes are used as well. The (simplified) half-cell reaction for the zinc electrode is the same in all electrolytes:

$$Zn \longleftrightarrow Zn^{2+} + 2e^- \tag{8.18}$$

This reaction may be followed by other (complex formation and/or precipitation) reactions which are independent of the electrode potential but determined by the nature and concentration of the electrolyte.

It is impossible to discuss all the problems related to zinc electrodes without looking at the electrolyte system and the kind of cell operation (primary or rechargeable). The only way to cover all the possible combinations is another mode of characterization or categorization, which is used in the subsequent sections: Sections 8.3.7.1, [(−) Zn/NH_4Cl, $ZnCl_2$/MnO_2 (+)]; 8.3.7.2, [(−) Zn/KOH/HgO, MnO_2, air (+)]; 8.3.7.3, [(−) Zn/KOH/NiOOH, AgO (+)]; 8.3.7.4, [(−) Zn/KOH/MnO_2, air (+)]; 8.3.7.5, [(−) Zn/H^+ or OH^-/Br_2, Cl_2, ferricyanide (+)]; and 8.3.7.6.

8.3.7.1 Zinc Electrodes for 'Acidic' (Neutral) Primaries

The 'classical' Leclanché cell uses zinc sheet formed into a cylindrical can serving simultaneously as the anode and as the cell container (AB1C1). The cathode is a mixture of MnO$_2$ and graphite wrapped into a piece of separator and contacted by a central carbon rod. The can dissolves slowly when the cell is not in use and faster when the cell delivers electrical energy.

The reaction following the primary electrochemical zinc dissolution (Equation 8.18) leads, in the case of an ammonium chloride electrolyte, to a zinc diammine cation:

$$Zn^{2+} + 2NH_4^+ + 2OH^- \rightarrow [Zn(NH_3)_2]^{2+} + 2H_2O \tag{8.19}$$

The (similar) corrosion reaction is:

$$Zn + 2NH_4^+ \rightarrow [Zn(NH_3)_2]^{2+} + H_2 \uparrow \tag{8.20}$$

If aqueous zinc chloride solution serves as electrolyte ('heavy duty' types), the hydrate of a basic zinc chloride is formed instead of the product in Equation 8.19:

$$5Zn^{2+} + 2Cl^- + 8OH^- + H_2O \rightarrow ZnCl_2.4ZnO.5H_2O \downarrow \tag{8.21}$$

The construction of the anode (sheet, cell container) is responsible for the two major problems:

- The relatively small specific surface (cm^2 g^{-1}) does not allow higher currents;
- The corrosion reactions usually proceed until the can starts leaking and the electrolyte spillage corrodes the electrical or electronic device powered by the cell.

The corrosion reactions may be slowed down by using zinc alloys (with lead and cadmium also improving the mechanical properties of zinc to simplify the production process) instead of the pure metal or by amalgamating the inner surface of the can by adding a small amount of a mercury compound to the electrolyte.

Only a few studies dealing with corrosion kinetics [122] or with substitutes for mercury as corrosion inhibitors [123] are available.

Other ('leak-proof') cells use an additional steel can on the outside to prevent any electrolyte loss caused by perforation of the inner zinc beaker when the cell is exhausted.

The problem of low specific surface (which, however, has a beneficial effect on the corrosion rate) cannot be solved so easily. This was one important reason for the development of the alkaline MnO$_2$/zinc cell known as '*Alkaline*' or 'PAM' (primary alkaline manganese dioxide).

8.3.7.2 Zinc Electrodes for Alkaline Primaries

The alkaline version of the MnO$_2$/zinc cell follows a different concept because it turns the construction of the Leclanché cell completely around: now the cathode (MnO$_2$ + carbon) forms a hollow cylinder contacting the inner wall of the cell container (steel) along its outer surface. The inner cavity has to accommodate anode, electrolyte, separator, and current collector. Usually, the separator forms a basket which is automatically inserted and prevents direct contact of the anode material with the cathode and the bottom of the cell container.

Anodic active species and electrolyte are provided as a gel consisting of zinc powder, aqueous KOH solution (7–9 mol L^{-1}), gelling agents, and additives. Finally the current collector (a brass nail spot-welded to the metallic part of the cell top) is introduced when the cell top is placed and the can is crimped to give a gas-tight closure.

Cells of cylindrical geometry are produced mainly in four sizes: D (LR-20), C (LR-14), AA (LR-6), and AAA (LR-03).

The two other alkaline cells of this section (using HgO or an oxygen electrode as cathode) were almost exclusively produced as small button cells. Larger zinc/air batteries gained some attention as candidates for electric vehicle propulsion [124].

The change from zinc sheet to zinc powder improved the high-current performance of the cell significantly but it increased the corrosion problems (a larger specific surface means a higher corrosion rate). Consequently, evaluation of the performance of zinc alloy powders in gelled electrolyte [125], the search for particularly active (high-rate) zinc anodes [126], and for methods to fabricate porous zinc anodes [127] were initiated. The electro-dissolution of zinc in alkaline solutions [128] and zinc corrosion (gassing) after partial discharge [129] have been thoroughly studied.

The discharge reactions now include formation of hydroxo complexes, preferably:

$$Zn^{2+} + 4OH^- \rightarrow [Zn(OH)_4]^{2-} \tag{8.22}$$

Depending on electrolyte saturation and KOH concentration, subsequent precipitation reactions may follow:

$$[Zn(OH)_4]^{2-} \rightarrow Zn(OH)_2 + 2OH^-, \text{ or:} \tag{8.23}$$

$$[Zn(OH)_4]^{2-} \rightarrow ZnO \downarrow + 2OH^- + H_2O. \tag{8.24}$$

The equilibrium concentration of zincate [130] as well as zinc oxide morphology and distribution in discharged Zn/MnO$_2$ batteries [131] became matters of particular interest.

In competition with the electrochemical discharge reaction and consequently diminishing the shelf life of the cell, chemical dissolution (corrosion) of zinc is more or less active:

$$Zn + 2OH^- + 2H_2O \rightarrow [Zn(OH)_4]^{2-} + H_2 \uparrow . \tag{8.25}$$

The loss of active zinc and the evolution of hydrogen, causing an intolerable rise of internal cell pressure, were retarded by amalgamation of zinc particles. For quite a long time (up to about 1982) mercury contents of 6% (in some cases up to 8%) were regarded as normal [132]. Then a rapid decrease in the mercury content took place (Figure 8.1). The first step was a reduction of Hg to 3%, which was made possible by the application of new amalgamation methods (surface-amalgamation instead of or additional to volume amalgamation [133, 134]. This first step was followed by a further decrease in the mercury content to 1% when the amalgamation techniques were regarded as suitable for achieving this goal [135]. In the meantime it was found that extremely pure zinc powder (made from selected raw material by a gas

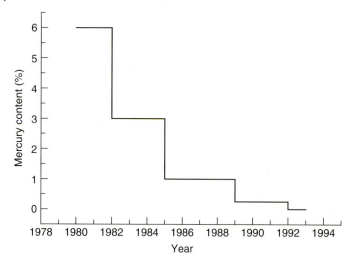

Figure 8.1 Decrease of mercury content of zinc powders within about 10 years.

atomization process) which is surface amalgamated may contain far less than 1% mercury without losing corrosion stability [132].

For a couple of years 0.25% Hg became the technical standard. It has to be pointed out that the mercury content of the metallic zinc has to be divided by a factor of roughly 10 to give the mercury content based on the total cell weight of an AA (LR-6) cell. These values are sometimes indicated on the cell labels.

With decreasing amalgamation, other corrosion inhibitors had to take over the role of mercury. There are numerous papers and patents claiming corrosion-inhibiting activities of elements like Al, In, Tl, Cd, Ga, Na, Ca, Co, Ni, Pb, Bi in single or combined application, with or without small amounts of mercury. The probably most important patents are cited in Ref. [132].

Finally, the research and development activities led to a zinc quality which is specified as 'no mercury added' (nobody dares to claim 'zero mercury'). The traded zinc powders frequently contain a combination of indium, lead, and bismuth in variable concentrations up to 500 ppm each [136]. New atomization techniques [137] made it possible to replace lead by nontoxic aluminum or calcium.

Some battery-producing companies prefer purchasing pure, nonamalgamated zinc powder to apply their own proprietary corrosion protection system based on organic inhibitors [138–141]. The general trend is to keep the anodes of all the consumer cells mercury-free (usually indicated by a 'green' label) and to make them disposable with the regular household trash. The exceptions to this rule are those cells where this makes no sense, for example, cells with mercuric oxide cathode (now nearly 100% withdrawn from consumer markets).

This trend also applies to the 'reusable' version of the manganese dioxide/zinc cell which came onto the consumer market in 1993 (Rayovac, USA). This type is treated in the next-but-one section (Section 8.3.7.4).

Zinc recovery from spent batteries [142] is only considered in exceptional cases.

8.3.7.3 Zinc Electrodes for Alkaline Storage Batteries

Battery systems of complex design and structure using – at least for one electrode – expensive materials are (for economical reasons) mainly conceived as storage batteries. Primary (and 'reserve') versions of the zinc/silver oxide battery [(−) Zn/KOH/AgO (+)] – as a first example – are only used in particular cases where the question of cost is not crucial, for example, for marine [143–146] and space applications [147].

The other example, called the *nickel/zinc battery* [(−) Zn/KOH/NiOOH (+)], has attracted more attention in two different versions from the viewpoints of application and cell design: one is the small cylindrical consumer cell [148, 149], the other is the flat-plate module for electrotraction [149, 150]. A very interesting review with an extensive collection of references was published in 1992 [151]. In 1996, an improved bipolar construction of this battery was presented [152]. The most recent version was described by Humble and co-authors [153]: a nickel/zinc microbattery developed for direct installation and use in autonomous microsystems.

The problems related to the zinc electrode grew significantly with the change to a rechargeable (reversible) system. Whereas the discharge of a zinc electrode in a primary cell is a simple electrochemical dissolution with little concern about the oxidation products, these may be of particular importance in a secondary (or storage) cell. The fact of starting with zinc oxide or hydroxide instead of metallic zinc had only minor influence (AB2C2). In any case, the solubility of zinc oxide or hydroxide in the KOH electrolyte was found to be a key parameter in a reversible zinc electrode [154]. The result of zinc migration ('shape change') was obvious when the electrodes of a flat-plate battery were inspected after a series of charge–discharge cycles. The active material was removed from the electrode edges and agglomerated toward the plate center. If the number of cycles was sufficiently high, the edge areas of the current collector were completely denuded of zinc.

Usually, this phenomenon limits the lifetime of a battery because the storage capacity falls under a reasonable lower limit. One reason for this zinc migration was identified by McBreen [155]: inhomogeneous current distribution makes the zinc move away from high current density areas. Another mechanism seems to be active as well: electrolyte convection induced by electro-osmosis through the separator [156]. Many attempts have been made to prevent or at least retard shape change, preferably by reducing the solubility of zincate in the electrolyte [157–163].

The consequences of shape change are densification and loss of electrode porosity, increased current density caused by loss of zinc surface area, and finally earlier passivation.

Two different forms of passivation can stop the discharge of a zinc electrode before the active material is exhausted. 'Spontaneous' passivation occurs at high current densities within a few seconds. 'Long-term' passivation may be observed after hours of continuous discharge in a current density range of 15–35 mA cm^{-2}. The effects are explained by the existence of supersaturated solutions of ZnO in KOH, which are normally quite stable, but if precipitation is induced by any means (nucleation) solid products form immediately and block the electrodes.

In rechargeable nickel/zinc and silver/zinc batteries this problem is partly compensated for by provision of a massive zinc reserve. The cells are cathode-limited and the amount of anode material exceeds the theoretically required mass by a factor between 2 and 3.

Another problem had to be solved when the zinc electrode was made reversible: in a battery with unstirred electrolyte or an electrolyte gel, dendritic growth of the electrolytically deposited metal takes place. The formation of dendrites cannot be fully suppressed by the use of current collectors with large surface areas (grids, wire fabrics). Chemical means may provide a significant contribution [49, 157, 158, 160, 164, 165]. However, by using improved separators combined in multi-layer arrangements, the danger of short-circuiting is reduced.

Design of zinc electrodes for storage batteries always has to find a balance between high-rate capability (this means high specific surface area [166] and good wettability by the electrolyte [167]), corrosion protection (with opposite requirements), and uniform zinc re-deposition upon the current collector [168–174].

8.3.7.4 Zinc Electrodes for Alkaline 'Low-Cost' Reusables

It was a reasonable idea to use the intensive research work in the fields of zinc, manganese dioxide, and oxygen electrodes on one hand, and on rechargeable metal oxide/zinc cells (the preceding section) on the other, to develop 'rechargeable' versions of the cells described in Section 8.3.7.2. The manganese dioxide/zinc system (for reason of low cost) and the zinc/air system (low cost and high energy density) were the most fascinating ones.

The specification 'rechargeable' is controversial: for many battery experts it requires the possibility of – at least – some hundred to a thousand full cycles to call a system 'rechargeable.' As a compromise, cells designed for 20 to about 200 cycles are designated 'reusable' or 'renewable' (e.g., RENEWAL™ of Rayovac, USA, for zinc/manganese dioxide, general name RAM™ of former Battery Technologies Inc. (BTI), Canada).

In the early stages of development the most significant difficulties seemed to come from the cathode side: it took a long time to convince people of the principle of rechargeability of manganese dioxide [175, 176] and to find suitable catalysts for the oxygen electrode showing a sufficiently high oxidation stability in the charging procedure [177, 178].

Soon it became evident that the zinc anode, working in both cases under capacity-limiting conditions, causes severe troubles too [179, 180].

Whereas in the zinc/air system the anode automatically limits the discharge (because access to oxygen from air is practically unlimited) the anode limitation in zinc/manganese dioxide cells has another reason: K. Kordesch and co-workers showed that the rechargeability of manganese dioxide (i.e., the number of available cycles) depends strongly on the depth of discharge (DOD) (Figure 8.2) [181].

It is possible to design a RAM (Rechargeable/Reusable Alkaline Manganese) cell either for high initial (first discharge) capacity (up to 1.8 Ah for AA size, close to the alkaline primary version) and a low cycle number, or for lower initial capacity (that means shallower discharge of MnO_2) but significantly higher cycle number.

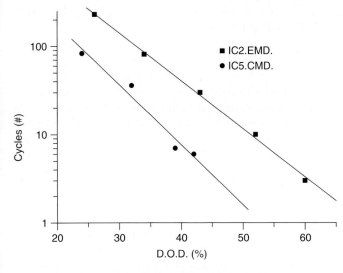

Figure 8.2 Influence of the DOD on the number of achievable cycles for CMD (Chemical Manganese Dioxide) and EMD (Electrolytic Manganese Dioxide) samples.

The commercially available products of the 1990s (RENEWAL™/USA, PURE ENERGY™/Canada, ALCAVA™/Korea, GRANDCELL™/Malaysia) followed the first principle.

Fortunately, shape-change effects are not important in cells of cylindrical geometry, which are preferred for making RAM products that are exchangeable with primaries. All flat-plate test cells and research batteries showed the same behavior as the nickel/zinc or silver/zinc batteries but without the possibility of oversizing the anode. Although small cylindrical cells cannot be manufactured economically with current collectors made from grid or mesh, it turned out that dendrite formation is not a critical feature if a laminated separator including a 'barrier' layer of regenerated cellulose is applied.

Most difficulties arose when – following the general trend – the RAM cell had to be made mercury-free. The final step especially, from about 0.15% mercury to 'no added mercury,' was quite challenging [182]. Many compounds, although evidently inhibiting zinc corrosion in primary cells, did not work so perfectly with the deposited zinc after the first charge and were only insufficient substitutes for mercury. However, some electrical properties of the first mercury-free test cells after shock or drop tests were inferior compared with those of cells using amalgamated zinc [183]. Apparently, the presence of mercury in the anode mixture had a positive effect on the adhesion of the zinc gel to the current collector, and on the discharge efficiency.

In connection with this observation, it should be noted that only a fraction (usually between 60 and 70%, depending on the discharge current and on the cut-off voltage) of the zinc powder provided in the anode gel can be used as active mass. The rest only acts as a metallic conductor or extension of the brass nail, which forms the current collector.

With primary cells the disposal of the remaining fraction of the zinc powder, or its collection for recycling, with the used battery is not troublesome. Zinc powder is not expensive and the rejected material has done its work as an electronically conductive component of the anode mass. By contrast, the rechargeable (reusable) version of the zinc/manganese dioxide cell requires – as stated before – a capacity-limiting anode with good and reproducible (!) discharge process efficiency. For this purpose it is necessary to establish a conductive matrix in the anode space of the cell by admixing lightweight materials with good surface conductivity (and not taking part in the redox reactions of the discharge and charge processes) to the anode gel. This conductive matrix is able to prevent the formation of isolated zinc particles out of contact with the current collector during discharge (a function of particular importance in mercury-free cells), but also serving as a three-dimensional substrate for the zinc deposition on charge [184].

For original equipment manufacturing (OEM) applications, that is, recharging strings of series-connected cells mounted inside the device, the possibility of utilizing the chemical oxygen–zinc reaction (to provide a certain overcharge capability) has been demonstrated in a modified version of the RAM cell [185].

The efforts made to obtain a rechargeable (reusable) version of the zinc/air battery had to overcome a series of other restrictions. This is not surprising because, in addition to the chemical differences, this system had to deal with a scale-up to sizes applicable to electric vehicle propulsion.

One of the first attempts was called '*mechanical recharge*.' It was a simple exchange of flat-plate anodes at a certain degree of oxidation (discharge) for new ones [186].

The next step was provision of the zinc electrode as a pumpable slurry as well as on-board and central recharge schemes [187, 188].

Finally, the development of stable ('bi-functional') air electrodes favored the construction of real 'in-cell' recharge systems [189–191] (see also Refs [177, 178]).

8.3.7.5 Zinc Electrodes for Zinc-Flow Batteries

There are three types of zinc-flow batteries (belonging in general to the group of flow or redox batteries) which have been studied intensively: two of them are similar with respect to the reactants involved, the zinc/chlorine [(−) Zn/HCl, $ZnCl_2/Cl_2$ (+)] and the zinc/bromine [(−) Zn/HBr, $ZnBr_2/Br_2$ (+)] batteries; the third one uses an alkaline electrolyte and potassium ferricyanide as active cathode material [(−) Zn/NaOH/$K_3[Fe(CN)_6]$(+)] [192].

The anodes of all three start with zinc provided as aqueous halide solutions (AB2C3).

While the zinc/chlorine battery was preferred for utility load-leveling applications [193], the zinc/bromine system was regarded as the more promising one for electric vehicle requirements [194, 195].

Both share more or less the same merits but also the same disadvantages. The beneficial properties are: high OCV (2.12 and 1.85 V respectively), flexibility in design (because the active chemicals are mainly stored in tanks outside the (usually bipolar) cell stack), no problems with zinc deposition in the charging cycle because it works under nearly ideal conditions (perfect mass transport by electrolyte convection, carbon substrates [196]), the possibility to ignore self-discharge by chemical attack of the acid on the deposited zinc because the stack runs dry in the standby mode, and use of relatively cheap construction materials (polymers) and reactants.

The most unpleasant drawbacks are: the requirement for tanks, tubes, and pumps for the electrolyte storage and transport system, increasing the volume of the battery and consuming energy (in the case of the zinc/chlorine battery an extra cooling device had to be provided); highly corrosive electrolytes; public distrust of storing halogens in any form (though it is frequently stated that organic polybromide complexes are quite harmless); and a certain imbalance of the electrochemistry involved (zinc ions form halide complexes of the type $[ZnX_4]^{2-}$ which behave – due to their charge – as anions in the electric field).

Despite the fact that the zinc/ferricyanide system employs an alkaline electrolyte, the electrode reactions are quite similar to those in zinc/halogen batteries, and battery constructions are usually bipolar too.

Zinc is electrodeposited from the sodium zincate electrolyte during charge. As in the zinc/bromine battery two separate electrolyte loops ('posilyte' and 'negalyte') are required. The only difference is the quality of the separator: the zinc/bromine system works with a microporous foil made from sintered polymer powder, but the zinc/ferricyanide battery needs a cationic exchange membrane in order to obtain acceptable coulombic efficiencies. The occasional transfer of solid sodium ferrocyanide from the negative to the positive tank, to correct for the slow transport of complex cyanide through the membrane, is proposed [197].

All in all, this system is more complicated than the other flow batteries, and this handicap is hindering wider application.

8.3.7.6 Zinc Electrodes for Printed Thin-Layer Batteries

Since the turn of the millennium, some companies in Israel (Power Paper, 2000), Finland (Enfucell, 2002), and the USA (Thin Battery Technology, 2002, meanwhile Blue Spark Technologies) are marketing thin and flexible zinc/manganese dioxide batteries produced by industrial printing, drying, and laminating techniques. They are preferably integrated in 'smart' cards for different applications or in active or semi-active tags. The user in general cannot even see the battery and may not have any idea about the nature of the power source operating his/her device.

The key technologies, for example, the production of proprietary printing inks, particular binder/solvent systems containing the active components in the form of very fine powders (e.g., zinc for anodes with an average particle diameter of 30 μm) are patented. Power Paper, as an example, is holding by more than 60 patents covering the applied materials and details of the production process.

Nevertheless, from time to time additional zinc electrode formulations for either zinc/nickel batteries [198] or for the zinc/manganese system [199] are presented.

Up to 2003, a joint project of Power Paper and the Institute for Chemical Technology of Inorganic Materials, Graz University of Technology, explored a rechargeable version of printed thin and flexible zinc/manganese dioxide batteries based on experience with RAM cells (see Section 8.3.7.4) [200].

References

1. Hasvold, O. et al. (1999) *J. Power Sources*, **80**, 254–260.
2. Yang, Sh. and Knickle, H. (2002) *J. Power Sources*, **112**, 162–173.
3. Li, Q. and Bjerrum, N.J. (2002) *J. Power Sources*, **110**, 1–10.
4. Shen, P.K., Tseung, A.C.C., and Kuo, C. (1994) *J. Appl. Electrochem.*, **24**, 145–148.
5. Tang, Y. et al. (2004) *J. Power Sources*, **138**, 313–318.
6. Paramasivam, M. et al. (2003) *J. Appl. Electrochem.*, **33**, 303–309.
7. Wang, X.Y. et al. (2005) *J. Appl. Electrochem.*, **35**, 213–216.
8. Wang, J.-B. et al. (2007) *J. Appl. Electrochem.*, **37**, 753–758.
9. Saidman, S.B. and Bessone, J.B. (1997) *J. Appl. Electrochem.*, **27**, 731–737.
10. El Shayeb, H.A. et al. (1999) *J. Appl. Electrochem.*, **29**, 601–609.
11. Flamini, D.O. and Saidman, S.B. (2008) *J. Appl. Electrochem.*, **38**, 663–668.
12. Nestoridi, M. et al. (2008) *J. Power Sources*, **178**, 445–455.
13. Reding, J.T. and Newport, J.J. (1966) *Mater. Prot.*, **5**, 15.
14. Sakans, T. et al. (1966) *Mater. Prot.*, **5**, 45.
15. Hunter, J. et al. (1991) in *Power Sources*, vol. 13 (eds T. Keily and W. Baxter), International Power Sources Symposium Committee, Leatherhead/England, p. 193.
16. Doche, M.L. et al. (1997) *J. Power Sources*, **65**, 197–205.
17. Zhuk, A.Z. et al. (2006) *J. Power Sources*, **157**, 921–926.
18. Ferrando, W.A. (2004) *J. Power Sources*, **130**, 309–314.
19. Zein el Abedin, S. and Saleh, A.O. (2004) *J. Appl. Electrochem.*, **34**, 331–335.
20. Jindra, J., Mrha, J., Micka, K., Zabransky, Z., Koudelka, V., and Malik, J. (1979) *J. Power Sources*, **4**, 227–237.
21. Seiger, H.N. (1984) *Proc. Electrochem. Soc.*, **84** (8), 50.
22. Barton, R.T. et al. (1986) *J. Power Sources*, **18**, 43–50.
23. Munshi, M.Z.A. and Tseung, A.C.C. (1986) *J. Power Sources*, **18**, 33–42.
24. Tamil Selvan, S. et al. (1991) *J. Appl. Electrochem.*, **21**, 646–650.
25. Paruthimal Kalaignan, G. et al. (1996) *J. Power Sources*, **58**, 29–34.
26. Simic, N. et al. (2001) *J. Power Sources*, **94**, 1–8.
27. Freitas, M.B.J.G., Penha, T.R., and Sirtoli, S. (2007) *J. Power Sources*, **163**, 1114–1119.
28. Chakkaravarthy, C., Periasamy, P., Jegannathan, S., and Vasu, K.I. (1991) *J. Power Sources*, **35**, 21–35.
29. Carlsson, L. and Ojefors, L. (1980) *J. Electrochem. Soc.*, **127**, 525.
30. Ojefors, L. and Carlsson, L. (1977/78) *J. Power Sources*, **2**, 287.
31. Buzzelli, E.S. (1978) *Proceedings of 28th Power Sources Symposium*, ECS Inc., Princeton, p. 160.
32. Bayles, G.A., Buzzelli, E.S., and Lauer, J.S. (1990) *Proceedings of 34th Power Sources Symposium*, IEEE Inc., New York, p. 312.
33. Berger, G. and Haschka, F. (1978) US Patent 4, 250,236, DE Patent 2, 837,980.
34. Caldas, C.A., Lopes, M.C., and Carlos, I.A. (1998) *J. Power Sources*, **74**, 108–112.
35. Souza, C.A.C. et al. (2004) *J. Power Sources*, **132**, 288–290.
36. Hang, B.T. et al. (2006) *J. Power Sources*, **155**, 461–469.
37. Hang, B.T. et al. (2007) *J. Power Sources*, **168**, 522–532.
38. Egashira, M. et al. (2008) *J. Power Sources*, **183**, 399–402.
39. Hang, B.T. et al. (2005) *J. Power Sources*, **143**, 256–264.
40. Hang, B.T. (2008) *J. Power Sources*, **178**, 393–401.
41. Soragni, E., Davolio, G., and Tarzia, G. (1993) in *Power Sources*, vol. 14 (eds A. Attewell and T. Keily), International Power Sources Symposium Committee, Leatherhead/England, p. 15.
42. Lexow, K.W., Krämer, G., and Oliapuram, V.A. (1981) in *Power*

Sources, vol. 8 (ed J. Thompson), Academic Press, London, New York, p. 389.
43. Bagshaw, N.E. (1995) *J. Power Sources*, **53**, 25–30.
44. Zhong, S. et al. (1997) *J. Power Sources*, **66**, 107–113.
45. Lakshimi, C.S., Manders, J.E., and Rice, D.M. (1998) *J. Power Sources*, **73**, 23–29.
46. Slavkov, D. et al. (2002) *J. Power Sources*, **112**, 199–208.
47. Jullian, E., Albert, L., and Caillerie, J.L. (2003) *J. Power Sources*, **116**, 185–192.
48. Prengaman, R.D. (2005) *J. Power Sources*, **144**, 426–437.
49. Dehmas, M. et al. (2006) *J. Power Sources*, **159**, 721–727.
50. Mukaitani, I. et al. (2006) *J. Power Sources*, **158**, 897–901.
51. Xu, J. et al. (2006) *J. Power Sources*, **155**, 420–427.
52. Sakai, M. et al. (2008) *J. Power Sources*, **185**, 559–565.
53. Li, H. et al. (2009) *J. Power Sources*, **191**, 111–118.
54. Jobst, K. et al. (1997) *J. Appl. Electrochem.*, **27**, 455–461.
55. Vinod, M.P. et al. (1997) *J. Appl. Electrochem.*, **27**, 462–468.
56. Boden, D.P., Arias, J., and Fleming, F.A. (2001) *J. Power Sources*, **95**, 277–292.
57. Francia, C., Maja, M., and Spinelli, P. (2001) *J. Power Sources*, **95**, 119–124.
58. Saez, F. et al. (2001) *J. Power Sources*, **95**, 174–190.
59. Ban, I. et al. (2002) *J. Power Sources*, **107**, 167–172.
60. Matrakova, M. et al. (2003) *J. Power Sources*, **113**, 345–354.
61. McNally, T. and Klang, J. (2003) *J. Power Sources*, **116**, 47–52.
62. Papazov, G., Pavlov, D., and Monahov, B. (2003) *J. Power Sources*, **113**, 335–344.
63. Valenciano, J., Trinidad, F., and Fernandez, M. (2003) *J. Power Sources*, **113**, 318–328.
64. Hirai, N. et al. (2006) *J. Power Sources*, **158**, 846–850.
65. Hirai, N. (2006) *J. Power Sources*, **158**, 1106–1109.
66. Petkova, G., Nikolov, P., and Pavlov, D. (2006) *J. Power Sources*, **158**, 841–845.
67. Hirai, N., Kubo, S., and Magara, K. (2009) *J. Power Sources*, **191**, 97–102.
68. Lam, L.T. et al. (2002) *J. Power Sources*, **107**, 155–161.
69. Wu, L., Chen, H.Y., and Jiang, X. (2002) *J. Power Sources*, **107**, 162–166.
70. Vermesan, H. et al. (2004) *J. Power Sources*, **133**, 52–58.
71. Karimi, M.A., Karami, H., and Mahdipour, M. (2006) *J. Power Sources*, **160**, 1414–1419.
72. Martha, S.K. et al. (2006) *J. Appl. Electrochem.*, **36**, 711–722.
73. Sawai, K. et al. (2006) *J. Power Sources*, **158**, 1084–1090.
74. Micka, K. et al. (2009) *J. Power Sources*, **191**, 154–158.
75. Newell, J.D., Patankar, S.N., and Edwards, D.B. (2009) *J. Power Sources*, **188**, 292–295.
76. Shiomi, M. et al. (1997) *J. Power Sources*, **64**, 147–152.
77. Calabek, M. et al. (2006) *J. Power Sources*, **158**, 864–867.
78. Moseley, P.T. (2009) *J. Power Sources*, **191**, 134–138.
79. Pavlov, D. et al. (2009) *J. Power Sources*, **191**, 58–75.
80. Huang, M. et al. (2006) *J. Electrochem. Soc.*, **153**, A631–A636.
81. D'Alkaine, C.V., Carubelli, A., and Lopes, M.C. (2000) *J. Appl. Electrochem.*, **30**, 585–590.
82. Yamaguchi, Y. et al. (2001) *J. Power Sources*, **93**, 104–111.
83. D'Alkaine, C.V. and G.A.D. and Brito (2009) *J. Power Sources*, **191**, 159–164.
84. Guo, Y., Wu, M., and Hua, S. (1997) *J. Power Sources*, **64**, 65–69.
85. Li, Z. et al. (2002) *J. Electrochem. Soc.*, **149**, A934–A938.
86. Petkova, G. and Pavlov, D. (2003) *J. Power Sources*, **113**, 355–362.
87. Kirchev, A. et al. (2009) *J. Power Sources*, **191**, 82–90.
88. Li, Y., Zhang, G., and Guo, Y. (2004) *J. Appl. Electrochem.*, **34**, 1113–1118.
89. Dai, C. et al. (2006) *J. Power Sources*, **158**, 885–890.
90. Das, K. and Mondal, A. (2000) *J. Power Sources*, **89**, 112–116.

91. Gyenge, E., Jung, J., and Mahato, B. (2003) *J. Power Sources*, **113**, 388–395.
92. Hariprakash, B. and Gaffoor, S.A. (2007) *J. Power Sources*, **173**, 565–569.
93. Yang, Y. II et al. (2006) *J. Power Sources*, **161**, 1392–1399.
94. Dai, J. et al. (2004) Proceedings of 41st Power Sources Conference, Army Communications-Electronics Command Fort Monmouth, New Jersey, pp. 295–298.
95. Ellis, K. et al. (2004) *J. Power Sources*, **136**, 366–371.
96. Lushina, M. et al. (2005) *J. Power Sources*, **148**, 95–104.
97. Soria, M.L. et al. (1999) *J. Power Sources*, **78**, 220–230.
98. Tabaatabaai, S.M. et al. (2006) *J. Power Sources*, **158**, 879–884.
99. Wang, J. et al. (2003) *J. Appl. Electrochem.*, **33**, 1057–1061.
100. Warlimont, H. and Hofmann, T. (2006) *J. Power Sources*, **158**, 891–896.
101. Pavlov, D., Petkova, G., and Rogachev, T. (2008) *J. Power Sources*, **175**, 586–594.
102. Beck, F. (1979) BMFT Forschungsbericht No. T79-142.
103. Pletcher, D. et al. (2008) *J. Power Sources*, **180**, 621–629.
104. Littauer, E.L. and Tsai, K.C. (1974) *Proceedings of 26th Power Sources Conference*, ECS Inc., Pennington, p. 570.
105. Littauer, E.L. and Tsai, K.C. (1976) *J. Electrochem. Soc.*, **123**, 964.
106. Pensado-Rodriguez, O., Urquidi-Macdonald, M., and Macdonald, D.D. (1999) *J. Electrochem. Soc.*, **146**, 1318–1325.
107. Pensado-Rodriguez, O. et al. (1999) *J. Electrochem. Soc.*, **146**, 1326–1335.
108. Shuster, N. (1990) *Proceedings of 34th Power Sources Symposium*, IEEE Inc., New York, p. 118.
109. Munichandraiah, N. (1999) *J. Appl. Electrochem.*, **29**, 463–471.
110. Hasvold, O. et al. (1997) *J. Power Sources*, **65**, 253–261.
111. Wilcock, W.S.D. and Kauffman, P.C. (1997) *J. Power Sources*, **66**, 71–75.
112. Hasvold, O. et al. (2004) *J. Power Sources*, **136**, 232–239.
113. Karpinski, A.P. et al. (2000) *J. Power Sources*, **91**, 77–82.
114. Renuka, R. (1997) *J. Appl. Electrochem.*, **27**, 1394–1397.
115. Sivashanmugam, A. et al. (2004) *J. Appl. Electrochem.*, **34**, 1135–1139.
116. Vuorilehto, K. (2003) *J. Appl. Electrochem.*, **33**, 15–21.
117. Renuka, R. (2000) *J. Appl. Electrochem.*, **30**, 483–490.
118. Sahoo, M. and Atkinson, J.T.N. (1982) *J. Mater. Sci.*, **17**, 3564.
119. Kan, J., Xue, H., and Mu, S. (1998) *J. Power Sources*, **74**, 113–116.
120. Kalaiselvi, D. and Renuka, R. (1999) *J. Appl. Electrochem.*, **29**, 797–803.
121. Rahmanifar, M.S. et al. (2004) *J. Power Sources*, **132**, 296–301.
122. Szczesniak, B., Cyrankowska, M., and Nowacki, A. (1998) *J. Power Sources*, **75**, 130–138.
123. Huang, C., Zhang, W., and Cao, X. (1997) *J. Appl. Electrochem.*, **27**, 695–698.
124. Goldstein, J., Brown, I., and Koretz, B. (1999) *J. Power Sources*, **80**, 171–179.
125. Huot, J.-Y. and Boubour, E. (1997) *J. Power Sources*, **65**, 81–85.
126. Huot, J.-Y. and Malservisi, M. (2001) *J. Power Sources*, **96**, 133–139.
127. Othman, R., Yahaya, A.H., and Arof, A.K. (2002) *J. Appl. Electrochem.*, **32**, 1347–1353.
128. Calabrese Barton, S., and West, A.C. (2001) *J. Electrochem. Soc.*, **148**, A490–A495.
129. Glaeser, W., Künzel-Keune, S., and Merkel, P. (1999) *J. Power Sources*, **80**, 72–77.
130. Kriegsmann, J.J. and Cheh, H.Y. (1999) *J. Power Sources*, **84**, 52–62.
131. Horn, Q.C. and Shao-Horn, Y. (2003) *J. Electrochem. Soc.*, **150**, A652–A658.
132. Glaeser, W. (1989) in *Power Sources*, vol. 12 (eds T. Keily and B.W. Baxter), International Power Sources Symposium Committee, Leatherhead/England, p. 265.
133. Myazaki K. et al. (1984) European Patent Application 0, 162,411.
134. Meeus M. et al. (1984) US Patent 4, 632,699.
135. Meeus, M., Strauven, Y., and Groothaert, L. (1986) in *Power Sources*, vol. 11 (ed. L.J. Pearce), International

Power Sources Symposium Committee, Leatherhead/England, p. 281.
136. Uemura T. et al. (1992) European Patent Application 0, 500,313 A2.
137. Strauven, Y. and Gay, B. (2008) US Patent 7, 374,840 B1.
138. Ein-Eli, Y., Auinat, M., and Starosvetsky, D. (2003) *J. Power Sources*, **114**, 330–337.
139. Ein-Eli, Y. and Auinat, M. (2003) *J. Electrochem. Soc.*, **150**, A1606–A1613.
140. Ein-Eli, Y. and Auinat, M. (2003) *J. Electrochem. Soc.*, **150**, A1614–A1622.
141. Auinat, M. and Ein-Eli, Y. (2005) *J. Electrochem. Soc.*, **152**, A1158–A1164.
142. de Souza, C.C.B.M., de Oliveira, D.C., and Tenorio, J.A.S. (2001) *J. Power Sources*, **103**, 120–126.
143. Giltner, L.J. (1996) Proceedings of 37th Power Sources Conference, Cherry Hill, p. 27.
144. Dallek, S., Cox, W.G., and Kilroy, W.P. (1996) Proceedings of 37th Power Sources Conference, Cherry Hill, p. 31.
145. James, S.D. and Serenyi, R. (1996) Proceedings of 37th Power Sources Conference, Cherry Hill, p. 398.
146. Skelton, J. and Serenyi, R. (1997) *J. Power Sources*, **65**, 39–45.
147. Adamedes, Z. (1996) Proceedings of 37th Power Sources Conference, Cherry Hill, p. 390.
148. Kanda, M. (1979) *Toshiba Rev.*, **124**, 3.
149. Jindra, J. (2000) *J. Power Sources*, **88**, 202–205.
150. McLarnon, F.R. and Cairns, E.J. (1991) *J. Electrochem. Soc.*, **138**, 645.
151. Jindra, J. (1992) *J. Power Sources*, **37**, 297.
152. Tassin, N., Bronoel, G., Fauvarque, J.F., and Millot, A. (1996) Proceedings of the 37th Power Sources Conference, Cherry Hill, p. 378.
153. Humble, P.H., Harb, J.N., and LaFollette, R.-. (2001) *J. Electrochem. Soc.*, **148**, A1357–A1361.
154. Jain, R., Adler, T.C., McLarnon, F.R., and Cairns, E.J. (1992) *J. Appl. Electrochem.*, **22**, 1039.
155. McBreen, J. (1972) *J. Electrochem. Soc.*, **119**, 1620.
156. Choi, K.W., Hamby, D., Bennion, D.N., and Newman, J. (1976) *J. Electrochem. Soc.*, **123**, 1628.
157. Chang, H. and Lim, C. (1997) *J. Power Sources*, **66**, 115–119.
158. Zhu, J., Zhou, Y., and Yang, H. (1997) *J. Power Sources*, **69**, 169–173.
159. Frackowiak, E. and Skowronski, J.M. (1998) *J. Power Sources*, **73**, 175–181.
160. Zhu, J. and Zhou, Y. (1998) *J. Power Sources*, **73**, 266–270.
161. Renuka, R., Ramamurthy, S., and Muralidharan, K. (1998) *J. Power Sources*, **76**, 197–209.
162. Zhang, C. et al. (2001) *J. Appl. Electrochem.*, **31**, 1049–1054.
163. Vatsalarani, J. et al. (2005) *J. Electrochem. Soc.*, **152**, A1974–A1978.
164. Zhu, J., Zhou, Y., and Gao, C. (1998) *J. Power Sources*, **72**, 231–235.
165. Wang, J.M. et al. (2001) *J. Power Sources*, **102**, 139–143.
166. Yang, C. and Lin, S. (2002) *J. Power Sources*, **112**, 174–183.
167. Yang, H. et al. (2004) *J. Power Sources*, **128**, 97–101.
168. Yano, M. et al. (1998) *J. Power Sources*, **74**, 129–134.
169. Yano, M., Fujitani, S., and Nishio, K. (1998) *J. Appl. Electrochem.*, **28**, 1221–1225.
170. Shivkumar, R., Paruthimal Kalaignan, G., and Vasudevan, T. (1998) *J. Power Sources*, **75**, 90–100.
171. Renuka, R., Ramamurthy, S., and Srinivasan, L. (2000) *J. Power Sources*, **89**, 70–79.
172. Wang, J.M. et al. (2000) *J. Appl. Electrochem.*, **30**, 113–116.
173. Renuka, R. (2001) *J. Appl. Electrochem.*, **31**, 655–661.
174. Yang, H. et al. (2004) *J. Electrochem. Soc.*, **151**, A389–A393.
175. Kordesch, K. and Gsellmann, J. (1979) in *Power Sources*, vol. 7 (ed. J.T. Thompson), Academic Press, New York, p. 557.
176. Kordesch, K., Gsellmann, J., Peri, M., Tomantschger, K., and Chemelli, R. (1981) *Electrochim. Acta*, **26**, 1495.

177. Müller, S., Holzer, F., Haas, O., Schlatter, C., and Comninellis, C. (1995) *Chimia*, **49**, 27.
178. Müller, S., Holzer, F., Haas, O., Schlatter, C., and Comninellis, C. (1996) *Electrochem. Soc. Proc.*, **95-14**, 135.
179. Shen, Y. and Kordesch, K. (2000) *J. Power Sources*, **87**, 162–166.
180. Sharma, Y. et al. (2001) *J. Power Sources*, **94**, 129–131.
181. Chemelli, R., Gsellmann, J., Körbler, G., and Kordesch, K. (1981) Rechargeability of manganese dioxide I.C. samples. 2nd International MnO_2 Symposium Tokyo, I.C.MnO_2 Sample Office, Cleveland.
182. Huot, J.-Y. (1993) in *Power Sources*, vol. 14 (eds A. Attewell and T. Keily), International Power Sources Symposium Committee, Leatherhead/England, p. 177.
183. Root, M.J. (1995) *J. Appl. Electrochem.*, **25**, 1057.
184. Taucher, W., Binder, L., and Kordesch, K. (1992) *J. Appl. Electrochem.*, **22**, 95.
185. Binder, L., Kordesch, K., and Urdl, P. (1996) *J. Electrochem. Soc.*, **143**, 13.
186. Witherspoon, R.R. (1969) Proceedings of the Automotive Engineering Conference, SAE, Paper No. 690 204, Detroit.
187. Foller, P.C. (1986) *J. Appl. Electrochem.* **16**, 527.
188. Jiricny, V. et al. (2000) *J. Appl. Electrochem.*, **30**, 647–656.
189. Holleck, G.L., Kon, A.B., and Morin, E.A. (1996) Proceedings of the 37th Power Sources Conference, Cherry Hill, p. 432.
190. Müller, S. et al. (1998) *J. Appl. Electrochem.*, **28**, 305–310.
191. Müller, S., Holzer, F., and Haas, O. (1998) *J. Appl. Electrochem.*, **28**, 895–898.
192. Clark, R.P., Chamberlin, J.L., Saxton, H.J., and Symons, P.C. (1983) in *Power Sources*, vol. 9 (ed. J. Thompson), Academic Press, New York, p. 271.
193. Carr, P., Warde, C.J., Lijoi, A., and Brummet, B.D. (1982) *Proceedings of the 30th Power Sources Symposium*, The Electrochemical Society Inc., Pennington, p. 99.
194. Kordesch, K. and Fabjan, C. (1984) DECHEMA Monographien, Bd.97, p. 255.
195. Tomazic, G. (1987) Österr. Zeitschrift für Elektrizitätswirtschaft, **40**, 13.
196. Iacovangelo, C.D. and Will, F.G. (1985) *J. Electrochem. Soc.*, **132**, 851.
197. Adams, G.B., Hollandsworth, R.P., and Littauer, E.L. (1981) Proceedings of the 4th US D.O.E. Battery Electrochemical Contractors Conference, p. 311.
198. Zhu, W.H., Flanzer, M.E., and Tatarchuk, B.J. (2002) *J. Power Sources*, **112**, 353–366.
199. Ghiurcan, G.A. et al. (2003) *J. Electrochem. Soc.*, **150**, A922–A927.
200. Klein, T.K. (2003) Doctoral dissertation, Graz University of Technology, Development of a thin-film battery produced by printing techniques.

9
Metal Hydride Electrodes
James J. Reilly

9.1
Introduction

Many metals and alloys reversibly absorb large quantities of hydrogen to form metal hydride (MH) phases. In most cases the volumetric hydrogen density in the hydride phase exceeds that of liquid hydrogen. While very few binary hydrides (phases consisting of the elemental metal and hydrogen) are suitable for hydrogen storage, many of them have properties that make them practical, convenient, and safe energy storage media.

For many years the focus of MH research was on the storage of hydrogen for use as a gaseous fuel, although the possibility of electrochemical applications was also recognized. In about 1980 that focus dramatically shifted toward electrochemical applications where a metal hydride (MH_x) electrode is used to replace the cadmium electrode in Cd/Ni batteries. The driving force for such replacement was the environmental problems associated with cadmium. An additional benefit was the higher energy density of the MH_x electrode. In the first edition this was concerned primarily with the electrochemical, thermodynamic, and structural properties of MHs that pertain to their use in Ni–MH_x batteries.

In the late 1990s the research focus again shifted, this time toward lithium ion batteries. In that same period the MH_x–Ni battery became an article of commerce, and consequently almost all government sponsored and academic research and development shifted toward Li based batteries employing nonaqueous electrolytes. Certainly research by private interests continues, but such efforts are considered proprietary and any significant advances are not likely to be published. Thus the original chapter appearing in the first edition has been little changed since, as noted above, it was concerned with basic properties which are little changed.

9.2
Theory and Basic Principles

Many metals and alloys reversibly absorb large quantities of hydrogen to form metal hydride (MH) phases, and for many years the focus of MH research was

---

Handbook of Battery Materials, Second Edition. Edited by Claus Daniel and Jürgen O. Besenhard.
© 2011 Wiley-VCH Verlag GmbH & Co. KGaA. Published 2011 by Wiley-VCH Verlag GmbH & Co. KGaA.

on the storage of hydrogen for use as a gaseous fuel, although the possibility of electrochemical applications was also recognized. In about 1980 that focus dramatically shifted toward electrochemical applications where a metal hydride (MH_x) electrode is used to replace the cadmium electrode in Cd/Ni batteries. The driving force for such replacement was the environmental problems associated with cadmium. An additional benefit was the higher energy density of the MH_x electrode, and in this chapter we discuss the electrochemical, thermodynamic, and structural properties of MHs that pertain to their use in Ni–MH_x batteries.

A *hydrogen–metal system* may be defined as consisting of an amorphous or crystalline metal phase containing dissolved hydrogen in interfacial contact with molecular, atomic, or ionized hydrogen. In many cases, depending on temperature and pressure, an MH phase will form of which there are three general categories: ionic, covalent, and metallic. Intermetallic hydrides are, of course, a sub-group of the latter class, where hydrogen occupies interstitial sites in the metal lattice and the hydride phase is crystalline. There are a large number of intermetallic compounds, many of which will form a hydride via the direct and reversible reaction with hydrogen. Consequently, even though most may not be of interest for practical applications, the sheer number of intermetallic hydride systems constitutes a great advantage over binary systems with respect to the formulation of attractive energy storage materials.

9.2.1
Thermodynamics

Flanagan and Oates have extensively reviewed the thermodynamics of intermetallic hydrides [1]; also recommended are the classic work of Libowitz [2] and the comprehensive text of Muller, Blackledge, and Libowitz [3] which treats the properties of binary hydrides. The properties of a metal–hydrogen system can be conveniently summarized by a pressure-temperature-composition (PTC) diagram of which an idealized version is shown in Figures 9.1 and 9.2. The former is

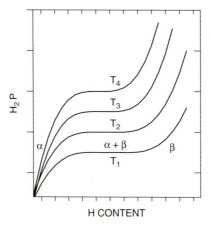

Figure 9.1 Ideal pressure-composition-isotherms showing the hydrogen solid solution phase, α, and the hydride phase, β. The plateau marks the region of the co-existence of the α and β phases. As the temperature is increased the plateau narrows and eventually disappears at some consolute temperature, T_c.

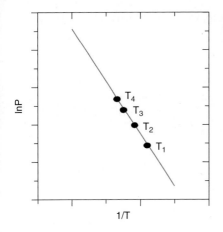

Figure 9.2 The enthalpy, ΔH, of the phase transformation can be calculated from the variation of $\ln P_{plateau}$ with reciprocal temperature in a van't Hoff plot.

essentially a phase diagram which consists of a family of isotherms that relate the equilibrium pressure of hydrogen to the H content of the metal. Initially the isotherm ascends steeply as hydrogen dissolves in the metal to form a solid solution which by convention is designated as the α phase. At low concentrations the behavior is ideal and the isotherm obeys Sievert's Law, that is,

$$H_{solid} = K_s P^{1/2} \tag{9.1}$$

where H_{solid} is the concentration of hydrogen in the metal, K_s is Sievert's constant, and P is the equilibrium hydrogen pressure. As the H content of the solid increases, the system departs from ideal behavior due to H-H attractive interactions primarily caused by elastic strain in the metal; this is reflected by a decreasing slope in the isotherm. When the terminal solubility of hydrogen in the α phase is exceeded, the hydride phase precipitates and is designated the β phase. Upon the appearance of the β phase the hydrogen pressure will remain constant and the isotherm forms a plateau as more hydrogen is added. The plateau is a consequence of the phase rule and will persist as long as the two solid phases coexist. When the phase conversion is complete the system regains a degree of freedom and the pressure again rises as a function of the hydrogen content. In this region of the diagram electronic factors become dominant as the limiting hydrogen concentration is approached. It is also possible that more than one hydride phase exists in which case a second plateau will appear. In many systems there is a significant hysteresis effect in the phase conversion process which is reflected by a higher isotherm plateau pressure for the $\alpha \Rightarrow \beta$ conversion than the reverse $\beta \Rightarrow \alpha$ process. The effect of increasing temperature is shown by the higher temperature isotherms T_2, T_3, and T_4 in Figure 9.1. Usually, the miscibility gap narrows as the temperature increases, eventually disappearing as the consolute temperature is reached.

The reaction of a metal with hydrogen gas may be written as

$$M + \frac{x}{2} H_2 \Leftrightarrow MH_x \tag{9.2}$$

Thermodynamic quantities for a system may be determined from the van't Hoff equation, which defines the equilibrium constant, K, in terms of the reaction enthalpy, ΔH, and the temperature, T.

$$\frac{d \ln K}{dT} = \frac{\Delta H}{RT^2} \tag{9.3}$$

For reaction (9.2),

$$K = \frac{a_{MH_x}}{a_M (f_{H_2})^{x/2}} \tag{9.4}$$

Under ideal conditions the activity of a solid may be taken as unity and the fugacity as the pressure, then Equation 9.3 may be rewritten as

$$\frac{d(\ln P_{H_2})^{-x/2}}{dT} = \left(\frac{\Delta H}{RT^2}\right) \tag{9.5}$$

which upon integration yields,

$$\ln P_{H_2} = \frac{2}{x}(\Delta H/RT) + C \tag{9.6}$$

The enthalpy of the phase conversion can be determined via Equation 9.6 by plotting the log of the absorption or desorption plateau pressure, P_{plateau}, vs the reciprocal temperature as indicated in Figure 9.2. When the solubility of hydrogen in the metal (α) phase is small, then $\Delta H_{\text{plat}} \approx \Delta H_f$, where ΔH_f is essentially the enthalpy of formation of the hydride from the metal [4]; the intercept, C, is equal to $2/x(\Delta S/R)$. Equation 9.6 is commonly presented as

$$\ln P_{\text{plateau}} = \frac{A}{T} + B \tag{9.7}$$

where the constants A and B are specified. Thermodynamic data for some representative compounds are given in Table 9.1.

9.2.2
Electronic Properties

Switendick was the first to apply modern electronic band theory to metal hydrides [11]. He compared the measured density of electronic states with theoretical results derived from energy band calculations in binary and pseudo binary systems. Recently, the band structures of intermetallic hydrides including $LaNi_5H_x$ and $FeTiH_x$ have been addressed; the results for these more complicated systems have been summarized in a review article by Gupta and Schlapbach [12]. All exhibit certain common features upon the absorption of hydrogen and the formation of a distinct hydride phase. These are (i) the density of states versus energy function is changed, (ii) new low-lying states having an s-like character appear and are associated with hydrogen, and (iii) to the extent that the hydrogen electrons cannot be accommodated in the new low-lying states they are inserted into empty states near the Fermi level, which in turn shifts.

Table 9.1 Thermodynamic data.

Alloy	Phase conversion	ΔH kJ/ mol H_2	ΔS JK^{-1} mol^{-1} H_2	References
MgH_2	$\beta \to \alpha$	77.1	137	[5]
TiH_2	$Ti \to TiH_2$	−123	−125	[3]
LaH_2	$\alpha \to \delta$	−206	−147	[3]
$PdH_{.6}$	$\beta \to \alpha$	40.9	91.1	[4]
Mg_2NiH_4	$\beta \to \alpha$	64.2	122	[5]
$FeTiH_x$	$\gamma \to \beta$	33.3	104	[6]
$FeTiH_x$	$\beta \to \alpha$	28.1	106	[6]
$LaNi_5H_x$	$\beta \to \alpha$	30.0	108	[7]
$LaNi_5H_x$[a]	$\alpha \to \beta$	−29.4		[1]
$LaNi_{4.6}Al_{.4}H_x$	$\beta \to \alpha$	36.3	109	[8]
$MmNi_5$	$\beta \to \alpha$	20.9	96	[9]
$Mm^b Ni_{3.55}Co_{.75}Mn_{.4}Al_{.3}$	$\beta \to \alpha$	29.7	100	[10]
$MmNi_{3.55}Co_{.75}Mn_{.4}Al_{.3}$	$\beta \to \alpha$	41.5	117	[10]

[a] Calorimetric measurement.
[b] Mm is mischmetal. See Table 9.2.

9.2.3
Reaction Rules and Predictive Theories

There have been numerous studies with the object of gaining an understanding of the factors that influence the stability, stoichiometry, and H site occupation in hydride phases. Stability has been correlated with cell volume [8] or the size of the interstitial hole in the metal lattice [13] and the free energy of the $\alpha \Rightarrow \beta$ phase conversion. This has been widely exploited to modulate hydride phase stability as discussed in Section 9.2.1.

Westlake developed a geometric model which is fairly successful in predicting site occupation in AB_5 and AB_2 hydride phases [14]. It involves two structural constraints; that the minimum hole size necessary to accommodate a H atom has a radius of 0.40 Å and that the minimum distance between two H occupied sites is 2.10 Å. The former criterion was empirically derived from a survey of known hydride structures while the latter was suggested by Switendick based on electronic [15] band structure calculations.

A relatively simple set of rules has been found to hold for all intermetallic hydrides useful for hydrogen storage [16]. They may be stated as follows:

1) In order for an intermetallic compound to react directly and reversibly with hydrogen to form a distinct hydride phase it is necessary that at least one of the metal components be capable of reacting directly and reversibly with hydrogen to form a stable binary hydride.
2) If a reaction takes place at a temperature at which the metal atoms are mobile, the system will assume its most favored thermodynamic configuration.

3) If the metal atoms are not mobile (as is the case in low-temperature reactions), only hydride phases can result in which the metal lattice is structurally very similar to the starting intermetallic compound because the metal atoms are essentially frozen in place. In effect the system may be considered to be pseudo binary as the metal atoms behave as a single component.

9.3
Metal Hydride–Nickel Batteries

The half cell reactions taking place in an MH_x–Ni battery may be written as follows:

$$MH_x + xOH^- \Leftrightarrow M + xH_2O + xe^- \tag{9.8}$$

$$Ni(OOH) + H_2O + e^- \Leftrightarrow Ni(OH_2) + OH^- \tag{9.9}$$

It is in effect a rocking-chair type battery in which hydrogen is transferred from one electrode to the other. It is also most convenient that the voltage is essentially the same as that in the conventional Nicad batteries. It is worthwhile noting that the NiOOH cathode has a maximum energy density, based on Equation 9.9, of 289 mAh g^{-1}. This may be compared with 300–400 mAh g^{-1} for current MH_x electrodes and >400 mAh g^{-1} projected for high-capacity MH_x electrodes which, though not yet developed, are certainly conceivable.

Two types of MH electrodes, comprising the AB_5 and AB_2 classes of intermetallic compounds, are currently of interest. The AB_5 alloys have the hexagonal $CaCu_5$ structure, where the A component comprises one or more rare earth elements and B consists of Ni, or another transition metal, or a transition metal combined with other metals. The paradigm compound of this class is $LaNi_5$, which has been well investigated because of its utility in conventional hydrogen storage applications. Unfortunately, $LaNi_5$ is too costly, too unstable, and too corrosion sensitive for use as a battery electrode. Thus commercial AB_5 electrodes use mischmetal, a low cost combination of rare earth elements, as a substitute for La. The B_5 component remains primarily Ni but is substituted in part with Co, Mn, Al, and so on. The partial substitution of Ni increases thermodynamic stability of the hydride phase [9] and corrosion resistance. Such an alloy is commonly written as MmB_5, where Mm represents the mischmetal component. The compositions of normal and cerium-free mischmetal are given in Table 9.2.

The other electrode type is usually referred to as the *AB_2 or Laves phase type* electrode and is discussed in Section 9.3. These electrodes are complicated, multiphase alloys with as many as nine metal components. Alloy formulation is primarily an empirical process where the composition is adjusted to provide one or more hydride-forming phases in the particle bulk but which has a surface that is presumed to be corrosion resistant because of the formation of semi-passivating oxide layers. Unlike the AB_5 alloys there are few systematic guidelines which can be used to predict alloy properties. Eventually AB_2 alloy electrodes may be more attractive than AB_5 electrodes in terms of cost and energy density, but that potential is not yet realized.

Table 9.2 Composition of mischmetal[a].

Rare earth	Normal (wt%)	Ce free (wt%)
Ce	56.9	0.13
La	20.5	58.2
Nd	14.7	29.9
Pr	5.5	7.6
Fe	0.42	0.07
O	0.058	0.47
C	0.032	0.104
N	0.002	0.162
Y	<0.01	0.04
Ca	0.13	0.27

[a] Analyzed at Materials Preparation Center, Ames Laboratory, Ames, Iowa.

There are also other intermetallic hydrides that have largely been ignored for battery applications because they are, or are perceived to be, too stable, electrochemically inactive or, most importantly, subject to severe corrosion in the battery environment. Alloys such as Mg_2Ni [5] and FeTi [6] have substantially greater hydrogen storage capacity than the conventional AB_5 and AB_2 alloys and are less costly. Unfortunately both are passivated rapidly in the electrochemical environment (Johnson, J.R. unpublished data). High-content Mg amorphous alloys, produced by mechanical alloying, have been demonstrated to have initial storage capacities of >400 mAh g^{-1} but they rapidly deteriorate upon cycling [17]. A novel vanadium based electrode, $TiV_3Ni_{.56}$, has also been shown to have a high initial storage capacity, 400–450 mAh g^{-1}, but also corrodes rapidly upon electrochemical cycling [18].

9.3.1
Alloy Activation

Essentially all metals and alloys which form metallic hydrides require an activation process before the metal will readily cycle between the hydride and metal phase. All the AB_5 and AB_2 alloys are quite brittle and during the activation procedure are pulverized to fine particles. This greatly enhances subsequent reaction rates. The activation process is considered to take place in two stages: the formation of a reactive surface and pulverization of the bulk solid to form fine particles. The surface composition of $LaNi_5$ after activation has been defined by X-ray photoemission spectroscopy, Auger electron spectroscopy, and magnetic susceptibility studies [19]. There is a surface enrichment of La to give a ratio of La/Ni $\approx$ 1; the La is associated with oxygen but Ni remains metallic and is present as surface clusters containing about 6000 atoms. Thus, the surface of cycled $LaNi_5$ appears to consist of islands of La associated with oxygen and Ni clusters. This is the mechanism by which catalytic metal surface sites are formed for chemical or electrochemical reactions. The activation procedure is straightforward; in gas/solid systems it merely consists of repeated formation and decomposition of the hydride phase [20]. Electrochemical activation also consists of repeated charge and discharge cycles which sometimes require extended cycling periods. Alloys may be activated via the gas–solid reaction and then used as an electrode; this can significantly shorten the period required for electrochemical activation.

9.3.2
AB_5 Electrodes

The use of $LaNi_5$ as an electrode was first reported by Justi in 1973 [21]. However, the capacity was less than 1/3 that of 372 mAh g^{-1}, which corresponds to the discharge of six hydrogens from $LaNi_5H_6$. This was primarily due to the fact that the dissociation pressure of $LaNi_5H_6$ exceeds 1 atm at 298 K. Thus, in an open cell most of the hydrogen is lost as H_2 gas. A few years later Percheron-Guegan and co-workers [22] substituted Al and Mn in part for nickel, thereby increasing hydride stability and charge capacity. However, this did not significantly affect the rapid corrosion of the electrode as observed with $LaNi_5$. In 1984 Willems [23] prepared the first multi-component AB_5 electrode that had an acceptable cycle life. He also reported the positive correlation between lattice expansion and electrode corrosion. Finally in 1987 an alloy of composition $MmNi_{3.55}Co_{0.75}Mn_{0.4}Al_{0.3}$ was shown to meet the minimum requirements for a practical battery with respect to cost, cycle life, and storage capacity [24]. Indeed, this composition is very similar to those currently used in commercial Ni–MH batteries with AB_5 hydride anodes. Ikoma et al. [25] describe an experimental electric vehicle (EV) battery having an energy density of 70 Wh kg^{-1} using an anode of composition $Mm(Ni,Co,Mn,Al)_5$. The electrochemical behavior of this alloy and its relation to small changes in alloy composition is of great practical interest and will be discussed at length in the following sections.

9.3.2.1 Chemical Properties of AB_5 Hydrides

In order to fully understand the electrochemical behavior of AB_5 hydrides a knowledge of their chemical properties is required. Van Vucht et al. [7] were the first to prepare $LaNi_5$ hydride, and this is arguably the most thoroughly investigated H storage compound. It reacts rapidly with hydrogen at room temperature at a pressure of several atmospheres above the equilibrium plateau pressure. PC isotherms for this system are shown in Figure 9.3. The nominal reaction may be written as

$$LaNi_5 + 3H_2 \Leftrightarrow LaNi_5H_6 \tag{9.10}$$

$LaNi_5$ has the $CaCu_5$ structure; space group P6/mmm [26]; the hexagonal metal lattice is shown in Figure 9.4. The crystal structure of $LaNi_5D_7$ has been determined [27, 28] and is illustrated in Figure 9.5. There are three types of interstitial D sites,

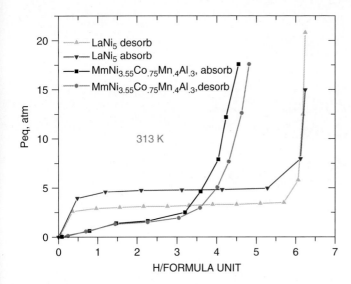

Figure 9.3 PC isotherms for LaNi$_5$ and MmNi$_{3.55}$Co$_{.75}$Mn$_{.4}$Al$_{.3}$ [42].

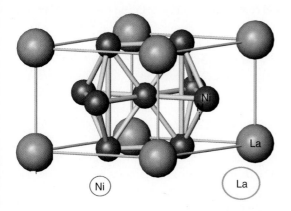

Figure 9.4 CaCu$_5$ structure of LaNi$_5$; space group P6/3mmm (no.191).

La$_2$Ni$_4$ octahedra, La$_2$Ni$_2$ tetrahedra, and Ni$_4$ tetrahedra. The unit cell is doubled along the c axis because of the formation of a super-lattice, which is a consequence of long range correlations between occupied and unoccupied Ni$_4$ tetrahedra.

With reference to rule 3 (Section 9.2.3) regarding metal atom mobility, we note that ΔG_f values at 298 K for LaNi$_5$ and LaH$_2$ are about -67 and -171 kJ respectively. Thus the following disproportionation reaction is highly favored [29],

$$\text{LaNi}_5 + \text{H}_2 \Rightarrow \text{LaH}_2 + 5\text{Ni} \quad \Delta G_{298K} = -104 \text{ kJ} \tag{9.11}$$

whereas for reaction (9.10) $\Delta G_{298} \approx 0$, but at low temperatures such disproportionation does not take place.

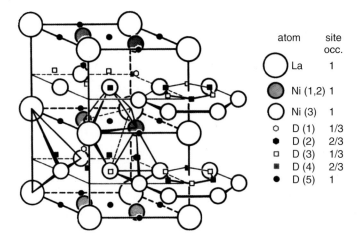

Figure 9.5 Structure of $LaNi_5D_7$. The unit cell is doubled along the c axis; space group $P6_3mc$ [27].

However, disproportionation on the surface of polycrystalline $LaNi_5$ occurs readily at room temperature and constitutes the alloy activation process described in Section 9.3.1.

The kinetics of the formation and decomposition of $LaNi_5$ hydride have been widely studied with just as widely varying results [30]. In most cases the investigations were done using static beds of metal/MH particles in contact with gaseous hydrogen. Such systems have inherently poor heat transport and exchange characteristics and, since reaction rates are high, isothermal conditions are difficult if not impossible to maintain. Consequently the data are difficult to interpret, which is the likely cause of the disparity in reported results. When kinetic experiments were carried out isothermally or nearly so, the kinetics was well described by a shrinking core model [31, 32]. In this model the rate-limiting process is the solid-state transformation taking place at the interface between the α and β phases. In hydride formation a growing product layer of ß $LaNi_5H_x$ proceeds inward from the surface while in hydride decomposition the reaction also proceeds inward from the surface but now the growing product layer is α $LaNi_5$ as illustrated in Figure 9.6 [33].

A particular advantage of the AB_5 hydride family is that the properties of the alloy–hydrogen system can be varied almost at will by substituting, in whole or in part, other metals for lanthanum and Ni. For example, mischmetal when substituted for La in $LaNi_5$ forms a hydride having about the same hydrogen content but is much more unstable [34]. Lundin et al. [13] carried out a systematic study of such substitutional alloys and correlated the free energy of formation (plateau region) with the change of the interstitial hole size caused by the substituted metal component. Gruen et al. [8] have taken a similar approach, but rather correlate the cell volume with $\ln P_{plateau}$, as shown in Figure 9.7.

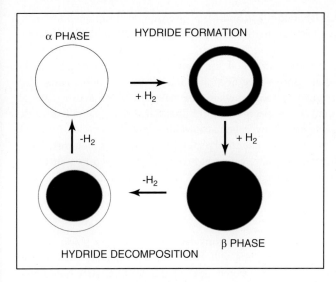

Figure 9.6 Schematic representation of the reaction paths for hydriding and dehydriding LaNi$_5$ [33].

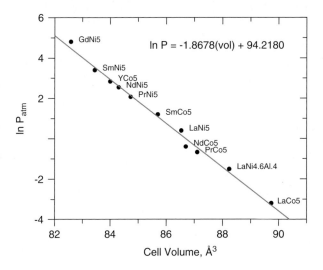

Figure 9.7 Alloy cell volume vs Ln $P_{plateau}$ for various AB$_5$-type hydrides [8].

9.3.2.2 Preparation of AB$_5$ Electrodes

Electrode behavior is strongly influenced by alloy microstructure, metal stoichiometry, and composition. Thus, an understanding of the physical metallurgy [35] of a particular system as well as a knowledge of its phase diagram are highly desirable if one wishes to prepare alloys having reproducible properties. While there are no phase diagrams for these multicomponent alloys, those having the AB$_5$ stoichiometry still behave similarly to LaNi$_5$. Percheron-Guegan and Welter have described both laboratory and industrial preparation techniques for many intermetallic hydride formers, particularly emphasizing LaNi$_5$ and its substituted analogs [36].

All AB$_5$ alloys are very brittle and are pulverized to fine particles in the hydriding–dehydriding process (see Section 9.3.1, Alloy Activation). Thus, electrodes must be designed to accommodate fine powders as the active material. There are several methods of electrode fabrication: Sakai *et al.* [35] pulverize the alloy by subjecting it to several hydrogen absorption–desorption cycles, then coat the resulting particles with Ni by chemical plating. The powder is then mixed with a Teflon dispersion to get a paste, which is finally roller pressed to a sheet and then hot pressed to an expanded nickel mesh. The fabrication of a simple paste electrode suitable for laboratory studies is reported by Petrov *et al.* [37].

9.3.2.3 Effect of Temperature

Ni–MH batteries are currently under consideration for use as power sources for automotive propulsion and thus will be required to operate over a large ambient temperature range. The stated goal of the USABC (US Automotive Battery Consortium) program [38] is to develop a battery which can operate satisfactorily over a range extending from -30 to 65 °C. However, hydride stability is a logarithmic function of the temperature and must be taken into account when choosing an electrode composition. For example, the equilibrium plateau pressure (decomposition) of LaNi$_5$H$_x$ at 65 °C $\approx$ 10 atm – much too high for use as a battery electrode. Van't Hoff plots [10] for LaNi$_5$H$_x$, MmNi$_{3.55}$Co$_{0.75}$Mn$_{.4}$Al$_{.3}$H$_x$ and Mm*Ni$_{3.55}$Co$_{0.75}$Mn$_{.4}$Al$_{.3}$H$_x$ (Mm* = cerium free mischmetal, see Table 9.2) are shown in Figure 9.8. At 65° the absorption plateau pressure of Mm*B$_5$ would be 0.5 atm, whereas that of MmB$_5$ is 5.0 atm. Thus, even though both mischmetal electrodes have similar electrochemical properties at room temperature, only the former would be suitable for use at higher temperatures.

9.3.3
Electrode Corrosion and Storage Capacity

Deterioration of electrode performance due to corrosion of electrode components is a critical problem. The susceptibility of MH$_x$ electrodes to corrosion is essentially determined by two factors, surface passivation due to the presence of surface oxides or hydroxides and the molar volume of hydrogen, V_H, in the hydride phase. As pointed out by Willems and Buschow [39], V_H is important since it governs alloy expansion and contraction during the charge–discharge cycle. Large volume changes increase the flushing action of the electrolyte through the pores and micro-cracks of the electrode during each charge and discharge cycle, thereby increasing the rate of contact of the alloy surface with fresh electrolyte and, consequently, the corrosion rate. Thus, when examining the effect of various substituents upon electrode corrosion the question always arises whether an observed change is due to a change in lattice expansion or to a change in surface passivation, for example, the formation of a corrosion-resistant oxide layer.

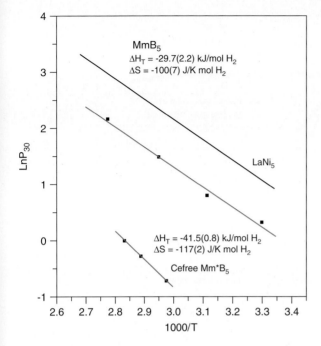

Figure 9.8 van't Hoff plots for $LaNi_5H_x$, $MmNi_{3.55}Co_{.75}Mn_{.4}Al_{.3}H_x$ and $Mm*Ni_{3.55}Co_{.75}Mn_{.4}Al_{.3}H_x$ [10].

While the partial substitution of Ni by other metals has ameliorated the corrosion problem it has also resulted in a reduced storage capacity and high alloy costs (because of the incorporation of Co). None of the substituted multicomponent hydrides approach the storage capacity of $LaNi_5H_x$ because it is reduced by the partial substitution of Ni. Percheron-Guegan et al. [40] noted this with the binary alloy $LaNi_{5-x}M_x$ with M = Al, Mn, Si, or Cu. Thus, although the cycle life of substituted AB_5 electrodes is greatly extended over that of $LaNi_5$, a severe penalty in storage capacity is exacted for this improvement as illustrated by the PCT diagram in Figure 9.3. It is also of interest to note that while $LaNi_5$ exhibits a significant hysteresis effect $MmNi_{3.55}Co_{.75}Mn_{.4}Al_{.3}$ does not. The small or even complete lack of hysteresis in multicomponent AB_5 hydrides is not unusual, but it is almost always present in less complex systems.

9.3.4
Corrosion and Composition

The long life of MmB_5 battery electrodes raises the question: why do such electrodes behave so differently than other more simple formulations? Such differences are very apparent in plots of charge capacity vs charge–discharge cycles as reported by Adzic et al. [41]. In Figure 9.9 four different electrodes are compared, Mm(or

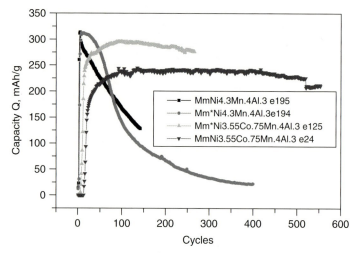

Figure 9.9 Charge capacity, Q, vs charge–discharge cycles for four mischmetal AB5 electrodes. Note high decay rate in charge capacity for Co free electrode (Johnson, J.R. unpublished data).

Table 9.3 Effect of Co in various MmB_5 electrodes [41].

Alloy	V_H (Å^3)	Q_{max} (mAh g^{-1})	n, H atoms/ unit cell	% $\Delta V/V$	Corrosion (wt%/cycle)
$MmNi_{3.55}Co_{0.75}Mn_4Al_3$	3.13	247	3.90	14.3	0.001
$Mm^*Ni_{3.55}Co_{0.75}Mn_4Al_3$[a]	3.05	295	4.64	16.0	0.041
$MmNi_{4.3}Mn_4Al_3$	3.51	314	4.96	20.1	0.354
$Mm^*Ni_{4.3}Mn_4Al_3$	3.14	314	4.94	17.7	1.029

[a] Mm^* = Ce free mischmetal.

$Mm^*)Ni_{3.55}Co_{0.75}Mn_4Al_3$ (Mm^* refers to Ce free mischmetal), and two Co-free electrodes, $Mm(\text{or } Mm^*)Ni_{4.3}Mn_4Al_3$. Both Co-free electrodes rapidly corrode and would not be suitable for battery applications. Obviously alloy composition is responsible for the observed behavior, and this is discussed in the following sections. The results of these experiments are summarized in Table 9.3.

While cycle life may vary dramatically, inspection of cycle life plots reveals a common behavior which is found in almost all MH_x electrodes. There is an initial steep increase in capacity in the first few cycles, which comprises the activation process. After activation a maximum in electrochemical storage capacity, Q_{max}, is reached. This is usually followed by an almost linear decrease in capacity which may be termed *capacity decay*. It is defined as the slope of the capacity vs cycle curve, that is, the $-dQ/d$ cycle.

Table 9.4 Crystallographic parameters and V_H of selected alloys.

Composition	a (Å)	c (Å)	Cell volume (Å^3)	V_H (Å^3)	References
NdNi$_{3.55}$Co$_{.75}$Mn$_{.4}$Al$_{.3}$	4.9992	4.0221	87.05	2.66	Johnson, J.R. (unpublished data)
La$_{.25}$Nd$_{.75}$Ni$_{3.55}$Co$_{.75}$Mn$_{.4}$Al$_{.3}$	5.0138	4.0254	87.63	2.74	Johnson, J.R. (unpublished data)
LaNi$_{3.55}$Co$_{.75}$Mn$_{.4}$Al$_{.3}$	5.0642	4.0325	89.56	2.93	[42]
La$_{.65}$Pr$_{.35}$Ni$_{3.55}$Co$_{.75}$Mn$_{.4}$Al$_{.3}$	5.0368	4.0206	88.33	2.97	Johnson, J.R. (unpublished data)
LaNi$_{3.5}$Co$_{.75}$Mn$_{.4}$Al$_{.3}$	5.0699	4.0392	89.91	3.00	[42]
Mm$_{.3}$Mm*$_{.7}$Ni$_{3.55}$Co$_{.75}$Mn$_{.4}$Al$_{.3}$	5.0234	4.0434	88.36	3.00	Johnson, J.R. (unpublished data)
LaNi$_{3.95}$Co$_{.75}$Al$_{.3}$	5.0378	4.0107	88.15	3.02	[43]
MmNi$_{3.55}$Co$_{.75}$Mn$_{.4}$Al$_{.3}$ Ce free	5.0318	4.0309	88.38	3.05	[43]
LaNi$_{3.55}$Co$_{.75}$Mn$_{.4}$Al$_{.3}$	5.0615	4.0298	89.40	3.06	[42]
La$_{.5}$Nd$_{.5}$Ni$_{3.55}$Co$_{.75}$Mn$_{.4}$Al$_{.3}$	5.0315	4.0259	88.26	3.07	Johnson, J.R. (unpublished data)
LaNi$_{3.55}$Co$_{.75}$Mn$_{.3}$Al$_{.3}$	5.0662	4.0321	89.70	3.07	[43]
LaNi$_{4.1}$Co$_{.2}$Mn$_{.4}$Al$_{.3}$	5.0609	4.0361	89.52	3.09	[41]
LaNi$_{3.9}$Co$_{.4}$Mn$_{.4}$Al$_{.3}$	5.0629	4.0349	89.57	3.09	[41]
MmNi$_{3.55}$Co$_{.75}$Mn$_{.4}$Al$_{.3}$^{a}	4.9890	4.0545	87.39	3.10	[42]
MmNiNi$_{3.55}$Co$_{.75}$Mn$_{.4}$Al$_{.3}$	4.9626	4.0560	86.50	3.13	[42]
La$_{.25}$Ce$_{.75}$Ni$_{3.55}$Co$_{.75}$Mn$_{.4}$Al$_{.3}$	4.9538	4.0559	86.19	3.15	[42]
La$_{.5}$Ce$_{.5}$Ni$_{3.55}$Co$_{.75}$Mn$_{.4}$Al$_{.3}$	4.9934	4.0446	87.33	3.15	[42]
La$_{.65}$Nd$_{.35}$Ni$_{3.55}$Co$_{.75}$Mn$_{.4}$Al$_{.3}$	5.0324	4.0211	88.19	3.15	Johnson, J.R. (unpublished data)
LaNi$_{3.55}$Co$_{.75}$Mn$_{.14}$Al$_{.3}$	5.0509	4.0321	89.08	3.16	[43]
LaNi$_{3.85}$Co$_{.75}$Mn$_{.38}$	5.0526	4.0195	88.86	3.20	[43]
La$_{.8}$Ce$_{.2}$Ni$_{3.55}$Co$_{.75}$Mn$_{.4}$Al$_{.3}$	5.0380	4.0416	88.84	3.21	[42]
MmNi$_{3.5}$Co$_{.75}$Mn$_{.4}$Al$_{.3}$	4.9623	4.0456	86.27	3.23	[42]
LaNi$_{4.3}$Mn$_{.4}$Al$_{.3}$	5.0591	4.0370	89.48	3.26	[41]
La$_{.65}$Ce$_{.35}$Ni$_{3.55}$Co$_{.75}$Mn$_{.4}$Al$_{.3}$	5.0168	4.0451	88.16	3.24	[42]
LaNi$_{3.85}$Co$_{.75}$Mn$_{.04}$	5.0494	4.0034	88.39	3.35	[43]
LaNi$_{4.7}$Al$_{.3}$	5.0195	4.0076	87.44	3.47	[42]
MmNi$_{4.3}$Mn$_{.4}$Al$_{.3}$	4.9652	4.0453	86.37	3.51	[41]

^{a}Synthetic mischmetal, that is, La$_{.26}$Ce$_{.52}$Pr$_{.06}$Nd$_{.16}$.

In order to elucidate the relationship between corrosion rate and composition it is necessary to quantitatively determine lattice expansion. This requires the determination of V_H, which is listed in Table 9.4 for a number of alloys.

In order to quantitatively determine electrode corrosion, Adzic et al. [42] used the following approach. The H content of the charged electrode, expressed as the number of H atoms, n, per formula unit, was calculated from Q_{max} via the Faraday equation,

$$n = e^- = \frac{3600}{9.65 \times 10^7}(mw)(Q_{max}) \qquad (9.12)$$

where mw is the molecular weight of the alloy and the units of Q are mAh g^{-1}. They assumed that after activation the remaining uncorroded alloy in each subsequent charge–discharge cycle is hydrided and dehydrided to the same degree and n is constant. The percent lattice expansion of the unit cell in each electrochemical cycle was calculated via the equation

$$\% \frac{\Delta V}{V} = \frac{V_H}{V} n \times 100 \qquad (9.13)$$

where ΔV is the actual volume change of the unit cell in Å^3 in each charge or discharge cycle, V is the initial unit cell volume, and n is the number of H atoms inserted into the unit cell and subsequently discharged.

Finally, the loss of electrochemical capacity is directly proportional to the loss of the AB$_5$ alloy by oxidation and readily calculated as follows;

$$\frac{\% \text{ wt loss}}{\text{cycle}} = \frac{-dQ}{\text{cycle}}(Q_{max}) \times 100 \qquad (9.14)$$

The effects of Ce, Co, Al, and Mn upon the properties and performances of A(NiCoMnAl)$_5$ electrodes employing the above equations are discussed in the following sections.

9.3.4.1 Effect of Cerium

The rare earth composition of commercial electrodes is also related to electrode corrosion. This was noted by Sakai et al. [44], who found that the presence of Nd and Ce inhibited corrosion when substituted in part for La in La$_{1-x}$Z$_x$(NiCoAl)$_5$ (Z = Ce or Nd) electrodes. However no explanation for the effect was noted. Willems [23] prepared an electrode having the composition of La$_{.8}$Nd$_{.2}$Ni$_{2.5}$Co$_{2.4}$Si$_{.1}$, which retained 88% of its storage capacity after 400 cycles. He attributed its long cycle life to a low V_H of 2.6 Å^3.

The case of cerium is of particular interest. Adzic et al. [42] examined the properties of a homologous series of alloys with a composition corresponding to La$_{1-x}$Ce$_x$Ni$_{3.55}$Co$_{.75}$Mn$_{.4}$Al$_{.3}$ and measured their comparative performance as battery electrodes. A PCT diagram for this system is shown in Figure 9.10. Note that at $x > 0.2$ there is a decrease in the H storage capacity and thermodynamic stability until at $x = 1$ the decreases in both parameters are marked. This reduced stability is not unexpected as the unit cell volume decreases with Ce content (see Figure 9.7).

Cycle life plots for the La$_{1-x}$Ce$_x$B$_5$ electrodes are illustrated in Figure 9.11. The decreased charge capacity found in all La$_{1-x}$Ce$_x$B$_5$ alloys with $x > 0.35$ conforms to the shorter and higher plateau pressures of the isotherms depicted in Figure 9.10. The extremely low electrochemical capacity of CeB$_5$ is a consequence of the high dissociation pressure of the hydride phase.

The corrosion rates for the La$_{1-x}$Ce$_x$B$_5$ electrodes are listed in Table 9.5. The results are summarized graphically in Figure 9.12, which plots lattice expansion,

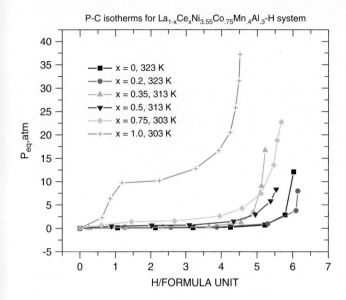

Figure 9.10 P-C isotherms for $La_{1-x}Ce_xNi_{3.55}Co_{.75}Mn_{.4}Al_{.3}$ – H system [43].

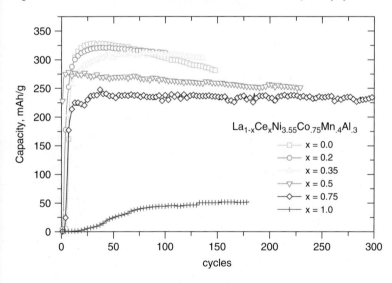

Figure 9.11 Charge capacity, Q, vs charge–discharge cycles for $La_{1-x}Ce_xNi_{3.55}Co_{.75}Mn_{.4}Al_{.3}$ electrodes [43].

corrosion rate, and H content (n) vs Ce content. The plot clearly shows the anomalous correlation of lattice expansion with corrosion; thus one concludes that the corrosion inhibition stemming from the presence of Ce is due to a surface effect. This conclusion is supported by previous work reporting that a film of CeO_2 on metal surfaces inhibits corrosion [45]. XAS (X-ray absorption spectroscopy) studies discussed in Section 9.3 confirm the corrosion inhibition effect of Ce [46].

Table 9.5 Effect of Ce in $La_{1-x}Ce_xNi_{3.55}Co_{.75}Mn_{.4}Al_{.3}$ electrodes [42].

x value	V_H (Å^3/atom)	n, H atoms per unit cell	%$\Delta V/V$	Q_{max} (mAh g^{-1})	Corrosion (wt%/cycle)
1.0	1.6[a]	0.8	1.4625	51	0
0.75	3.15	3.8	13.919	241	0.003
0.5	3.15	4.4	15.917	278	0.04
0.20	3.21	4.8	17.534	305	0.042
0.20	3.21	4.6	16.547	293	0.047
0.50	3.15	4.0	14.634	260	0.054
0.20	3.21	4.6	16.659	293	0.051
0.20	3.21	5.0	18.122	318	0.057
0.35	3.24	5.0	18.376	318	0.057
0.20	3.21	5.0	18.112	318	0.066
0.0	2.99[b]	4.8	15.96	305	0.15
0.0	2.99[b]	5.2	17.33	331	0.139
0.0	2.99[b]	5.1	17.002	325	0.145
$LaNi_{4.7}Al_{.3}$[c]	3.47	4.5	17.943	285	0.291

[a] α phase.
[b] Average.
[c] Included for comparison.

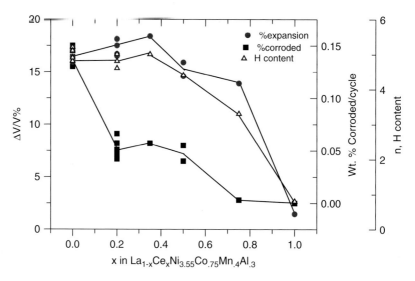

Figure 9.12 % $\Delta V/V$, wt% corroded/cycle, and H content vs Ce content, x, in $La_{1-x}Ce_xNi_{3.55}Co_{.75}Mn_{.4}Al_{.3}$ electrodes [42].

9.3.4.2 Effect of Cobalt

Cobalt is invariably present in commercial MH_x battery electrodes. It tends to increase hydride thermodynamic stability and inhibit corrosion. However, it is also expensive and substantially increases battery costs; thus, the substitution of Co by a lower cost metal is desirable. Willems and Buschow [39] attributed reduced corrosion in $LaNi_{5-x}\,Co_x$ ($x = 1–5$) to low V_H. Sakai et al. [47] noted that $LaNi_{2.5}Co_{2.5}$ was the most durable of a number of substituted $LaN_{5-x}\,Co_x$ alloys but it also had the lowest storage capacity.

The results of a systematic study of the effect of Co in an alloy series corresponding to $LaNi_{4.3-x}\,Co_x\,Mn_{.4}\,Al_{.3}$ is shown in Figures 9.13 and 9.14 and summarized in Table 9.6. The correlation between expansion and corrosion is rather weak; for example, even though the H content increases at $x = 0.2–0.4$ corrosion is decreased while expansion is unchanged. It is thus likely that that corrosion inhibition by Co is also due to a surface effect, as with Ce. In this connection Kanda et al. [48] found evidence that Co suppresses the transport of Mn to the surface where it is readily oxidized causing rapid electrode deterioration. Recent XAS results also suggest that Co inhibits corrosion via a surface process by suppressing Ni oxidation [49].

9.3.4.3 Effect of Aluminum

Aluminum appears to be present in all commercial AB_5 electrodes. Sakai et al. [50] noted that the incorporation of Al in $La(NiCoAl)_5$ alloys substantially reduced electrode corrosion; they attributed this to the formation of protective surface oxides. The corrosion-inhibiting effect of Al is clearly shown in Figure 9.15, which plots storage capacity versus cycle life for $LaNi_{3.85-x}\,Co_{.75}\,Mn_{.4}\,Al_x$ ($x = 0, 0.1, 0.2, 0.3$)

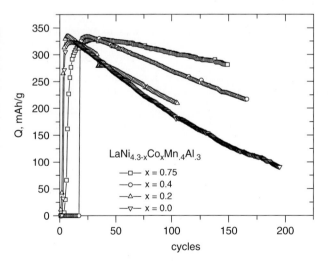

Figure 9.13 Charge capacity, Q, vs charge–discharge cycles for $LaNi_{4.3-x}Co_xMn_{.4}Al_{.3}$ electrodes [41].

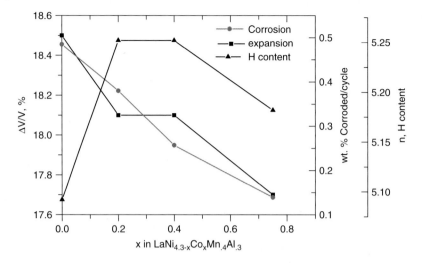

Figure 9.14 % $\Delta V/V$, wt.% corroded/cycle, and H content vs Co content, x, in LaNi$_{4.3-x}$Co$_x$Mn$_{.4}$Al$_{.3}$ electrodes [43].

Table 9.6 Effect of Co in LaNi$_{4.3-x}$Co$_x$Mn$_{.4}$Al$_{.3}$ electrodes [41].

x value	V_H (Å^3)	Q_{max} (mAh g^{-1})	n, H atoms unit cell	$\Delta V/V\%$	Corrosion (wt%/cycle)
0.75	2.99	330	5.18	17.3	0.139
0.40	3.09	334	5.25	18.1	0.257
0.20	3.09	334	5.25	18.1	0.380
0.0	3.26	324	5.09	18.5	0.485

electrodes [43]; the Al-free electrode corrodes at a greatly increased rate. As illustrated in Table 9.7 and Figure 9.16, the presence of even a small amount of Al substantially decreases V_H and n and, consequently, both lattice expansion and corrosion.

9.3.4.4 Effect of Manganese

Manganese is also present in most commercial electrodes. In a series of experiments examining the cycle lives of the homologous alloys LaNi$_{5-x}$ M$_x$ (M = Mn, Cu, Cr Al, and Co) Sakai et al. [24] noted that Mn was the least effective. In the more complex alloy examined by Adzic et al. [25] the function of Mn is still open to question. The cyclic behavior of a series of electrodes of varying Mn content is shown in Figure 9.17. It apparently increases V_H (Table 9.8) slightly, and, although the correlation between lattice expansion, n, and corrosion rate is fairly strong, they are not a function of Mn content, as shown in Figure 9.18.

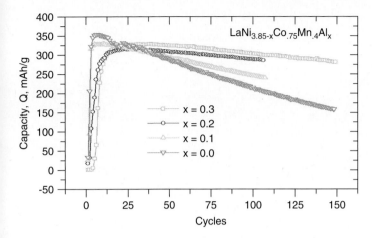

Figure 9.15 Charge capacity, Q, vs charge–discharge cycles for $LaNi_{3.85-x}Co_{.75}Mn_{.4}Al_x$ electrodes [43].

Table 9.7 Effect of Al in $LaNi_{3.85-x}Co_{.75}Mn_{.4}Al_x$ electrodes [43].

x Value	V_H (Å^3)	Q_{max} (mAh g^{-1})	n, H atoms per unit cell	% $\Delta V/V$	Corrosion (wt%/cycle)
0.2	3.01	314	4.98	16.66	0.1274
0.3	2.99	330	5.18	17.33	0.1394
0.1	3.01	327	5.22	17.58	0.2905
0.0	3.20	353	5.66	20.39	0.4079
0.0	3.35	366	5.88	22.30	0.4126

9.4
Super-Stoichiometric AB$_{5+x}$ Alloys

Notten et al. [51, 52] reported that the electrochemical cycling stability can be improved dramatically when using nonstoichiometric $La(NiCu)_{5+x}$ alloys. They attributed such improvement to an alteration of the crystal structure in which the excess of B-type atoms is accommodated in the AB$_5$ lattice by the occupation of empty A sites (La) with dumbbell pairs of Ni atoms oriented along the c-axis, although the hexagonal P6/mmm space group is preserved.

Recently Vogt et al. have shown that the structure of $La_{.9}Ni_{4.54}Sn_{.32}$ (a stoichiometry of AB$_{5.40}$ when La is normalized) also compensates for La deficiency by also inserting Ni dumbbells in empty A sites [53]. A representation of the $La(Ni, Sn)_{5+x}$ lattice is shown in Figure 9.19. Here too the presence of Ni dumbbells greatly improves the cycle life compared to two stoichiometric alloys as illustrated in Figure 9.20. Indeed it is remarkable that the super-stoichometric alloy, $LaNi_{4.84}Sn_{0.32}$, performs better than the simulated commercial alloy containing Co.

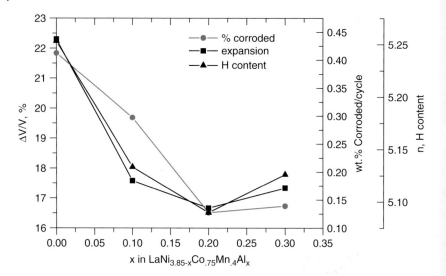

Figure 9.16 % $\Delta V/V$, wt % corroded/cycle, and H content vs Al content, x, in $LaNi_{3.85-x}Co_{.75}Mn_{.4}Al_x$ electrodes [43].

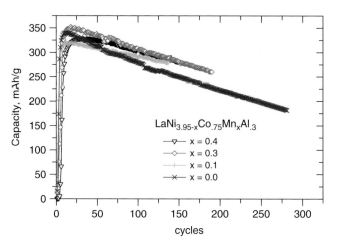

Figure 9.17 Charge capacity, Q, vs charge–discharge cycles for $LaNi_{3.95-x}Co_{.75}Mn_xAl_{.3}$ electrodes [43].

The $La(Ni, Sn)_{5+x}$ materials are most interesting as they constitute a new class of compounds which may provide alloys that could be fabricated into low-cost, corrosion-resistant electrodes. We also note that the existence of transition metal dumbbells is not unique. They exist in $LaNi_{5+x}$ [54] and $RE_2Fe_{17}C_x$, whose A_2B_{17} type structure is an ordered superstructure of the $CaCu_5$ lattice and AB_7 compounds where B is Cu or Ni [55, 56].

Table 9.8 Effect of Mn in $LaNi_{3.95-x}Co_{.75}Mn_xAl_{.3}$ electrodes [43].

x Value	V_H (Å³)	Q_{max} (mAh g⁻¹)	n, H atoms per unit cell	%ΔV/V	Corrosion (wt%/cycle)
0.14	3.16	320	4.87	17.27	0.106
0.40	2.99	330	5.18	17.33	0.1394
0.0	3.02	340	5.37	18.38	0.1676
0.30	3.07	353	5.48	18.75	0.150

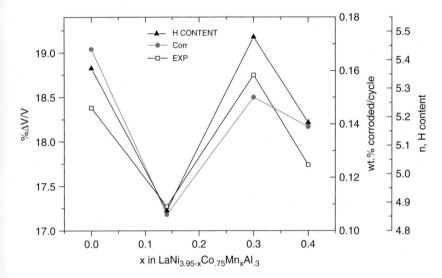

Figure 9.18 % ΔV/V, wt% corroded/cycle, and H content vs Mn content, x, in $LaNi_{3.95-x}Co_{.75}Mn_xAl_{.3}$ electrodes [43].

9.5
AB₂ Hydride Electrodes

The active materials in these electrodes are Laves phase alloys. These have close-packed structures in which the radii of the A and B atoms must lie within a certain range based on a hard-sphere packing model. The ideal ratio r_a/r_b is 1.225, but known Laves phases have ratios ranging from 1.05 to 1.68. There are three structural types, the hexagonal C14 ($MgZn_2$), the cubic C15 ($MgCu_2$), and the hexagonal C36 ($MgNi_2$). The C14 and C15 structures are common, and many form hydride phases [57]. However, the alloys used in battery applications are very complicated and may contain as many as three distinct bulk phases [58]. Ovshinsky et al. [59] describe the properties of a series of alloys containing V, Ti, Zr, Ni, Cr, Co, and Fe in various proportions; they qualitatively discuss the how AB₂ alloy properties are influenced by various elemental constituents. Gifford et al. [60]

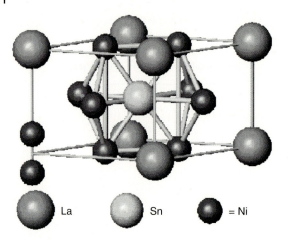

Figure 9.19 Representation of the unit cell of La(Ni, Sn)$_{5+x}$ structure showing substitution of a Ni dumbbell for La on the A site [53].

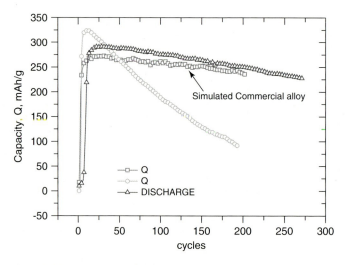

Figure 9.20 Comparison of charge capacity, Q, vs charge–discharge cycles for LaNi$_{4.84}$Sn$_{0.32}$, △ vs simulated commercial, □ and a Co-free electrode, ○ [53].

describe an experimental EV battery incorporating an AB$_2$ anode having an energy density of 80 Wh kg^{-1}. The battery lost 18% of its original charge capacity after 800 cycles at 80% depth of discharge.

PCT diagrams of AB$_2$(electrode alloys)/H systems reflect multiphase or nonideal behavior. This is illustrated in Figure 9.21, which plots both the equilibrium pressure and the open-circuit equilibrium voltage, E_r, for Zr$_{.5}$Ti$_{.5}$V$_{.5}$Ni$_{1.1}$Fe$_{.2}$Mn$_{.2}$. The pressure was calculated from E_r using the Nernst equation [61]. The use

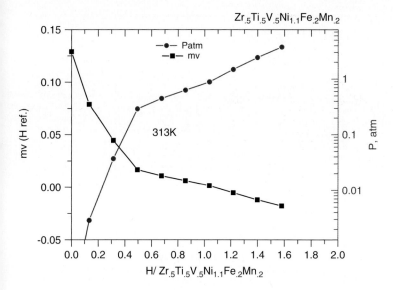

Figure 9.21 Electrochemical isotherm for $Ti_{.5}Zr_{.5}V_{.5}Ni_{1.1}Fe_{.2}Mn_{.2}$. The P_{eq} is calculated from the equilibrium voltage, E_r, via the Nernst equation [61].

of the electrochemical technique is more convenient to measure such equilibria than the conventional gas/solid method [20], when equilibrium pressures are <<1 atm over a significant portion of the H content range. The isotherm is highly sloping with no plateau and reflects nonideal behavior, that is, the presence of inhomogeneities, defects, and so on. However, unlike applications which involve the storage of hydrogen for subsequent evolution as a gas, battery applications do not require flat, wide plateaus because the pressure is a logarithmic function of the voltage. Of course, if the isotherm is too steep, a portion of the H storage capacity will not be electrochemically accessible, either because the voltage will become too anodic and corrosion will ensue or the equilibrium H_2 pressure will become excessive.

The cycle lives of several AB_2 electrodes are illustrated in Figure 9.22. In some cases AB_2 alloys require many charge–discharge cycles to become fully activated; pre-activation via the direct reaction with H_2 gas is helpful in this regard. Some pertinent properties and results are given in Table 9.9. It is of interest to note that V_H in the hydride phase is significantly less than that in AB_5 hydrides. Consequently, lattice expansion is also significantly reduced. However, the corrosion rates of the electrodes in Table 9.9 are still appreciable. Indeed for electrode $x = 0.25$ the corrosion rate is very high in spite of a small lattice expansion. Obviously this material is quite sensitive to corrosion and indicates that a moderately high Zr content is necessary to inhibit corrosion by surface passivation, as suggested by Zuttel et al. [62].

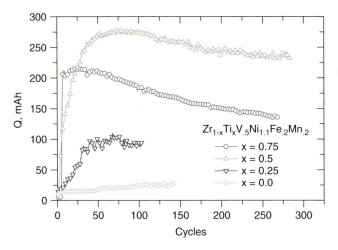

Figure 9.22 Charge capacity, Q, vs charge–discharge cycles for $Zr_{1-x}Ti_xV_{.5}Ni_{1.1}Fe_{.2}Mn_{.2}$ electrodes [61].

Table 9.9 Properties of the $Ti_{1-x}Zr_xV_{0.5}Ni_{1.1}Fe_{0.2}Mn_{0.2}$ electrodes [61].

x value	V_H (Å^3)	Q_{max} (mAh g^{-1})	n, H atoms[a] per unit cell	% $\Delta V/V$	Corrosion (wt%/cycle)
0.25	1.95	215	5.48	6.45	0.214
0.5	2.76	299	8.12	13.1	0.097
0.5	2.76	278	7.56	12.3	0.083
0.75	–	95	0.7	–	0.0
1.0	–	27	0.2	–	0.0

[a]There are four formula units in the hexagonal C14 unit cell.

9.6
XAS Studies of Alloy Electrode Materials

The availability of high-intensity, tunable X-rays produced by synchrotron radiation has resulted in the development of new techniques to study both bulk and surface materials properties. Both *in situ* and *ex situ* XAS methods have been applied to determine electronic and structural characteristics of electrodes and electrode materials [63, 64]. XAS combined with electron yield techniques can be used to distinguish between surface and bulk properties. In the latter procedure, X-rays are used to produce high energy Auger electrons [65] which, because of their limited escape depth (~150–200 Å), can provide information regarding near surface composition.

The element-specific nature of XAS makes it particularly useful for the study of complex AB_5 and AB_2 MH electrode materials. Mukerjee *et al.* [46] examined

$La_{0.8}Ce_{0.2}Ni_{4.8}Sn_{0.2}$ and $LaNi_{4.8}Sn_{0.2}$ electrodes using *in situ* XAS. It was determined by analysis of the X-ray absorption near-edge structure (XANES) that the presence of Ce reduced Ni corrosion – a finding which confirmed previous cycle life experiments [42]. This was done by quantitatively determining the amount of oxidized Ni (assumed to be $Ni(OH)_2$) in cycled electrodes as a function of Ce content. It is of interest to note the 001 peak of α $Ni(OH)_2$ was weakly observed in an electrode after 500 cycles using conventional X-ray diffraction (XRD). While this is to be expected, since the nickel hydroxide formed is somewhat amorphous, it illustrates an important advantage of XAS over XRD, since the former probes short range order and thus can provide quantitative information regarding amorphous or partly amorphous materials. Tryk *et al.* have similarly examined $LaNi_5$ [66] and $MmNi_{3.5}Co_{0.8}Mn_{.3}Al_{.4}$ [67] electrodes and noted the electronic transitions taking place in the metal lattice as a function of charge and the strong interaction of absorbed H with Ni. This is not unexpected as hydrogen occupies an Ni tetrahedral site in $LaNi_5H_6$ (Figure 9.5).

XAS studies have also been carried out on C14 Laves phase alloys $Ti_{0.5}Zr_{0.5}M_2$ and $Ti_{0.75}Zr_{0.25}M_2$ ($M = V_{0.5}Ni_{1.1}Fe_{0.2}Mn_{0.2}$) [61]. The XANES spectra at the Ni K edge indicates that, unlike the AB_5 alloys, there is very little interaction between hydrogen and Ni but rather strong interactions with Ti, V, and Zr. The hydrogen is presumably located in tetrahedra that contain large fractions of these three elements, whereas the Ni-rich sites are probably empty. Thus the function of Ni in AB_2 alloys may be primarily to serve as a catalyst for the electrochemical and hydriding reactions.

9.7 Summary

This survey presents an overview of the chemistry of metal-hydrogen systems which form hydride phases by the reversible reaction with hydrogen. The discussion then focuses on the AB_5 class and, to a lesser extent, the AB_2 class of MHs, both of which are of interest for battery applications. A new section has been introduced on super-stoichiometric $La(Ni, Sn)_{5+x}$ electrodes, which have a higher storage capacity and cycle life than commercial-type electrodes containing Co.

Electrode corrosion is the critical problem associated with the use of MH anodes in batteries. The extent of corrosion is essentially determined by two factors: alloy expansion and contraction in the charge–discharge cycle and chemical surface passivation via the formation of corrosion resistant oxides or hydroxides. Both factors are sensitive to alloy composition, which can be adjusted to produce electrodes having an acceptable cycle life. In AB_5 alloys the effects of Ce, Co, Mn, and Al upon cycle life in commercial type AB_5 electrodes are correlated with lattice expansion and charge capacity. Ce was shown to inhibit corrosion even though lattice expansion increases. Co and Al also inhibit corrosion. XAS results indicate that Ce and Co inhibit corrosion via surface passivation.

There are few systematic guidelines which can be used to predict the properties of AB_2 MH electrodes. Alloy formulation is primarily an empirical process where the composition is designed to provide a bulk hydride-forming phase (or phases) but which will form, *in situ*, a corrosion-resistant surface of semi-passivating oxide (hydroxide) layers. Lattice expansion is usually reduced relative to the AB_5 hydrides because of a lower V_H. Pressure–composition isotherms of complex AB_2 electrode materials indicate nonideal behavior.

Acknowledgment

The author wishes to acknowledge the support of Brookhaven National Laboratory operating under contract No. DE-AC02-98CHI-886 with the Department of Energy. Further thanks are due to John R. Johnson and Claire Reilly for proof-reading the manuscript and for offering many helpful suggestions.

References

1. Flanagan, T.B. and Oates, W.A. (1988) in *Hydrogen in Intermetallic Compounds I*, Topics in Applied Physics, Vol. 63 (ed. L. Schlapbach), Springer-Verlag, New York, p. 49.
2. Libowitz, G.G. (1965) *The Solid State Chemistry of Binary Metal Hydrides*, W. A. Benjamin, New York.
3. Mueller, W.M., Blackledge, J.P., and Libowitz, G.G. (1968) *Metal Hydrides*, Academic Press, New York.
4. Wicke, E., Brodowsky, H., and Zuchner, H. (1978) in *Hydrogen in Metals I*, Topics in Applied Physics, Vol. 28 (eds G. Alefeld and J. Voklkl), Springer-Verlag, New York, p. 101.
5. Reilly, J.J. and Wiswall, R.H. (1968) *Inorg. Chem.*, **7**, 2254.
6. Reilly, J.J. and Wiswall, R.H. (1974) *Inorg. Chem.*, **13**, 218.
7. van Vucht, J.H.N., Kuipers, F.A., and Bruning, H.C.A.M. (1970) *Philips Res. Rep.*, **25**, 133.
8. Gruen, D.N., Mendelsohn, M.H., and Dwight, A.E. (1979) *J. Less-Common Met.*, **63**, 193.
9. Reilly, J.J. (1979) *Z. Phys. Chem. N. F.*, **117**, 155.
10. Adzic, G.D., Johnson, J.R., Mukerjee, S., McBreen, J., and Reilly, J.J. (1996) *Meeting Abstracts of the 189th Meeting of the Electrochemical Society, Los Angeles, 1996*, vol. 96-1, The Electrochemical Society, Pennington, NJ, Abstract # 65.
11. Switendick, A.C. (1978) in *Hydrogen in Metals I*, Topics in Applied Physics, Vol. 28 (eds G. Alefeld and J. Voklkl), Springer-Verlag, New York, p. 101.
12. Gupta, M. and Schlapbach, L. (1988) in *Hydrogen in Intermetallic Compounds I*, Topics in Applied Physics, Vol. 63 (ed. L. Schlapbach), Springer-Verlag, New York, p. 139.
13. Lundin, C.E., Lynch, F.E., and Magee, C.B. (1977) *J. Less-Common Met.*, **56**, 19.
14. Westlake, D.G. (1983) *J. Less Common Met.*, **91**, 1.
15. Switendick, A.C. and Phsik, Z. (117) *Chem. N.F.*, **1979**, 89.
16. Reilly, J.J. (1978) in *Proceedings International Symposium on Hydrides for Energy Storage, Gielo, Norway* (eds A.F. Andresen and A.J. Maeland), Pergamon Press, New York, p. 301.
17. Lei, Y., Wu, Y., Yanf, Q., Wu, J., and Wang, Q. (1994) *Z. Phys. Chem.*, **Bd. 183**, S. 379.
18. Tsukahara, M., Takahashi, K., Mishima, T., Miyamura, H., Sakai, T., Kuriyama, N., and Uehara, I. (1995) *J. Alloys Compd.*, **231**, 616.
19. Siegmann, H.C., Schlapbach, L., and Brundle, C.R. (1978) *Phys. Rev. Lett.*, **40**, 547.
20. Reilly, J.J. (1983) in *Inorganic Syntheses* (ed. S.L. Holt), John Wiley & Sons, Inc., New York, p. 90.

21. Justi, E.W., Ewe, H.H., and Stephan, H. (1973) *Energy Convers.*, **13**, 109.
22. Percheron-Guegan, A., Achard, J.C., Sarradin, J., and Bronoel, G. (1978) in *Proceedings of the International Symposium on Hydrides for Energy Storage, Gielo, Norway* (eds A.F. Andersen and A.J. Maeland), Pergamon Press, New York, p. 485.
23. Willems, J.J.G. (1984) *Philips J. Res.*, **39** (Suppl. 1), 55–70.
24. (a) Ikowa, M., Kawano, H., Matsumoto, I., and Yanagihara, N. (1987) Eur. Patent Appl. # 0271043; (b) Ogawa, H., Ikowa, M., Kawano H., and Matsumoto, I. (1988) *Power Sources*, **12**, 393.
25. Ikoma, M., Hamada, S., Morishita, N., Hoshina, Y., Ohta, K., and Kimura, T. (1996) in *Proceedings of the Symposium on Hydrogen and Metal Hydride Batteries*, vol. 94-27 (eds P.D. Bennett and T. Sakai), The Electrochemical Society, Pennington, NJ, p. 370.
26. Wernick, J.H. and Geller, S. (1959) *Acta Cryst.*, **12**, 662.
27. Thompson, P., Reilly, J.J., Corliss, L.M., Hastings, J.M., and Hempelmann, R. (1986) *J. Phys. F. Met. Phys.*, **16**, 679.
28. Lartigue, C., Percheron-Guegan, A., Achard, J.C., and Soubeyoux, J.L. (1985) *J. Less Common Met.*, **113**, 127.
29. Buschow, K.H.J. and Medima, A.R. (1978) in *Proceedings of the International Symposium on Hydrides for Energy Storage, Gielo, Norway* (eds A.F. Andersen and A.J. Maeland), Pergamon Press, New York, p. 235.
30. Goodell, P.D. and Rudman, P.S. (1983) *J. Less Common Met.*, **89**, 117.
31. Reilly, J.J., Josephy, Y., and Johnson, J.R. (1989) *Z. Phys. Chem. N. F.*, Bd. **164**, S. 1241.
32. Miyamoto, M., Yamaji, K., and Nakata, Y. (1989) *J. Less Common Met.*, **89**, 111.
33. Reilly, J.J. (1992) in *Proceedings of the Symposium on Hydrogen Storage Materials, Batteries, and Electrochemistry*, vol. 92-5 (eds D.A. Corrigan and S. Srinivasan), Electrochemical Society, Pennington, NJ, p. 24.
34. Reilly J.J. and Wiswall, R.H. Jr. Hydrogen Storage and Purification Systems, U. S. Atomic Energy Commission, BNL-17136, Brookhaven National Laboratory, Upton, NY, August 1972.
35. Sakai, T., Yoshinaga, H., Miyamura, H., Kuriyama, N., and Ishikawa, H. (1992) *J. Alloys Compd.*, **180**, 37.
36. Percheron-Guegan, A. and Welter, J.-M. (1988) in *Topics in Applied Physics, Hydrogen in Intermetallic Compounds I*, vol. 63 (eds L. Schlapbach), Springer-Verlag, New York, p. 11.
37. Petrov, K., Rostami, A.A., Visintin, A., and Srinivasan, S. (1994) *J. Electrochem. Soc.*, **141** (7), 1747.
38. Adams, W.A. (1996) in *Symposium Proceedings of "Exploratory Research and Developpment of Batteries for Electric and Hybrid Vehicles"*, vol. 96-14 (eds W.A. Adams, A.R. Landgrebe, and R. Scrosati), Electrochemical Society, Penninton, NJ, p. 1.
39. Willems, J.J.G. and Buschow, K.H.J. (1987) *J. Less Common Met.*, **129**, 13.
40. Latroche, M., Percheron-Guegan, A., Chabre, Y., Bouet, J., Pannetier, J., and Ressouche, E. (1995) *J. Alloys Compd.*, **231**, 537.
41. Adzic, G., Johnson, J.R., Mukerjee, S., McBreen, J., and Reilly, J.J. (1997) *J. Alloys Compd.*, **253–254**, 579.
42. Adzic, G., Johnson, J.R., Reilly, J.J., McBreen, J., Mukerjee, S., Kumar, M.P.S., Zhang, W., and Srinivasan, S. (1995) *J. Electrochem. Soc.*, **142**, 3429.
43. Adzic, G., Johnson, J.R., Mukerjee, S., McBreen, J., and Reilly, J.J. (1996) in *Symposium Proceedings of Exploratory Research and Developpment of Batteries for Electric and Hybrid Vehicles*, vol. 96-14 (eds W.A. Adams, A.R. Landgrebe, and R. Scrosati), Electrochemical Society Penninton, NJ, p. 189.
44. Sakai, T., Hazama, T., Miyamura, H., Kuriyama, N., Kato, A., and Ishikawa, H. (1991) *J. Less Common Met.*, **172–174**, 1175.
45. Davenport, A.J., Isaacs, H.S., and Kendig, M.W. (1991) *Corros. Sci.*, **32** (5/6), 653.
46. Mukerjee, S., McBreen, J., Reilly, J.J., Johnson, J.R., Adzic, G., Petrov, K., Kumar, M.P.S., Zhang, W., and Srinivasan, S. (1995) *J. Electrochem. Soc.*, **142** (7), 2278.

47. Sakai, T., Oguro, K., Miyamura, H., Kuriyama, N., Kato, A., Ishikawa, H. et al. (1990) *J. Less Common Met.*, **161**, 193.
48. Kanda, M., Yamamoto, M., Kanno, K., Satoh, Y., Hayashida, H., and Suzuki, M. (1987) *J. Less Common Met.*, **129**, 13.
49. Mukerjee, S., McBreen, J., Adzic, G.D., Johnson, J.R., and Reilly, J.J. (1996) The function of cobalt in AB_5H_x metal hydride electrodes as determined by x-ray absorption spectroscopy. Extended Abstracts, National Meeting of the Electrochemical Society, San Antonio, Texas, October, Vol. 96-2, Abstract # 48.
50. Sakai, T., Miyamura, H., Kuriyama, N., Kato, A., Oguro, K., and Ishikawa, H. (1990) *J. Less Common Met.*, **159**, 127.
51. Notten, P.H.L., Einerhand, R.E.F., and Daams, J.L.C. (1994) *J. Alloys Compd.*, **210**, 221.
52. Notten, P.H.L., Einerhand, R.E.F., and Daams, J.L.C. (1994) *J. Alloys Compd.*, **210**, 233.
53. Vogt, T., Reilly, J.J., Johnson, J.R., Adzic, G.D., and McBreen, J. (1999) *Electrochem. Solid State Lett.*, **2** (3), 111.
54. Buschow, K.H.J. and Van Mal, H.H. (1972) *J. Less Common Met.*, **29**, 203.
55. Coene, W., Hakkens, F., Jacobs, T.H., and Buschow, K.H.J. (1991) *J. Solid State Chem.*, **92**, 191.
56. Givord, D., Lemaire, R., Moreau, J.M., and Roudaut, E. (1972) *J. Less Common Met.*, **29**, 361.
57. Ivey, D.G. and Northwood, D.O. (1986) *Z. Phys. Chem. N.F.*, **147**, 191.
58. Huot, J., Akiba, E., and Ishido, Y. (1995) *J. Alloys Compd.*, **231**, 85.
59. Ovshinsky, S.R., Fetcenko, M.A., and Ross, J. (1993) *Science*, **260**, 176.
60. Gifford, P.R., Fetcenko, M.A., Venkatesan, S., Corrigan, D.A., Holland, A., Dhar, S.K., and Ovshinsky, S.R. (1996) in *Proceedings of the Symposium on Hydrogen and Metal Hydride Batteries*, vol. 94-27 (eds P.D. Bennett and T. Sakai), The Electrochemical Society, Pennington, NJ, p. 353.
61. Johnson, J.R., Mukerjee, S., Adzic, G.D., Reilly, J.J., and McBreen, J. (1996) In situ XAS studies on AB_2 type metal hydride alloys for battery applications. Poster Presented at the The International Symposium on Metal Hydrogen Systems, Fundamentals and Applications, Les Diablerets, Switzerland August 1996.
62. Zuttel, A., Meli, F., and Schlapbach, L. (1995) *J. Alloys Compd.*, **231**, 645.
63. McBreen, J. (1996) in *Symposium Proceedings of "Exploratory Research and Developpment of Batteries for Electric and Hybrid Vehicles"*, vol. 96-14 (eds W.A. Adams, A.R. Landgrebe, and R. Scrosati), Electrochemical Society, Penninton, NJ, p. 162.
64. Scherson, D.A. (1996) *Interface*, **5** (3), 34.
65. Mansour, A.N., Melandres, C.A., Poon, S.J., He, Y., and Shiflet, G.J. (1987) *J. Electrochem. Soc.*, **143**, 219.
66. Tryk, D.A., Bae, I.T., Hu, Y., Kim, S., Antonio, M.R., and Sherson, D.A. (1995) *J. Electrochem. Soc.*, **142** (3), 824.
67. Tryk, D.A., Bae, I.T., Sherson, D.A., Antonio, M.R., Jordan, G.W., and Huston, E.L. (1995) *J. Electrochem. Soc.*, **142** (5), L76.

Further Reading

Adams, W.A. (1996) in *Symposium Proceedings of Exploratory Research and Development of Batteries for Electric and Hybrid Vehicles*, vol. 96-14 (eds W.A. Adams, A.R. Landgrebe, and R. Scrosati), Electrochemical Society, Pennington, NJ, p. 1.

Reilly, J.J., Vogt, T., Johnson, J.R., Adzic, G.D., and McBreen, J. (2001) US Patent 6, 238, 823 B1, May 29.

10
Carbons
Kimio Kinoshita

10.1
Introduction

Solid carbon materials are available in a variety of crystallographic forms, typically classified as diamond, graphite, and amorphous carbon. More recently another structure of carbon was identified, namely the fullerenes, whose structure resembles that of a soccer ball (C_{60}). In this chapter the discussion will focus on graphites and amorphous carbons, which are practical materials for use in aqueous batteries.

Carbonaceous materials serve several functions in electrodes and other cell components for aqueous-electrolyte batteries, and these are summarized in Table 10.1.

Of practical importance is the contribution that is made by carbonaceous materials as an additive to enhance the electronic conductivity of the positive and negative electrodes. In other electrode applications, carbon serves as the electrocatalyst for electrochemical reactions and/or the substrate on which an electrocatalyst is located. In addition, carbonaceous materials are fabricated into solid structures which serve as the bipolar separator or current collector. Clearly, carbon is an important material for aqueous-electrolyte batteries. It would be very difficult to identify a practical alternative to carbon-based materials in many of their battery applications. The attractive features of carbon in electrochemical applications are its high electrical conductivity, acceptable chemical stability, and low cost. These characteristics are important for the widespread acceptance of carbon in aqueous electrolyte batteries.

10.2
Physicochemical Properties of Carbon Materials

10.2.1
Physical Properties

The crystal structure of graphite and amorphous carbon is illustrated by the schematic representations given in Figure 10.1.

Handbook of Battery Materials, Second Edition. Edited by Claus Daniel and Jürgen O. Besenhard.
© 2011 Wiley-VCH Verlag GmbH & Co. KGaA. Published 2011 by Wiley-VCH Verlag GmbH & Co. KGaA.

Table 10.1 Application of carbon in aqueous batteries.

Battery	Application of carbonaceous material
Lead-acid	Bipolar current collector, electrode additive
Metal/air	Air electrode, electrocatalyst support
Redox flow	Positive electrode, negative electrode substrate, electrocatalyst support, current collector, bipolar separator
Metal hydride/NiOOH	Electrode additive
Hydrogen/NiOOH	Electrode additive, electrocatalyst support
Cd/NiOOH	Electrode additive
Zn/NiOOH	Electrode additive
ZnAgO and Zn/Ag$_2$O	Electrode additive
Zn/HgO	Electrode additive
Alkaline Zn/MnO$_2$	Electrode additive
Zinc/carbon (Leclanché cell)	Electrode additive, current collector

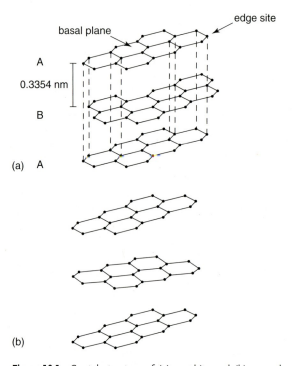

Figure 10.1 Crystal structure of (a) graphite and (b) amorphous carbon.

The structure consists of carbon atoms arranged in hexagonal rings that are stacked in an orderly fashion in graphite (see Figure 10.1a). Only weak van der Waals bonds exist between these layer planes. The usual stacking sequence of the carbon layers is ABABA... for hexagonal graphite. The stacking sequence ABCABC... is found less frequently (i.e., in a few percent of the solid) and is called *rhombohedral graphite*. The $d(0\ 0\ 2)$ interplanar spacing in graphite is 0.3354 nm in the C-axis direction (perpendicular to the layer planes), while the C–C bond distance in the A-axis direction (parallel to the layer planes) is 0.142 nm. It is apparent in Figure 10.1a that graphite has two distinct surfaces present, the basal plane and the edge sites. Furthermore, the physical properties of graphite are highly anisotropic because of this crystallographic structure. For instance, the electrical conductivity in the direction parallel to the basal plane is about 100 times higher than in the perpendicular direction.

Amorphous carbons (see Figure 10.1b) also consist of hexagonal carbon rings, but the number of these rings that constitutes a crystallite is much less than for graphite. In addition there is very little order between the layers. Instead, the layers are rotated with respect to each other, but they are parallel to each other (i.e., the material is turbostratic) and there is no three-dimensional ordering. The layer spacing of carbon blacks is typically >0.350 nm, and the crystallite sizes are typically 1.0–2.0 nm for L_a (crystallite size in the direction parallel to the basal plane) and L_c (crystallite size in the direction perpendicular to the basal plane). In contrast, L_a and L_c for graphites can be >100 nm. The surface area of graphite and amorphous carbon can be <10 to >1000 $m^2\ g^{-1}$ respectively. The densities of these carbonaceous materials are 2.25 g cm^{-3} for graphite and usually <1.80 g cm^{-3} for amorphous carbon. Further details on the physical properties can be found in the extensive discussion by Kinoshita [1] and in review articles [2–5].

The lattice plane images of carbonaceous materials, which were obtained by high-resolution transmission electron microscopy (HRTEM), are reviewed by Millward and Jefferson [6]. Examples of HRTEM of carbon blacks are presented in Figure 10.2 to illustrate the difference in the structure of an amorphous carbon and a graphitized carbon. The electron micrographs show a distinct difference in the structure of the carbon particles. The amorphous carbon ($d(0\ 0\ 2)$ spacing of 0.352 nm) shows little evidence for long-range order of the basal planes. On the other hand, the graphitized carbon black ($d(0\ 0\ 2)$ spacing of 0.344 nm) has well-defined layer planes that follow the surface contours of the carbon particles. Despite heat treatment at 2700 °C, the $d(0\ 0\ 2)$ spacing is much higher than that

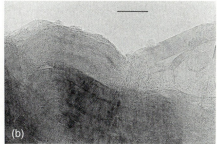

Figure 10.2 High-resolution transmission electron micrographs of carbon black (Sterling R. Cabot Corp.): (a) as-received and (b) heat-treated at 2700 °C. Scale marker 10 nm.

of graphite, and there are regions in the carbon particles which appear to be amorphous. These observations are typical for carbon blacks which are heat-treated at graphitizing temperatures. The particle size restricts the motion of the layer planes and the stresses that result inhibit the formation of a highly graphitized structure that is similar to that of pure graphite.

A terminology to identify carbons that are graphitizable or those that are nongraphitizable by heat treatment has been adopted. Hard carbons are those carbons that are nongraphitizable and are mechanically hard – hence the name. In contrast, soft carbons are mechanically soft and can be graphitized. Hard carbons are obtained by carbonizing precursors such as thermosetting polymers (e.g., phenol-formaldehyde resins), furfuryl alcohol, divinylbenzene-styrene copolymer, cellulose, charcoal, and coconut shells. These carbons are usually formed by solid-state transformation during the carbonization steps. One explanation for the inability of hard carbons to form a graphitic structure by heat treatment is the presence of strong sp^3 crosslinking bonds which impede movement and reorientation of the carbon atoms to form the ordered layer structure of graphite. Soft carbons are formed by carbonizing precursors such as petroleum coke, oil, and coal-tar pitch. In these materials the formation of carbon proceeds through an intermediate liquid-like phase (referred to as a *mesophase*), which facilitates the three-dimensional ordering that is necessary to create a graphite-like structure. Besides the discussion by Kinoshita [1], an extensive review that describes the formation of carbonaceous materials and their physical properties is presented by researchers from Japan [7].

Natural graphite is classified as flake, vein, or microcrystalline (amorphous), depending on the crystallite size and particle shape. Major sources of natural graphite are found in Mexico, China, and Brazil. Flake graphite is anisotropic and has a crystallinity similar to that of single-crystal graphite. One problem with many sources of natural graphite is their ash content (e.g., Fe, Si), which can be as high as 25%. Much of this ash can be removed by leaching in concentrated acid or exposure to halogen gases. Synthetic or artificial graphite is produced by heat treatment of a precursor carbon such as petroleum coke to temperatures in the region of 2800 °C or higher. Solid graphite structures for bipolar separators or electrode substrates for batteries are obtained by extrusion or molding of blended mixtures of petroleum coke and a binder of coal-tar pitch which are heat-treated to graphitization temperatures [8].

A variety of amorphous carbons such as carbon black, active carbon, and glassy carbon is available. With the exception of glassy carbon, these amorphous carbons generally have high surface area, high porosity, and small particle size. Carbon blacks, for example, are available with surface areas that are $>1000\,m^2\,g^{-1}$, particle size <50 nm, and density much less than the theoretical value for graphite (2.25 g cm^{-3}). In addition, the morphology of carbon blacks may resemble individual spheres of about 250 nm diameter (i.e., thermal blacks) or a cluster of fused carbon particles of <50 nm diameter (i.e., furnace blacks). The morphology of carbon black particles has been the subject of much discussion [9–12]. Active carbons are typically granular carbons which are produced by carbonizing materials such as wood (charcoal), coconut and other fruit shells, and low-rank coals. The resulting carbon is activated by treatment with gas (steam activation) or chemical processing [7]. The end product is a carbon material with high surface area ($>1000\,m^2\,g^{-1}$) and extensive micropores (pore size $<2\,nm$); these properties were analyzed by Kaneko et al. [13]. It is this microporosity that contributes to the high adsorption properties of these carbons.

10.2.2
Chemical Properties

The surface of carbonaceous materials contains numerous chemical complexes that are formed during the manufacturing step by oxidation or introduced during post-treatment. The surface complexes are typically chemisorbed oxygen groups such as carbonyl, carboxyl, lactone, quinone, and phenol (see Figure 10.3).

In addition, carbon–hydrogen bonds are present, particularly in carbonaceous materials obtained by carbonizing polymers at low temperatures, typically $<1000\,°C$. Detailed discussions on the types of surface groups and their surface concentrations are presented by Boehm [14] and Rivin [15]. These surface groups exhibit different thermal stabilities, with the functional groups that contain two oxygen atoms (carboxyl and lactone) desorbing as CO_2, generally at $<500\,°C$. Functional groups that contain one oxygen atom (phenol, quinone) evolve CO at temperatures of about $600\,°C$ and higher. The thermal analysis by Rivin [15] with an oil furnace black (surface area $122\,m^2\,g^{-1}$) indicated that surface concentrations of about $1 \times 10^{-10}\,mol\,cm^{-2}$ (CO_2) and $5 \times 10^{-10}\,mol\,cm^{-2}$ (CO) were present. However after oxidation in air for 2 h at $420\,°C$, the surface concentrations increased by about two and threefold, respectively, and the surface area increased about 2.5-fold. Hydrogen is frequently found in carbon blacks and other carbonaceous materials. Analysis of carbon blacks indicates that the hydrogen content is in the range 0.01–0.7%. The hydrogen that is bonded to carbon is relatively stable, commencing evolution at about $700\,°C$ and reaching a maximum at about $1100\,°C$. Other common heteroatoms such as nitrogen and sulfur are also found in carbon.

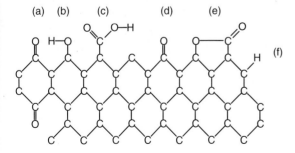

Figure 10.3 Schematic representation of the common functional groups that are present on carbon: (a) quinone; (b) phenol; (c) carboxyl; (d) carbonyl; (e) lactone; and (f) hydrogen.

Nitrogen is usually present in minor amounts, but sulfur can be present in high concentrations, >1%, depending on the precursor that is used to manufacture the carbonaceous material. Besides sulfur that is bonded to carbon, other forms such as elemental sulfur, inorganic sulfate, and organosulfur compounds may be present. The carbon–sulfur surface compounds on carbon blacks are relatively stable, but they desorb as H_2S when carbon is heat-treated in H_2 between 500 and 1000 °C.

The surface oxide groups on carbon play a major role in its surface properties; for example, the wettability in aqueous electrolytes, work function, and pH in water are strongly affected by the presence of surface groups on the carbonaceous material. Typically, the wettability of carbon blacks increases as the concentration of surface oxides increases [16]. The pH of an aqueous slurry of carbon decreases as the volatile or oxygen content of the carbon increases [17]. The work function of carbon blacks shows a minimum at a pH near 6 [18].

The physicochemical properties of carbonaceous materials can be altered in a predictable manner by different types of treatments. For example, heat treatment of soft carbons, depending on the temperature, leads to an increase in the crystallite parameters, L_a and L_c and a decrease in the $d(0\,0\,2)$ spacing. Besides these physical changes in the carbon material, other properties such as the electrical conductivity and chemical reactivity are changed. A review of the electronic properties of graphite and other types of carbonaceous materials is presented by Spain [3].

10.3
Electrochemical Behavior

10.3.1
Potential

Several significant electrode potentials of interest in aqueous batteries are listed in Table 10.2; these include the oxidation of carbon and oxygen evolution/reduction reactions in acid and alkaline electrolytes. For example, for the oxidation of carbon in alkaline electrolyte, $E°$ at 25 °C is −0.780 V vs SHE (standard hydrogen electrode) or −0.682 V (vs Hg/HgO reference electrode) in 0.1 mol L^{-1} CO_3^{2-} at pH = 14. Based on the standard potentials for carbon in aqueous electrolytes, it is thermodynamically stable in water and other aqueous solutions at a pH less than about 13, provided no oxidizing agents are present.

The typical products that form during oxidation of carbon in acid and alkaline electrolytes are CO_2 and carbonate species, respectively. Additional details of the thermodynamic stability of carbon in aqueous electrolytes, and the electrode potentials for reactions involving carbon, are presented in the review by Randin [19].

The standard oxidation potentials suggest that carbon has a limited stability domain in aqueous electrolytes. As noted in Table 10.2 the oxidation (corrosion) of carbon should occur at potentials much lower than the reversible potential for oxygen evolution/reduction. To illustrate this point further, take the example of an

Table 10.2 Standard potentials for reactions of carbon materials in batteries containing aqueous electrolytes.

Electrochemical reaction	Standard potential (V vs SHE)	Electrolyte
$C + 2H_2O \rightarrow CO_2 + 4H^+ + 4e^-$	0.207	Acid
$C + 6OH^- \rightarrow CO_3^{2-} + 3H_2O + 4e^-$	−0.780	Alkaline
$O_2 + 4H^+ + 4e^- \rightarrow 2H_2O$	1.229	Acid
$O_2 + 2H^+ + 2e^- \rightarrow H_2O_2$	0.682	Acid
$H_2O_2 + 2H^+ + 2e^- \rightarrow 2H_2O$	1.776	Acid
$O_2 + 2H_2O + 4e^- \rightarrow 4OH^-$	0.401	Alkaline
$O_2 + H_2O + 2e \rightarrow HO_2^- + OH^-$	−0.076	Alkaline
$HO_2^- + H_2O + 2e^- \rightarrow 3OH^-$	0.878	Alkaline

air electrode (for instance in a metal/air battery) that utilizes carbon. In an acid electrolyte, for instance, a typical potential at which oxygen reduction occurs is about 0.7 V, whereas in alkaline electrolyte the reaction may take place at about 0.1 V. At these operating potentials, the overpotential for carbon oxidation is high in both electrolytes (i.e., 0.5 V in acid and 0.9 V in alkaline electrolytes). Furthermore, the overpotential is much higher in alkaline electrolyte, which suggests that carbon oxidation should be much greater at high pH. In rechargeable alkaline metal/air batteries that utilize carbon in the bifunctional air electrodes, corrosion during charge is a major problem that has not been resolved satisfactorily. The net result is that practical rechargeable metal/air batteries are not available because of their limited cycle life.

10.3.2
Conductive Matrix

Perhaps the first practical application of carbonaceous materials in batteries was demonstrated in 1868 by Georges Leclanché in cells that bear his name [20]. Coarsely ground MnO_2 was mixed with an equal volume of retort carbon to form the positive electrode. Carbonaceous powdered materials such as acetylene black and graphite are commonly used to enhance the conductivity of electrodes in alkaline batteries. The particle morphology plays a significant role, particularly when carbon blacks are used in batteries as an electrode additive to enhance the electronic conductivity. One of the most common carbon blacks which is used as an additive to enhance the electronic conductivity of electrodes that contain metal oxides is acetylene black. A detailed discussion on the desirable properties of acetylene black in Leclanché cells is provided by Bregazzi [21]. A suitable carbon for this application should have characteristics that include: (i) low resistivity in the presence of the electrolyte and active electrode material, (ii) absorption and retention of a significant amount of electrolyte without reduction of its capability of mixing with the active material, (iii) compressibility and resilience in the cell, and (iv) only low contents of impurities. Graphite has higher electrical conductivity than acetylene black but it is not capable of retaining the same amount of electrolyte or demonstrating the same mechanical properties in the cell. Acetylene black has a well-developed chain structure, and it is this characteristic which provides the capability to retain a significant amount of electrolyte. In addition, acetylene black is produced with a low ash content and it does not contain surface groups. The results obtained by Bregazzi [21] indicate that acetylene black is capable of retaining over three times as much electrolyte (cubic centimeter electrolyte/gram carbon) as graphite. The capacity of Leclanché cells is dependent on the amount and type of carbon black that is used. Generally about 55 vol% carbon black mixed with MnO_2 yields the maximum capacity [22]. This composition agrees closely with the minimum in the electrical resistivity of the electrode mixture.

A carbon rod is used as a current collector for the positive electrode in dry cells. It is made by heating an extruded mixture of carbon (petroleum coke, graphite) and pitch which serves as a binder. A heat treatment at temperatures of about 1100 °C is used to carbonize the pitch and to produce a solid structure with low resistance. For example, Takahashi [23] reported that heat treatment reduced the specific resistance from 1 to 3.6×10^{-3} Ω cm and the density increased from 1.7 to 2.02 g cm^{-3}. Fischer and Wissler [24] derived an experimental relationship (Equation 10.1) between the electrical conductivity, compaction pressure, and properties of graphite powder:

$$\log \rho = K - 0.45 \log L_c - 0.43 \log d_{50} - 0.54 \log p \qquad (10.1)$$

where ρ is the electrical resistivity, K is a constant, L_c is the crystallite size in the direction perpendicular to the basal plane, d_{50} is the mean particle diameter, and p is the compaction pressure. This relationship indicates that the electrical resistivity decreases as the crystallite size increases, and with a given average particle size and compaction pressure. When graphite is mixed with MnO_2 in an electrode structure, the conductivity increases with a decrease in the particle size of graphite. In addition, the conductivity increases dramatically when the graphite concentration increases above about 10%.

Another example of the use of a graphite as an additive to improve the electronic conductivity of an electrode can be found in the discussion of the Fe/NiOOH cell developed by Edison in the early 1900s [25]. The positive electrode which contained graphite (20–30% graphite flake) degraded rapidly during charge because of oxidation and swelling. This experience led to the development of electrolytic nickel flakes and eventually to the porous nickel plaque for use in NiOOH electrodes.

Composite structures that consist of carbon particles and a polymer or plastic material are useful for bipolar separators or electrode substrates in aqueous batteries. These structures must be impermeable to the electrolyte and electrochemical reactants or products. Furthermore, they must have acceptable electronic conductivity and mechanical properties. The physicochemical properties of carbon blacks, which are commonly used, have a major effect on the desirable properties of the conductive composite structures. Physicochemical properties such as the surface area, structure, and volatile content (oxygen surface groups) influence the electronic conductivity of the composite structure. Typically the electronic conductivity is significantly lower when less than about 30 wt% carbon black is incorporated in the composite structure. Carbon blacks with higher surface area, and usually smaller particle size, are desirable because the interparticle distance is shorter and electron tunneling can occur more easily. The structure of carbon blacks is conventionally defined by the amount of adsorption of dibutyl phthalate (DBP); higher adsorption means a higher structure. For example, acetylene black has DBP absorption of 200–250 cm^3 g^{-1} and is a high-structure black, whereas a low structure has a DBP absorption of <100 cm^3 g^{-1}. A higher oxygen or volatile content is not desirable because its presence renders the carbon black less conductive. Further details are available in the publication by Kinoshita [1].

10.3.3
Electrochemical Properties

A comprehensive review which discusses the surface properties and their role in the electrochemistry of carbon surfaces was written by Leon and Radovic [26]. This review provides a useful complement to the following discussion on the role of carbon in aqueous batteries. Four key parameters that are important for carbonaceous materials in batteries, which were identified by Fischer and Wissler [24], are:

1) chemical purity
2) crystalline structure
3) particle size distribution, and
4) porosity.

These parameters are critical to the operation of alkaline batteries. Evaluation of graphite additives in the positive electrode for Cd/NiOOH cells by Veres and Csath [27] showed that the state of oxidation of graphite and the level of impurities strongly influence the electrode capacity. The capacity of the active material decreased with an increase in the amount of impurities and the degree of oxidation of the graphite.

The studies by Biermann *et al.* [28] indicate that the carbon blacks used as the conductive matrix in Leclanché cells remain chemically inert, that is, they do not undergo oxidation during storage or discharge of the cell. However, Caudle *et al.* [29] found evidence that the ion-exchange properties of carbon black, which exist because of the presence of surface redox groups, are responsible for electrochemical interactions with MnO_2. The extent of MnO_2 reduction to MnOOH depends on the carbon black (e.g., furnace black > acetylene black).

10.3.4
Electrochemical Oxidation

In acid electrolytes, carbon is a poor electrocatalyst for oxygen evolution at potentials where carbon corrosion occurs. However, in alkaline electrolytes carbon is sufficiently electrocatalytically active for oxygen evolution to occur simultaneously with carbon corrosion at potentials corresponding to charge conditions for a bifunctional air electrode in metal/air batteries. In this situation, oxygen evolution is the dominant anodic reaction, thus complicating the measurement of carbon corrosion. Ross and co-workers [30] developed experimental techniques to overcome this difficulty. Their results with acetylene black in 30 wt% KOH showed that substantial amounts of CO in addition to CO_2 (carbonate species) and O_2, are produced at 550–600 mV (vs Hg/HgO reference electrode) and temperatures up to 65 °C. Evidence for the formation of an organic species (appearance of a deep reddish-brown color in the solution) was found at 65 °C but the composition was not identified. However, Thiele [31] and Heller [32] concluded that the organic species was probably a humic acid.

The major oxidation reactions of acetylene black in an alkaline electrolyte (30 wt% KOH + 2 wt% LiOH) are strongly dependent on the potential (vs Hg/HgO) and temperature [30]:

- 500 mV and <50 °C: carbonate formation (CO_2) is the dominant reaction;
- 500–600 mV and <50 °C: carbonate formation and O_2 evolution rates are comparable;
- 600 mV or >60 °C and >450 mV: O_2 evolution and CO formation are the dominant reactions.

Other experiments by Ross and co-workers [30] clearly indicate that the common metal (Co, Ni, Fe, Cr, Ru) oxides that are used for oxygen electrocatalysts also catalyze the oxidation of carbon in alkaline electrolytes.

The surface structure has a strong influence on the corrosion rate of carbon in both acid and alkaline electrolytes. Studies by Kinoshita [33] clearly showed that the specific corrosion rate (milliamperes per square centimeter of carbon black in 96 wt% H_3PO_4 at 160 °C) was affected by heat treatment. A similar trend in the corrosion rate in alkaline electrolyte was observed by Ross [30c], as shown in Figure 10.4. It is evident that the corrosion rates of the nongraphitized carbons are higher than those of the corresponding graphitized carbons. Their study further indicated that some types of carbon blacks (e.g., semi-reinforcing furnace blacks) showed a larger decrease in the corrosion rate after graphitization than others that were evaluated. The decrease in the corrosion rate is attributable to the change in the surface microstructure after heat treatment. The surface layers rearrange to form a graphitic structure with basal planes that are exposed to the electrolyte. This surface is more resistant to corrosion than the edge plane sites, and experiments by Ross [30c] indicate that the nongraphitic surface sites, which are capable of adsorbing iodine from solution, are the likely corrosion sites.

10.3.5
Electrocatalysis

Carbon shows reasonable electrocatalytic activity for oxygen reduction in alkaline electrolytes, but it is a relatively poor oxygen electrocatalyst in acid electrolytes.

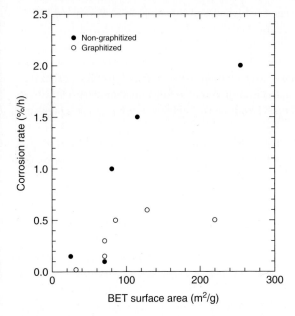

Figure 10.4 Corrosion of carbon blacks at 550 mV (vs. Hg/HgO) in 35 wt% KOH at 55 °C. From Ross [30c].

A detailed discussion on the mechanism of oxygen reduction and evolution on carbon was presented by Kinoshita [1]. The experimental studies suggest that oxygen reduction in alkaline electrolytes is first order in O_2 concentration. There is evidence that the reaction mechanism is not the same on different carbon electrodes, as illustrated by Equations 10.2–10.7 for graphite and carbon black.

Graphite:

$$O_2 \rightarrow O_2(ads) \tag{10.2}$$

$$O_2(ads) + e^- \rightarrow O_2^-(ads) \tag{10.3}$$

[rate-determining step]

$$2O_2^-(ads) + H_2O \rightarrow O_2 + HO_2^- + OH^- \tag{10.4}$$

where O_2^- is a superoxide radical ion.

Carbon black:

$$O_2 + e^- \rightarrow O_2^- \tag{10.5}$$

$$O_2^- + H_2O \rightarrow HO_2^- + OH \tag{10.6}$$

[rate-determining step]

$$OH + e^- \rightarrow OH^- \tag{10.7}$$

The rate and mechanism are different on the basal plane and edge sites of carbon. The reactions involving oxygen are 2–3 orders of magnitude slower on the basal plane than on the edge sites because of the weak adsorption of oxygen molecules on the basal plane surface [34].

The overpotentials for oxygen reduction and evolution on carbon-based bifunctional air electrodes for rechargeable Zn/air batteries are reduced by utilizing metal oxide electrocatalysts. Besides enhancing the electrochemical kinetics of the oxygen reactions, the electrocatalysts serve to reduce the overpotential to minimize carbon oxidation during charge (oxygen evolution). An example of the polarization curves for oxygen reduction and evolution on a bifunctional air electrode with an electrocatalyst of cobalt and nickel oxides on a graphitized carbon black is presented in Figure 10.5.

These results were obtained in a 1.5 Ah Zn/air cell with 12 mol L^{-1} KOH at 27 °C by Ross [35]. The reversible potential for the electrochemical reactions of oxygen is 0.303 V (vs Hg/HgO, OH^-), and the corresponding reversible potential for the oxidation of carbon is -0.682 V in alkaline electrolyte. Based on these reversible potentials and the polarization curves in Figure 10.5, it is apparent that oxygen reduction and evolution occur at high overpotentials. For example, at 10 mA cm^{-2} the electrode potentials for oxygen reduction (discharge) are -0.130 V in air and 0.638 V for oxygen evolution (charge); these correspond to overpotentials of 0.433 and 0.335 V, respectively. These results indicate several of the technical problems facing the viability of a rechargeable Zn/air battery which utilizes carbon-based bifunctional air electrodes. That is, the overpotentials for the electrochemical oxygen reactions must be reduced to improve energy efficiency, and the potential of the electrode during charge must be lowered to protect the carbon from electrochemical oxidation. As mentioned above, electrocatalysts such as cobalt and nickel oxides enhance the kinetics for the oxygen reactions, but they are also catalysts for carbon oxidation. Thus, the challenge is to identify electrocatalysts which are beneficial

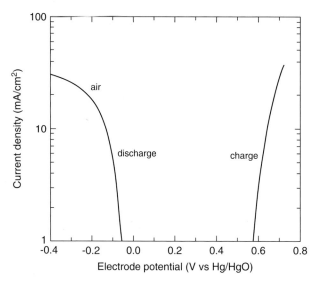

Figure 10.5 Polarization curves for bifunctional air electrode in 1.5 Ah Zn/air cell with 12 KOH at 27 s. From Ross [35].

for the electrochemical reactions of oxygen and at the same time do not promote carbon oxidation.

In redox flow batteries such as Zn/Cl_2 and Zn/Br_2, carbon plays a major role in the positive electrode where reactions involving Cl_2 and Br_2 occur. In these types of batteries, graphite is used as the bipolar separator, and a thin layer of high-surface-area carbon serves as an electrocatalyst. Two potential problems with carbon in redox flow batteries are: (i) slow oxidation of carbon and (ii) intercalation of halogen molecules, particularly Br_2, in graphite electrodes. The reversible redox potentials for the Cl_2 and Br_2 reactions (Equations 10.8 and 10.9)

$$Cl_2 + 2e^- \rightarrow 2Cl^- \tag{10.8}$$

$$Br_2 + 2e^- \rightarrow 2Br^- \tag{10.9}$$

are 1.358 and 1.066 V, respectively. These potentials are considerably higher than the reversible potential for the C/H_2O reaction (see Table 10.2), which suggests that carbon is susceptible to oxidation at the redox potentials for the Cl_2 and Br_2 reactions.

In the Zn/Cl_2 battery, carbon is utilized in both electrodes, serving as a flow-through positive electrode and a substrate for the zinc negative electrode. The requirements are listed below. Chlorine, flow-through electrode:

- relatively narrow pore size distribution for uniform flow characteristics;
- uniform porosity and permeability for good electrolyte flow distribution;
- low resistivity to minimize IR drop in the electrode;
- capability to accept activation treatment;
- no distortion in flowing electrolyte;
- adequate physical strength to permit press-fitting of electrode into the intercell busbar.

Graphite substrate for zinc deposit:

- low surface porosity and fine grain-size for attaining an adherent and uniform zinc deposit;
- low exchange current for hydrogen evolution;
- good physical strength for press-fitting of the electrode into the intercell bus-bar;
- easily machined into thin electrodes, about 1 mm thick.

Jorne et al. [36] investigated the reactivity of graphites in acidic solutions that are typically used for Zn/Cl_2 cells. The degradation of porous graphite is attributed to oxidation to CO_2. The rate of CO_2 evolution gradually decreased with oxidation time until a steady state was reached. The decline in the CO_2 evolution rate is attributed to the formation of surface oxides on the active sites.

A composite consisting of a mixture of carbon particles (e.g., carbon black or graphite) and a polymer binder such as polyethylene or polypropylene with a surface layer of a carbon-black or carbon-felt flow-through structure serves as the Br_2 electrode in Zn/Br_2 batteries. Because of the low surface area of the carbon-polymer surface, an additional layer of carbon is necessary to obtain higher reaction rates. The mechanical deterioration of graphite-polymer composite electrodes (e.g., 50 wt% high-density polyethylene, 35 wt% graphite, 15 wt% carbon black) in Br_2-containing electrolytes was investigated by Futamata and Takeuchi [37]. The intercalation of Br_2 in graphite and the reaction of Br_2 with polyethylene resulted in mechanical degradation of the composite electrode.

Another type of redox flow battery that utilizes carbon electrodes and soluble reactants involving vanadium compounds in H_2SO_4 is under evaluation [38, 39]:

$$\text{Positive electrode (discharge): } VO_2^+ + 2H^+ + e^- \rightarrow VO^{2+} + H_2O \quad (10.10)$$

$$\text{Negative electrode (discharge): } V^{2+} \rightarrow V^{3+} + e^- \quad (10.11)$$

Electrodes consisting of carbon-reinforced graphite or carbon fibers were investigated with the redox reactions of soluble vanadium ions. The former material showed evidence for the intercalation of H_2SO_4 at concentrations >5 mol L^{-1}; however, a similar reaction was not observed with the carbon fibers. Skyllas-Kazacos and co-workers [39] noted that the electrochemical activity of graphite-polymer composite electrodes in the vanadium redox battery was enhanced by a chemical activation treatment involving strong inorganic acids (H_2SO_4, HNO_3). The increase in electrochemical activity is attributed to the increase in the concentration of surface functional groups containing C–O and C=O, which could behave as active sites.

Activation by electrochemical or gas-phase oxidation can alter the performance of carbon electrodes for redox reactions. The two major changes that occur to the carbon electrodes as a result of these treatments are an increase in the surface area of the carbon and the formation of surface functional groups on the surface. Jorne and Roayaie [40] reported that electrochemical activation (applying a current density of 33 mA cm^{-2} for 5 h in 0.975 mol L^{-1} H_2SO_4 at 40 °C) of porous graphite electrodes produced an increase in the surface area of nearly an order of magnitude, and this is mainly responsible for the improved kinetics of the Cl^-/Cl_2 redox reaction. On the other hand, gas-phase oxidation of highly oriented pyrolytic graphite in air at 600 °C is reported to enhance the surface area and form acidic surface oxides which help to increase the kinetics of the redox reactions involving both Cr^{3+}/Cr^{2+} and Fe^{3+}/Fe^{2+} [41].

10.3.6
Intercalation

Highly ordered graphite serves as a host for intercalation of ions such as HSO_4^-, ClO_4^-, and BF_4^- in aqueous electrolytes. Graphite intercalation compounds in H_2SO_4 containing HNO_3 have shown some encouraging results [42]. In lead–acid batteries, graphite in the positive electrode is beneficial because the formation of an intercalation compound $C_nHSO_4 \cdot 2.5H_2SO_4$ expands the electrode structure [43]. This expansion increases the porosity and the amount of electrolyte available in the electrode to improve the discharge performance. More recently, carbon has played a pivotal role in the success of Li-ion batteries, serving as the host material for lithium storage in the negative electrode. In this application, the high electronic conductivity of carbon and its ability to intercalate and/or adsorb lithium ions are critical to the success of the Li-ion battery. A detailed review of carbon in the negative electrode of Li-ion batteries is discussed in Part III, Chapter 15.

10.4
Concluding Remarks

The element carbon has many desirable characteristics which have prompted its use in aqueous batteries; they include its low cost, acceptable corrosion stability, high electronic conductivity, compatibility with processing conditions used in porous electrode structures, availability in a range of particle sizes and shapes, and reasonable electrochemical activity. It would be difficult to find an alternative material which could match these advantageous features. In this review, the electrochemical behavior of amorphous carbons and graphitic materials is discussed. Carbon can be tailored to meet the electrochemical requirements in many battery applications because of the wide range of properties that are available with its various forms extending from amorphous carbon to graphite.

References

1. Kinoshita, K. (1988) *Carbon Electrochemical and Physicochemical Properties*, John Wiley & Sons, Inc., New York, p. 20.
2. Hess, W. and Herd, C. (1993) in *Carbon Black* (eds J. Donnet, R. Bansal, and M. Wang), 2nd edn, Marcel Dekker, New York, p. 89.
3. Spain, I. (1981) in *Chemistry and Physics of Carbon*, vol. 16 (eds P. Walker and P. Thrower), Marcel Dekker, New York, p. 119.
4. Delhaés, P. and Carmona, F. (1981) in *Chemistry and Physics of Carbon*, vol. 17 (eds P. Walker and P. Thrower), Marcel Dekker, New York, p. 89.
5. Dannenberg, E. (1978) *Kirk-Othmer: Encyclopedia of Chemical Technology*, vol. 4, John Wiley & Sons, Inc., New York, p. 631.
6. Millward, G. and Jefferson, D. (1978) in *Chemistry and Physics of Carbon*, vol. 14 (eds P. Walker and P. Thrower), Marcel Dekker, New York, p. 1.
7. Ishikawa, T. and Nagaoki, T. (1983) *Recent Carbon Technology Including Carbon and SiC Fibers*, JEC Press, Cleveland, OH.
8. Allera, R. and Ruopp, P. (1993) *Am. Ceram. Soc. Bull.*, **72**, 99.
9. (a) Medalia, A., Richards, L., and Heckman, F. (1972) *J. Coll. Sci.*, **40**, 223; (b) Medalia, A., Richards, L., and Heckman, F. (1971) *J. Coll. Sci.*, **36**, 173; (c) Medalia, A., Richards, L., and Heckman, F. (1970) *J. Coll. Sci.*, **32**, 115.
10. (a) Hess, W., Ban, L., McDonald, G., and Urban, E. (1977) *Rubber Chem. Technol.*, **50**, 842; (b) Hess, W., Ban, L., McDonald, G., and Urban, E. (1973) *Rubber Chem. Technol.*, **46**, 204; (c) Hess, W., Ban, L., McDonald, G., and Urban, E. (1969) *Rubber Chem. Technol.*, **42**, 1209.
11. Ban, L. and Hess, W. (1969) in *Petroleum Derived Carbons*, ACS Symposium Series (eds M. Deviney and T. O'Grady), American Chemical Society, Washington, DC, p. 358.
12. Ehrburger-Dolle, F. and Misono, S. (1992) *Carbon*, **30**, 31.
13. Kaneko, K., Ishii, C., Ruike, M., and Kuwabara, H. (1992) *Carbon*, **30**, 1075.
14. Boehm, H. (1994) *Carbon*, **32**, 759.
15. Rivin, D. (1971) *Rubber Chem. Technol.*, **44**, 307.
16. Kinoshita, K. and Bett, J. (1975) *Carbon*, **13**, 405.

17. Wiegand, W. (1937) *Ind. Eng. Chem.*, **29**, 953.
18. Fabish, T. and Schleifer, D. (1984) *Carbon*, **22**, 19.
19. Randin, J.-P. (1976) in *Encyclopedia of Electrochemistry of the Elements*, vol. 7 (ed. A. Bard), Marcel Dekker, New York, p. 1.
20. Vinal, G.W. (1950) *Primary Batteries*, John Wiley & Sons, Inc., New York, p. 20.
21. Bregazzi, M. (1967) *Electrochem. Technol.*, **5**, 507.
22. Lahaye, J., Wetterwald, M., and Messiet, J. (1984) *J. Appl. Electrochem.*, **14**, 545.
23. Takahashi, K. (1980) *Prog. Batt. Solar Cells*, **3**, 140.
24. (a) Fischer, F. and Wissler, M. (1984) in *Battery Material Symposium*, Brussels 1983, vol. 1 (eds A. Kozawa and M. Nagayama), International Battery Material Association, Cleveland, OH, p. 115; (b) (1985) *New Mater. New Proc.*, **3**, 268.
25. Tuomi, D. (1987) in *Proceedings of the Symposium on History of Battery Technology* (ed. A. Salkind), The Electrochemical Society, Pennington, NJ, p. 21.
26. Leon, L. and Radovic, L. (1994) in *Chemistry and Physics of Carbon*, vol. 24 (ed. P. Thrower), Marcel Dekker, New York, p. 213.
27. Veres, A. and Csath, G. (1986) *J. Power Sources*, **18**, 305.
28. Biermann, J., Wetterwald, M., Messiet, J., and Lahaye, J. (1981) *Electrochim. Acta*, **26**, 1237.
29. Caudle, J., Summer, K., and Tye, F. (1977) in *Power Sources*, vol. 6 (ed. D. Collins), Academic Press, New York, p. 447.
30. (a) Ross, P., Staud, N., and Sokol, H. (1984) *J Electrochem. Soc.*, **131**, 1742; (b) Ross, P., Staud, N., and Sokol, H. (1986) *J. Electrochem. Soc.*, **133**, 1079; (c) Ross, P., Staud, N., and Sokol, H. (1988) *J Electrochem. Soc.*, **135**, 1464 (d) Ross, P., Staud, N., and Sokol, H. (1989) *J Electrochem. Soc.*, **136**, 3570.
31. Thiele, H. (1938) *Trans. Faraday Soc.*, **34**, 1033.
32. Heller, H. (1945) *Trans. Electrochem. Soc.*, **87**, 501.
33. Kinoshita, K. (1984) in *Proceedings of the Workshop on the Electrochemistry of Carbon* (eds S. Sarangapani, J. Akridge, and B. Schumm), The Electrochemical Society, Pennington, NJ, p. 273.
34. Morcos, I. and Yeager, E. (1970) *Electrochim. Acta*, **15**, 953.
35. Ross, P. (1986) *Proceedings of the 21st Intersociety Energy Conversion Engineering Conference*, American Chemical Society, Washington, DC, p. 1066.
36. Jorne, J., Roayaie, E., and Argade, S. (1988) *J. Electrochem. Soc.*, **135**, 2542.
37. Futamata, M. and Takeuchi, T. (1992) *Carbon*, **30**, 1047.
38. Kaneko, H., Nozaki, K., Wada, Y., Aoki, T., Negishi, A., and Kamimoto, M. (1991) *Electrochim. Acta*, **36**, 1191.
39. (a) Skyllas-Kazacos, M., Kazacos, M., Zhong, S., and Sun, B. (1992) *Electrochim. Acta*, **37**, 2459; (b) Kazacos, M., and Skyllas-Kazacos, M. (1989) *J. Electrochem. Soc.*, **136**, 2759; (c) Zhong, S., Kazacos, M., Burford, R.P., and Skyllas-Kazacos, M. (1991) *J. Power Sources*, **36**, 29; (d) Skyllas-Kazacos, M., Kazacos, M., Zhong, S., and Sun, B. (1992) *Electrochim. Acta*, **39**, 1.
40. Jorne, J. and Roayaie, E. (1986) *J. Electrochem. Soc.*, **133**, 696.
41. Hollax, E. and Cheng, D. (1985) *Carbon*, **23**, 655.
42. (a) Iwashita, N., Shiogama, H., and Inagaki, M. (1995) *Synth. Met.*, **73**, 33; (b) Inagaki, M., Tanaika, O., and Iwashita, N. (1995) *Synth. Met.*, **73**, 83; (c) Inagaki, M. and Iwashita, N. (1994) *J. Power Sources*, **52**, 69.
43. (a) Tokunaga, A., Tsubota, M., Yonezu, K., and Ando, K. (1987) *J. Electrochem. Soc.*, **134**, 525; (b) Tokunaga, A., Tsubota, M., Yonezu, K., and Ando, K. (1989) *J. Electrochem. Soc.*, **136**, 33.

11
Separators
Werner Böhnstedt

11.1
General Principles

The separator – the distance-keeping component between the positive and the negative electrode of a galvanic cell – does not directly participate in the electrochemical processes of electricity storage. As a 'passive' element it has naturally attracted little scientific interest; its significance lies in the technical challenge to build batteries ever more compact and long-lasting. A decisive breakthrough was achieved only in the second half of the twentieth century by the development of sufficiently stable synthetic materials. The know-how of the chemical industry in selecting suitable plastics and their processing was combined with the experience of the battery industry regarding the unique conditions of use; an independent separator industry developed which, since the late 1960s, from the combination of these two aspects, has given essential impulses to the advancement of batteries.

A comprehensive modern survey of separators for electrochemical power sources exists only in incomplete parts [1–3], and textbooks on batteries treat this important element only as a side aspect [4–11]. This section is an attempt to describe, besides some fundamental aspects, the development history of the battery separator, competing systems of the present day with their advantages and weaknesses, and also future development trends.

11.1.1
Basic Functions of the Separators

Separators serve two primary functions: while having to keep the positive electrode physically apart from the negative in order to prevent any electronic current passing between them, they also have to permit an ionic current with the least possible hindrance. These two opposing requirements are best met by a compromise: a porous nonconductor.

The electronic insulation – the origin of the term '*separator*') – has to be durable, that is, it must be effective over many years over a wide range of temperatures and in a highly aggressive medium. Under these conditions no substance harmful to the electrochemical reactions may be generated.

The unhindered ionic charge transfer requires many open pores of the smallest possible diameter to prevent electronic bridging by deposition of metallic particles floating in the electrolyte. Thus the large number of microscopic pores form immense internal surfaces, which inevitably are increasingly subject to chemical attack.

Not only the electrolyte, but also the electrodes directly or indirectly exert a chemical attack, either by an oxidation or reduction potential of the electrode material itself or by the generation of soluble oxidizing or reducing substances.

The requirements for the separator properties are generally lower in primary cells, that is, in nonrechargeable systems. This results from the lack of problematic phenomena accompanying any charging of a battery, such as recrystallization of active materials or the generation of oxidizing species during overcharge. Within the framework of this chapter, therefore, separators mainly for secondary cells will be described.

In the older battery literature the term 'separator' is frequently used very loosely, to include all nonmetallic solid components between the electrodes, such as supporting structures for active materials (tubes, gauntlets, glass mats), spacers, and separators in a narrow sense. In this section, only the last of these, the indispensable separating components in secondary cells, will be termed 'separators,' distinguished from the others by their microscopically small pores, that is, with a mean diameter significantly below 0.1 mm.

11.1.2
Characterizing Properties

Some terms and properties common to all separators are defined and discussed below.

11.1.2.1 Backweb, Ribs, and Overall Thickness

Separator backweb refers to the porous separating membrane. It is of uniform thickness and has a macroscopically uniform pore distribution. Only in this way can an overall uniform current density be ensured during the operation of the storage battery, achieving a uniform charging and discharging of the electrodes and thus a maximum utilization of the electrode materials.

The lead–acid battery has a peculiarity: the electrolyte sulfuric acid not only serves as ion conductor (as charge-transport medium), but it actively participates in the electrochemical reaction:

$$Pb + PbO_2 + 2H_2SO_4 \leftrightarrow 2PbSO_4 + 2H_2O \tag{11.1}$$

During charging at the positive electrode one additional water molecule is consumed per electron converted, which is regenerated during discharging.

In practice the desired electrolyte distribution is achieved by distance-maintaining ribs on the porous backweb; this in addition has the advantage of maintaining a maximum distance between the origin of oxidizing substances located at the positive electrode and the highly porous separating membrane, sensitive due to its large inner surface.

The total or overall thickness thus comprises the backweb thickness and the rib height. For achieving a uniform current distribution the thickness is normally specified very precisely and is acceptable only within rather narrow tolerances. Besides technical difficulties in the production, this also presents a problem in measurement: since all separator materials are more or less compressible, a specified measuring pressure has to be used. Moreover, the measuring area is also significant; one can easily imagine an extended area touching only the microscopic elevations of the separator, whereas a measuring tip may very well hit 'valleys.'

11.1.2.2 Porosity, Pore Size, and Pore Shape

Porosity of a separator is defined as the ratio of void volume to apparent geometric volume. High porosity is desirable for unhindered ionic current flow.

The pores of the separating membrane are to be most uniformly distributed and of minimum size to avoid deposition of metallic particles and thus electronic bridging. One distinguishes between macroporous and microporous separators, the latter having to show pore diameters below 1 micron (μm), that is, below one-thousandth of a millimeter. Thus the risk of metal particle deposition and subsequent shorting is quite low, since active materials in storage batteries usually have particle diameters of several microns.

However, even these small pores cannot prevent the formation of so-called 'microshorts,' arising by metal deposition (e.g., dendrites) from the solution phase. The pores of modern separators have a diameter of about 0.1 μm, equal to 100 nm, while metal ions have a diameter of few angstroms, equal to 0.5–1 nm. On an atomic scale even micropores are barn doors!

Micropores are invisible to the naked human eye; thus for outsiders it is always surprising that separators of typically 60% porosity (i.e., 60% void volume, 40% solid material) present the impression of a compact, hole-free, nontransparent sheet.

In a first approximation the average size of pore diameter has no effect on porosity, even though a superficial view leads to other conclusions. A mental experiment may be of assistance: imagine a pore and its outside wall, decrease both to identical scale, then the ratio of void to outside volume remains constant. Of course the requirements as to pore sizes and their uniform distribution increase with decreasing separator backweb thickness. The risk of defects also increases; so-called 'pinholes' can originate, for example, by bubble inclusion within the separator membrane during the production process.

Pores generally are not of a hose-like configuration of constant diameter in a straight-line direction from one electrode to the other. In practice, a separator's pores are formed as void between fibers (Figure 11.1) or spherical bodies in amorphous agglomerates (Figure 11.2), thus being very different in their form and size. Statements of any pore diameter are always to be viewed with the above in mind. Figures 11.1 and 11.2 represent macroporous systems, whereas Figures 11.3

Figure 11.1 Microfiber glass fleece separator (SEM).

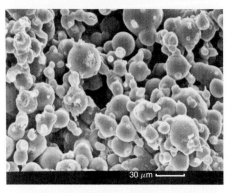

Figure 11.2 Sintered PVC separator (SEM).

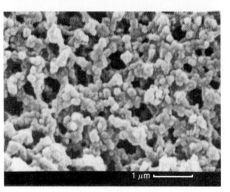

Figure 11.3 Phenol–formaldehyde resin–resorcinol separator (SEM).

and 11.4 show microporous separators. It should be noted that the latter figures have a 50-fold larger magnification!

The path taken by an ion from one electrode to the other will not be a straight one, as it has to evade the solid structures by making detours. The ratio of the mean actual path in comparison with the direct distance is called the *tortuosity factor T*. For plastic bodies consisting essentially of spherical, interconnected particles with voids in between, with a porosity of about 60%, this value is roughly 1.3; for higher porosities it decreases to approach a value of 1.0 at very high porosities.

Figure 11.4 Microporous polyethylene separator (SEM).

11.1.2.3 Electrical Resistance

The *electrical resistance exerted by a separator on the ionic current* is defined as the total resistance of the separator filled with electrolyte minus the resistance of a layer of electrolyte of equal thickness, but without the separator. The separator resistance has to be considered as an increment over the electrolyte resistance.

$$R(\text{separator}) = R(\text{electrolyte} + \text{separator}) - R(\text{electrolyte}) \qquad (11.2)$$

$$R = \frac{1}{\sigma} \times \frac{l}{q} \qquad (11.3)$$

where l is the length of the ion path and q the area of the ionic flow; σ is the specific electrolytic conductivity, the reciprocal of the specific resistance ρ of the electrolyte, and is a temperature-dependent material constant.

The tortuosity factor T of a separator can be expressed by

$$T = \frac{l_s}{d} \qquad (11.4)$$

with l_s, the ion path through the separator and d, the thickness of the separating layer.

The porosity of the separator is defined thus:

$$P = \frac{\text{void separator volume}}{\text{geometric separator volume}}$$

$$= \frac{q_s l_s}{q d} \qquad (11.5)$$

with q_s as the 'open' area of the separator.

A transformation results in

$$q_s = q \cdot \frac{P}{T}$$

and substitution into Equation 11.2 gives:

$$R(\text{separator}) = \frac{1}{\sigma} \cdot \frac{l_s}{q_s} - \frac{1}{\sigma} \frac{d}{q}$$

$$= \frac{1}{\sigma}\left(dT \cdot \frac{T}{qP}\right) - \frac{d}{q}$$

$$= \frac{1}{\sigma}\frac{d}{q}\left(\frac{T^2}{P} - 1\right)$$

$$= R(\text{electrolyte})\left(\frac{T^2}{P} - 1\right)$$

$$R_{sep} = R_{el}\left(\frac{T^2}{P} - 1\right)$$

with $\quad R_{el} = \frac{1}{\sigma} \cdot \frac{d}{q}$ \hfill (11.6)

This formula shows the factorial effect of the separator on the electrical resistance; the measured resistance of the electrolyte-filled separator is the (T^2/P)-fold multiple of the electrolyte resistance without the separator; by definition, $T^2/P \geq 1$.

With increasing tortuosity factor T and lower porosity P, R increases sharply. The electrical resistance of a separator is proportional to the thickness d of the membrane and is subject to the same dependence on temperature or concentration as the electrolyte itself.

For sulfuric acid (H_2SO_4) of specific density $1.28\,\text{g cm}^{-3}$ at $25\,°C$, the specific resistance $(1/\sigma)$ is $1.26\,\Omega$ cm; using this value in the Equation 11.6 and selecting values typical for polyethylene starter battery separators at $d = 0.25$ mm, $P = 0.6$, and $T = 1.3$, the electrical resistance for 1 cm² of separators area results in

$$R_{sep} = 1.26\,\Omega\text{cm}\frac{0.025\,\text{cm}}{1\,\text{cm}^2}\left(\frac{1.3^2}{0.6} - 1\right)$$

$$= 0.057\,\Omega$$

Usually the electrical resistance of a separator is quoted in relation to area; in the above case it is 57 mΩ cm². In order to quote it for other areas, because of the parallel connection of individual separator areas, Kirchhoff's law has to be taken into account:

$$\frac{1}{R_{total}} = \frac{1}{R_1} + \frac{1}{R_2} + \ldots$$

or, as all R_i are equal,

$$R_{total} = \frac{1}{n}R_n$$

Applying this to the above example for an area of 1 in² = 6.45 cm², the result is $R = 8.8$ mΩ in². Taking an example from starter-lighting-ignition (SLI) battery practice, one cell with six positive and seven negative electrodes of typical 114 mm × 147 mm size with the above separator shows a resistance of $28.3 \times 10^{-6}\,\Omega$ at $25\,°C$, or close to $75 \times 10^{-6}\,\Omega$ at $-18\,°C$. For a cold crank current of 320 A and six cells in series in

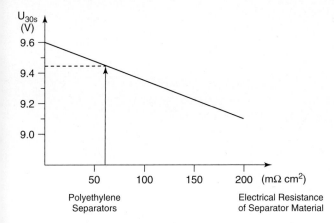

Figure 11.5 Cold crank voltage as a function of separator electrical resistance. Reprinted from W. Bohnstedt, Automotive lead/acid battery separators: a global overview. *J. Power Sources*, 1996, **59**, 45–50, with kind permission from Elsevier Science S.A., Lausanne [3].

a 12 V battery, the voltage drop due to the separator resistance amounts to ∼0.15 V; Figure 11.5 shows this correlation.

The dependence of separator electrical resistance on porosity for the selected SLI battery separator (0.25 mm backweb thickness) and the practical approximation $T^2 P = 1$ can be seen in Figure 11.6.

Other characterizing separator properties are either application-related or product-specific; they will therefore be discussed with the individual separator types.

11.1.3
Battery and Battery Separator Markets

There are no indications, or only vague ones, of the size of the various battery separator markets in the literature [3]. A rough estimate can be deduced from the sales figures for battery systems by a rule of thumb: the sales value of separators is roughly 2–5% of the sales for the battery producers. Even the data for battery markets are not uniformly gathered, however, and contain considerable uncertainties.

The total sales value for battery systems worldwide in 1997 may amount to US $25.5 billion and the sales value for battery separators correspondingly to US $600 million; Table 11.1 gives an estimate of the work battery market, split according to the different battery systems.

From this – albeit rather rough – overview, the proportions become clear: around 45% of all battery sales worldwide and thus also separator sales worldwide are in lead–acid batteries and a further 13% in the rechargeable alkaline battery sector.

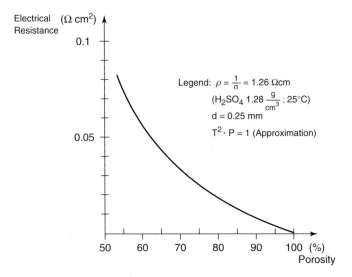

Figure 11.6 Electrical resistance as a function of porosity. Reprinted from W. Bohnstedt, Automotive lead/acid battery separators: a global overview. *J. Power Sources*, 1996, **59**, 45–50, with kind permission from Elsevier Science S.A., Lausanne [3].

Table 11.1 World battery markets 1998 (US $ million, estimate).

Lead–acid batteries	
Automotive batteries	8200
Industrial batteries	2300
VRLA batteries	1000
Total	11 500
Alkaline batteries	
Vented	500
Sealed	2800
Total	3300
Lithium-ion batteries	1300
Consumer batteries	9400
Total markets	25 500

The remaining 40% or more are split among the recently introduced lithium-ion batteries as well as a multitude of primary systems in the portable battery sector. This distribution of battery production is not geographically uniform; whereas in Europe and the USA automotive and industrial batteries are in the lead, in the Asia-Pacific area consumer batteries are more strongly represented. In this section separators for mainly those rechargeable batteries which have aqueous electrolyte will be discussed individually, whereas separators for batteries with nonaqueous electrolyte, which have attained a commercial breakthrough in recent years, will be the subject of a separate chapter.

11.2
Separators for Lead–Acid Storage Batteries

11.2.1
Development History

11.2.1.1 Historical Beginnings

The historical development of the separator and of the lead–acid storage battery are inseparably tied together. When referring to lead–acid batteries today one primarily thinks of starter batteries or forklift traction batteries, but the original applications were quite different.

The very first functioning lead–acid battery was presented by Gaston Planté in 1860: spirally wound lead sheets served as electrodes, separated by a layer of felt – the first separator of a lead–acid battery [12]. This assembly in a cylindrical vessel in 10% sulfuric acid had only a low capacity, which prompted Planté to undertake a variety of experiments resulting in many improvements that are still connected with his name.

Until about 1880 the lead–acid battery was exclusively the subject of scientific study. Possible commercial utilization lacked suitable charging processes; secondary cells had to be charged by means of the primary cells already known at that time.

Only with the discovery of the dynamoelectric effect and its rapid commercialization after 1880 did the industrial use of lead–acid storage battery begin. Here the development of pasted plates by Camille Faure was essential for significantly raising the amount of stored energy; they were separated by layers of parchment and felt [13]. These batteries served predominantly for illumination and, later – beginning around 1890, also as stationary batteries for peak power load leveling in power plants. Glass rods frequently sufficed as spacers, or these batteries were even built without separators at all, simplifying the frequent removal of anode mud from the containers.

The development of the tubular plate during the last decade of the nineteenth century required oxidation- and acid-stable porous material. Of the natural materials only a few are moderately stable in sulfuric acid: glass, asbestos, rubber, and cellulose. All have been tested, singly or in combination. Asbestos fabrics as tubular material for positive electrodes, textiles for fixing the negative mass, and rubber rods as spacers were in the first batteries for driving electric vehicles, an application becoming popular in that period. These vehicles, however, required increased energy densities, that is, the electrode distance had to be decreased. After many trials, extruded hard rubber tubes prevailed for the positive electrodes, with finely sawn cross-slits to allow ion migration. At that time the first wooden veneers

[14] were used for separating the electrodes (the first separator in the narrow meaning of the word), and for about 60 years the most successful material.

11.2.1.2 Starter Battery Separators

In the competition between the systems (electric motor versus combustion engine) for vehicles, one essential disadvantage of the latter was the tedious and demanding process of starting by muscular power. Only the development of the electric starter by Kettering in 1911, and the battery accompanying it, changed this situation suddenly. It is an irony of history that the starter battery contributed essentially to the downfall of the early electric car. A rapid development of the car industry, and of the battery industry in parallel with this, followed. Wooden veneer became the standard separation of the lead–acid storage battery, be it in double separation as wood veneer with rubber spacers or later as ribbed wood veneer alone, when the electrodes became thinner; thus the required acid supply and also the distance between electrodes decreased in order to increase the energy and power density.

Wood veneers were produced preferentially from Port Orford cedar, primarily domiciled in Oregon (USA). Trials with other types of wood, for example, poplar, remained makeshift measures. The preparation of the wood veneer, that is, the sawing and slicing of the trees, the dissolving of the lignin to achieve porosity, and the almost complete leaching of resins which would otherwise accelerate the corrosion, was quite difficult [15]. Wood veneer separators could be stored and transported only when wet; dry-charged starter batteries could not be built using them. Nevertheless, wood veneers remained the predominant separators until about 1960!

In the meantime another development had decisively altered the outset situation; plastics had been discovered and synthesized, among them also some acid-stable ones such as phenol–formaldehyde resin and poly(vinyl chloride) (PVC). These opened up new possibilities: cellulose papers could be impregnated with phenol–formaldehyde resin solution and thus rendered sufficiently acid-stable, and sintered sheets from PVC powder were developed. Independent producers of separators were founded, combining knowledge of the chemical industry with experience of the battery industry and thus accelerating the development process.

During the first trials with synthetic separators around 1940 it had already been observed that some of the desired battery characteristics were affected detrimentally. The cold crank performance decreased and there was a tendency toward increased sulfation and thus shorter battery life. In extended test series, these effects could be traced back to the complete lack of wooden lignin, which had leached from the wooden veneer and interacted with the crystallization process at the negative electrode. By a dedicated addition of lignin sulfonates – so-called organic expanders – to the negative mass, not only were these disadvantages removed, but an improvement in performance was even achieved.

Larger vehicles with bigger engines required even higher cold crank performance. In order to meet the resulting requirements for separators with lower electrical resistance, around 1970 the polyethylene separator [16] and more or less at the same time glass separators were developed and introduced. Glass separators are very similar to the cellulosic separators already mentioned; they do not require special machinery for processing and offer a very low electrical resistance. They succeeded in the USA in largely displacing sintered PVC and cellulosic separators, before they themselves were supplanted by a completely new technology, polyethylene pocket separation. In Europe this intermediate step – glass separators – did not occur; the transition from PVC or cellulosic leaf separators to the polyethylene pockets proceeded directly, albeit some 5 or 10 years later than in the USA.

The decisive advantage of the polyethylene pocket is its flexibility and sealability as well as its very small pore diameters. In the course of their efforts at efficiency improvement, around 1970 Delco–Remy had designed a completely novel starter battery concept: electrodes of expanded lead strip were to replace cast grids, promising a significant saving in weight and thus in cost. Only lead–calcium alloys proved to be suitable for expansion, which had a tendency, however, toward increased mud shedding during battery operation. Against this the microporous polyethylene separator film, developed by W.R. Grace & Co. in 1966 [16], promised to be a remedy. While the mud could be accommodated by the formation of a three-sided sealed pocket around the electrode, at the same time the micropores eliminated the risk of penetration through the separator. A largely automated starter battery production was the positive result of these developments. The commercial advantages of pocketed starter batteries as well as their technical ones, such as freedom from maintenance, high cold crank power, and prolonged battery life, have led to a victorious worldwide advance of this technology [3].

Alternative separator materials such as organic fibers or sintered PVC did not succeed as pocket materials due to their excessive pore size. The pore diameter of 10–20 μm typical for such materials is insufficient to effectively prevent the preferential formation of shorts by mass particles at the folding edge of the pocket.

A completely separate development took place in Japan. Traditionally, very porous positive active materials are used in that country requiring support glass mats to avoid premature capacity loss and shorting in cycling service. Since the introduction of the cellulosic separator a version of this kind has been used in Japan without distance ribs, achieving the total thickness of the separator by means of a thick glass mat instead. Certainly this is an expensive type of separator, but it can be balanced by savings in active material. In Japan, meanwhile, the fleece of organic fibers, which was unsuitable for pocketing, has replaced the impregnated cellulosic sheet. It distinguishes itself by favorable electrical resistance data and good acid and oxidative stability. The transition to microporous polyethylene pockets proceeds more slowly than in the USA or in Europe, because it requires a simultaneous change in formulation of the positive mass.

Starting with the development of sealed lead–acid cells by The Gates Rubber Co. in 1972 [17], this principle of internal oxygen transfer was transferred to starter batteries around 1980. The electrolyte is absorbed in a microfiber glass mat; this has to leave electrolyte-free channels through which the oxygen generated at the positive electrode can diffuse to the negative, where it is reduced. Thus one has a theoretically maintenance-free battery without any water consumption.

Table 11.2 Automotive lead–acid battery production 1997 (million kilowatt-hour, estimate).

	Polyethylene pocket separators	Sintered PVC/rubber separators	Cellulosic/glass separators	Synthetic pulp/GM separators	VRLA SLI batteries	Total
USA–Canada	56.5	1.2	0.6	–	0.7	59.0
Europe	30.0	11.0	3.4	–	0.1	45.4
Asia–Pacific	16.3	11.4	4.5	9.8	0.3	42.3
Latin America	11.7	0.9	3.9	–	–	16.5
Total (million kWh)	115.4	24.5	12.4	9.8	1.1	163.2
(%)	70.7	15.0	7.6	6.0	0.7	100.0

Despite additional indisputable advantages such as spill-proofness or flexibility of position in the car, this design – mostly for reasons of cost – has not yet made a breakthrough in starter battery applications.

After this stroll through history, let us consider the current markets, split according separation systems in the various geographic areas (Table 11.2).

The microporous polyethylene pocket has succeeded worldwide; more than 70% of all starter batteries use this form of separation. Whereas in the USA and Western Europe the transition is essentially complete, a similar development in the Asia-Pacific area and Latin America, and in the medium term also in Russia and China, is expected [3].

11.2.1.3 Industrial Battery Separators

11.2.1.3.1 Stationary Battery Separators   As already mentioned, at the beginning of the twentieth century the electric power supply was still very susceptible to load changes, requiring the use of stationary lead batteries for load leveling. As more powerful generators were developed this application diminished, but from the increasing dependence on general supply of electricity the need for emergency power batteries developed, for example, for emergency lights. From the start, the telephone systems required huge battery installations in float service, on the one hand as buffer batteries filtering interferences from the alternating current circuits and on the other hand permitting (as least for limited periods) an uninterrupted service during power outages. The batteries in these applications are charged continuously with a low current to counteract self-discharge and to allow discharging at comparably high currents when required. Due to the level of maintenance necessary, these batteries were initially built in an open construction, the required electrolyte reservoir being supplied with wide electrode spacing, frequently without any separators. Later, spacers of hard rubber – initially rods and then corrugated spacers – were used. With the invention of PVC separators and their low-cost industrial production process, sintered PVC separators have been used since around 1950 and some are still employed today.

In the second half of the 1960s, at the same time but independently, three basically different plastic separators were developed. One was the polyethylene separator [16] already referred to in starter batteries, used only rarely in stationary batteries, but successful in traction batteries. The others were the microporous phenolic resin separator (DARAK) [18] and a microporous PVC separator [19], both of which became accepted as the standard separation for stationary batteries. They are distinguished by high porosity (about 70%) and thus very low electrical resistance and very low acid displacement, both important criteria for stationary batteries.

The desire for maintenance-free service, for example, in decentralized single emergency lights for panic lighting, had led, around 1960, to the development of small, sealed, lead–acid batteries with gelled electrolyte [20, 21]. An idea that was already known – the gelling of electrolyte – became applicable to the sulfuric acid electrolyte with the industrial availability of silica types of very high surface area. These fumed silicas have an internal surface of, say, $200 - 300$ m^2 g^{-1} and convert sulfuric acid into a thixotropic gel. By vigorous stirring, the electrolyte is liquefied and can be filled into the battery cell, where it gels again, rendering it leakproof and serviceable in all positions. Microporous separators, for example, of phenolic resin or PVC like the ones referred to above are required for maintaining the spacing and preventing shorts.

However, because of the viscous electrolyte, the charging gases can no longer escape! The application of the principle of sealed nickel–cadmium batteries, known since around 1933 [22], letting the oxygen generated at the positive electrode diffuse to the negative electrode for reduction with partial discharge there, was successful. After an initial water loss, the gel starts to dry out, thereby forming cracks and allowing the oxygen to find a path. Unfortunately this technology is rather costly and therefore can only be justified for special applications.

Uninterrupted power supply for computers, initially only for central units, grew significantly with the introduction of personal computers (PCs): the batteries became smaller and, since they had to be located in offices, had to be in a sealed version, since no aggressive charging gases were permitted to escape. This advanced the breakthrough of the development of so-called recombination or valve-regulated batteries, a version with an absorptive glass mat and an internal oxygen cycle. At about the same time British Telecom phased them out because of the high maintenance, especially the water replenishment required for large stationary batteries with liquid electrolyte, and also started using recombination batteries. The liquid electrolyte is completely absorbed by a microfiber glass fleece, although some channels have to remain free from electrolyte to permit oxygen transfer. The electrolyte absorption occurs on the surface of the microfibers, which – as known – increases steeply with decreasing diameter. Frequently such separators are therefore also characterized by means of their internal surface area (~ 1 m^2 g^{-1}!). The pore distribution is anisotropic, which is desirable; in the plane between the electrodes – because of the fiber diameters – very small pores are formed effecting a large capillary force, while perpendicular to this plane, due to the high porosity of $>90\%$, pores of 10–20 µm diameter are found, which are necessary to ensure the oxygen transport [23–25]. A number of producers of specialized papers started to manufacture and develop these microfiber glass mats further. Fibers below 1 μm in diameter are expensive, and due to their shortness (~1 mm) contribute little to the tensile strength. Binder may be omitted, however, to achieve good wettability; the addition of longer glass fibers of large diameter is required to improve the processability of such separators. A microfiber content of 20–30% has proven sufficient largely to optimize the desired characteristics [26].

The market for sealed stationary batteries has greatly increased since 1980, both by the growth of the PC market as well as by the decentralization of emergency power supplies and telephone exchanges, even though this conversion has not remained undisputed [27]. Table 11.3 gives an estimate of the present situation; these figures also include small consumer lead–acid batteries, which are constructed similarly. More than 60% of all stationary batteries are currently being produced in the sealed version, with the total market growing by roughly 5–10% annually.

11.2.1.3.2 Traction Battery Separators   Electric road vehicles have been reduced to insignificance, as mentioned already, by vehicles with combustion engines. Another electric vehicle – the electrically driven submarine – presented a continuous challenge to lead–acid battery separator development since the 1930s and 1940s. The wood veneers originally used in electric vehicles proved too difficult to handle, especially if tall cells had to be manufactured. Therefore much intense effort took place to develop the first plastic separators. In this respect the microporous hard rubber separator, still available today in a more advanced version, and a microporous PVC separator (Porvic I) merit special mention [28]. For the latter, a molten blend of PVC, plasticizer, and starch was rolled into a flat product. In a lengthy process the starch was subsequently leached out, leaving voids interconnected through holes in their walls. This resulted in an extremely high porosity (levels of up to 85% were reported), but due to a high tortuosity factor of about 1.7 there was also a relatively high electrical resistance.

Table 11.3 World lead–acid stationary and consumer battery production 1997 (million watt-hour, estimate).

	Polyethylene separators	Phenol–formaldehyde–resorcinol separators	PVC separators	Rubber separators	Microfiber glass mat separators	Total
USA–Canada	620	3.50	60	120	3900	5050
Europe	210	1520	510	180	2350	4770
Asia–Pacific	150	50	420	410	2600	3630
Latin America	80	40	120	140	100	480
Total (million Wh)	1060	1960	1110	850	8950	13 930
(%)	7.6	14.1	8.0	6.1	64.2	100.0

Table 11.4 World lead–acid traction battery production 1997 (million watt-hour, estimate).

	Polyethylene separators	Phenol–formaldehyde–resorcinol separators	PVC separators	Rubber separators	Microfiber glass mat separators	Total
USA–Canada	4150	80	50	350	150	4780
Europe	3700	800	1100	950	50	6600
Asia–Pacific	950	150	500	900	50	2550
Latin America	20	50	50	100	–	220
Total (million Wh)	8820	1080	1700	2300	250	14 150
(%)	62.3	7.6	12.0	16.3	1.8	100.0

In Europe, with the economic upswing after 1950, forklifts with batteries came into use – a development which met less acceptance in the USA for various reasons, among them low fuel cost. In this application rubber separators and microporous PVC (Porvic I) were finally able to replace wood veneers, until from around 1975 they again met strong competitors in the new separators already mentioned made of phenolic resin (DARAK), PVC, and mainly polyethylene (Daramic). Today this market is dominated by the polyethylene separator, as is shown in Table 11.4. The annual growth of this market is 2–3%, but with large fluctuation based on prevailing economic conditions.

Sealed batteries have made little entry into this market with heavy cycling service, since the lead–calcium alloys required for these versions tend toward premature capacity loss, a phenomenon intensively investigated in recent years and possibly close to a solution.

11.2.1.3.3 Electrical Vehicle Battery Separators

Although electric vehicles are only a special application for traction batteries, the general interest in them may justify their own separate section.

Electric vehicles are around only in a few surviving niches, electric baggage carts at German railway stations, postal delivery trucks, and milk delivery vans in the UK being the best-known examples. Based on a growing consciousness of decreasing natural resources and especially on the oil crisis around 1970 there were intensive efforts to develop electric propulsion further, but they focused mainly on high-energy battery systems such as sodium–sulfur. The serious difference in energy density between a fuel tank of around 12 000 Wh kg^{-1} and the batteries of 30–40 Wh kg^{-1} actually available was insurmountable; even when considering all efficiencies involved, there remains a factor in the order of magnitude of 100; the electric vehicle returned to the background. Only since about 1990, prompted by the California Clean Air Act and by considerable research grants from the US Advanced Battery Consortium (USABC) – a joint activity mainly of the three major US car manufacturers – have increased efforts on electric vehicles been resumed.

USABC has set the goal so high that lead–acid batteries have been put out of the question for this application [29]. This led to an initiative by the lead–acid battery industry and their suppliers to set up the Advanced Lead–Acid Battery Consortium (ALABC) with the goal of fostering development of the lead–acid battery for use in electric vehicles, at least for an interim period until more powerful batteries with higher energy density will become available. Here a series of complex technical problems have to be solved [30]. Of course, such electric vehicle batteries have to be maintenance-free, that is, of sealed construction; the resulting use of lead–calcium alloys and thus the premature capacity loss have already been touched on.

For the separators of such batteries, gel construction and microfiber glass fleece separators again compete: because of the deep discharge cycles, the gel construction with its lower tendency to acid stratification and to penetration shorts has advantages; for the required power peaks, microfiber glass fleece construction would be the preferred solution. The work on reduction of premature capacity loss with lead–calcium alloys has shown that considerable pressure (e.g., 1 bar) on the positive electrode is able to achieve a significantly better cycle life [31–36]. Pressure on the electrodes produces counter pressure on the separators, which is not unproblematic for both separation systems. New separator developments have been presented with the goal of their being only a little deformed even at high pressure despite high porosity, be they of ceramics [37] or highly filled polymer [38]. Because of the power requirements the trend is clearly toward thinner electrodes and thus thinner separators, which would render a microporous pore size structure indispensable.

11.2.2
Separators for Starter Batteries

11.2.2.1 Polyethylene Pocket Separators

11.2.2.1.1 Production Process The term *'polyethylene separator'* is somewhat misleading, since this separator consists mainly of agglomerates of precipitated silica held within a network of extremely long-chained, ultrahigh-molecular-weight polyethylene (UHMW PE) molecules. The raw materials, precipitated silica (SiO_2 – about 60%), UHMW PE (about 23%), a mineral process oil (about 15%) – all percentages are relative to the final product – and some processing aids (e.g., antioxidants) together with an additional considerable excess of mineral oil, are mixed intensively and fed into an extruder. Here, by the effects of heat and mechanical shear, a viscous melt is formed which is extruded through a slit die 1 m wide into a sheet 1–2 mm thick, which is then formed between the two profiling rolls of a calendar into the desired separator profile. Generally this is characterized by a backweb of about 0.2 mm, which on one side has continuous ribs 0.6–1 mm high in the machine direction at a distance of about 10 mm. At this point the separator material is oil-filled and thus shiny black. In a subsequent step the mineral oil serving as pore-former is largely extracted in a solvent bath [16]. Some producers use trichloroethylene as the solvent; it is easy to handle processwise, but as a chlorinated hydrocarbon it carries environmental risks.

11.2.2.1.2 Mixing and Extrusion The alternative is hexane, which because of the explosion hazard requires a more expensive type of extractor construction. After the extraction the product is dull gray. The continuous sheet is slit to the final width according to customer requirements, searched by fully automatic detectors for any pinholes, wound into rolls of about 1 m diameter (corresponding to a length of 900–1000 m), and packed for shipping. Such a continuous production process is excellently suited for supervision by modern quality assurance systems, such as statistical process control (SPC). Figures 11.7–11.9 give a schematic picture of the production process for microporous polyethylene separators.

11.2.2.1.3 Properties Filled polyethylene is the only separator pocket material that has been able to meet all requirements of a starter battery reliably [39–48].

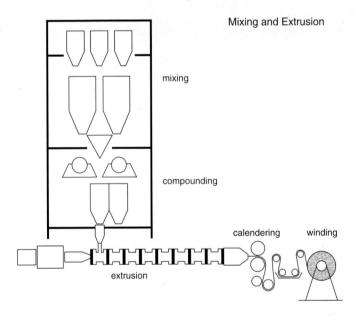

Figure 11.7 Polyethylene separator production process (I) mixing and extrusion.

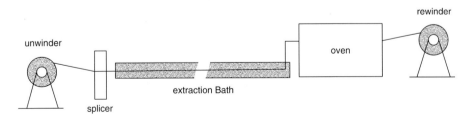

Figure 11.8 Polyethylene separator production process (II) extraction.

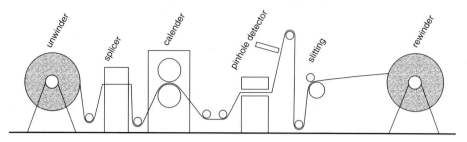

Figure 11.9 Polyethylene separator production process (III) slitting.

Figure 11.10 Starter battery with pocketed plates. Reprinted from W. Böhnstedt, Automotive lead/acid battery separators: a global overview, *J. Power Sources*, 1996, **59**, 45–50, with kind permission from Elsevier Science S.A., Lausanne [3].

It is flexible and weldable into three-sided closed pockets, making the previously usual mud room at the bottom of a starter battery redundant; the consequent increase in grid size of 8% can crank up performance and energy density results (cf. Figures 11.10 and 11.11) [3]. It is microporous, that is, its pore diameters are significantly below 1 µm, which durably prevents penetration by lead particles. Only in this way has the use of lead–calcium alloys in electrodes, with their increased tendency to shedding, become possible, together with a reduction in water consumption over the life of the battery, allowing today's batteries to be properly called *maintenance-free*.

The thin backweb, typically 0.2 mm thick with a porosity of 60%, yields excellent electrical resistance values of ~50 mΩ cm^2, permitting further optimization of high-performance battery constructions. These require very thin electrodes due to the overproportionally increasing polarization effects at higher current densities and consequently also low distances: most modern versions have separators only 0.6 mm thick. Such narrow spacings enforce microporous separation!

Practical experience has shown polyethylene pocket separators only in very exceptional cases to be considered as a cause of failure in starter batteries

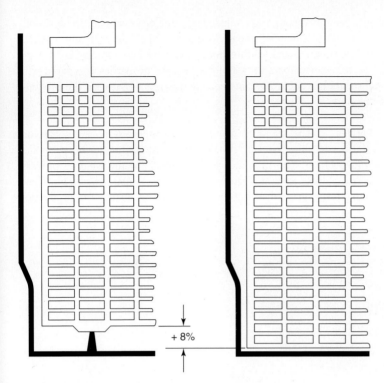

Figure 11.11 Grid comparison: conventional vs pocket construction. (Courtesy: VARTA Batterie AG.) Reprinted from W. Böhnstedt, Automotive lead/acid battery separators: a global overview, *J. Power Sources*, 1996, **59**, 45–50, with kind permission from Elsevier Science S.A., Lausanne [3].

[40, 49–51]. Here it has usually been the case of an atypical application, for example, a power supply in seasonal use on a boat or long-term deep discharges resulting in penetration shorts from the solution phase. Under extreme temperature conditions, as in the famous Las Vegas taxicab service, the battery life is severely reduced, but again the predominant failure modes are corrosion or worn-out positive electrodes and expander deterioration. One has to concede that under such extreme conditions the separators also approach their limits of stability [40] and less oxidation-stable versions can begin to shorten the battery life. The prevailing cost pressure has led to increasing use of thinner backweb, for example, 150 μm, in order to reduce raw material costs; this calls for a thorough evaluation of the limitations mentioned above [41].

In this connection the remaining oil in the separator plays an important role. At first glance, to increase the porosity a total extraction of the oil would be expedient, but certain oil components have been shown to exert a protective action on the polyethylene. Oil content and its distribution, as well as selection of the oil, thus gain particular significance [41, 52–54].

For problem-free processing, high tensile and puncture strengths are desirable. Especially when using expanded metal electrodes, sharp edges or points which may puncture the backweb and lead to shorts have to be taken into account. Even though the polyethylene separator is unique in these properties compared with conventional separators, these are considerable differences between the products of various suppliers.

11.2.2.1.4 Profiles The standard profile for microporous pocket separators exhibits continuous longitudinal ribs 10–12 mm apart, which determine the total thickness (Figure 11.12). The margin area, used later for the welding process, generally has ribs of lower height (Figure 11.13) or only a thicker backweb. These measures facilitate the mass distribution in calendering during the production

Figure 11.12 Polyethylene separator: pockets.

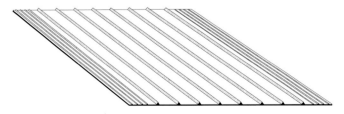

Figure 11.13 Polyethylene separator: standard profile.

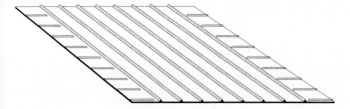

Figure 11.14 Polyethylene separator: cross-ribs in the margin area.

process, and apart from this they protect the particularly exposed edges of the pockets during the life of the battery. One noteworthy version of a profile has recently been presented by providing cross-ribs within the margin area (Figure 11.14) to keep the backweb in this area always at a safe distance from the grid edge of the positive electrode; the oxidizing substances originating there will thus do less damage [55, 56]. A similar protection is offered by profiles with a continuous rib pattern, that is, extending also into the welding zone, be it as narrow vertical ribs or especially as sinusoidal ribs [57].

The narrow tolerances to be maintained for the total separator thickness are tightened even further by the trend toward high-performance batteries with many thin electrodes and therefore many separators also. One can easily calculate that for, say, 10 or more electrodes and an equal number of separators per cell, the permitted tolerances become very small for fitting the electrodes/separators stack into the cell container. With electrodes and separators being produced continuously, that is, the thickness of consecutive individual pieces all having the same tendency, this means that if they are too thick, the stack does not fit into the cell without great pressure; if they are too thin, there is the danger of the electrode stack suffering in service due to vibration. As one solution, compressible ribs have been proposed [58] with groups of three ribs of which the middle one which is not back-to-back, on the opposite side of the backweb, generates a spring effect, balancing the tolerances and fixing the electrode stack within the container by its resilience (Figure 11.15).

The desire for cost savings starts with utilization of material. Is the continuous vertical rib necessary? Interrupted rib versions [56] or so-called dimples [47] have been proposed repeatedly, but they have not succeeded because production or

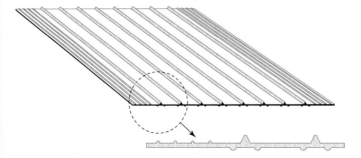

Figure 11.15 Polyethylene separators: compressible rib design.

Figure 11.16 Polyethylene separator: intermediate vertical ribs.

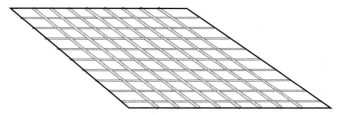

Figure 11.17 Polyethylene separator: cross-rib design.

processing problems in practice could not be justified by the minor cost advantages. The trend toward thinner backwebs has already been mentioned several times; it leads to a significant decrease in separator stiffness and thus possibly to processing problems. This loss in stiffness concerns the cross-direction more seriously than the longitudinal one, which is supported essentially by its high vertical ribs [41]. To regain stiffness, additional small longitudinal ribs between the main ones have been proposed (Figure 11.16) [47], and, for the far more seriously affected cross-stiffness, additional flat ribs across the main rib direction can help (Figure 11.17) [59]. These enable the cross-stiffness of a backweb of almost twice the thickness to be maintained without obstructing the escape of charging gases.

The selection of suitable profiles improves the efficiency of processing on pocketing machines. Experience has shown that, even with an optimum machine adjustment, the pocketing material, by its profile design and the strict adherence to tolerances, contributes essentially to the quantity and quality of the pockets produced.

11.2.2.1.5 Product Comparison Table 11.5 shows typical values for polyethylene pocket materials; of course, for the various producers [60–64] (Grace Gmbh (unpublished results)) they vary slightly owing to differences in formulation and process. An exact comparison is also difficult, since not all producers stating tolerances respectively clarify their statistical base.

11.2.2.2 Leaf Separators

The term '*leaf separator*' characterizes the customary stiff version of a starter battery separator that can be inserted individually between the electrodes on automatic stackers, in contrast to pocket separators. This processing requires considerably

Table 11.5 Microporous polyethylene pocket separators.

Brand name	Backweb thickness[a] (mm)	Electrical resistance (mΩ cm^2)	Oil content (%)	Porosity (%)	Puncture strength (N)	Supplier
Daramic II	0.20–0.25	60	12 ± 3	60	9	Daramic, Inc [60, 61]
Daramic High Performance	0.15–0.20	50	12 ± 3	60	6	Daramic, Inc [60, 61]
EMARK	0.20–0.25	55	12 ± 3	60	6	Entek Int. LLC [62, 63]
RhinoHide	0.15	50	13 ± 3	60	6	Entek Int. LLC [62, 63]
Junfer PE Separator SLI	0.25	90	13 ± 3	60	n.a.[b]	Jungfer GmbH & Co. KG [64]

[a] Other backweb thicknesses request.
[b] n.a., not available.

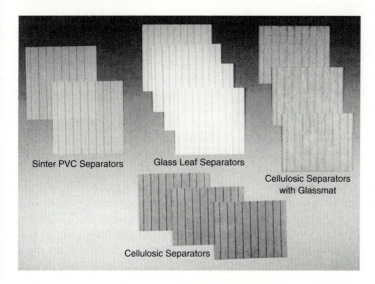

Figure 11.18 Leaf type separators.

higher bending stiffness than for pocket separators, calling for thicker backwebs, typically 0.4–0.6 mm (Figures 11.18 and 11.19).

11.2.2.2.1 Sintered PVC Separators The first synthetic separator is still in use today in some geographical areas, for two reasons: this separator is unchallenged in its low raw-material and production costs and it shows good stability against oxidation at elevated temperatures and vibrations. However, its processing is difficult

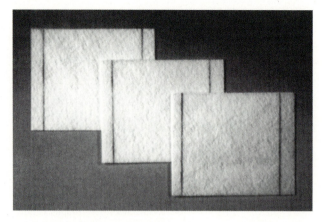

Figure 11.19 'Japanese separators.'

and its brittleness leads to higher scrap rates. Tender treatment, preferentially by manual labor, and an increased quality control effort may be justifiable at low labor rates. Sintered PVC separators arc thus still widely used in China, India, Russia, and some Asia-Pacific countries as well as around the Mediterranean Sea.

Sintered PVC is not at all one uniform product; large differences in properties and quality are possible. Experience has shown that premium qualities require significantly higher production costs.

The production process is comparatively simple, even though – of course – the respective know-how is also decisive. The equipment for the production of sintered PVC separators is suitable in size and production capacity to be operated on its own by individual, medium-sized, starter battery plants, in contrast to the far larger units required for the production of polyethylene pocket material.

Fine-grained PVC powder is spread onto a flat steel transport belt and, by means of a doctor knife, brought into the desired profile, which is generally quite a thin sheet of 0.3–0.6 mm thickness with vertical ribs. While passing through a sintering oven the surface of the PVC grains is just barely molten, causing neighboring particles to stick together (cf. Figure 11.2); the remaining void spaces within this spherical packing are the resulting porosity. Finally the product is slit and chopped into the dimensions required. In an alternative version of the process, the thin, sintered sheet produced initially is embossed in a second step between heated calender rolls to achieve the requisite total thickness.

Whereas a maximum number of contact points between PVC grains is desired to achieve mechanical stability, this prevents higher porosities. Typical values for porosity are 30–35%; therefore the electrical resistance is rather high, that is, 170 mΩ cm^2, despite thin 0.3 mm backwebs for top qualities. As mentioned, the range is very wide – even considerably higher electrical resistances are sometimes acceptable, for example, in areas where cold crank performance is of no significant importance.

Typical pore size distributions result in mean pore diameters of around 15 μm. Even long and intensive efforts did not succeed in decreasing this value decisively in order to enable production of microporous pocketing material resistant to penetration (Grace GmbH (unpublished results)) [66]. In practice, PVC separators prove themselves in starter batteries in climatically warmer areas, where the battery life is, however, noticeably reduced because of increased corrosion rates at elevated temperature and vibration due to the road condition. The failure modes are similar for all leaf separator versions: shedding of positive active mass fills the mud room at the bottom of the container and leads to bottom shorts there, unless – which is the normal case – the grids of the positive electrodes are totally corroded beforehand.

In many countries starter batteries are almost 100% recycled; PVC separators can cause some problems here [67]. A prior separation of PVC from other battery components, which is quite tedious, would be desirable, because a PVC content decreases the recycling purity of the container polypropylene and makes further processing of this plastic more difficult. Also, any chlorine compounds liberated can form environmentally hazardous products with other substances; the usual remedy is to install costly filter stations, with the residues representing possibly toxic wastes requiring special disposal methods.

Sintered PVC separators are frequently produced only for captive consumption; beyond that there are specialized producers for these separators [64, 67, 68] and for equipment for their production [64]. The data compiled in Table 11.6 comprise only premium products of independent producers.

11.2.2.2.2 Cellulosic Separators

Cellulosic separators, the closest relatives to the wood veneer, surprisingly have retained some of its properties, which differentiates these separators from purely synthetic ones: primarily a positive effect in reducing the water loss in starter batteries [39, 69–71]. This impact tends to decrease as the antimony content in the alloys is lowered, but it still represents an advantage over other leaf separators, unless a microporous pocket is required by the alloy anyway.

A voluminous, highly porous, special paper of cotton linters, or other premium α-cellulose fibers is passed through an immersion bath of aqueous phenol–formaldehyde resin solution and dried. A different process combines the production of the paper directly with its impregnation. Common to both processes

Table 11.6 Sintered PVC separators.

Brand name	Backweb thickness (mm)	Electrical resistance[a]	Porosity (%)	Pore size (average) (μm)	Supplier
Accuma PVC	0.20–0.30	170	30	30	Accuma S.p.A. [68]
ICS LR type	0.30	170	37	<25	ICS S.p.A. [69]
Jungfer LJF	0.25–0.30	170	33	<30	Jungfer GmbH & Co. KG [64]

[a] Electrical resistance of a separator of 1.3 mm overall thickness.

is the coating of cellulose fibers with a very thin layer of phenol–formaldehyde resin, largely protecting them from acid or oxidative attack. The pore structure of the separator is predetermined by the paper; to increase the porosity further, glass fibers may be mixed in during paper production. In a second step, inside a curing oven the phenolic resin is crosslinked at elevated temperature, and finally ribs of thermoplastic polymers are applied to achieve the desired total thickness. Some versions have the backweb embossed with longitudinal corrugations with plastic coated surfaces for better oxidation stability, since they are placed in the battery directly against the positive electrode. Slitting and chopping processes set the required dimensions for the product. Cellulosic separators show a high porosity (70–75%) and thus also low electrical resistances (100–150 mΩ cm^2 according to quality and producer). It is the preferred leaf separator in climatically more moderate areas, where cold crank performance is of importance. The water-loss-reducing properties of the cellulosic separator have made it possible to meet the criteria of the standards for maintenance-free starter batteries, even when using easy-to-cast antimony alloys (1.8–2.5%).

A thorough study (cf. Ref. [40]) of failure modes in practice has shown that with this form of separation also the cause of failure has not been the separator; the usual failure modes for leaf-type separators, as they have been described for sintered PVC separators, apply here as well.

Table 11.7 shows typical values for different qualities of cellulosic separators from various producers.

11.2.2.2.3 Glass Fiber Leaf Separators

Glass fiber leaf separators in the USA – especially at one large manufacturer – were for over a decade, between 1980 and 1995, an intermediate in the transition from conventional leaf separator to microporous pocket. The web is produced from glass fibers of suitable quality (C-glass) and of various diameters (mainly from 3 μm to around 10 μm) on a special paper machine. Even though an impregnation for protecting the fibers is not required, a small quantity of phenolic or acrylate resin is nevertheless applied to achieve the desired bending strength. A thermoplastic rib is added in the usual way. Glass fiber leaf separators are distinguished by very high porosity (80–85%)

Table 11.7 Cellulosic separators.

Brand name	Backweb thickness (mm)	Electrical resistance (mΩ cm^2)	Porosity (%)	Pore size (average) (μm)	Supplier
Arniorib-L	0.55	140	70	25	Daramic, Inc. [60]
Darak 101[a]	0.50	110	75	22	Daramic, Inc. [60]
Axohm A 428	0.60	210	70	20	Iydall Axohm [72]
Axohm A 438[a]	0.55	140	75	23	Iydall Axohm [72]

[a] These types consist of a cellulosic/glass fiber blend paper.

Table 11.8 Glass fiber leaf separators.

Brand name	Backweb thickness (mm)	Electrical resistance (mΩ cm^2)	Porosity (%)	Pore size (average) (μm)	Supplier
Axohm 10 G+	0.70	65	85	27	Iydall Axohm [72]

and very low electrical resistance (65 mΩ cm^2). The battery performance meets expectations: very good cold crank data attract attention, and the water loss is comparable with that of PVC but exceeds that of cellulosic separators significantly; the separator is not a cause of battery life limitation.

There is at present only one producer of this type of separator left; typical data are shown in Table 11.8.

11.2.2.2.4 Leaf Separators with Attached Glass Mat Even though this version is not a distinct type of separator, this section is dedicated to it. To all leaf-type separators described, a glass mat can be applied on the side directed toward the positive electrode, which is usually fixed by an adhesive coated onto the ribs (cf. Figure 11.18). This raises the cost of the separator and is usually not required for starter batteries used under normal service conditions, but it holds the positive active mass better inside the electrode and thus prevents premature shedding. It is especially important for batteries subject to severe vibrations or encountering frequent deep discharges. Typical applications include construction machinery batteries or the area bordering on cycling applications, such as marine batteries, truck and off-road vehicle batteries, electric lawn mowers, golf carts, or other small traction batteries.

11.2.2.2.5 'Japanese' Separators The development of the starter battery in Japan has taken an independent course (see Section 11.2.1.2), visibly expressed by the separator's thick glass mat and its lack of spacing ribs (cf. Figure 11.19). The cellulosic backweb impregnated with phenolic resin, generally in use until around 1980 and largely identical to the separator of the same type already mentioned has been completely replaced by thin ($\sim$0.3 mm) fleece materials made of organic fibers.

Since the glass mat supplies sufficient stiffness, high backweb thickness was no longer needed! These fleeces are made of organic fibers (polyester and polypropylene, as well as so-called 'synthetic pulp,' i.e., fibrillated polypropylene) on paper machines.

The basic materials are sufficiently stable in sulfuric acid not to require the expensive phenolic resin impregnation. Traces of adhesive are applied to hold the glass mat in order to achieve the total thickness. This separation system may be expensive to manufacture, a fact certainly largely balanced by savings in positive active mass, but it also has some indisputable advantages.

The electrical resistance, at 60–90 mΩ cm^2, is astonishingly low, because the backweb is only 0.25–0.30 mm thick and the glass mat with its porosity in excess of 90% also contributes only a little. In some types of construction the low electrical resistance cannot be fully utilized, however, due to a tendency for gas to be trapped within the glass mat.

The oxidative stability is excellent. Direct contact between the glass mat and the positive electrode produces a far lower tendency to shed active mass; thus as a general rule the failure mode is positive grid corrosion.

It is rather difficult to compare constructions differing to such an extent, since in the course of development the standards and also the electric layout of vehicles have been adapted to accommodate to the products available. Table 11.9 shows typical data for 'Japanese' separators. From the above it can be clearly seen that a direct comparison or even an exchange for other leaf separators is almost impossible.

11.2.2.2.6 Microfiber Glass Separators

Even though this separation system has not yet entered the starter battery field, it should be discussed here as a possible option for the future.

Microfiber glass fleece mats are typically produced from a blend of 20–30% glass microfibers <1μm in diameter, with the balance of the glass fibers thicker (3–10 μm) and longer (cf. Figure 11.1), on a specialized paper machine (Foudrinier), since this is the only way of achieving the desired tensile strength without binder. The material is supplied in roll form, even though it is normally not processed into pockets, which are not required due to the absence of free electrolyte. The classification here as a leaf separator should be seen in this sense.

The microfiber glass separators have to fill the space between the electrodes completely; the backweb thickness is thus identical to the total thickness. Due to the high compressibility of such porous glass mats, a standard measuring pressure of 2 or 10 kPa (BCI method) is generally used; during assembly they are compressed by an additional 25% of their nominal thickness to make it possible to match the volume changes of the electrodes during charging and discharging; otherwise dry spots could be formed causing performance losses. Characterization

Table 11.9 Synthetic pulp–glass mat separators ('Japanese' separators).

Brand name	Backweb thickness (mm)	Electrical resistance (mΩ cm^2)	Pore size (average) (μm)	Supplier
GSK – CV	0.30	90[a]	18	GS KASEI KOGYO K.K. [73]
GSK – MS	0.25	30[a]	23	GS KASEI KOGYO K.K. [73]
GSK – TC	0.25	60[a]	10	GS KASEI KOGYO K.K. [73]
GSK – SI	0.50	150[a]	25	GS KASEI KOGYO K.K. [73]

[a] Electrical resistance without glass mat, which adds approx. 30 mΩ cm^2.

of microfiber glass separators by their thickness alone has proven to be ambiguous; therefore the preferred method is by area weight. For a typical separation thickness of 1 mm, glass fiber mats of 200 g m^{-2} are used. Resulting from the extremely high porosity of more than 90% the measured electrical resistance is extremely low, but the difference in the pore spectrum inside the battery due to compression during installation has to be taken into account; moreover, not all pores must be acid-filled, in order not to block the oxygen transfer. The actual electrical resistance 'experienced' by the battery is in the order of magnitude of other modern separation systems (50–70 mΩ cm^2). An excellent description of these relationships exists in the literature [23].

Despite all the efforts over many years to establish this 'sealed' construction in starter batteries, field results have been published only sparingly and they have not always been satisfactory. Cold crank results are very good; despite using lead–calcium, the cycle life – at least in laboratory tests – has been found to be very good [74], probably due to the mat support of the positive active mass. In the day-today practice, other influences, such as insufficient recharging and microshorts as a result of deep discharges or valve leaks, appear to lead to premature sulfation of the negative electrode and eventually to capacity deterioration. Improved constructions are continually being presented and tested on a large scale. Besides some open technical questions, the cost structure also has prohibited a wider introduction to date: sturdier containers and more precise electrode geometry, voluminous separators, and reliable valves, an expensive filling process, and last but not least temperature-controlled charging management could only be justified with difficulty in times of cost trimming within the automotive industry.

The range of microfiber glass mat separators offered by the leading producers is presented in Section 11.2.3.3 with typical data in connection with their predominant application in sealed stationary batteries.

11.2.2.3 Comparative Evaluation of Starter Battery Separators

The individual starter battery separator systems have been described; here they are evaluated comparatively. There are no standards for evaluating separators! Therefore the comparison will be concentrated primarily on the effects on the performance of the starter battery, with other decisive criteria such as cost structure and effects on productivity indicated.

Cold crank performance, battery life expectancy, and freedom from maintenance are generally co-affected by the separators, whereas ampère-hour capacity remains largely unaffected at a given separator thickness. The properties of the different leaf and pocket separators are compared in Table 11.10. These typical separator properties (lines 1–4) are reflected in the electrical results of battery tests (lines 5–8). The data presented here are based on the 12 V starter battery standard DIN 43 539–02; tests based on other standards lead to similar results.

The cold crank voltage is directly affected by the separator electrical resistance (cf. Section 11.1.2.3; Figure 11.5), but to a much smaller extent than is normally assumed. Nevertheless the effect of low electrical resistance of the separator is not

Table 11.10 Comparative evaluation of starter battery separators.

	Polyethylene pocket separators	Sintered PVC separators	Cellulosic separators	Cellulosic/glass mix separators	Glass fiber leaf separators
Backweb thickness (mm)	0.25	0.30	0.55	0.50	0.70
Pore size (average) (µm)	0.1	15	25	22	27
Acid displacement ($cm^3 m^{-2}$)	120	210	170	140	110
Electrical resistance ($m\Omega\ cm^2$)	60	160	140	100	65
Test results (DIN 43 539-02) (PbSb 1.6/PbSb 1.6)					
Cold crank voltage (V)	9.40	9.20	9.25	9.30	9.35
Cycle life test (weeks)	>10	>10	>10	>10	>10
Accelerated cycle life test (weeks) (DIN 43 539 E-1980)	>10	5	6	6	7
Water consumption[a] ($g\ Ah^{-1}$)	2–4	4	2	2	4

[a] 504 h overcharge with 14.4 V at 40 °C.

to be underestimated, since it often presents the only way of meeting the acceptance criteria of high-performance batteries reliably.

The life expectancy of a starter battery, according to DIN 43 539–02, is not affected by the separator. Experience has shown that all batteries with modern standard separators in cycle life tests not only last the required 5 weeks of cycling, but mostly 10 weeks or more. The cause of failure is typically positive grid corrosion – in good agreement with practice [49]. This fact supports this standard, which is supposed to reflect reality, but on an accelerated time scale.

A modification of the test conditions (DIN 43 539-02 E) did prove to show more about the effect of the separator on battery life expectancy. In Figure 11.20 the weekly cycling regimes of these two standards are compared. DIN 43 539-02 E, requiring more discharges at a higher temperature, has been shown to uncover significant differences between the individual types of separators. Macroporous separators with average pore sizes of 10–30 µm like PVC, cellulosic, or glass separators, just meet the required cycle life. Microporous pocket separators with pore sizes distinctly below 1 µm prevent not only penetration through the separator, but also – because of the pocket – bottom or side shorts, and this to such an extent that even under these aggravated test conditions the separator does not limit the cycle life duration.

Surprisingly the water consumption of a starter battery, provided that it contains antimonial alloys, is affected by the separator. Some cellulosic separators as well

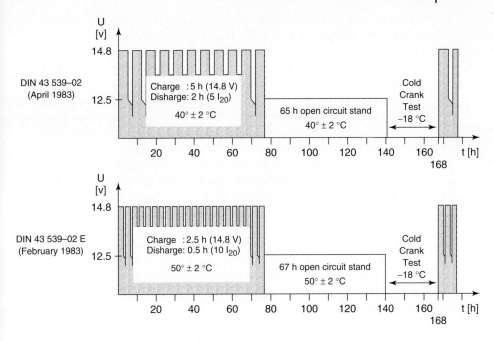

Figure 11.20 DIN standards: weekly cycling regimes.

as specially developed polyethylene separators (e.g., DARAMIC V [75]) are able to decrease the water consumption significantly. The electrochemical processes involved are rather complex, and a detailed description is beyond the scope of this chapter. Briefly, the basic principle behind the reduction of water loss by separators is their continuous release of specific organic molecules, for example, aromatic aldehydes, which are selectively adsorbed at antimonial sites of the negative electrode, inhibiting there the catalytic effect of antimony on hydrogen evolution and thus lowering the water consumption [69, 70]. The current trend toward low-antimony or lead–calcium alloys – primarily for productivity reasons – reduces the importance of these effects; nevertheless, they remain decisive in many instances.

The above comparative evaluation of starter battery separators refers to moderate ambient temperatures: the standard battery tests are performed at 40 or 50 °C. What happens, however, on going to significantly higher temperatures, such as 60 or 75 °C? This question cannot be answered without considering the alloys used: batteries with antimonial alloys show a water consumption that rises steeply with increasing temperature [40], leaving as the only possibilities for such applications either the hybrid construction, that is, positive electrode with low-antimony alloy, negative electrode lead–calcium, or even both electrodes lead–calcium.

Because of the increased shedding with these alloys, pure leaf separation is hardly suitable. Separations with supporting glass mats or fleeces as well as microfiber glass mats provide technical advantages, but are expensive and can be justified only in special cases. Also under these conditions of use the microporous polyethylene pocket offers the preferred solution [40]. Lower electrical properties at higher temperatures, especially decreased cold crank duration, are battery-related; the choice of suitable alloys and expanders gains increased importance.

However, it has to be conceded that after battery life cycle tests at such temperatures polyethylene separators also reach their limits (although this fact does not yet reflect in failure-mode studies [49]) even in locations with extreme ambient temperatures. The tendency toward using ever-thinner backwebs cannot be continued, however, without seeking protective measures. Suitable provisions have to be made, especially with respect to the separator's oxidative stability at elevated temperature. The leading producers of polyethylene separators have recently presented solutions [41, 47] which, even at 150 µm backweb, provide for oxidative stability and puncture strength in excess of that for the standard product at 250 µm backweb [41].

Without any doubt, the microporous polyethylene pocket will meet all requirements of modern starter batteries for the foreseeable future. Whether and to what extent other constructions, such as valve-regulated lead–acid (VRLA) batteries, other battery systems, or even supercapacitors, will find acceptance, depends – besides the technical aspects – on the emphasis which is placed on the ecological or economical factors.

11.2.3
Separators for Industrial Batteries

11.2.3.1 Separators for Traction Batteries

Traction batteries are the workhorses among batteries; day in, day out they have to perform reliably, that is, for years. They are discharged to about 80% by their nominal capacity, typically during an 8 h shift of a forklift, and are recharged during the remaining hours of the day. A life of 1500 cycles or five years is taken for granted, with concession regarding the life expectancy only made under extreme conditions.

It can be stated generally that requirements for traction battery separators in respect to mechanical properties and chemical stability are considerably higher than those for starter battery separators. This is due to the fact that a forklift battery is typically operated for about 40 000–50 000 h in charge–discharge service, whereas a starter battery for only about 2000 h. The requirements for electrical resistance are lower because of the typically lower current densities for traction batteries. These differences are of course reflected in the design of modern traction battery separator material.

11.2.3.1.1 Polyethylene Separators
A detailed description of the production process and the properties of polyethylene separators can be found in Section 11.2.2.1, so only the modifications which are important for traction battery separators are covered here.

Industrial battery separators are often supplied in cut-piece form, that is, they have to have a certain stiffness and robustness in order to withstand the assembly into cells, with electrodes weighing several kilograms, without damage. Modern polyethylene traction battery separators have backwebs of about 0.50–0.65 mm, that is, about three times the backweb thickness of starter battery separators with the respective effect on stiffness, electrical resistance, and also production process line speed. The larger backweb thickness combined with a higher oil content (~15–20%) gives the separator the required oxidative stability, which to a first approximation is proportional to the product of backweb thickness and its oil percentage. A somewhat lower porosity and thus lower acid availability are the consequences.

The microporosity is also important for this application, in order not to allow shorts through the backweb during battery life. Bottom shorts are avoided by a mud room of sufficient dimensions, and side shorts by plastic edge protectors on the frames of the negative electrode. Some manufacturers have switched to using sleeves of polyethylene separator material, rendering an edge protection superfluous. The use of three-side-sealed separator pocket in traction batteries should be avoided, because experience has shown this can lead to increased acid stratification, subsequent sulfation, and thus capacity loss.

The choice of a suitable oil has special importance. Besides beneficial effects of the oil on the oxidative stability of the separator, other consequences have to be considered. From the chemical mixture of which an oil naturally consists, polar substances may migrate into the electrolyte. Being of lower density than the electrolyte, they accumulate on its surface and may interfere, for instance, with the proper float function of automatic water refilling systems. Some oils which fully meet both of the above requirements have been identified, that is, they provide sufficient oxidation stability without generating black deposits [53].

An effect similar to the water loss in starter batteries is characterized as top-of-charge performance in traction batteries. Antimony is dissolved from the alloy of the positive electrode, migrates through the electrolyte, and is deposited on the negative electrode, where – because of its far lower hydrogen overvoltage than lead – it catalyzes hydrogen evolution, thus reducing the charging voltage at constant current during the overcharge period [76]. From long experience it is known that some separators are able to influence this behavior [77–80]. Many hypotheses have been proposed, examined, and discarded again; for the current status of the discussion reference should be made to the literature [69, 70]. Suitable additives, such as uncrosslinked natural rubber [81] or Voltage Control Additive (VCA [82]) allow significant improvement of the top-of-charge performance of batteries, helping polyethylene separators to gain acceptance in the great majority of applications.

Traction batteries are assembled either with pasted and glass mat-wrapped positive electrodes, as is the case predominantly in the USA, or with tubular positive plates, which prevail in Europe. The former electrodes place no particular requirement on the separator profile; vertical ribs on the positive side are standard. The construction with tubular positive electrodes preferably uses a diagonal (Figure 11.21) or sinusoidal (Figure 11.22) rib pattern. Insufficiently narrowly spaced supporting contact points between tube, rib, and separator backweb have

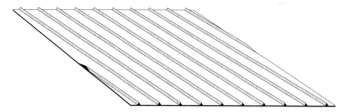

Figure 11.21 Polyethylene separator: diagonal industrial battery profile.

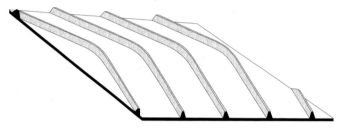

Figure 11.22 Polyethylene separator: sinusoidal profile.

shown the latter to yield to expansion of the negative electrode during cycling. Capacity deterioration by overexpansion and gas trapping results; thus a narrower rib spacing is desirable, but is limited by increased acid displacement. Interrupted ribs or even dotted spacers (dimples) on the backweb are under discussion.

Flexible polyethylene separators have facilitated a novel cell construction: the separator material, supplied in roll form, is wound so that it meanders around electrodes of alternating polarity (Figure 11.23), requiring ribs in the cross machine direction; such profiles are available commercially [60].

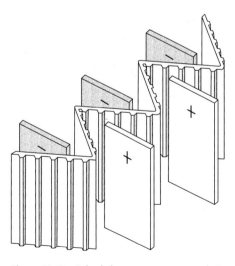

Figure 11.23 Polyethylene separator: meandering separation.

Finally, one development results from returning to a basic idea from the dawn of the lead–acid battery, wherein the functions of support for the positive active material and of the separator are combined into one component: the gauntlet separator [83], consisting of a coarsely porous, flexible support structure coated with microporous polyethylene material for separation. The future has to show whether this approach will be able to meet all demands.

Characteristic data for polyethylene separators in comparison with competing systems are discussed later in this section (Table 11.11).

11.2.3.1.2 Rubber Separators A thin layer of a mix of natural rubber, sulfur, precipitated silica, water, and some additives, such as carbon black and vulcanizing agents, is extruded on a paper support belt, calendered, and vulcanized as a roll in an autoclave under elevated pressure and temperature ($\approx 180\,°C$). A modified process extrudes and calenders a ribbed profile and crosslinks the rubber separator by irradiation.

Rubber separators have a relatively low porosity (~ 50–55%) and thus high acid displacement and electrical resistance. Furthermore, they are brittle and for this reason difficult to handle in larger sizes. In order to balance this disadvantage, an adjustment to a lower degree of crosslinking has been attempted; the result was a corresponding increase in susceptibility to oxidative attack.

Table 11.11 Separators for lead–acid traction batteries.

	Polyethylene separators	Rubber separators	Phenol–formaldehyde–resorcinol separators	Microporous PVC separators
Supplier	Daramic, Inc. [60, 61][a]	Amerace [86]	Daramic, Inc. [60, 61]	AMER-SILs[a] [86]
Brand name	DARAMIC Industrial CL	Ace-Sil	DARAK 5000	AMEK-SIL Standard
Backweb thickness (mm)	0.6	0.8	0.6	0.5
Pore size (average) (μm)	0.1	0.2	0.5	0.5
Acid displacement ($cm^3\,m^{-2}$)	320	450	235	250
Electrical resistance ($m\Omega\,cm^2$)	280	240	120	200
Initial capacity[b]	+	0	++	++
Handling properties[b]	++	0	+	+
Water consumption[b]	+	++	0	0

[a] Polyethylene industrial separators are also available from ENTEK International [62].
[b] Ranking: ++, very good; +, good; 0, acceptable; −, poor.

These disadvantages have led to an extensive displacement of rubber separators by polyethylene separators. Nevertheless, a few market segments exist, such as golf cart batteries – which for statistical reasons due to their construction are shown under SLI batteries – and traction batteries in severe heavy-duty service, especially at elevated temperatures, where rubber separators continue to be used. The reason is the top-of-charge behavior of traction batteries referred to above. Rubber separators are able to delay the process of antimony poisoning significantly. Its mechanism is based on uncrosslinked rubber components inhibiting hydrogen evolution [69, 82] in a similar manner to that described for the water loss of starter batteries. With a constant-voltage charging regime, this leads to a lower increase in charging current and lower water consumption [79]. A comparative tabulation of rubber separator properties can be found in Table 11.11 (see also Section 11.2.3.1.5 below).

11.2.3.1.3 Phenol–Formaldehyde–Resorcinol Separators (DARAK 5000)

An aqueous solution of phenol–formaldehyde resin and resorcinol (~70 vol% water) is crosslinked by means of organic catalysts [18] at ~95–98 °C between two continuous Teflon belts. With growing molecular weight, the water solubility of the phenolic resin decreases and a phase separation occurs. A three-dimensional phenol resin structure is generated, in the interconnected cavities of which the water accumulates which is evaporated in a subsequent process step, thus generating the porosity of some 70%. The Teflon belts mentioned contain grooves, which determine the rib geometry of the separators. Phenolic resin is a duroplast and thus brittle; a polyester fleece is incorporated into the separator backweb during the production process, resulting in a sufficient reinforcing effect. In contrast to the macroporous (phenolic-resin-impregnated) cellulosic separator, the pore size of the present microporous phenolic resin–resorcinol system is determined exclusively by process conditions and not by the reinforcing fleece.

A stiff, microporous separator is formed with a very narrow pore size distribution with an average of 0.5 µm – about 90% of all pores being between 0.3 and 0.7 µm in diameter!

The porosity, at 70%, is excellent, and it also achieves strikingly good values for acid displacement and electrical resistance for an industrial battery separator (0.12 Ω cm^2).

The stiffness of the duroplast is helpful in counteracting the tendency of the negative active material to expand during cycling, even at larger rib spacing. The prevailing profiles have vertical or diagonal ribs on the positive side, and on the negative side a low rib is frequently added for better gas release from the negative electrode.

Further details are given in the systems comparison under Section 11.2.3.1.5 below.

11.2.3.1.4 Microporous PVC Separators

A mixture of powdered PVC, cyclohexanone as solvent, silica, and water is extruded and rolled in a calender into a profiled separator material. The solvent is extracted by hot water, which is evaporated in an oven, and a semiflexible, microporous sheet of very high porosity (~70%) is formed [19]. Further developments up to 75% porosity have been reported [84, 85], but these materials suffer increasingly from brittleness. The high porosity results in excellent values for acid displacement and electrical resistance. For profiles, the usual vertical or diagonal ribs on the positive side and, as an option, low ribs on the negative side, are available [85].

11.2.3.1.5 Comparative Evaluation of the Traction Battery Separators

Which separator properties are important for use in traction batteries? For this aspect primarily the highly predominant application, namely forklift traction batteries, is to be considered: chemical resistance against attacks by acid and oxidation, mechanical stability for problem-free assembly, stiffness to counteract overexpansion of the negative active material, and low acid displacement are particularly desirable. Delay in antimony poisoning, absence or near-absence of oily deposits in the cells, and – last but not least – a low electrical resistance complete the requirement profile.

In Table 11.11 an attempt is made to include the above criteria in the form of quantitative data or qualitative evaluations.

Polyethylene separators offer the best balanced property spectrum: excellent mechanical and chemical stability as well as good values for acid availability and electrical resistance have established their breakthrough to be the leading traction battery separator. Rubber separators, phenolic resin–resorcinol separators, and microporous PVC separators are more difficult to handle than polyethylene separators; their lack of flexibility does not allow folding into sleeves or use in a meandering assembly; in addition they are more expensive.

Special applications are often governed by different priorities: as already discussed in relation to golf carts, the low water loss and the delay in antimony poisoning in heavy-duty service of a forklift are of eminent importance, with the result that rubber separators remain the preferred product there. Submarine batteries offer a different picture: the number of cycles to be reached is far lower (~500) and, due to the slow (~100 h) but very deep discharge, the acid availability becomes the decisive criterion, which favors, for example, the phenolic resin–resorcinol separator. Such requirements are already similar to the application in open stationary cells.

11.2.3.2 Separators for Open Stationary Batteries

Stationary batteries serve predominantly as an emergency power supply, that is, they are on continuous standby in order to be discharged for brief periods and sometimes deeply, up to 100% of nominal capacity, in the rare case of need. The following profile of requirements for the separator thus arises: very low electrical resistance, low acid displacement, no leaching of substances harmful to float-service, as well as an excellent mechanical and chemical stability, especially against oxidation at continuous overcharge, because such batteries have a life expectancy of 20–30 years.

11.2.3.2.1 Polyethylene Separators The production process for polyethylene separators (Section 11.2.1.1) as well as the characteristic properties (see Sections 11.2.2.1 and 11.2.3.1) have already been described in detail above. Deviating therefrom, the desire for low acid displacement has to be added for separators in open stationary batteries. This can be met either by decreasing the backweb thickness or by increasing the porosity; the latter, however, is at the expense of separator stability.

Stationary batteries, moreover, often have transparent containers, historically, probably to allow observation of the electrolyte level or the extent of shedding. Deposits of oily substances accumulating at the electrolyte surface due to their stickiness could gather lead particles and produce an unpleasantly dirty rim, which can be avoided by careful selection of suitable oils [53].

11.2.3.2.2 Phenol–Formaldehyde Resin–Resorcinol Separators (DARAK 2000/ 5005) The production process and the principal properties of this system have been described in detail in the section on traction battery separators (see Section 11.2.3.1). The outstanding properties, such as excellent porosity (70%) and resulting very low acid displacement and electrical resistance, come into full effect when applied in open stationary batteries. Due to the good inherent stiffness the backweb may even be reduced to 0.4 mm, reducing acid displacement and electrical resistance to low levels that are not achievable by any other system. Furthermore, the phenolic resin–resorcinol separator neither generates any harmful substances nor is it attacked chemically or by oxidation. The sum of these properties has made it the preferred separator for open stationary batteries.

11.2.3.2.3 Microporous PVC Separators Much of the above also holds true for the application of microporous PVC separators (see Section 11.2.3.1) in open stationary batteries. Very high porosity and thus low acid displacement and electrical resistance are also offered by this system. The relevant properties are compiled in Table 11.12.

Since the early days of using PVC separators in stationary batteries, there has been a discussion about the generation of harmful substances: caused by elevated temperatures or other catalytic influences, a release of chloride ions could occur which, oxidized to perchlorate ions, form soluble lead salts resulting in enhanced positive grid corrosion. Since this effect proceeds by self-acceleration, the surrounding conditions such as temperature and the proneness of alloys to corrosion as well as the quality of the PVC have to be taken carefully into account.

11.2.3.2.4 Sintered PVC Separators Sintered PVC separators for open stationary batteries are produced in the same way as the corresponding starter battery version (Section 11.2.2.2). Their brittleness and thus difficult processability are

Table 11.12 Separators for flooded lead–acid stationary batteries.

	Polyethylene separators	Phenol–formaldehyde–resorcinol separators	Microporous PVC separators	Sintered PVC separators	Rubber separators
Supplier	Daramic, Inc. [60, 61][a]	Daramic, Inc. [60, 61]	AMER-SIL s.a. [86]	Jungfer GmbH [64]	AMERACE Microporous Products [87]
Brand name	DARAMIC Industrial CL	DARAK 5005	AMER-SIL HP	Sintered PVC	Micropor-Sil M3
Backweb thickness (mm)	0.5	0.5	0.5	0.5	0.50
Pore size (average) (µm)	0.1	0.5	0.5	15	0.25
Acid displacement ($cm^3\,m^{-2}$)	280	200	220	350	n.a.[c]
Electrical resistance ($m\Omega\,cm^2$)	240	110	150	300	150
Initial capacity[b]	0	++	++	0	++
Handling properties[b]	++	++	+	0	+
Black deposits[b]	+	++	++	++	++

[a] Polyethylene industrial separators are also available from ENTEK International 1621.
[b] Ranking: ++, very good; +, good; 0, acceptable; –, poor.
[c] n.a.: not available.

disadvantages, as is their relatively low porosity; the concerns about release of chloride ions and subsequent increased corrosion are to be considered here as well. On the other hand, they are unrivalled in low cost, even up to extreme overall thickness (up to 5 mm). Since at these thicknesses electrical resistance and acid displacement by the back-web have a relatively low impact, there is a remaining niche for the application of sintered PVC separators.

11.2.3.2.5 Comparative Evaluation of Separators for Open Stationary Batteries

Table 11.12 shows the physicochemical data of separators used in open stationary batteries. Since the emphasis is on low acid displacement, low electrical resistance, and high chemical stability, the phenolic resin–resorcinol separator is understandably the preferred system, even though polyethylene separators, especially at low backweb, are frequently used. For large electrode spacing and consequently high separation thickness, microporous as well as sintered PVC separators also find use.

11.2.3.3 Separators for Valve Regulated Lead–Acid (VRLA) Batteries

11.2.3.3.1 Batteries with Absorptive Glass Mat

VRLA batteries are frequently also somewhat misleadingly called *sealed* or *recombinant* batteries. Their operating principle is – as mentioned already – based on oxygen, which is generated during charging at the positive electrode and is able to reach the negative electrode internally and be reduced there again. The negative electrode thus becomes partially discharged, so that it does not enter the overcharge phase, that is, it does not lead to hydrogen evolution. No water consumption occurs; viewed externally, the total charging current is transformed into heat. For a more detailed description of the system, the literature [7, 23–27, 87–97] should be consulted.

What requirements are placed by this construction on the separator? First, the free mobility of the electrolyte has to be hampered in order to maintain tiny open channels for oxygen transfer from the positive to the negative electrode. One solution to this problem is the use of highly porous microfiber glass mats as separators. This glass mat has to fill the space between the electrodes completely and absorb a maximum amount of electrolyte. These requirements imply extremely high porosity (>90%), large internal surface area, and good wettability to assure a high absorption for the electrolyte. Starting from a fibrous structure, a large internal surface means a fiber diameter as small as possible: glass fibers of 0.5 μm reach around $3 \, m^2 \, g^{-1}$, whereas 10 μm fibers have only some $0.15 \, m^2 \, g^{-1}$ of surface. The good wettability of glass fibers suffers if binder is used. Of course, the separator has to have long-term resistance against various kinds of chemical and electrochemical attack inside a lead–acid battery, and its susceptibility increases with the internal surface! It must not generate substances that increase the gassing rate, corrosion, or self-discharge. Finally, it has to be mechanically robust enough to be handled during the battery production process. Sharp corners or edges should not be able to penetrate it. This last demand competes, of course, with the desire for the least possible binder content.

These generally defined requirements are met quite comprehensively by microfiber glass fleeces. These are blends of C-glass fibers of various diameters, which are processed in the usual way on a Foudrinier paper machine into a voluminous glass mat. The blending ratio gains special importance since cost aspects have to be balanced against technical properties. The expensive microfibers below 1 μm in diameter (∼20–30% share) give a large internal surface and the desired pore size distribution, but do not contribute substantially to the mechanical properties. Fibers of significantly larger diameter increase the tensile strength and thus the processability, but tend to break more easily when the glass mat is under compression, as it needs to be to maintain at all times sufficient contact with the electrodes as they contract and expand during charging and discharging respectively.

The conventional requirements of a separator are met fully by microfiber glass mats: the extreme porosity guarantees – in spite of the free volume for oxygen transfer of about 10 vol% – that acid displacement and also electrical resistance remain very low, even though they are significantly higher (about double [23]) than is indicated by values measured on fully soaked samples. The chemical and oxidative stability is very good. The dimensional stability of absorptive microfiber glass fleeces is a critical parameter. On one hand, during the production of these fleeces the thickness (i.e., weight per unit area and fiber distribution/cloudiness) has to be maintained within narrow limits in order to assure a uniform distribution of electrolyte and subsequently of the depth of discharge at the assembly pressure. On the other hand, the resilience arising from the assembly pressure must not be noticeably reduced by fiber fracture or drying-out.

The small pore size and the uniform distribution result in capillary forces which should allow wicking heights and thus battery heights of up to 30 cm. Due to the cavities required for gas transfer and under the effect of gravity, the electrolyte forms a filling profile, that is, fewer cavities remain at the bottom than at the top. Therefore with absorptive glass mats a rather flat battery construction is preferred. Another reason for this is acid stratification: since the electrolyte is still liquid, and acid of higher density formed, for example, during charging will diffuse downwards – even at a delayed pace – this may detrimentally affect especially any deep-cycling service. Furthermore, due to the severe acid limitation of such cells during deep discharge, lead sulfate will dissolve increasingly, and during recharge – and thus at higher acid density – it is again precipitated and can lead after reduction to microshorts. This effect is partially counteracted by the addition of sodium sulfate to the electrolyte. Nevertheless, sealed batteries with microfiber glass fleece separation are therefore predominantly used in service that rarely involves deep-discharge cycles. A special development, the addition of a low percentage of organic fibers to microfiber glass fleeces [98], allegedly simplifies the acid filling; excess acid is removed simply by dumping. Due to their hydrophobicity, the organic fibers facilitate the oxygen transfer, and they should suffice to weld such fleeces into pockets. Reports of practical experience have not yet been published.

Developments to produce such absorptive mats totally from organic fibers even go one step further. Only recently success came in achieving a suitable fiber diameter and permanent hydrophilization [99]. Such materials are not yet commercially available, however, and field experience has not been reported as yet.

Table 11.13 compares the specification data of microfiber glass fleeces from various manufacturers.

11.2.3.3.2 Batteries with Gelled Electrolyte

An advanced solution to the problem of decreasing the free mobility of the electrolyte in sealed batteries is its gel formation. By adding some 5–8 wt% of pyrogenic silica to the electrolyte, a gel structure is formed due to the immense surface area (~ 200–$300\,m^2\,g^{-1}$) of such

Table 11.13 Separators for valve-regulated lead–acid batteries (liquid electrolyte).

	Absorptive microfiber glass mat separators[a] (100% glass fibers[b])					
Supplier	B. Dumas S.A. [100]	Hollingsworth & Vose Co. [101]	Lydall Axohm [72]	Nippon Glass Fiber Co., Ltd. [102]	Technical Fibre Prod. Ltd. [103]	Whatman Int. Ltd. [104]
Brand name	2133 XP	05 series	AXΩMAT	MS type	40101 series	SLA 1250
Thickness at 10 kPa (mm)	1.33	1.35	1.30	1.25	n.a.	1.35
Grammage (g m^{-2})	210	200	200	200	n.a.	200
Tensile strength (N/15 mm)	n.a.	n.a.	7.65	7.5	9.0	11.2
Porosity (%)	94.5	n.a.	>93	n.a.	>94	n.a.
Pore size (average) (μm)	5.5	n.a.	7.5	10	n.a.	n.a.

[a] AMER-SIL. s. a. [86] has recently introduced a microfiber glass fleece separator ('AMER-GLASS').
[b] Dumas ('Series 6000'), H & V ('Hovosorb II'), Technical Fiber Products ('Polymer Reinforced Sealable Separator'), Nippon Glass Fiber ('MFC'), and Whatman also offer products with organic fibers and/or binders.

silicas, which fixes the sulfuric acid solution molecules by van der Waals bonds within a lattice. These gels have thixotropic properties; that is, by mechanical stirring they can be liquefied and used to fill into the battery cells, where they gel again within a few minutes. Initially such batteries suffer some water loss during overcharge. The gel dries to some extent and forms cracks, allowing the oxygen to transfer to the negative electrode where the internal oxygen consumption occurs, which avoids further water loss and gel drying [20, 21, 105, 106].

Batteries with gelled electrolyte have been shown to require a separator, in the conventional sense, to secure spacing of the electrodes as well as to prevent any electronic shorts; the latter is achieved by microporous separators. An additional important criterion is minimal acid displacement, since these batteries – in comparison with batteries with liquid electrolyte – lack the electrolyte volume share taken up by gelling and by the cracks.

Among the separator varieties described, the phenol–formaldehyde–resorcinol separator (DARAK 2000) [60] as well as the microporous PVC separator [85] have proven effective for this construction. For applications without deep discharges, concessions may be made with the respect to porosity and pore sizes of the separator; therefore polyethylene separators or a special version of glass leaf separators with attached glass mat [72] are occasionally used in such cases.

Table 11.14 compares the most important physicochemical data of separators used in batteries with gelled electrolyte.

Table 11.14 Separators for valve-regulated lead–acid batteries (gelled electrolyte).

	Phenol–formaldehyde–resorcinol separators	Microporous PVC separators	Glass fiber/polyester fiber separators	Rubber separators
Supplier	Daramic, Inc. [60, 61]	AMER-SIL s.a. [86]	Lydall Axohm [72]	AMERACE Microporous Products [86]
Brand name	Darak 2003 Ind. With glassmat	DGT 200 HP	Standard D. S. R.	Micropor-Sil
Backweb thickness (mm)	0.3	0.55	0.7	0.4
Porosity (%)	70	71	85	70
Pore size (average) (μm)	0.5	0.2	27.5	0.1
Acid displacement ($cm^3\,m^{-2}$)	145	260	n.a.	n.a.
Electrical resistance ($m\Omega\,cm^2$)	120	160	170	140

11.3
Separators for Alkaline Storage Batteries

11.3.1
General

In acidic electrolytes, only lead, because it forms passive layers on the active surfaces, has proven sufficiently chemically stable to produce durable storage batteries. In contrast, in alkaline media there are several substances basically suitable as electrode materials: nickel hydroxide, silver oxide, and manganese dioxide as positive active materials may be combined with zinc, cadmium, iron, or metal hydrides. In each case potassium hydroxide is the electrolyte, at a concentration – depending on battery systems and application – in the range of 1.15–1.45 g cm^{-3}. Several electrochemical couples consequently result, which are available in a variety of constructions and sizes, with an even larger variety of separators, of course.

For alkaline storage batteries, requirements often exceed by far those for lead storage batteries. The reason is that the suitable materials for the positive electrode are very expensive (silver oxide, nickel hydroxide), and thus the use of these storage batteries is only justified where requirements as to weight, number of cycles, or temperature range prohibit other solutions. Besides a few standardized versions – mainly for nickel–cadmium batteries – this has led to the existence of a large diversity of constructions for special applications [4–6, 107, 108].

In order to classify this diversity from the viewpoint of the separator, the basic requirements for separators in alkaline cells are discussed below and an attempt at structuring them accordingly is made.

The prime requirements for the separators in alkaline storage batteries are on the one hand to maintain durably the distance between the electrodes, and on the other to permit the ionic current flow in as unhindered a manner as possible. Since the electrolyte participates only indirectly in the electrochemical reactions, and serves mainly as an ion-transport medium, no excess of electrolyte is required; that is, the electrodes can be spaced closely together in order not to suffer unnecessary power loss through additional electrolyte resistance. The separator is generally flat, without ribs. It has to be sufficiently absorbent, and it also has to retain the electrolyte by capillary forces. The porosity should be at a maximum to keep the electrical resistance low (see Section 11.1.2.3); the pore size is governed by the risk of electronic shorts. For systems where the electrode substance does not dissolve or is only slightly soluble (e.g., nickel hydroxide, cadmium), separators which prevent a deposit of particles of the active materials and subsequent shorting are sufficient, whereas for electrodes that dissolve (e.g., zinc), effective ion-selective barriers are desirable, delaying the onset of penetration from the solution phase. Positive electrodes (e.g., silver oxide) whose ions are dissolved – even sparingly – and deposited on the negative electrode, form local elements there, and thus increase self-discharge, also requiring separators with ion-segregating properties. Ion separation means, however, pore sizes on an atomic scale; this leads empirically to higher electrical resistance and especially to chemical susceptibility. The optimizations achieved to date toward increasing the service life of alkaline storage batteries are still unsatisfactory; this presents a particular challenge to the further development of separators.

11.3.2
Primary Cells

Primary cells generally do not place high demands on the separator, so these are not covered exhaustively here; the lack of a charging process avoids undesirable electrochemical deposits (e.g., dendrites) as well as generation of oxidizing substances. Thus low-priced, alkali-resistant sheets are used as separators; generally cellulosic papers, fleeces, or woven fabrics of poly-amide, poly(vinyl alcohol) (PVA) or polypropylene fibers meet this requirement satisfactorily [4]. It is generally sufficient for them to absorb and retain as much as possible of the electrolyte without decomposition and to be resistant against the substance of the positive electrode under the conditions of use to be expected. The fleeces of organic fibers are also used in alkaline secondary cells and will be explained in more detail in that context (cf. Section 11.3.5).

11.3.3
Nickel Systems

11.3.3.1 Nickel–Cadmium Batteries

11.3.3.1.1 Vented Construction   The first practical alkaline storage batteries were developed in the 1890–1910 period by Waldemar Jungner in Sweden and almost simultaneously by Thomas Alva Edison in the USA [10]. These nickel–iron batteries, because of their high self-discharge rates due to iron poisoning of the nickel electrodes, have been replaced almost completely by the nickel–cadmium batteries also developed by W. Jungner. The original construction with so-called pocket plates is still available today, with only little change. The active material powders are held in pockets of perforated (nickel-coated) steel sheets. In the simplest case the pocket electrodes are kept at a spacing of about 1–3 mm by PVC rod plates ('ladders'), and occasionally also by extruded PVC ribs or perforated, corrugated PVC spacers, according to the designed electrical power performance. Since, as mentioned, no soluble ions cause any interferences in a nickel–cadmium pocket plate battery, a separator in the narrow sense is not required.

For increased power requirements, electrode constructions have been developed which bring the electronic conductors into closer contact with the active material particles: first, around 1930, the sinter electrode [109], recently in sealed cells largely replaced by the nickel–foam electrode, and then, around 1980, the fiber structure electrode [110]. In order to take full advantage of their increased performance, the electrodes have to be as close together as possible; that is, a uniformly thin, highly porous separator is required with sufficiently small pores to prevent any penetration even at narrow spacing. For medium electrical performance – that is, electrode spacings of about 1 mm – ribbed or corrugated sintered PVC separators are used. They largely correspond to the product used in lead–acid batteries and have been described in that context in detail (cf. Section 11.2.2.2). This separator is good value, but it is rather brittle and thus difficult to handle, and it has relatively large pores (−15 μm).

For higher current loads, especially for sinter electrodes, smaller separator pores are desired; such materials are mostly sensitive, frequently requiring multiple layers performing different duties. Both electrodes are wrapped in a relatively open fleece or woven fabric of polyamide ('nylon') or, if higher temperatures apply, of polypropylene fibers, which provide sufficient electrolyte at the electrode surface to keep the electrical resistance low. Between these, an ion-semipermeable membrane, typically regenerated cellulose ('cellophane') [112], serves as a gas barrier to prevent the generated oxygen from reaching the negative electrode. In wet a condition, where it swells and achieves the desired pore sizes and properties, cellophane is mechanically very sensitive; the aforementioned nylon fleeces offer the required support from both sides.

Better mechanical stability can be expected from irradiated polyethylene or microporous polypropylene ('Celgard') membranes, but these account for increased electrical resistance values.

One version of the microporous, filled polyethylene separator ('PowerSep') [113], which is so successful in the lead–acid battery, is also being tested in nickel–cadmium batteries. This separator is manufactured largely in the same way and also has similar properties to those described in Section 11.2.2.1. Of course, silica cannot be used as a filler, but has to be replaced by an alkali-resistant substance, for example, titanium dioxide. The resulting separator membrane excels, with very small pore sizes and low electrical resistance as well as outstanding mechanical properties. A comprehensive presentation of the different separation materials follows in Section 11.3.5.

The microporous or semipermeable separators serve, as explained, to avoid oxygen transfer and thus increased self-discharge. In special cases of severe cycling service without extended standing periods, this oxygen transfer is actually desired, in order to suppress – by means other than constructional techniques – hydrogen generation and consequently water consumption. Batteries for electric vehicles are such a case, in which freedom from maintenance is the primary goal. As separators, several layers of macroporous fleeces of either polyamide, polyethylene, or polypropylene fibers and blends thereof, as well as spun fleece (melt-blown) of polypropylene, are used. This construction ('partial recombination') is already a transition stage to sealed batteries.

11.3.3.1.2 Sealed Construction The working principle of sealed nickel–cadmium batteries is based on internal oxygen consumption. The negative electrodes have a larger capacity than the positive ones; therefore, during the charging step the latter reach their fully charged status earlier and start to evolve oxygen, which migrates through voids in the electrolyte to the negative electrode to discharge cadmium, which was already charged. As a prerequisite the separator has to be permeable to gaseous oxygen; this is achieved by separator pores being of a specific minimum size and not all of them being filled with electrolyte at the same time, so as to leave some gas channels. For this application the fleeces of polyamide, polyethylene, or polypropylene fibers mentioned above have proven themselves. With their porosity they can absorb sufficient electrolyte, and due to their pore size distribution they can simultaneously bind electrolyte and allow oxygen transfer.

Mechanical strength becomes an important criterion, because wound cells (spiral-type construction), in which a layer of separator material is spirally wound between each pair of electrodes, are manufactured automatically at very high speed. Melt-blown polypropylene fleeces, with their excellent tensile properties, offer an interesting option. Frequently, two layers of the same or different materials are used, to gain increased protection against shorts; for button cells the use of three layers, even, is not unusual. Nevertheless the total thickness of the separation does not exceed 0.2–0.3 mm. For higher-temperature applications (up to about 60 °C) polypropylene fleeces are preferred since they offer better chemical stability, though at lower electrolyte absorption [113].

11.3.3.2 Nickel–Metal Hydride Batteries

Cadmium presents an environmental risk. Since small nickel–cadmium cells are often not separately disposed of, they may enter municipal garbage incinerators. The search for alternative materials for the negative electrode led to metal hydrides, which not only are regarded as environmentally less critical, but also allow higher energy density than cadmium. This is especially important for use in portable equipment, such as cellular phones or laptop computers, where the nickel–metal hydride system is especially successful. Only in applications requiring high current densities are they second to nickel–cadmium. The requirements for the separators are largely identical with those for the sealed nickel–cadmium cells; therefore mostly the same separator materials are used. They are described in Section 11.3.5.

11.3.4
Zinc Systems

11.3.4.1 Nickel–Zinc Storage Batteries

Electrochemical systems with zinc as the negative electrode material in alkaline electrolyte promise high energy and power densities. The nickel–zinc storage battery especially is being discussed as a candidate for the power source of electric vehicles, last but not least because zinc – compared with the above-mentioned metal hydrides – is of low cost and available in sufficient quantity. Even though this system has been studied and developed since 1930 [114], no success has yet been achieved in reaching a sufficient number of cycles, so no commercial utilization has resulted; 200–300 cycles are still considered to be the limit today, although recently laboratory cells are reported to have reached 600 cycles [115].

The reason for this limited cycle life is the high solubility of the zinc electrode in alkaline electrolyte; the zincate ions formed are deposited again during the subsequent charging in the form of dendrites, that is, of fernlike crystals. They grow in the direction of the counterelectrode and ultimately cause shorts.

A remedy could be achieved by a decrease in the zinc solubility in the electrolyte or by suppression of dendrite formation; oxides of cadmium, lead, or bismuth, as well as calcium hydroxide or aluminum hydroxide, have been added to the zinc electrode or the electrolyte for this purpose, but not with long-lasting effectiveness.

Thus in this system, in addition to the usual requirements, the separator has the task of delaying penetration for as long as possible. A membrane would be regarded as perfect which lets hydroxyl ions pass, but not the larger zincate ions. This requirement is best met by regenerated cellulose ('cellophane') [10, 11], which in swollen condition shows such ion-selective properties but at the same time is also chemically very sensitive and allows only a limited number of cycles; the protective effects of additional fleeces of polyamide or polypropylene have already been taken into account.

Chemically more stable systems with microporous properties, such as stretched polypropylene films ('Celgard'), irradiated, coated polyethylene, or filled polyethylene separators ('PowerSep') offer a compromise: smaller pore diameters have been shown to increase the number of cycles to penetration. However, a different failure mode occurs at an earlier stage, namely 'shape change' of the negative electrode [116]. If the off-diffusion of zincate ions into the bulk electrolyte is obstructed, for example, by small separator pores, concentration gradients on the electrode surface cause a shifting of the zinc deposit from the edges toward the center of the electrode [117]. In summary it may be noted that these opposing effects have prevented a breakthrough of the nickel–zinc system, as yet.

11.3.4.2 Zinc–Manganese Dioxide Secondary Cells

The system known as *primary alkaline manganese cells* has been further developed since 1975 into secondary cells [118, 119]. The above-mentioned problems of the zinc electrode apply here as well, although safety is assured for these sealed cells by constructional measures. Depending on the depth of discharge, between 20 and 200 cycles can be attained, which may be sufficient for many applications, for example, as low-cost rechargeable power sources for children's toys. The described combination of a few layers of fleece of polyamide or polypropylene fibers with an ion-selective film of regenerated cellulose ('cellophane') is being used as a means of separation to prevent shorting by dendrites. A further development of the separator has been achieved by impregnation of a polyamide fleece with regenerated cellulose in order to obtain a single, stable, ion-semipermeable separator layer.

11.3.4.3 Zinc–Air Batteries

A completely different way has been taken to render zinc–air elements of very high energy density rechargeable for the use in electric vehicles [120]. In the vehicle they are used exclusively as primary cells to be 'mechanically' recharged at a central depot. The zinc electrodes are removed from the discharged battery; then they are mechanically crushed, chemically dissolved, and electrolytically deposited, again compacted, and supplied with a separator pocket before being reinstalled in the battery. A woven fabric of polyamide ('nylon') fibers serves a separator, which is sufficient to prevent shorts during discharge.

11.3.4.4 Zinc-Bromine Batteries

Even though zinc–bromine batteries operate with a slightly acidic electrolyte (pH 3), they are discussed here briefly, because they offer another way of escaping the problems of zinc deposition. At this pH value both zinc corrosion as well as the tendency toward dendrite formation are low; the latter, furthermore, is prevented by electrolyte circulation [121]. The separator, besides meeting the usual requirements, has to perform an additional duty: although it must permit the charge transfer of zinc and bromide ions, it should suppress the transfer of dissolved bromine, of polybromide ions, or of the complex phase. Due to mechanical and chemical susceptibility, ion-selective membranes did not prove effective. Microporous polyethylene separators are usually used; in their manufacture and properties they are quite similar to those described in Section 11.2.3.1.

11.3.4.5 Zinc–Silver Oxide Storage Batteries

Zinc–silver oxide batteries as primary cells are known both as *button cells*, for example, for hearing aids, watches, or cameras, and for military applications, usually as *reserve batteries*. Since the latter after activation have only a very short life (a few seconds to some minutes), a separation by cellulosic paper is generally sufficient.

Rechargeable zinc–silver oxide batteries have to struggle against the same problems as the zinc electrode, which have been described in detail for the nickel–zinc systems. To make matters even worse, the silver oxide electrode contributes an additional problem: silver ions – even to a small extent – dissolve, deposit on the negative electrodes, and poison them by forming local corrosion elements and causing self-discharge with hydrogen evolution. In order to prevent this, several layers of semipermeable cellophane membranes are used [122], among other methods. The beneficial effect is caused by a sacrificial action: the silver ions migrate through the electrolyte and oxidize (i.e., they thus destroy) cellophane film sites, simultaneously being reduced to metallic silver and thereby becoming less harmful. The life of the cellophane is therefore limited; together with wetting fleeces to prevent also direct contact with the silver oxide electrode, this is fully sufficient for primary cells. For rechargeable batteries, cycle lives of 10–100 cycles are quoted [123, 124], depending on type of separation and depth of discharge; in special cases of very shallow discharges of only a few percent, however, 3000 cycles and three years of life have been reported.

Advanced development of ion-selective films has been attempted by radiation grafting of methacrylic acid onto polyethylene films, and combinations of this with cellophane are also being tested. Polyamide fleece impregnated with regenerated cellulose is another option for zinc-silver oxide batteries.

Occasionally the zinc electrode is wrapped in a polypropylene fleece filled with inorganic substances, such as potassium titanate, in order to reduce the solubility of zinc, since the problem of dendrite growth is aggravated even by the metallization of the cellophane separator due to the aforesaid silver reduction and its promoting the generation of shorts.

After these comments it is understandable that this expensive and life-limited system could succeed only in a few special applications where the high energy and power density could not be achieved by other systems.

11.3.5
Separator Materials for Alkaline Batteries

In the product range of alkaline power sources, each manufacturer has developed for each special application an optimum type of separator. Generally, however, these consist of the combination of a relatively small variety of proven materials. They are presented here jointly, even though they can hardly be compared with each other. They may be divided into three groups, depending on their application: macroporous wetting fleeces (Table 11.15), microporous separators (Table 11.16), and ion-semipermeable membranes (Table 11.17).

Of all possible manufacturing processes for macroporous separators to be employed in alkaline batteries, the wet-fleece process using paper machines is the predominant one [125]; it permits a very uniform ('cloud-free') production of such material and the use of different types of fibers as well as of short and very thin fibers, thus achieving a uniform structure of small pores (Table 11.15).

Whereas PVA fleeces are used only in primary cells, polyamide fleeces compete with polyolefin, preferably polypropylene fleeces. The latter are more stable at higher temperatures and do not contribute to electrolyte carbonation, but they wet only after a pretreatment either by fluorination [126] or by coating and crosslinking with hydrophilic substances (e.g., polyacrylic acid [127]) on the surface of the fiber.

Only very recently, the production of melt-blown polypropylene fleeces with considerably thinner fiber diameter became possible [99], thus making it possible to achieve attractive properties with regard to small pore size and excellent tensile performance for use in highly automated assembly processes, provided that a low-cost hydrophilization is available.

Very different microporous separators for alkaline batteries are included in Table 11.16. The very thin ($\sim$25 µm) films of stretched polypropylene ('Celgard') are generally employed in combination with fleeces, while separators of sintered PVC or filled UHMW PE find use also in single separation of alkaline industrial batteries. Their production process corresponds to the analogous version for lead–acid batteries and is described in detail in Sections 11.2.2.1 and 11.2.2.2 respectively.

As ion-semipermeable membranes, which – despite good permeability to hydroxyl ions – hinder the transfer of zincate and silver ions, essentially only regenerated cellulose is being used in alkaline batteries. In a complex conversion process, a pulp of wood cellulose, primarily from eucalyptus trees, is dissolved

Table 11.15 Nonwoven materials for alkaline batteries.

	Wet-laid poly(vinyl alcohol) (PVA) fiber fleece	Wet-laid polyamide (PA) fiber fleece	Wet-laid polyolefin (PP/PE) fiber fleece	Wet-laid grafted polypropylene fiber fleece	Melt-blown grafted polypropylene fiber fleece
Supplier[a]	C. Freudenberg [128]	C. Freudenberg [128]	C. Freudenberg [128]	Sci MAT Ltd. [129]	Sci MAT Ltd. [129]
Brand name	Viledon FS 2183	Viledon FS 2117	Viledon FS 2123 WI	Sci MAT 700/25	Sci MAT 700/35
Weight[b] (g m^{-2})	70	70	72	70	60
Thickness (mm)	0.35	0.33	0.26	0.23	0.18
Tensile strength (N/15 mm)	≥260	≥21	≥45	45	45
Air permeability (L s^{-1} m^{-2})	450	700	400	n.a.	n.a.
KOH absorption (g m^{-2})	≥350	≥300	150	200	120
Porosity (%)	84	81	71	n.a.[c]	n.a.
Pore size (average) (μm)	36	40	40	20	20
Electrical resistance (KOH) (mΩ cm^2)	40	40	70	n.a.	n.a.

[a]Other suppliers include PGI Nonwovens Chicopee, Inc. [130], Kuraray Co., Ltd. [131], and Hollingsworth and Vose Comp. [101].
[b]Representative examples of a variety of weight/thickness.
[c]n.a.: not available.

in caustic solution, under reaction with carbon disulfide, to be precipitated as cellulose xanthate. The xanthate is then dissolved in a sodium hydroxide solution, thickened ('viscose'), then extruded and regenerated with sulfuric acid to give cellophane. 'Cellophane' was originally a registered trade name and in some countries still is to this day. This material is macromolecular and swells in potassium hydroxide solution forming pores in the order of magnitude of 25–50 Å. The molecules are attacked and oxidized by potassium hydroxide, by oxygen, but particularly by silver ions. This process can be slowed down by a pre-treatment with silver salt solution [133]: the precipitated silver atoms, by comproportionation with the 'dangerous' silver(II) ions, form silver(I) ions, which are less soluble and precipitate on

Table 11.16 Microporous separators for alkaline batteries.

	Microporous polypropylene film		Microporous filled UHMW polyethylene separator	Sintered PVC separator
Supplier	Hoechst Celanese Corp. [132] Celgard 3401	Hoechst Celanese Corp. [132] Celgard 3501	Daramic, Inc [60, 61] PowerSep	Jungfer GmbH [64] Sintered PVC
Thickness (mm)	0.025	0.025	1.3	1.3
Backweb thickness (mm)	0.025	0.025	0.2	0.3
Porosity (%)	38	45	40	33
Pore size (average) (μm)	<1	<1	0.2	15
Electrical resistance (KOH) (mΩ cm^2)	80	4	200	300

Table 11.17 Semipermeable membranes for alkaline batteries.

	Regenerated cellulose membrane		Regenerated cellulose membrane (silver-treated)	Polyamide nonwoven impregnated with regenerated cellulose
Supplier	UCB Cellophane Ltd. [134]	Viskase Corp. [135]	Yardney Techn. Prod., Inc. [136]	Viskase Corp. [135]
Brand name	350 POO	SEPRA-CEL	C 19	SEPRA-CEL
Thickness (mm)	0.025	0.025	0.025	0.175
Pore size (nm)	2.5	2.5–5	2.5	n.a.
KOH absorption (g m^{-2})	n.a.[a]	60	55	275
Electrical resistance (KOH) (mΩ cm^2)	40	n.a.	<120	90

[a] n.a.: not available.

the semipermeable membrane without being able to reach the negative electrode. However, this pre-poisoning contributes additionally to the metal content of the separator.

The development of stable, effectively ion-separating membranes for rechargeable alkaline batteries remains a persistent challenge, in which the separator can provide decisive contributions to the advancement of storage batteries of high power and energy density.

Acknowledgments

I thank many colleagues in the battery industry for critical discussions, for suggestions, and for supplying me with references, as well as Daramic, Inc., Burlington, MA (USA) for the permission to publish. My special appreciation goes to Dr. F. Theubert and Mrs. A. Hayen for their untiring assistance in the preparation of the manuscript.

References

1. Lander, J.J. (1970) *Proceedings of the Symposium on Battery Separators*, The Electrochemical Society, Columbus, OH, p. 4.
2. Selsor, J.Q. (1985) *Battery Man*, 9.
3. Böhnstedt, W. (1996) *J. Power Sources*, 59, 45.
4. Linden, D. (1995) *Handbook of Batteries*, 2nd edn, McGraw-Hill, New York.
5. Barak, M. (1980) *Electrochemical Power Sources*, The Institution of Electrical Engineers, London and P. Peregrinus Ltd, Stevenage.
6. Wiesener, K., Garche, J., and Schneider, W. (1980) *Elektrochemische Stromquellein*, Akademie-Verlag, Berlin.
7. Berndt, D. (1997) *Maintenance–Free Batteries*, 2nd edn, Research Studies Press Ltd., Taunton, Somerset.
8. Bode, H. (1977) *Lead–Acid Batteries*, John Wiley & Sons, Inc., New York.
9. Vinal, G.W. (1955) *Storage Batteries*, John Wiley & Sons, Inc., New York.
10. Falk, S.U. and Salkind, A.J. (1969) *Alkaline Storage Batteries*, John Wiley & Sons, Inc., New York.
11. Fleischer, A. and Lander, J.J. (1971) *Zinc–Silver Oxide Batteries*, John Wiley & Sons, Inc., New York.
12. Schallemberg, R.H. (1982) *Bottled Energy*, American Philosophical Society, Memoirs Series, Vol. 148, G. H. Buchanan, Philadelphia, PA, p. 27.
13. Schallemberg, R.H. (1982) *Bottled Energy*, American Philosophical Society, Memoirs Series, Vol. 148, G. H. Buchanan, Philadelphia, PA, p. 58.
14. Schallemberg, R.H. (1982) *Bottled Energy*, American Philosophical Society, Memoirs Series, Vol. 148, G. H. Buchanan, Philadelphia, PA, p. 269.
15. Moll, P.J. (1952) *Die Fabrikation von Bleiakkurnulatoren*, 2nd edn, Akademischer Verlag Geest/Portig K.-G., Leipzig, p. 131.
16. Larsen, D.W. and Kehr, C.L. (1966) US Patent 3 351 495.
17. McClelland, D.H. and Devitt, J.L. (1972) US Patent 3 862 861.
18. Decker, E. (1965) German Patent 1 279 795; (1969) US Patent 3 475 355.
19. Tudor, A.B. (1966) British Patent 1 183 470.
20. Jache, O. (1958) German Patent 1 194 015.
21. Jache, O. (1967) German Patent 1 671 693.
22. Lange, A.E., Langguth, E., Breuning, E., and Dassler, A. (1933) German Patent 674 829.

23. Culpin, B. and Hayman, J.A. (1986) in *Power Sources*, vol. 11 (ed. L.J. Pearce), International Power Sources Symposium Committee, Leatherhead, p. 45.
24. Peters, K. (1993) *J. Power Sources*, **42**, 155.
25. Zguris, G.C., Klauber, D.W., and Lifshutz, N.L. (1991) in *Power Sources*, vol. 13 (eds T. Keily and B.W. Baxter), International Power Sources Symposium Committee, Leatherhead, p. 45.
26. Harris, F.J.T. (1982) US Patent 4 465 748.
27. Feder, D. (1994) Performance measurement and reliability of VRLA batteries. Paper presented at Battery Council International, 106th Convention (Technical Committee Workshop), SanDiego.
28. Pritchett & Gold and E.P.S. Co. Ltd. (1942) British Patent 565 022.
29. Kalhammer, F.R., Kozawa, A., Moyer, C.B., and Owens, B.B. (1995) *Performance and Availability of Batteries, for Electric Vehicles*, California Air Resource Board, El Monte, CA. Report of the Battery Technical Advisory Panel.
30. Moseley, P.T. (1997) *J. Power Sources*, **67**, 115.
31. Alzieu, J. and Robert, J. (1984) *J. Power Sources*, **13**, 93.
32. Alzieu, J., Koechlin, N., and Robert, J. (1987) *J. Electrochem. Soc.*, **134**, 1881.
33. Takahashi, K., Tsubota, M., Yonezu, K., and Ando, K. (1983) *J. Electrochem. Soc.*, **130**, 2144.
34. Hollenkamp, A.F., Constanti, K.K., Koop, M.J., and McGregor, K. (1995) *J. Power Sources*, **55**, 269.
35. Hollenkamp, A.F. (1996) *J. Power Sources*, **59**, 87.
36. Hollenkamp, A.F. and Newham, R.H. (1997) *J. Power Sources*, **67**, 27.
37. Stempin, J.L., Steward, R.L., and Wexell, D.R. (1995) European Patent EPO 750 357 A2.
38. Böhnstedt, W., Deitcs, J., Ihmels, K., and Ruhoff, J. (1997) German Patent Application DE 197 02 757.
39. Böhnstedt, W. and Weiß, A. (1992) *J. Power Sources*, **38**, 103.
40. Böhnstedt, W. (1993) *J. Power Sources*, **42**, 211.
41. Böhnstedt, W. (1997) *J. Power Sources*, **67**, 299.
42. Reitz, J.W. (1987) *J. Power Sources*, **19**, 181.
43. Reitz, J.W. (1988) *J. Power Sources*, **23**, 109.
44. Weighall, M.J. (1991) *J. Power Sources*, **34**, 257.
45. Weighall, M.J. (1992) *J. Power Sources*, **40**, 195.
46. Weighall, M.J. (1995) *J. Power Sources*, **53**, 273.
47. Weighall, M.J. (1996) 5th European Lead Battery Conference, Barcelona.
48. Schneider, J. (1988) *J. Power Sources*, **23**, 113.
49. Hoover, J. (1995) Failure modes of batteries removed from service. Paper presented at Batteries Council International, 107th Convention, Orlando.
50. Fischer, K. (1996) LABAT/96 Conference, Varna, p. 171.
51. Kung, J. (1994) *Batteries Int.*, **1**, 46.
52. Sugarman, N. (1979) German Patent DE 30 04 659, British Patent 2 044 516.
53. Bünsch, H., Ihmels, K.H., and Teubert, F.O. (1989) German Patent DE 39 28 468; European Patent EP 0 425 784.
54. Choi, W.M. and Böhnstedt, W. (1983) US Patent 5 384 211.
55. Nakano, K. (1991) European Patent EP 0 541 124.
56. Knauer, D.J. (1995) US Patent S 558 952.
57. O'Rell, D.D., Lin, N.-J., Gordon, G., and Gillespie, J.H. (1981) British Patent GB 2 097 174.
58. Böhnstedt, W. and Lindenstruth, W. (1988) German Patent DE 38 30 728; US Patent 4 927 722.
59. Böhnstedt, W. (1994) German Patent DE 44 14 723; World Patent WO 95/29508.
60. Daramic, Inc. Postfach 1509, D-22805 Norderstedt, Germany.
61. Daramic, Inc. 20 Burlington Mall Road, Burlington, MA 01803, USA.
62. ENTEK International LLC P.O. Box 127, 250 North Hansard Avenue, Lebanon, OR, 97355, USA.

63. ENTEK International Ltd. Mylord Crescent, Camperdown Industrial Estate, Killingworth, Newcastle upon Tyne NE 12 OXG, UK.
64. Jungfer GmbH & Co. KG, A-9181 Feistritz, Austria.
65. Engelmann, M., Kraus, H., and Plewan, O. (1992) German Patent DE 43 37 429.
66. Müller, R. (1993) The lead market towards the 21st century. Paper presented at EUROBAT Convention, Prague.
67. Accuma S.p.A. Via Medici, 14, 1 20052 Monza, Italy.
68. I.C.S. Industria Composizioni Stampate S.p.A. I −24040 Canonica d'Adda, Bergamo, Italy.
69. Böhnstedt, W., Radel, C., and Scholten, F. (1987) *J. Power Sources*, **19**, 301.
70. Döring, H., Radwan, M., Dietz, H., Garche, J., and Wiesener, K. (1989) *J. Power Sources*, **28**, 381.
71. Cerný, J. and Koudelka, V. (1990) *J. Power Sources*, **31**, 183.
72. Axohm Industries S.A., Lydall, Inc. Saint Rivalain, F −56310 Melrand, France.
73. GS Kasei Kogyo K.K. lnoguchi Takatsuki−Cho Ikagun, Siga Prefecture, Japan 529−02.
74. Nann, E. (1991) *J. Power Sources*, **33**, 93.
75. Yaacoub, C.M. (1989) German Patent DE 39 22 160; European Patent EP 0 409 363.
76. Dawson, J.L., Gillibrand, M.I., and Wilkinson, J. (1971) in *Power Sources*, vol. 3 (ed. D.H. Collins), Oriel Press, Newcastle upon Tyne, p. 1.
77. Herrmann, W. and Pröbstl, G. (1957) *Z. Elektrochem.*, **61**, 1154.
78. Zehender, E., Herrmann, W., and Leibssle, H. (1964) *Electochim. Acta*, **9**, 55.
79. Goldberg, B.S., Hausser, A.G., and Le, B.T. (1983) *J. Power Sources*, **10**, 137.
80. Paik, S.L. and Terzaghi, G. (1995) *J. Power Sources*, **53**, 283.
81. Böhnstedt, W. and Radel, C. (1991) German Patent DE 41 08 176; European Patent EP 0 507 090.
82. Palmer, N.I., Rooney, J.L., and Sugarman, N. (1974) German Patent DE 25 20 961; British Patent 1512 997.
83. Choi, W.M. and Schmidt, I.W. (1995) US Patent 5 478 677.
84. Ferreira, A.L. and Lingscheidt, H.A. (1997) *J. Power Sources*, **67**, 291.
85. AMER-SIL sa., Zone Industrielle L-8287 Kehlen, Luxembourg.
86. AMERACE Microporous Products L. P., 596 Industrial Park Road, Piney Flats, TN 37686, USA.
87. Nelson, R.F. (1991) in *Power Sources*, vol. 13 (eds T. Keily and B.W. Baxter), International Power Sources Symposium Committee, Leatherhead, p. 13.
88. Nelson, R.F. (1993) *J. Power Source.s*, **46**, 159.
89. Crouch, D.A. and Reitz, J.W. Jr. (1990) *J. Power Sources*, **31**, 125.
90. Culpin, B. (1995) *J. Power Sources*, **53**, 127.
91. Miura, H. and Hosono, H. (1994) *J. Power Sources*, **48**, 233.
92. Zguris, G.C. (1996) *J. Power Sources*, **59**, 131.
93. Nguyen, T.V., White, R.E., and Gu, H. (1990) *J. Electrochem. Soc.*, **137**, 2998.
94. Nguyen, T.V. and White, R.E. (1993) *Electrochem. Acta*, **38**, 935.
95. Bernardi, D.M. and Carpenter, M.K. (1995) *J. Electrochem. Soc.*, **142**, 2631.
96. May, G.J. (1995) *J. Power Sources*, **53**, 111.
97. Zguris, G.C. and Harmon, F.C. Jr. (1992) US Patent 5 336 275.
98. Badger, J.P. (1987) US Patent 4 908 282.
99. Zucker, J. (1997) Application, Inc., US Patent, (not yet published).
100. Dumas, B. S.A., P.O. Box 3, Creysse, F −24100 Bergerac, France.
101. Hollingsworth & Vose Co., 112 Washington Street, East Walpole, MA 02032, USA.
102. Nippon Glass Fiber Co., Ltd., 2 Chitose-Cho, Yokkaichi, Mie Prefecture, Japan.
103. Technical Fibre Products, Burneside Mills, Kendal, Cumbria LA9 6P2, UK.
104. Whatman International Ltd. Springfield Mill, James Whatman Way, Maidstone, Kent ME 14 2LE, UK.

105. Tuphorn, H. (1988) *J. Power Sources*, **23**, 143.
106. Tuphorn, H. (1993) *J. Power Sources*, **46**, 361.
107. Crompton, T.R. (1990) *Battery Reference Book*, Butterworth, London.
108. Jaksch, H.D. (1993) *Batterie-Lexikon*, Pflaum Verlag, München.
109. Pfleiderer, G., Spoun, F., Gmelin, P., and Ackermann, K. (1928) German Patent 491 498.
110. Benczür-Ürmössy, G., Berger, G., and Haschka, F. (1983) *Etz*, **104**, 1098.
111. Danko, T. (1995) *Proceedings of the 10th Annual Battery Conference on Applications and Advances, Long Beach, CA*, Institute of Electrical and Electronic Engineers (IEEE), New York, p. 261.
112. Lundquist, J.T., Jr. and Lundsager, C.B. (1979) US Patent 4 287 276.
113. Bennett, J. and Choi, W.M. (1995) *Proceedings of the 10th Annual Battery Conference on Applications and Advances, Long Beach, CA*, Institute of Electrical and Electronic Engineers (IEEE), New York, p. 265.
114. Drumm, J.J. (1930) British Patent 365 125.
115. Coates, D., Ferreira, E., and Charkey, A. (1997) *Proceedings of the 12th Annual Battery Conference on Applications and Advances, Long Beach, CA*, Institute of Electrical and Electronic Engineers (IEEE), New York, p. 23.
116. Lundquist, J.T. Jr. (1983) *J. Membr. Sci.*, **13**, 337.
117. Klein, M. and McLarnon, F. (1995) in *Handbook of Batteries* (ed. D. Linden), 2nd edn, McGraw-Hill, New York, p. 29.4.
118. Kordesch, K. and Gsellmann, J. (1979) *Power Sources*, vol. 7 (ed. J. Thompson), Academic Press, London, p. 557.
119. Kordesch, K. and Gsellmann, J. (1983) US Patent 4 384 029.
120. Hamlen, R. (1995) in *Handbook of Batteries* (ed. D. Linden), 2nd edn, McGraw-Hill, New York, p. 38.18.
121. Bellows, R.J., Eustace, D.J., Grimes, P., Shropshirc, J.A., Tsien, H.C., and Venero, A.P. (1979) in *Power Sources*, vol. 7 (ed. J. Thompson), Academic Press, London, p. 301.
122. André, H. (1941) *Bull. Soc. Franc. Electriciens*, **6**, 132.
123. Lewis, H.L., Grun, C., and Salkind, A. (1995) *Electrochem. Soc. Proc.*, **14**, 68.
124. Serenyi, R., Kuklinski, J., Williams, D.C., and Thompson, A.F. (1995) *Electrochem. Soc. Proc.*, **14**, 84.
125. Hoffmann, H. (1995) *Ratrevies Int.*, **10**, 44.
126. Schwöbel, R.P., Hoffmann, H., Keil, A., and Möller, B. (1995) German Patent DE 195 23 231.
127. Singleton, R. and Barker, D. (1996) *Batteries Int.*, **7**, 35.
128. Freudenberg, C. D-69465 Weinheirn, Germany (or licensee: Japan Vilene Co., Ltd., Tokyo, Japan).
129. Sci MAT Ltd. Dorcan 200, Murdock Road, Dorcan, Swindon SN3 5HY, UK.
130. PGI Nonwovens, Chicopee, Inc. 2351 US Route 130, Dayton, NJ 08810-1004, USA.
131. Kuraray Co., Ltd. 1-12-39, Umeda, Kita-ku, Osaka 530, Japan.
132. Hoechst Celanese Corporation 13800 South Lakes Drive, Charlotte, NC 28273, USA.
133. Mendelsohn, M. (1957) US Patent 2 816 154.
134. UCB Cellophane Ltd. Bath Road, Bridgwater, Somerset TA6 4PA, UK.
135. Viskase Corp. 6855 W. 65th St., Chicago, IL 60638, USA.
136. Yardney Technical Products, Inc. 82 Mechanic St., Pawcatuck, CT 06379, USA.

Part III
Materials for Alkali Metal Batteries

Handbook of Battery Materials, Second Edition. Edited by Claus Daniel and Jürgen O. Besenhard.
© 2011 Wiley-VCH Verlag GmbH & Co. KGaA. Published 2011 by Wiley-VCH Verlag GmbH & Co. KGaA.

12
Lithium Intercalation Cathode Materials for Lithium-Ion Batteries
Arumugam Manthiram and Theivanayagam Muraliganth

12.1
Introduction

Lithium-ion batteries have revolutionized the portable electronics market, and they are being intensively pursued for vehicle applications including hybrid electric vehicles (HEVs), plug-in hybrid electric vehicles (PHEVs), and electric vehicles (EVs). They are also being seriously considered for the efficient storage and utilization of intermittent renewable energies like solar and wind. The attractiveness of lithium-ion batteries for these applications is due to their much higher energy density compared to the other rechargeable systems. This higher energy density is a result of our ability to achieve at least 4 V per cell in practical systems by means of using nonaqueous electrolytes, the figure for aqueous electrolyte-based systems normally being <2 V [1]. The use of nonaqueous electrolytes also offers an added advantage of a wider operating temperature. However, the overall performance of lithium-ion batteries, including energy density, power capability (charge-discharge rate), safety, and cost, is related to the properties and characteristics of the various components (anodes, cathodes, and electrolytes) used in the cells. Of these, the cathode materials play a critical role in terms of cost, safety, energy, and power, and this chapter provides an overview of the developments in lithium insertion cathode materials for lithium-ion batteries.

12.2
History of Lithium-Ion Batteries

The concept of rechargeable lithium batteries was first illustrated with a transition metal sulfide TiS_2 as the cathode, metallic lithium as the anode, and a nonaqueous electrolyte [2]. During discharge, the Li^+ ions from the metallic lithium anode are inserted into the empty octahedral sites of the layered TiS_2 cathode, this being accompanied by a reduction of the Ti^{4+} ions to Ti^{3+} ions. During charge,

Handbook of Battery Materials, Second Edition. Edited by Claus Daniel and Jürgen O. Besenhard.
© 2011 Wiley-VCH Verlag GmbH & Co. KGaA. Published 2011 by Wiley-VCH Verlag GmbH & Co. KGaA.

exactly the reverse reaction occurs. The layered structure of TiS$_2$ is maintained during the charge–discharge (lithium extraction/insertion) process, resulting in good reversibility. Following this, several other sulfides and chalcogenides with high capacities were investigated as cathodes during the 1970s and 1980s [3]. However, most of these exhibited a low cell voltage of <2.5 V versus a metallic lithium anode. This limitation in cell voltage is due to the overlap of the higher-valent M^{n+}:d band with the top of the nonmetal:p band. Figure 12.1, for example, illustrates the overlap of the Co^{3+}:3d band with the top of the S^{2-}:3p band in cobalt sulfide. Such an overlap results in an introduction of holes or removal of electrons from the S^{2-}:3p band and the formation of molecular ions such as S_2^{2-}. This results in an inaccessibility of the higher oxidation states of the M^{n+} ions in a sulfide like M_yS_z, leading to a limitation in cell voltage to <2.5 V.

Recognizing this difficulty with chalcogenides, Goodenough's group at the University of Oxford focused on oxide cathodes during the 1980s [4–6]. The larger Madelung energy in an oxide compared to that in a sulfide as well as the positioning of the top of the O^{2-}:2p band below that of the S^{2-}:3p band make the higher-valence states accessible in oxides. For example, while Co^{3+} can be readily stabilized in an oxide, it is difficult to stabilize Co^{3+} in a sulfide since the $Co^{2+/3+}$ redox couple lies within the S^{2-}:3p band, as seen in Figure 12.1. Accordingly, several transition metal oxide hosts (e.g., LiCoO$_2$ and LiMn$_2$O$_4$) providing ~4 V vs Li/Li$^+$ have been identified as lithium intercalation cathodes during the past three decades. Although the cell voltage could be raised significantly with the oxide cathodes, rechargeable lithium cells based on a metallic lithium anode could not be commercialized because of the safety problems associated with metallic lithium [7, 8]. The inherent safety problem of the metallic lithium anode and the dendrite formation during the charge–discharge cycling eventually forced the use of intercalation compounds as anodes. This led to the commercialization of the lithium-ion battery technology by Sony in 1990, with LiCoO$_2$ as the cathode and graphite as the anode.

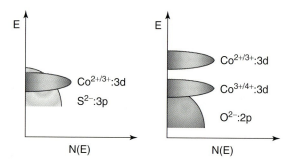

Figure 12.1 Relative energies of metal:d (e.g., Co:3d) and nonmetal:p in a sulfide and an oxide.

12.3
Lithium-Ion Battery Electrodes

Rechargeable lithium batteries involve a reversible insertion/extraction of lithium ions into/from a host electrode material during the charge–discharge process. The lithium insertion/extraction process, which occurs with a flow of ions through the electrolyte, is accompanied by an oxidation/reduction (redox) reaction of the host matrix assisted by a flow of electrons through the external circuit (Figure 12.2).

The open-circuit voltage V_{oc} of such a lithium cell is given by the difference in the lithium chemical potential between the cathode (μ_C) and the anode (μ_A) as

$$V_{oc} = (\mu_A - \mu_C)/F$$

where F is the Faraday constant ($F = 96485\,\mathrm{C\,mol^{-1}}$). Figure 12.3 gives a schematic energy diagram of a lithium-ion cell at open circuit. The cell voltage is determined by the energies involved in both the electron transfer and the Li$^+$-ion transfer. While the energy involved in electron transfer is related to the redox potential of the ion involved in the cathode and anode, the energy involved in Li$^+$-ion transfer is determined by the crystal structure and the coordination geometry of the site into/from which the Li$^+$ ions are inserted/extracted [9]. The energy separation E_g between the lowest unoccupied molecular orbital (LUMO) and the highest occupied molecular orbital (HOMO) of the electrolyte defines the stability window of the electrolyte. Therefore, thermodynamic stability considerations require the redox energies of the cathode (μ_C) and anode (μ_A) to lie within the band gap E_g of the electrolyte, as shown in Figure 12.3. An anode with a μ_A above the LUMO will reduce the electrolyte, and a cathode with a μ_C below the HOMO will oxidize the electrolyte unless an appropriate solid electrolyte interfacial (SEI) layer is formed to prevent such reactions. Thus, the electrochemical stability requirement imposes a limitation on the cell voltage as given by

$$V_{oc} = (\mu_A - \mu_C)/F \leq E_g$$

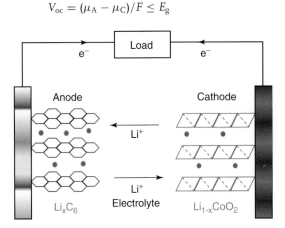

Figure 12.2 Illustration of the charge–discharge process involved in a lithium-ion cell consisting of graphite as the anode and layered LiCoO$_2$ as the cathode.

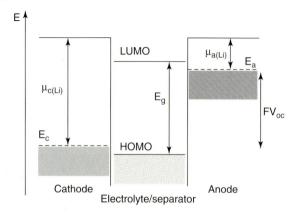

Figure 12.3 Schematic energy diagram of a lithium cell at open circuit. HOMO and LUMO refer, respectively, to the HOMO and LUMO in the electrolyte.

The key requirements for a successful cathode material in a lithium-ion battery are given below:

- The intercalation cathode $Li_xM_yX_z$ (X = anion) should have a low lithium chemical potential, and the intercalation anode should have a high lithium chemical potential to maximize the cell voltage. This implies that the transition metal ion M^{n+} should have a high oxidation state in the cathode and a low oxidation state in the anode. The chemical potential or redox energies of the cathode and anode could also be tuned by counter cations as illustrated by an increase in voltage on going from an oxide to a polyanion cathode with the same oxidation state for the transition metal ions.
- The intercalation compound should allow for insertion/extraction of a large number of lithium ions per formula unit to maximize cell capacity. This depends on the number of available lithium sites and the accessibility of multiple valence states for M in the insertion host.
- The lithium insertion/extraction reaction should be reversible, with minimal or no change in structure, leading to good cycle life.
- The intercalation compound should support mixed conduction. It should have good electronic conductivity and lithium-ion conductivity to minimize polarization losses during the charge–discharge process, thereby supporting fast charge–discharge rates and power density. The lithium-ion and electronic conductivities depend on the crystal structure, arrangement of the MX_n polyhedral geometry, interconnection of lithium sites, electronic configuration, and relative positions of the M^{n+} and X^{n-} energies.
- The redox energies of the cathode and anode should lie within the band gap of the electrolyte.
- The intercalation compound should be inexpensive, environmentally benign, and thermally and chemically stable.

12.4
Layered Metal Oxide Cathodes

Oxides with the general formula $LiMO_2$ (M = V, Cr, Co, and Ni) crystallize in a layered structure in which the Li^+ and M^{3+} ions occupy the alternate (111) planes of the rock salt structure to give a layered sequence of –O–Li–O–M–O– along the c axis. The Li^+ and M^{3+} ions occupy the octahedral interstitial sites of the cubic close-packed oxygen array as shown in Figure 12.4. This structure is also called the *O3 layered structure*, since the Li^+ ions occupy the octahedral sites (O referring to octahedral) and there are three MO_2 sheets per unit cell. This structure with covalently bonded MO_2 layers allows a reversible extraction/insertion of lithium ions from/into the lithium planes. The lithium-ion movement between the MO_2 layers provides fast two-dimensional lithium-ion diffusion [10], and the edge-shared MO_6 octahedral arrangement with a direct M-M interaction provides good electronic conductivity. As a result, the $LiMO_2$ oxides have become attractive cathode candidates for lithium-ion batteries.

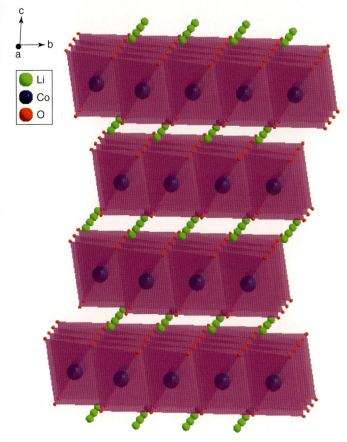

Figure 12.4 Crystal structure of layered $LiCoO_2$.

12.5
Layered LiCoO$_2$

LiCoO$_2$ is the most commonly used transition metal oxide cathode in commercial lithium-ion batteries because of its high operating voltage (~4 V) (Figure 12.5), ease of synthesis, and good cycle life. LiCoO$_2$, synthesized by conventional high-temperature procedures at $T > 800\,°C$, adopts the O3 layered structure shown in Figure 12.4, with an excellent ordering of the Li$^+$ and Co^{3+} ions on the alternate (111) planes of the rock salt lattice. The ordering is due to the large charge and size differences between the Li$^+$ and Co^{3+} ions. The highly ordered structure exhibits good lithium-ion mobility and electrochemical performance. The direct Co-Co interaction with a partially filled t_{2g}^{6-x} band associated with the Co$^{3+/4+}$ couple leads to high electronic conductivity (metallic) for Li$_{1-x}$CoO$_2$ (10^{-3} S cm^{-1}) [11]. In addition, a strong preference of the low-spin Co^{3+} and Co^{4+} ions for the octahedral sites, as evident from the high octahedral-site stabilization energy (OSSE), as seen in Table 12.1, provides good structural stability. In contrast, synthesis at low temperatures (~400 °C) results in a considerable disordering of the Li$^+$ and Co^{3+} ions, leading to the formation of a lithiated spinel-like phase with a cation distribution of [Li$_2$]$_{16c}$[Co$_2$]$_{16d}$O$_4$, which exhibits poor electrochemical performance [12–14].

Even though one Li$^+$ ion per formula unit can be theoretically extracted from LiCoO$_2$ with a capacity of ~274 mAh g^{-1}, only 50% (~140 mAh g^{-1}) of its theoretical capacity can be utilized in practical lithium-ion cells because of structural and chemical instabilities at deep charge ($x > 0.5$ in Li$_{1-x}$CoO$_2$) [15, 16]. Extraction of more than 0.5 Li$^+$ ions from LiCoO$_2$ leads to chemical instability due to the overlap of the Co$^{3+/4+}$:t_{2g} band with the top of the O^{2-}:2p band as shown in Figure 12.6. The removal of a significant amount of electron density from the O^{2-}:2p band will result in an oxidation of O^{2-} ions and a slow loss of oxygen and cobalt from the lattice during repeated cycling [16, 17]. However, a strong covalent mixing or hybridization of the Co$^{3+/4+}$:3d orbitals with the top of the O^{2-}:2p band prevents

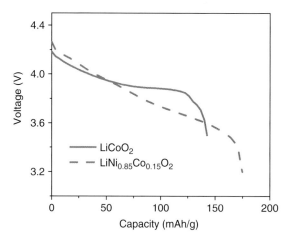

Figure 12.5 Typical discharge curves of LiCoO$_2$ and LiNi$_{0.85}$Co$_{0.15}$O$_2$.

Table 12.1 Crystal field stabilization energies (CFSEs) and octahedral site stabilization energies (OSSE) of some 3d transition metal ions.

Ion	Octahedral coordination		Tetrahedral coordination		OSSE[d] (Dq)
	Configuration[a]	CFSE[b] (Dq)	Configuration[a]	CFSE[h,c] (Dq)	
$V^{3+}:3d^2$	$t_{2g}^2 e_g^0$	−8	$e^2 t_2^0$	−5.33	−2.67
$Cr^{3+}:3d^3$	$t_{2g}^3 e_g^0$	−12	$e^2 t_2^1$ (HS)	−3.56	−8.44
$Mn^{3+}:3d^4$	$t_{2g}^3 e_g^1$ (HS)	−6	$e^2 t_2^2$ (HS)	−1.78	−4.22
$Fe^{3+}:3d^5$	$t_{2g}^3 e_g^2$ (HS)	0	$e^2 t_2^3$ (HS)	0	0
$Co^{3+}:3d^6$	$t_{2g}^6 e_g^0$ (LS)	−24	$e^3 t_2^3$ (HS)	−2.67	−21.33
$Ni^{3+}:3d^7$	$t_{2g}^6 e_g^1$ (LS)	−18	$e^4 t_2^3$ (HS)	−5.33	−12.67

[a] LS and HS refer, respectively, to low-spin and high-spin configurations.
[b] Pairing energies are neglected for simplicity.
[c] Obtained by assuming $\Delta_t = 0.444 \Delta_o$; Δ_t and Δ_o refer, respectively, to tetrahedral and octahedral splittings.
[d] Obtained by taking the difference between the CFSE values for octahedral and tetrahedral coordinations.

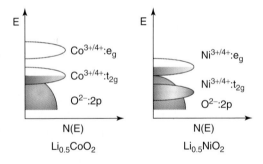

Figure 12.6 Comparison of the qualitative energy diagrams of $Li_{0.5}CoO_2$ and $Li_{0.5}NiO_2$.

a quick loss of oxygen during the first charge itself; instead, the oxygen loss from the lattice or a consequent slow reaction with the electrolyte may occur slowly over a number of cycles involving the loss of cobalt ions as well from the lattice. This results in a severe formation of SEI layer and consequent increase in impedance, resulting in poor cycle performance.

One way to suppress the chemical instability or reactivity with the electrolyte is to coat or modify the surface of the cathode with other inert oxides. In fact, surface modification of the layered $LiCoO_2$ cathode with nanostructured oxides like Al_2O_3, TiO_2, ZrO_2, SiO_2, MgO, ZnO, and MPO_4 (M = Al and Fe) has been found to increase the reversible capacity of $LiCoO_2$ from ∼140 to ∼ 200 mAh g^{-1}, which corresponds to a reversible extraction of ∼0.7 lithium per formula unit of

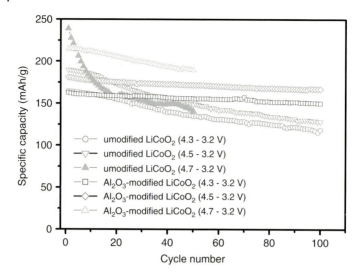

Figure 12.7 Comparison of the cyclability data of unmodified and Al$_2$O$_3$-modified LiCoO$_2$ cathodes in different Voltage ranges at C/5 rate.

LiCoO$_2$ [18–25]. Figure 12.7 compares the cyclability of Al$_2$O$_3$-modified LiCoO$_2$ to various cutoff voltages [23]. The surface modification with Al$_2$O$_3$ suppresses the reaction of the LiCoO$_2$ surface with the electrolyte and improves the capacity retention. This clearly demonstrates that the limitation in practical capacity of LiCoO$_2$ is primarily due to the chemical instability at deep discharge and is not due to the structural (order–disorder) transition at $(1 - x) = 0.5$. In addition to the improvement in electrochemical properties, nano-coating of AlPO$_4$ on LiCoO$_2$ has also been found to improve the thermal stability and safety of LiCoO$_2$ cathodes [26]. However, the long-term performance of these nano-oxide-coated cathodes will rely on the robustness of the coating. Even though the LiCoO$_2$ cathode is widely used in small-format lithium-ion batteries employed in portable electronic devices, the safety concerns arising from the chemical instability at deep charge as well as the high cost and toxicity of Co necessitate the development of alternative cathode materials for vehicle and stationary storage applications. In this regard, other layered LiMO$_2$ oxides with M = Ni and Mn and their solid solutions with LiCoO$_2$ have become attractive in recent years because of the lower cost and better chemical stability of Mn and Ni.

12.6
Layered LiNiO$_2$

LiNiO$_2$ is isostructural with LiCoO$_2$ and offers a cell voltage of ~3.8 V (Figure 12.5). Ni is less expensive and less toxic than Co. The operating voltage of the Ni$^{3+/4+}$ couple is slightly lower than that of the Co$^{3+/4+}$ couple in LiMO$_2$, in spite of

Ni being more electronegative than Co and lying to the right of Co in the Periodic Table. This is because while the redox reaction with $Ni^{3+}:t_{2g}^2 e_g^1$ involves the upper-lying, σ-bonding e_g band, that with $Co^{3+}:t_{2g}^2 e_g^0$ involves the lower-lying, π-bonding t_{2g} band. However, it is difficult to synthesize $LiNiO_2$ as a well-ordered stoichiometric material with all Ni^{3+} because of the difficulty of stabilizing Ni^{3+} at the high synthesis temperatures and the consequent volatilization of lithium [27–29]. It invariably forms as $Li_{1-x}Ni_{1+x}O_2$ with some Ni^{2+}, which results in a disordering of the cations in the lithium and nickel planes due to smaller charge and size differences between Li^+ and Ni^{2+} and consequently poor electrochemical performance. In addition, charged $Li_{1-x}NiO_2$ suffers from a migration of Ni^{3+} ions from the octahedral sites of the nickel plane to the octahedral sites of the lithium plane via the neighboring tetrahedral sites, particularly at elevated temperatures, due to a lower OSSE associated with the low-spin $Ni^{3+}:t_{2g}^2 e_g^1$ ions compared to that of the low-spin $Co^{3+}:t_{2g}^2 e_g^0$ ions (Table 12.1) [30, 31]. While a moderate OSSE allows the Ni^{3+} ions to migrate through the tetrahedral sites under mild heat, the stronger OSSE of Co^{3+} hinders such a migration. Moreover, $LiNiO_2$ also suffers from Jahn–Teller distortion (tetragonal structural distortion) associated with the low-spin $Ni^{3+}:3d^7$ ($t_{2g}^2 e_g^1$) ion. Also $Li_{1-x}NiO_2$ electrodes in their charged state are thermally less stable than the charged $Li_{1-x}CoO_2$ electrodes, an indication that Ni^{4+} ions are reduced more easily than Co^{4+} ions [32, 33]. As a result, $LiNiO_2$ is not a promising material for lithium-ion cells.

However, partial substitution of Co for Ni has been shown to suppress the cation disorder and Jahn–Teller distortion. For example, $LiNi_{0.85}Co_{0.15}O_2$ has been found to show a reversible capacity of $\sim$180 mAh g^{-1} (Figure 12.5) with excellent cyclability [34, 35]. The increase in the capacity of $LiNi_{0.85}Co_{0.15}O_2$ compared to that of $LiCoO_2$ can be understood by considering the qualitative band diagrams for the $Li_{1-x}CoO_2$ and $Li_{1-x}NiO_2$ systems, as shown in Figure 12.6. With a low-spin $Co^{3+}:3d^6$ configuration, the t_{2g} band is completely filled and the e_g band is empty ($t_{2g}^2 e_g^0$) in $LiCoO_2$. Since the t_{2g} band overlaps with the top of the $O^{2-}:2p$ band, deep lithium extraction with $(1-x) < 0.5$ in $Li_{1-x}CoO_2$ results in the removal of a significant amount of electron density from the $O^{2-}:2p$ band and consequent chemical instability, limiting its practical capacity. In contrast, the $LiNiO_2$ system with a low-spin $Ni^{3+}:t_{2g}^2 e_g^1$ configuration involves the removal of electrons only from the e_g band. Since the e_g band barely touches the top of the $O^{2-}:2p$ b and, $Li_{1-x}NiO_2$, and $LiNi_{1-y}Co_yO_2$ exhibit better chemical stability [15] than $LiCoO_2$, resulting in higher capacity values. In addition, nanocoating of AlF_3 on doped $LiNi_{0.85}Co_{0.15}Al_{0.05}O_2$ has been shown to improve the cycle performance and thermal stability of the cathodes in its oxidized (charged) state. This benefit is ascribed to the AlF_3 coating layer protecting the oxidized cathode from attack by hydrogen fluoride in the electrolyte [36].

Recent studies have shown that partial substitution of manganese in $LiNi_{0.5}Mn_{0.5}O_2$ not only provides high capacities ($\sim$200 mAh g^{-1}), but also results in a significant improvement in thermal stability compared to $LiNiO_2$ [37]. The increase in capacity and thermal stability is associated with the substitution of chemically more stable Mn^{4+} ions for Ni^{3+}. Recently, the mixed layered oxide

$LiMn_{1/3}Ni_{1/3}Co_{1/3}O_2$ has become an attractive cathode material because of its high capacity, better thermal stability, and stable cycle performance [38, 39]. In these mixed layered oxides, Ni, Mn, and Co exist as, respectively, Ni^{2+}, Mn^{4+}, and Co^{3+}. However, only $Li_{1-x}CoO_2$ becomes metallic on charging, because of the partially filled t_{2g} band, while $Li_{1-x}NiO_2$ and $Li_{1-x}MnO_2$ remain as semiconductors during charging as the e_g band is redox active and not the t_{2g} band in the edge-shared MO_6 lattice.

12.7
Layered LiMnO$_2$

Layered $LiMnO_2$ is attractive from an economical and environmental point of view, since manganese is inexpensive and environmentally benign compared to cobalt and nickel. However, $LiMnO_2$ synthesized at high temperatures adopts an orthorhombic structure instead of the layered O3-type structure, resulting in poor electrochemical performance [40]. Ion-exchange of Na^+ by Li^+ in layered α-$NaMnO_2$ has been shown to form layered, monoclinic $LiMnO_2$ [41, 42]. The reduction in crystal symmetry from trigonal in $LiCoO_2$ to monoclinic in $LiMnO_2$ is attributed to the crystallographic Jahn–Teller distortion induced by the Mn^{3+} ions. However, the layered $LiMnO_2$ synthesized by the ion-exchange method exhibits poor cycling performance because of the transformation of the charged $Li_{1-x}MnO_2$ into spinel $LiMn_2O_4$ during electrochemical charge–discharge cycling. This is because of the low OSSE value of Mn^{3+} ions and the consequent easy migration of the Mn^{3+} ions from the octahedral sites of the Mn planes to the octahedral sites of the Li planes via the neighboring tetrahedral sites [43].

12.8
Li[Li$_{1/3}$Mn$_{2/3}$]O$_2$ - LiMO$_2$ Solid Solutions

Recently, solid solutions between $Li[Li_{1/3}Mn_{2/3}]O_2$ (commonly known as Li_2MnO_3) and $LiMO_2$ (M = $Mn_{0.5}Ni_{0.5}$, Co, Ni, and Cr) have become attractive as they exhibit a high reversible capacity of around 250 mAh g^{-1} with lower cost and better safety than $LiCoO_2$ cathodes [44–51]. Li_2MnO_3 has the layered structure similar to $LiCoO_2$, but one third of the transition metal planes are occupied by Li^+ ions. Although Li_2MnO_3 is electrochemically inactive at 3–4 V vs Li/Li$^+$, subsequent studies showed that it can be made electrochemically active by acid leaching [52] or charging to high voltages [53]. Although $xLi[Li_{1/3}Mn_{2/3}]O_2 - (1-x)LiMO_2$ can be considered as solid solutions on a macroscopic scale, more detailed investigations with high-resolution transmission electron microscopy (TEM) and NMR have shown nanodomains consisting of layered Li_2MnO_3-like phases and layered $LiMO_2$ phases [50, 51].

Figure 12.8a shows a typical first charge–discharge curve of the $xLi[Li_{1/3}Mn_{2/3}]O_2 - (1-x)LiMO_2$ cathodes. Following the initial sloping region corresponding to the oxidation of the transition metal ions, a plateau region around 4.5 V

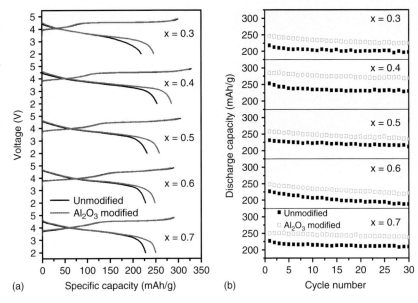

Figure 12.8 (a) First charge–discharge and (b) cycle life of pristine and Al_2O_3 coated $(1 − x)Li[Li_{1/3}Mn_{2/3}O_2]$ - $xLi[Mn_{1/3}Co_{1/3}Ni_{1/3}O_2]$.

is found which corresponds to an irreversible loss of oxygen from the lattice during first charge. The high discharge capacity of these cathodes is due to the irreversible loss of oxygen from the lattice and removal of Li as Li_2O above 4.5 V during the first charge, as revealed by *in-situ* X-ray diffraction [54] and electrochemical mass spectroscopy [46] followed by a lowering of the oxidation state of the transition metal ions in the subsequent discharge compared to that in the initial material. Based on electrochemical mass spectroscopy and powder neutron diffraction data, it was also suggested that the oxygen vacancies formed at the end of first charge are eliminated to give a defect-free layered oxide lattice. Since the elimination of oxygen vacancies will also result in an elimination of a corresponding number of lithium sites from the lattice, a large difference between the first charge and discharge capacity values (irreversible capacity loss) occurs, as part of the lithium extracted could not be put back into the layered lattice.

However, it has subsequently been found that the irreversible capacity loss in the first cycle can be reduced significantly by coating these layered oxide cathodes with nanostructured Al_2O_3 or $AlPO_4$ [55, 56]. Figure 12.9 shows the TEM images of the $AlPO_4$-coated layered oxide, in which the thickness of the $AlPO_4$ coating is around 5 nm. Figure 12.8a,b compares the first charge–discharge profiles and the corresponding cyclability data of layered $0.6[LiMn_2O_3]–0.4Li[Ni_{1/3}Mn_{1/3}Co_{1/3}]O_2$ before and after surface modification with nanostructured Al_2O_3 [47]. The surface-modified samples exhibit lower irreversible capacity loss and higher discharge capacity values than the pristine layered oxide samples. Remarkably,

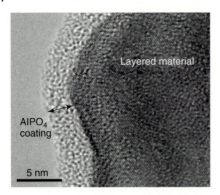

Figure 12.9 TEM image of 4 wt% nano AlPO$_4$ modified Li[Li$_{0.2}$Mn$_{0.54}$Ni$_{0.13}$Co$_{0.13}$]O$_2$ cathode.

the surface-modified samples exhibit a high discharge capacity, ~280 mAh g^{-1}, which is twice that of LiCoO$_2$. This improvement in surface-modified samples has been explained on the basis of the retention of a higher number of oxygen vacancies in the layered lattice after the first charge compared to that in the unmodified samples. The bonding of the nano-oxides to the surface of the layered oxide lattice suppresses the diffusion of oxygen vacancies and their elimination. A careful analysis of the first charge and discharge capacity values of the pristine sample also reveals that part of the oxygen vacancies are retained in the material after the first charge [57]. Moreover, the surface-modified cathodes have been found to exhibit higher rate capability than the unmodified samples despite the electronically insulating nature of the coating materials like Al$_2$O$_3$ and AlPO$_4$ [57]. This is believed to be due to the suppression of the formation of thick SEI layers, as the coating material minimizes the direct reaction of the cathode surface with the electrolyte at the high charging voltages.

However, these high-capacity layered oxide cathodes have to be charged up to about 4.8 V, so more stable, compatible electrolytes need to be developed to fully exploit their potential as high-energy-density cathodes. Moreover, oxygen is lost irreversibly from the lattice during first charge, and it may have to be vented appropriately during cell manufacture. Also, the long-term cyclability of these high-capacity cathodes at elevated temperatures needs to be fully assessed.

12.9
Other Layered Oxides

LiVO$_2$ is isostructural with LiCoO$_2$ and has the O3 layered structure. However, in de-lithiated Li$_{1-x}$VO$_2$ with $(1 - x) < 0.67$, the vanadium ions migrate from the octahedral sites of the vanadium layer into the octahedral sites of the lithium layer because of the low OSSE of the vanadium ions [58]. Therefore, the kinetics of lithium transport and the electrochemical performance is very poor with LiVO$_2$, making it an unattractive cathode material. LiCrO$_2$ can also be prepared in the O3 structure, but it has been shown to be electrochemically inactive for lithium insertion/extraction. Layered LiFeO$_2$, like LiMnO$_2$, is thermodynamically unstable at high temperatures and has to be prepared by an ion exchange of layered NaFeO$_2$ [59] with Li$^+$. However, the O3-type LiFeO$_2$ also exhibits poor electrochemical performance due to structural instabilities since the high-spin Fe^{3+}:3d^5, with an OSSE value of zero, can readily migrate from the octahedral sites to the tetrahedral sites.

12.10
Spinel Oxide Cathodes

Oxides with the general formula LiM$_2$O$_4$ (M = Ti, V, and Mn) crystallize in the normal spinel structure (Figure 12.10), in which the Li$^+$ and the M$^{3+/4+}$ ions occupy, respectively, the 8a tetrahedral and 16d octahedral sites of the cubic close-packed oxygen array. A strong edge-shared octahedral [M$_2$]O$_4$ array permits reversible extraction of the Li$^+$ ions from the tetrahedral sites without collapsing the three-dimensional [M$_2$]O$_4$ spinel framework. While an edge-shared MO$_6$ octahedral arrangement with direct M–M interaction provides good hopping electrical conductivity, the interconnected interstitial (lithium) sites via the empty 16c octahedral sites in the three-dimensional structure provide good lithium-ion conductivity.

12.11
Spinel LiMn$_2$O$_4$

Spinel LiMn$_2$O$_4$ has become an attractive cathode, as Mn is inexpensive and environmentally benign compared to Co and Ni involved in the layered oxide

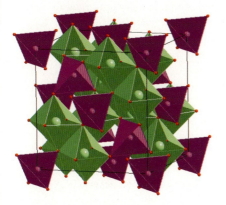

Figure 12.10 Crystal structure of spinel LiMn$_2$O$_4$.

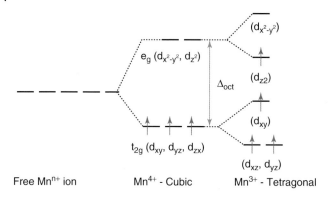

Figure 12.11 Illustration of Jahn–Teller distortion in manganese oxides.

cathodes. The extraction/insertion of two lithium ions from/into the LiMn$_2$O$_4$ spinel framework occurs in two distinct steps [9]. The lithium extraction/insertion from/into the 8a tetrahedral sites occurs around 4 V with the maintenance of the initial cubic symmetry, while that from/into the 16c octahedral sites occurs around 3 V by a two-phase mechanism involving the cubic spinel LiMn$_2$O$_4$ and the tetragonal lithiated spinel Li$_2$Mn$_2$O$_4$. A deep energy well for the 8a tetrahedral Li$^+$ ions and the high activation energy required for the Li$^+$ ions to move from one 8a tetrahedral site to another via an energetically unfavorable neighboring 16c site lead to a higher voltage of 4 V. On the other hand, the insertion of an additional lithium into the empty 16c octahedral sites occurs at 3 V. Thus, there is a 1 V jump on going from octahedral-site lithium to tetrahedral-site lithium with the same Mn$^{3+/4+}$ redox couple, reflecting the contribution of site energy to the lithium chemical potential and the overall redox energy. The Jahn–Teller distortion associated with the single electron in the e$_g$ orbitals of a high spin Mn^{3+}:3d^4 (t$_{2g}^3$e$_g^1$) ion results in the cubic-to-tetragonal transition (Figure 12.11) on going from LiMn$_2$O$_4$ to Li$_2$Mn$_2$O$_4$. The cubic-to-tetragonal transition is accompanied by a 6.5% increase in unit cell volume, which makes it difficult to maintain structural integrity during discharge–charge cycling and results in rapid capacity fade in the 3 V region.

Therefore, LiMn$_2$O$_4$ can only be used in the 4 V region with a limited practical capacity of around 120 mAh g^{-1}, which corresponds to an extraction/insertion of 0.8 Li$^+$ ion per formula unit of LiMn$_2$O$_4$ (Figure 12.12). However, LiMn$_2$O$_4$ tends to exhibit capacity fade even in the 4 V region as well, particularly at elevated temperatures (55 °C) (Figure 12.13). Several factors, such as Jahn–Teller distortion under conditions of nonequilibrium cycling [60, 61], manganese dissolution into the electrolyte [62, 63], formation of two cubic phases in the 4 V region, loss of crystallinity [64], and development of micro-strain [65] during cycling, have been suggested to be the source of capacity fade. Among these, dissolution of manganese is believed to be the main cause for capacity fade, especially at elevated temperatures. Manganese dissolution is due to the disproportionation of Mn^{3+} into Mn^{4+} (remains in the solid) and Mn^{2+} (leaches out into the electrolyte) in the

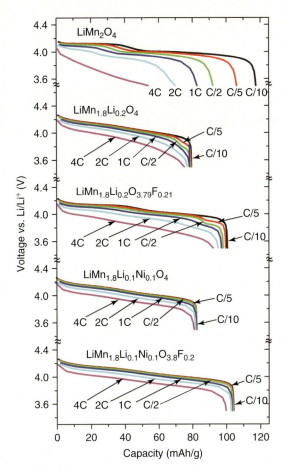

Figure 12.12 Comparison of the discharge profiles at Various C rates for $LiMn_2O_4$, $LiMn_{1.8}Li_{0.2}O_4$, $LiMn_{1.8}Li_{0.2}O_{3.79}F_{0.21}$, $Li_{1.1}Mn_{1.8}Ni_{0.1}O_4$, and $Li_{1.1}Mn_{1.8}Ni_{0.1}O_{3.8}F_{0.2}$.

presence of trace amounts of HF that is produced by a reaction of trace amounts of water in the electrolyte with the $LiPF_6$ salt. The Mn disproportionation reaction is given below as

$$2Mn^{3+} \rightarrow Mn^{2+} + Mn^{4+} \tag{12.1}$$

Several strategies have been pursued to overcome the capacity fade of $LiMn_2O_4$, for example, reducing the surface area by tuning the particle morphology and increasing the oxidation state of Mn ions via cationic substitutions in $LiMn_{2-y}M_yO_4$ [66]. The most significant among them is cationic substitutions to give $LiMn_{2-y}M_yO_4$ (M = Li, Cr, Co, Ni, and Cu) to suppress Jahn–Teller distortion and Mn dissolution since Mn^{4+}:$3d^3$ ($t_{2g}^3 e_g^0$) has a cubic octahedral coordination and does not disproportionate [67–71]. Recently, it has been shown that appropriate cationic substitutions,

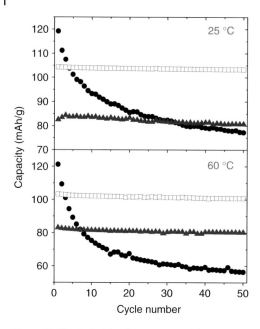

Figure 12.13 Cycle life of LiMn$_2$O$_4$ (solid circles), Li[Mn$_{1.8}$Ni$_{0.1}$Li$_{0.1}$]O$_4$ (solid triangles), and Li[Mn$_{1.8}$Ni$_{0.1}$Li$_{0.1}$]O$_{3.8}$F$_{0.2}$ (open squares) spinel cathodes at 25 and 60 °C.

as in LiMn$_{1-2y}$Li$_y$Ni$_y$O$_4$, reduce both the lattice parameter difference Δa between the two cubic phases formed during the charge–discharge process and Mn dissolution, thereby significantly improving the cyclability at elevated temperatures (Figure 12.13) [72, 73]. However, such substitutions increase the oxidation state of Mn and lower the capacity to less than 100 mAh g^{-1}. In this regard, fluorine doping at the oxygen site has been shown to be effective in increasing the capacity of the doped spinels. Therefore, dual cationic and anionic substitutions are an attractive strategy to improve the electrochemical performance of spinel oxides [74–81].

Although the partial substitution of some of the O^{2-} ions by F$^-$ ions is appealing to increase the capacity, it is difficult to incorporate large amounts of fluorine into the spinel lattice because it tends to volatilize during the conventional high-temperature (800 °C) reactions with LiF as a fluorine source. To overcome this difficulty, a low-temperature procedure has been developed in which the already synthesized cation-substituted spinel oxides such as LiMn$_{2-2y}$Li$_y$Ni$_y$O$_4$ are heated with NH$_4$HF$_2$ at 450 °C to give LiMn$_{2-2y}$Li$_y$M$_y$O$_{4-x}$F$_{2x}$ (M = Co, Zn, and Fe). Accordingly, LiMn$_{1.8}$Li$_{0.1}$Ni$_{0.1}$O$_{3.8}$F$_{0.2}$ synthesized by the low-temperature fluorination method was found to deliver a high capacity (104 mAh g^{-1}) with good cycle life compared to the 84 mAh g^{-1} obtained for pristine LiMn$_{1.8}$Ni$_{0.1}$O$_4$ (Figures 12.12 and 12.13). The fluorine doping was also found to offer an important advantage of higher tap density and better thermal stability, which may be particularly useful to improve the volumetric energy density and safety, an important criterion for large-scale electric vehicle applications. Another strategy that has been pursued to improve the cyclability of the LiMn$_2$O$_4$ spinel is surface modification or coating of its surface with other oxides such as LiCoO$_2$, ZrO$_2$, SiO$_2$, V$_2$O$_5$, Al$_2$O$_3$, or MgO with the aim of minimizing the contact of the cathode surface with the electrolyte and thereby suppressing the dissolution of manganese. In fact, the surface oxides coated on LiMn$_2$O$_4$ have been found to suppress Mn dissolution from the spinel lattice in contact with the electrolyte and improve the capacity retention [82–84].

12.12
5 V Spinel Oxides

Initially, cation-substituted LiMn$_{2-x}$M$_x$O$_4$ spinel oxides were studied to improve the capacity retention in the 4 V region, as discussed earlier. However, such substitutions to give LiMn$_{2-x}$M$_x$O$_4$ (M = Ni, Fe, Co, and Cr) lead to a 5 V plateau in addition to the 4 V plateau, which was first recognized by Amine et al. [85] and Dahn et al. [86] in 1997. The 4 V region in LiMn$_{2-x}$M$_x$O$_4$ corresponds to the oxidation of Mn^{3+} to Mn^{4+}, while the 5 V region corresponds to the oxidation of M^{3+} to M^{4+} or the oxidation of M^{2+} to M^{3+} and then to M^{4+} (Figure 12.14). It is interesting to note that while the M = Co$^{3+/4+}$ and Ni$^{3+/4+}$ couples offer around 4 V, corresponding to the extraction/insertion of lithium from/into the octahedral sites of the layered LiMO$_2$, they offer 5 V corresponding to the extraction/insertion of lithium from/into the tetrahedral sites of the spinel LiMn$_{2-x}$M$_x$O$_4$. The 1 V difference is due to the differences in the site energies between octahedral and tetrahedral sites, as discussed earlier.

With a higher operating voltage and theoretical capacities of around 145 mAh g^{-1}, LiMn$_{1.5}$Ni$_{0.5}$O$_4$ has emerged as an attractive cathode candidate. In comparison to LiMn$_2$O$_4$, here Mn predominantly remains in the +4 oxidation state during cycling, avoiding the normal Jahn–Teller distortions of Mn^{3+} ions, while Ni^{2+} first oxidizes to Ni^{3+} and then to Ni^{4+}. However, the LiMn$_{1.5}$Ni$_{0.5}$O$_4$

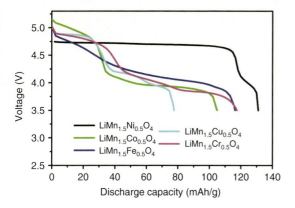

Figure 12.14 Discharge curves of the 5 V spinel oxides Li$_1$Mn$_{1.5}$M$_{0.5}$O$_4$ (M = Ni, Fe, Co, Cr, and Cu).

spinel encounters the formation of NiO impurity during synthesis, and the ordering between Mn^{4+} and Ni^{2+} leads to inferior performance compared to the disordered phase [87]. It has been found that the formation of the NiO impurity phase and ordering can be suppressed by appropriate cationic substitutions, as in $LiMn_{1.5}Ni_{0.42}Zn_{0.08}O_4$ and $LiMn_{1.42}Ni_{0.42}Co_{0.16}O_4$ [88].

One major concern with the spinel $LiMn_{1.5}Ni_{0.5}O_4$ cathode is the chemical stability in contact with the electrolyte at the higher operating voltage of 4.7 V. To overcome this difficulty, surface modification of $LiMn_{1.42}Ni_{0.42}Co_{0.16}O_4$ cathodes with oxides like $AlPO_4$, ZnO, Al_2O_3, and Bi_2O_3 have been carried out, as shown in Figure 12.15 [89]. The surface modified cathodes exhibit better cyclability (Figure 12.16) and rate performance compared to the pristine unmodified samples. The surface coating not only acts as a protective shell between the active cathode material surface and the electrolyte, but also offers fast lithium-ion and electron diffusion channels compared to the SEI layer formed by a reaction of the cathode surface with the electrolyte, resulting in enhanced cycle life and rate performance.

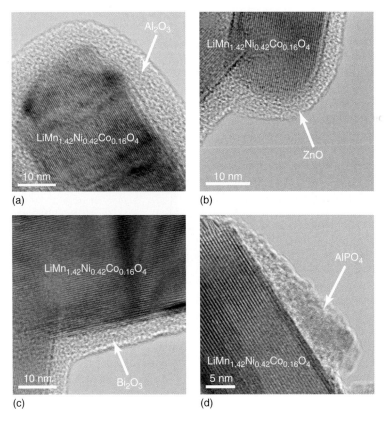

Figure 12.15 High-resolution TEM images of 2 wt% (a) Al_2O_3, (b) ZnO, (c) Bi_2O_3, and (d) $AlPO_4$-coated $LiMn_{1.42}Ni_{0.42}Co_{0.16}O_4$.

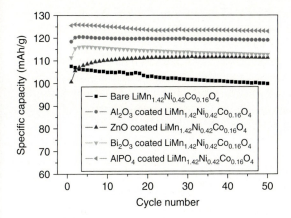

Figure 12.16 Cycling performances of pristine unmodified ('bare') LiMn$_{1.42}$Ni$_{0.42}$Co$_{0.16}$O$_4$ compared with the same material coated with 2 wt% Al$_2$O$_3$, ZnO, Bi$_2$O$_3$, and AlPO$_4$.

X-ray photoelectron spectroscopic (XPS) analysis has shown that the surface modification indeed suppresses the formation of thick SEI layers and thereby improves the rate capability [89].

12.13
Other Spinel Oxides

Both LiTi$_2$O$_4$ and LiV$_2$O$_4$ crystallize in the normal spinel structure (Li)$_{8a}$[M$_2$]$_{16d}$O$_4$ (M = Ti and V) and are metallic as a result of the direct M-M interactions with a partially filled t$_{2g}$ band. LiTi$_2$O$_4$ can insert an additional lithium into the empty 16c octahedral sites to give the lithiated spinel Li$_2$Ti$_2$O$_4$, which occurs with a flat discharge profile at a much lower voltage of around 1.5 V [90]. Accordingly, (Li)$_{8a}$[Ti$_{1.67}$Li$_{0.33}$]$_{16d}$O$_4$ or Li$_4$Ti$_5$O$_{12}$, which is much easier to synthesize than LiTi$_2$O$_4$ due to the fully oxidized Ti^{4+}, has become appealing as an anode. (Li)$_{8a}$[Ti$_{1.67}$Li$_{0.33}$]$_{16d}$O$_4$ and the corresponding lithiated spinel (Li$_2$)$_{16c}$[Ti$_{1.67}$Li$_{0.33}$]$_{16d}$O$_4$ differ in unit cell volume by only 0.1%, which is attractive to maintain good electrode integrity and capacity retention, unlike LiMn$_2$O$_4$. However, the higher voltage (1.5 V) and lower capacity (175 mAh g^{-1}) of Li$_2$Ti$_5$O$_{12}$ make it uncompetitive compared to the currently available carbon anodes (~0.1 V and 372 mAh g^{-1}). LiV$_2$O$_4$ also inserts an additional lithium into the 16c sites. The lithium ions could also be extracted from the 8a tetrahedral sites of LiV$_2$O$_4$ [91]. However, LiV$_2$O$_4$ suffers from vanadium-ion migration during these processes, which leads to poor capacity retention.

Spinels LiCr$_2$O$_4$ and LiFe$_2$O$_4$ are not known and have not been investigated. Although LiCo$_2$O$_4$ cannot be made by conventional high-temperature methods, it can be synthesized in the spinel structure by chemically extracting 50% of the lithium with aqueous acid or NO$_2$PF$_6$ in an acetonitrile medium from the low-temperature form of $LiCoO_2$, which has a lithiated spinel structure [30]. The extraction of lithium from the 8a sites occurs at around 3.9 V, and the insertion of additional lithium into 16c sites occurs at around 3.5 V. However, the system suffers from a huge polarization loss, as indicated by a large separation between the charge and discharge profiles. Attempts to make $LiNi_2O_4$ spinel by chemically extracting 50% of lithium from $LiNiO_2$ followed by heating at 200 °C results in a spinel-like cubic phase, but with the $Ni^{3+/4+}$ ions occupying both the 16c and 16d sites. Heating at $T > 200\,°C$ results in a disproportionation of cubic $LiNi_2O_4$ into $LiNiO_2$ and NiO [92].

12.14
Polyanion-containing Cathodes

Although simple oxides such as $LiCoO_2$, $LiNiO_2$, and $LiMn_2O_4$ with highly oxidized redox couples ($Co^{3+/4+}$, $Ni^{3+/4+}$, $Mn^{3+/4+}$, respectively) were able to offer high cell voltages of ~4 V in lithium-ion cells, they are prone to release oxygen from the lattice in the charged state at elevated temperatures because of the chemical instability of highly oxidized species such as Co^{4+} and Ni^{4+}. One way to overcome this problem is to work with lower-valent redox couples like $Fe^{2+/3+}$. However, a decrease in the oxidation state raises the redox energy of the cathode and lowers the cell voltage. Recognizing this, and to keep the cost low, oxides containing polyanions such as XO_4^{2-} (X = S, Mo, and W) were proposed as lithium insertion hosts in the 1980s by Manthiram and Goodenough [92, 93]. Although the $Fe^{2+/3+}$ couple in a simple oxide like Fe_2O_3 would normally operate at a voltage of <2.5 V vs Li/Li^+, surprisingly the polyanion-containing $Fe_2(SO_4)_3$ host was found to exhibit 3.6 V vs Li/Li^+, while both $Fe_2(MoO_4)_3$ and $Fe_2(WO_4)_3$ were found to operate at 3.0 V vs Li/Li^+ (Figure 12.17). The remarkable increase in cell voltage on going from

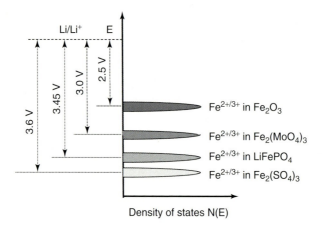

Figure 12.17 Positions of the $Fe^{2+/3+}$ redox energies relative to that of Li/Li^+ in various Fe-containing lithium insertion hosts and consequent changes in cell voltages, illustrating the role of polyanions.

a simple oxide such as Fe_2O_3 to polyanion hosts like $Fe_2(XO_4)_3$ and a difference of 0.6 V between the isostructural $Fe_2(SO_4)_3$ and $Fe_2(MoO_4)_3$ polyanion hosts, all operating with the same $Fe^{2+/3+}$ couple, were attributed to the influence of inductive effect and consequent differences in the location of the $Fe^{2+/3+}$ redox levels relative to the Li/Li^+ redox level, as seen in Figure 12.17. In the Nasicon-related $Fe_2(SO_4)_3$ and $Fe_2(MoO_4)_3$ hosts with corner-shared FeO_6 octahedra, XO_4 tetrahedra, and Fe–O–X–O–Fe (X = S, Mo, or W) linkages, the strength of the X–O bond can influence the Fe–O covalence and thereby the relative position of the $Fe^{2+/3+}$ redox energy. The stronger the X–O bonding, the weaker the Fe–O bonding, and consequently the lower the $Fe^{2+/3+}$ redox energy relative to that in a simple oxide like Fe_2O_3. The net result is a higher cell voltage on going from Fe_2O_3 to $Fe_2(MoO_4)_3$ or $Fe_2(SO_4)_3$. Comparing $Fe_2(MoO_4)_3$ and $Fe_2(SO_4)_3$, the S–O covalent bonding in $Fe_2(SO_4)_3$ is stronger compared to the Mo–O bonding in $Fe_2(MoO_4)_3$, leading to a weaker Fe–O covalence in $Fe_2(SO_4)_3$ than that in $Fe_2(MoO_4)_3$, resulting in a lowering of the $Fe^{2+/3+}$ redox energy in $Fe_2(SO_4)_3$ compared to that in $Fe_2(MoO_4)_3$ and a consequent increase in cell voltage by 0.6 V. Thus, the replacement of simple O^{2-} ions by XO_4^{n-} polyanions was recognized as a viable approach to tune the position of redox levels in solids and consequently to realize higher cell voltages with chemically more stable, lower-valent redox couples like $Fe^{2+/3+}$.

12.15
Phospho-Olivine LiMPO$_4$

Although the above findings in the late 1980s demonstrated an important fundamental concept in tuning the redox energies in solids, the cathode hosts pursued did not contain any lithium, so they could not be combined with the carbon anode in a lithium-ion cell. Following this initial concept, several phosphates have been investigated [94–96], and, in 1997, Goodenough's group identified LiFePO$_4$, crystallizing in the olivine structure (Figure 12.18), as a facile lithium extraction/insertion host that could be combined with a carbon anode in lithium-ion cells [94]. They also identified other olivines of the general formula LiMPO$_4$ (M = Mn, Co, and Ni) as lithium insertion/extraction hosts. Since its identification as a potential cathode, LiFePO$_4$ has been the subject of extensive studies from both scientific and technological points of view.

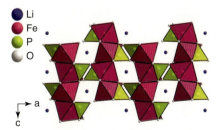

Figure 12.18 Crystal structure of olivine LiFePO$_4$.

In the initial work, fewer than 0.7 lithium ions were extracted per formula unit of LiFePO$_4$ even at very low current densities, which corresponds to a reversible capacity of <120 mAh g^{-1} [94]. The lithium extraction/insertion occurred via a two-phase mechanism with LiFePO$_4$ and FePO$_4$ as end members without much solid solubility. The limitation in capacity was attributed to the diffusion-limited transfer of lithium across the two-phase interface and poor electronic conductivity due to the corner-shared FeO$_6$ octahedra. Nevertheless, because Fe is abundant, inexpensive, and environmentally benign, olivine LiFePO$_4$ attracted immense interest as a potential cathode. Recognizing that the limited reversible capacity and low rate capability may be linked to the poor electronic conductivity, researchers investigated the possibility of improving the electronic conductivity by coating the LiFePO$_4$ powder with conductive carbon [97]. However, LiFePO$_4$ is a one-dimensional lithium-ion conductor with the lithium-ion diffusion occurring along edge-shared LiO$_6$ chains (b axis). It was therefore suggested that both intimate contact with conductive carbon and particle size minimization are necessary to optimize the electrochemical performance [98, 99]. Consequently, with a reduction in particle size and coating with conductive carbon, reversible capacity values of ~160 mAh g^{-1} were realized. Also, doping of LiFePO$_4$ with supervalent cations like Ti^{4+}, Zr^{4+}, Nb^{5+}, and organometallic precursors of the dopants was reported to increase the electronic conductivity by a factor of 10^8 [100]. Although this report attracted significant interest, subsequent investigations have suggested that the formation of a percolating nano-network of metallic iron phosphides may play a role in enhancing electronic conductivity [101].

Recognition of the importance of both the decrease in particle size and improvement in electronic conductivity has generated a flurry of activity in investigations into the solution-based synthesis of LiFePO$_4$ to minimize the particle size and coating the LiFePO$_4$ particles with conductive species such as carbon and conducting polymers [102–111]. Of these investigations, microwave-assisted hydrothermal and solvothermal approaches are appealing as they offer single-crystal LiFePO$_4$ with high crystallinity at significantly low temperatures of 230–300 °C within a relatively short reaction time of 5–15 min [108–111]. The products obtained by such approaches exhibit unique nanorod-like morphologies with excellent crystallinity (see the TEM fringe pattern), as seen in Figure 12.19, with the easy lithium diffusion direction (b axis) perpendicular to the long axis, which is beneficial for achieving high rate capability. In addition, the width and length of the nanorods depend on the synthesis conditions (e.g., reactant concentration), which could help to tune the rate capability and volumetric energy density.

While decreasing the particle size to the nanometer level has been successful in reducing the diffusion length of lithium-ions and overcoming the lithium-ion transport limitations in LiFePO$_4$, the pristine LiFePO$_4$ nanorods still suffer from poor electronic conductivity. In this context, addition of conducting polymers and nano-networking with multi-walled carbon nanotubes (MWCNTs) (Figures 12.20–12.22) have been found to offer significantly improved electrochemical performances [108, 110]. Figure 12.22 compares the discharge capacities

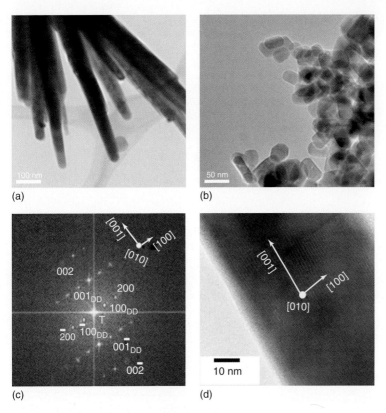

Figure 12.19 TEM images of (a) large LiFePO$_4$ nanorods, (b) small LiFePO$_4$ nanorods, and (c) Fast Fourier Transform (FFT), and (d) high-resolution TEM image of the LiFePO$_4$ nanorods.

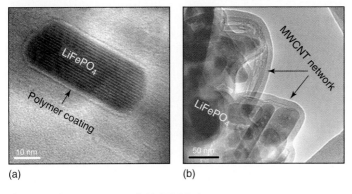

Figure 12.20 TEM images of (a) LiFePO$_4$/poly(3,4-ethylenedioxythiophene) (PEDOT) and (b) LiFePO$_4$/MWCNT nanocomposites.

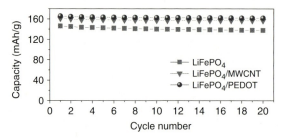

Figure 12.21 Cyclability of pristine LiFePO$_4$ prepared by the MW-ST method, after networking it with MWCNT, and after encapsulating it with p-toluene sulfonic acid (p-TSA) doped PEDOT.

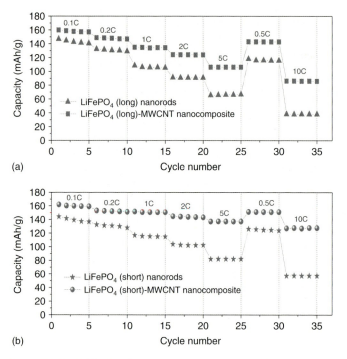

Figure 12.22 Cyclability of the (a) long LiFePO$_4$ nanorods and (b) short LiFePO$_4$ nanorods at different discharge rates from C/10 to 10 C before and after networking with MWCNT. The charging rate was kept constant at C/10 for all the samples.

at various C-rates for the long and short LiFePO$_4$ nanorods before and after networking with MWCNT. With both the pristine and MWCNT networked samples, the shorter LiFePO$_4$ nanorods exhibit higher discharge capacity at a given C-rate than the long nanorods due to a faster lithium-ion diffusion arising from a shorter diffusion length. With both the long and short nanorods, the LiFePO$_4$/MWCNT nanocomposites exhibit higher capacity at a given C-rate than the pristine LiFePO$_4$ due to the enhancement in electronic conductivity.

Although the initial work by Goodenough's group revealed a two-phase reaction mechanism with LiFePO$_4$ and FePO$_4$ as end members, subsequent investigations have indicated several interesting observations [112–115]. For example, the miscibility gap between the two phases has been found to decrease with increasing temperature [113], and the occurrence of a single-phase solid solution Li$_x$FePO$_4$ with $0 \leq x \leq 1$ has been reported at 450 °C. Similarly, the miscibility gap has been found to decrease with decreasing particle size, and complete solid solubility between LiFePO$_4$ and FePO$_4$ at room temperature has been reported for 40 nm size particles [114, 115]. Thus, what was originally found to be a two-phase reaction mechanism with micrometer-sized particles has now turned into a single-phase reaction mechanism with nano-sized particles. This is a clear demonstration of how nanoparticles can behave entirely differently from their micrometer-sized counterparts. Defects caused by the existence of cationic vacancies in the samples prepared at low temperatures have been suggested to contribute to the unique behavior of the nano-sized particles.

Replacing the transition-metal ion Fe^{2+} by Mn^{2+}, Co^{2+}, and Ni^{2+} increases the redox potential significantly from 3.45 V in LiFePO$_4$ to 4.1, 4.8, and 5.1 V, respectively, in LiMnPO$_4$, LiCoPO$_4$, and LiNiPO$_4$ because of the changes in the positions of the various redox couples (Figure 12.23). As we have seen earlier, the electronegativity of X and the strength of the X–O bond play a role in controlling the redox energies of metal ions in polyanion-containing samples. However in the case of LiMPO$_4$ cathodes, the polyanion PO$_4$ is fixed, so the shifts in the redox potential can only be associated with the changes in the M^{2+} cations. It is well known that the redox energies of transition metal M$^{2+/3+}$ couples decrease as we go from left to right on the periodic table because of the increase in the nuclear charge, the extra electrons being added to the same principal quantum number (e.g., 3d in

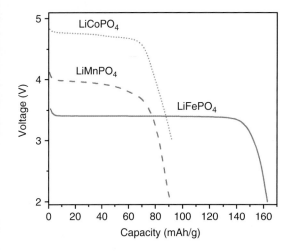

Figure 12.23 Discharge curves of LiMPO$_4$/C (M = Mn, Fe, and Co) nanocomposites.

the case of first row transition metals). However, LiFePO$_4$ exhibits a lower voltage (3.43 V) than LiMnPO$_4$ (4.13 V) despite Fe being to the right of Mn in the periodic table as the upper-lying t$_{2g}$ of Fe^{2+}:t$_{2g}^4$e$_g^2$ is the redox-active band (due to the pairing of the sixth electron in the t$_{2g}$ orbital) compared to the lower-lying e$_g$ of Mn^{2+}:t$_{2g}^3$e$_g^2$ (Figure 12.24). In addition, a systematic shift in the redox potential (open-circuit voltage) of the M$^{2+/3+}$ couples has been observed in the LiM$_{1-y}$M$_y$PO$_4$ (Mn, Fe, and Co) solid solutions compared to those of the pristine LiMPO$_4$. The potential of the lower-voltage couple increases, while that of the higher-voltage couple decreases in the LiM$_{1-y}$M$_y$PO$_4$ solid solutions compared to that of the pristine LiMPO$_4$. The shifts in the redox potentials have been explained by the changes in the M–O covalence (inductive effect) caused by the changes in the electronegativity of M or M–O bond length as well as by the influence of the M–O–M interactions in the solid solutions [116].

LiMnPO$_4$ is of particular interest because of the environmentally benign manganese and the favorable position of the Mn$^{2+/3+}$ redox couple at 4.1 V vs Li/Li$^+$, which is compatible with most of the electrolytes. However, it has been shown to offer low practical capacity even at low currents due to the wide band gap of ~2 eV and low electronic conductivity of ~10^{-14} S cm^{-1} compared to LiFePO$_4$, which has an electronic conductivity of ~10^{-9} S cm^{-1} and a band gap of ~0.3 eV [117, 118]. Optimizing the synthesis process and carbon coating has recently shown promising electrochemical performances for LiMnPO$_4$ nanoparticles [116, 119, 120]. On the other hand, the flat voltage profile at 4.8 V of LiCoPO$_4$ is desirable to increase the energy density, but its full theoretical capacity (~167 mAh g^{-1}) has not been

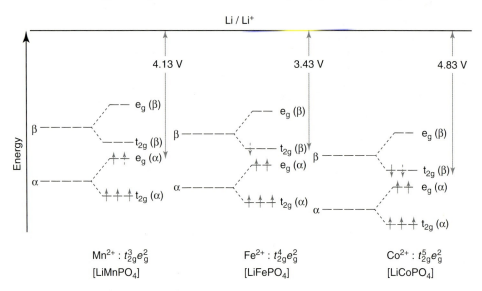

Figure 12.24 Crystal field splitting and 3d-orbital energy level diagram for the high-spin Mn^{2+}, Fe^{2+}, and Co^{2+} ions in olivine LiMPO$_4$. The electron involved in the redox reaction is shown with a dashed arrow.

realized in practical cells so far, and the cycle life of LiCoPO$_4$ is also poor because of the instability of the commonly used LiPF$_6$ in ethylene carbonate (EC):diethyl carbonate (DEC) electrolyte at these high operating voltages [111]. The redox potential of LiNiPO$_4$ is even higher at 5.1 V, which makes it difficult to analyze with the currently available electrolytes.

Recently, the mixed transition-metal ion systems have attracted considerable interest. Among these, the LiFe$_{1-y}$Mn$_y$PO$_4$ solid solution system has attracted particular attention as it exhibits higher energy density and improved redox kinetics due to improved electronic conductivity compared to LiMnPO$_4$ [116,121–124]. Alternatively, the LiMn$_{1-y}$Co$_y$PO$_4$ and LiFe$_{1-y}$Co$_y$PO$_4$ solid solutions are appealing because of their high operating voltages, which arise from the Co$^{2+/3+}$ redox couple. Even though LiMnPO$_4$, LiCoPO$_4$, and their solid solutions have higher theoretical energy densities than LiFePO$_4$, their experimental energy densities are lower than or equal to that of LiFePO$_4$ owing to their lower capacity values and high polarization losses. Further improvements in high-voltage electrolytes could allow efficient use of the high-voltage Co$^{2+/3+}$ redox couples and enhance the energy density.

12.16
Silicate Li$_2$MSiO$_4$

Even though LiFePO$_4$ has become successful because of its low cost and good safety characteristics, its energy density is limited because of the lower operating voltage (3.4 V), lower theoretical capacity (~170 mAh g^{-1}), and lower crystallographic density. In this regard, polyanion-containing frameworks that can accept two lithium ions per transition metal atom would help to realize high capacities and energy densities. Recently, a new class of silicates, Li$_2$MSiO$_4$ (M = Fe, Mn, and Co), has been introduced, which offers the possibility of reversibly extracting/inserting two lithium ions per formula unit with a theoretical capacity around ~333 mAh g^{-1} [125, 126]. Li$_2$MSiO$_4$ crystallizes in an orthorhombic β-Li$_3$PO$_4$ structure, with all the cations occupying tetrahedral sites.

Li$_2$FeSiO$_4$ has been shown to undergo an initial charge at around 3.1 V with a capacity of around 160 mAh g^{-1} and stable cycle life. The redox voltage of the Fe$^{2+/3+}$ couple is considerably lower than the 3.45 V in LiFePO$_4$ because of the lower electronegativity of Si compared to P. The stronger P–O covalent bonding in LiFePO$_4$ leads to a weaker Fe–O covalence and a consequent lowering of the Fe$^{2+/3+}$ redox energy in LiFePO$_4$ compared to the weaker Si–O covalent bonding in Li$_2$FeSiO$_4$, resulting in an increase in the cell voltage by ~0.3 V in LiFePO$_4$. It is important to note that the capacity corresponding to the Fe$^{3+/4+}$ couple has not been realized in Li$_2$FeSiO$_4$ at room temperatures [125]. Nevertheless, close to two lithium atoms per formula could be extracted in Li$_2$MnSiO$_4$ at ~4 V corresponding to the oxidation of Mn^{2+} to Mn^{3+} and then Mn^{3+} to Mn^{4+}. In contrast to Li$_2$FeSiO$_4$, the extraction of both the electrons occurs around the same potential range in Li$_2$MnSiO$_4$ because of the extraction of the electron from the same e$_g$ redox level for both the Mn$^{2+/3+}$ and Mn$^{3+/4+}$ redox couples.

However, Li_2MnSiO_4 suffers from poor cycle life, which is most likely caused by Jahn–Teller distortion and loss of crystallinity during cycling. In addition, these materials also suffer from poor electronic conductivity and the consequent slow reaction kinetics. Therefore, various synthetic routes such as sol-gel and microwave-solvothermal methods have been employed to prepare nanostructured materials, and coating with electronically conductive agents has been carried out to improve their electrochemical performance [127–129].

12.17
Other Polyanion-containing Cathodes

Polyanion-containing compounds other than $LiMPO_4$, such as $Li_3M_2(PO_4)_3$ (M = V, Fe, or Ti), $LiVPO_4F$, $Li_5V(PO_4)_3$, and $LiVOPO_4$, have also attracted a great deal of interest in recent years because of their high thermal stability and attractive electrochemical properties [130–134]. For example, $LiVPO_4F$ is isostructural with the mineral tavorite ($LiFePO_4OH$) and has been shown to exhibit a capacity of ~156 mAh g^{-1} with a flat voltage profile (two-phase reaction) at 4.2 V corresponding to the $V^{3+/4+}$ redox couple. The main drawbacks of these materials are their poor electronic conductivity and consequently slow reaction kinetics.

Recently, $LiFeSO_4F$ synthesized by a nonaqueous solvothermal process employing ionic liquids (iono-thermal) has been shown to deliver capacities of around 140 mAh g^{-1} at 3.6 V [135]. Even though the theoretical specific capacity of $LiFeSO_4F$ (151 mAh g^{-1}) is slightly lower than that of $LiFePO_4$ (170 mAh g^{-1}), the material is proposed to provide better ionic and electronic conductivities, which could eliminate the need for carbon coating or nanoparticles.

12.18
Summary

This chapter presents an overview of the structural characteristics, chemical stability, and electrochemical properties of various lithium-insertion cathode materials. The materials systems that are currently used are layered, spinel oxides, and polyanion-containing cathodes. Although only 50% of the lithium can be extracted from $LiCoO_2$, limiting its practical capacity to ~140 mAh g^{-1}, a recently discovered class of layered oxide solid solution cathodes belonging to the series xLi_2MnO_3–$(1 - x)LiMO_2$ deliver capacities of ~250 mAh g^{-1}. However, these cathodes release oxygen during first charge, and their adoption needs more robust electrolytes that can operate up to 4.8 V. Although the conventional $LiMn_2O_4$ spinel is plagued by severe capacity fade, spinel cathodes have been optimized by doping the cation sites, as in the case of $LiMn_{1.8}Ni_{0.1}O_4$, which provides more stable cycle life and higher rate performance. These characteristics make spinels an attractive option for electric vehicle applications, but they commonly have low energy densities. This can be mitigated by using the 5 V region in $LiMn_{1.5}Ni_{0.5}O_4$ to realize higher energy densities, but the long-term cycling stability of these high-voltage cathodes has yet to be established, and their successful application requires the development of more stable high-voltage electrolytes.

In recent years, the phospho-olivine LiFePO$_4$ cathode has attracted much attention because of its excellent cycle life, high chemical stability, and good safety characteristics compared to the layered and spinel oxide cathodes. However, its lower cell voltage (3.45 V) and crystallographic density limit the energy density. Although other olivines like LiCoPO$_4$ and LiNiPO$_4$ offer higher energy densities due to higher voltages (4.8–5.1 V), the instability of conventional electrolytes at these higher voltages limits their adoptability. In this regard, the challenge is to develop a polyanion-containing framework that can allow for extraction/insertion of more than one lithium ion per transition metal ion with a voltage around 4 V without undergoing any major structural changes during cycling.

Overall, there is immense interest in lowering the cost, improving the safety, and increasing the energy and power densities so that the lithium-ion technology can be successfully employed for vehicle and stationary storage applications. The development of (a) new materials that can allow the insertion/extraction of more than one lithium per transition metal ion without unduly increasing the insertion host weight or (b) more stable, robust electrolytes to utilize the high-voltage (>4.3 V) cathodes is needed to move the field forward.

Acknowledgments

This work was supported by the Welch Foundation grant **F-1254**, Office of Vehicle Technologies of the U.S. Department of Energy under Contract No. **DE-AC02-05CH11231**, and NASA contract NNC09CA08C.

References

1. Lithium Battery Energy Storage (LIBES) Publications, Technological Research Association, Tokyo (1994).
2. Whittingham, M.S. (1976) *Science*, **192**, 1126.
3. Whittingham, M.S. and Jacobson, A.J. (1982) *Intercalation Chemistry*, Academic Press, New York.
4. Mizushima, K., Jones, P.C., Wiesman, P.J., and Goodenough, J.B. (1980) *Mater. Res. Bull.*, **15**, 783.
5. Goodenough, J.B., Mizushima, K., and Takeda, T. (1983) *Jpn. J. Appl. Phys.*, **19**, 305.
6. Thackeray, M.M., David, W.I.F., Bruce, P.G., and Goodenough, J.B. (1983) *Mater. Res. Bull.*, **18**, 461.
7. Megahed, S. and Scrosati, B. (1995) *Interface*, **4** (4), 34.
8. Takehare, Z. and Kanamura, K. (1993) *Electrochim. Acta*, **38**, 1169.
9. Aydinol, M.K. and Ceder, G.J. (1997) *J. Electrochem. Soc.*, **144**, 3832.
10. Whittingham, M.S. (2004) *Chem. Rev.*, **104**, 4271.
11. Tukamoto, H. and West, A.R. (1997) *J. Electrochem. Soc.*, **144**, 3164.
12. Gummow, R.J., Thackeray, M.M., David, W.I.F., and Hull, S. (1992) *Mater. Res. Bull.*, **27**, 327.
13. Gummow, R.J., Liles, D.C., and Thackeray, M.M. (1993) *Mater. Res. Bull.*, **28**, 1177.

14. Choi, S. and Manthiram, A. (2002) *J. Electrochem. Soc.*, **149**, A162.
15. Chebiam, R.V., Prado, F., and Manthiram, A. (2001) *Chem. Mater.*, **13**, 2951.
16. Venkatraman, S., Shin, Y., and Manthiram, A. (2003) *Electrochem. Solid State Lett.*, **6**, A9.
17. Venkatraman, S. and Manthiram, A. (2002) *Chem. Mater.*, **14**, 3907.
18. Kim, J., Noh, M., Cho, J., and Kim, H. (2005) *J. Electrochem. Soc.*, **152**, A1142.
19. Kim, B., Kim, C., Kim, T., Ahn, D., and Park, B. (2006) *J. Electrochem. Soc.*, **153**, A1773.
20. Kim, Y.J., Cho, J., Kim, T.J., and Park, B. (2003) *J. Electrochem. Soc.*, **150**, A1723.
21. Liu, L., Wang, Z., Li, H., Chen, L., and Huang, X. (2002) *Solid State Ionics*, **152**, 341.
22. Cho, J., Kim, C.S., and Yoo, S.I. (2000) *Electrochem. Solid State Lett.*, **3**, 362.
23. Kannan, A.M., Rabenberg, L., and Manthiram, A. (2003) *Electrochem. Solid State Lett.*, **6**, A16.
24. Wang, Z., Huang, X., and Chen, L. (2003) *J. Electrochem. Soc.*, **150**, A199.
25. Fang, T., Duh, J., and Sheen, S. (2005) *J. Electrochem. Soc.*, **152**, A1701.
26. Cho, J., Kim, Y.W., Kim, B., Lee, J.G., and Park, B. (2003) *Angew. Chem. Int. Ed.*, **42**, 1618.
27. Dutta, G., Manthiram, A., and Goodenough, J.B. (1992) *J. Solid State Chem.*, **96**, 123.
28. Hirano, A., Kanno, R., Kawamoto, Y., Takeda, Y., Yamamura, K., Takeda, M., Ohyama, K., Ohashi, M., and Yamaguchi, Y. (1995) *Solid State Ionics*, **78**, 123.
29. Kanno, R., Kubo, H., Kawamoto, Y., Kamiyama, T., Izumi, F., and Takano, Y. (1998) *J. Solid State Chem.*, **140**, 145.
30. Choi, S. and Manthiram, A. (2002) *J. Electrochem. Soc.*, **149**, A1157.
31. Huheey, J.E. (1978) *Inorganic Chemistry*, 2nd edn, Harper and Row, New York, p. 348.
32. Dahn, J.R., Fuller, E.W., Obrovac, M., and Von Sacken, U. (1994) *Solid State Ionics*, **69**, 265.
33. Zhang, Z., Fouchard, D., and Rea, J.R. (2001) *J. Power Sources*, **70**, A49.
34. Li, W. and Curie, J. (1997) *J. Electrochem. Soc.*, **144**, 1773.
35. Im, D. and Manthiram, A. (2002) *J. Electrochem. Soc.*, **149**, A1001.
36. Kim, H.B., Park, B.C., Myung, S.T., Amine, K., Prakash, J., and Sun, Y.K. (2008) *J. Power Sources*, **179**, 347.
37. Yabuuchi, N., Kim, Y.T., Li, H.H., and S-Horn, Y. (2008) *Chem. Mater.*, **20**, 4936.
38. Choi, J. and Manthiram, A. (2005) *Electrochem. Solid State Lett.*, **8**, C102.
39. Choi, J. and Manthiram, A. (2005) *J. Electrochem. Soc.*, **152**, A1714.
40. Mishra, S.K. and Ceder, G. (1999) *Phys. Rev. B.*, **59**, 6120.
41. Armstrong, A.R. and Bruce, P.G. (1996) *Nature*, **381**, 499.
42. Capitaine, F., Gravereau, P., and Delmas, C. (1996) *Solid State Ionics*, **89**, 197.
43. Vitins, G. and West, K. (1997) *J. Electrochem. Soc.*, **144**, 2587.
44. Numata, K., Sakaki, C., and Yamanaka, S. (1999) *Solid State Ionics*, **117**, 257.
45. Lu, Z.H., Beaulieu, L.Y., Donaberger, R.A., Thomas, C.L., and Dahn, J.R. (2002) *J. Electrochem. Soc.*, **149**, A778.
46. Armstrong, A.R., Holzapfel, M., Novak, P., Johnson, C.S., Kang, S., Thackeray, M.M., and Bruce, P.G. (2006) *J. Am. Chem. Soc.*, **128**, 8694.
47. Wu, Y. and Manthiram, A. (2006) *Electrochem. Solid State Lett.*, **9**, A221.
48. Arunkumar, T.A., Wu, Y., and Manthiram, A. (2007) *Chem. Mater.*, **19**, 3067.
49. Tackeray, M.M., Johnson, C.S., Vaughey, J.T., Li, N., and Hackney, S.A. (2005) *J. Mater. Chem.*, **15**, 2257.
50. Johnson, C.S., Li, N., Vaughey, J.T., Hackney, S.A., and Thackeray, M.M. (2005) *Electrochem. Commun.*, **7**, 528.
51. Kim, J.S., Johnson, C.S., and Thackeray, M.M. (2002) *Electrochem. Commun.*, **4**, 205.
52. Rossouw, M.H. and Thackeray, M.M. (1991) *Mater. Res. Bull.*, **26**, 463.
53. Kalyani, P., Chitra, S., Mohan, T., and Gopukumar, S. (1999) *J. Power Sources*, **80**, 103.

54. Lu, Z.H. and Dahn, J.R. (2002) *J. Electrochem. Soc.*, **149**, A815.
55. Wu, Y., Vadivel Murugan, A., and Manthiram, A. (2008) *J. Electrochem. Soc.*, **155**, A635.
56. Wu, Y. and Manthiram, A. (2009) *Solid State Ionics*, **180**, 50.
57. Wang, Q.Y., Liu, J., Vadivel Murugan, A., and Manthiram, A. (2009) *J. Mater. Chem.*, **19**, 4965.
58. de Picciotto, L.A., Thackeray, M.M., David, W.I.F., Bruce, P.G., and Goodenough, J.B. (1984) *Mater. Res. Bull.*, **19**, 1497.
59. Shimode, M., Sasaki, M., and Mukaida, K.I. (2000) *Mol. Cryst. Liq. Cryst.*, **341**, 183.
60. Ohzuku, T., Kitagawa, M., and Hirai, T. (1990) *J. Electrochem. Soc.*, **137**, 769.
61. Thackeray, M.M., Shao-Horn, Y., Kahaian, A.J., Kepler, K.D., Skinner, E., Vaughey, J.T., and Hackney, S.A. (1998) *Electrochem. Solid State Lett.*, **1**, 7.
62. Jang, D.H., Shin, Y.J., and Oh, S.M. (1996) *J. Electrochem. Soc.*, **143**, 2204.
63. Inoue, T. and Sano, M. (1998) *J. Electrochem. Soc.*, **145**, 3704.
64. Xia, Y. and Yoshio, M. (1998) *J. Electrochem. Soc.*, **143**, 825.
65. Kannan, A.M. and Manthiram, A. (2002) *Electrochem. Solid State Lett.*, **5**, A167.
66. Scrosati, B. (2000) *Electrochim. Acta*, **45**, 2461.
67. Thackeray, M.M., Manusuetto, M.F., Dees, D.W., and Vissers, D.R. (1996) *Mater. Res. Bull.*, **31**, 133.
68. Gao, Y. and Dahn, J.R. (1996) *J. Electrochem. Soc.*, **143**, 1783.
69. Thackeray, M.M., Manusuetto, M.F., and Johnson, C.S. (1996) *J. Solid State Chem.*, **125**, 274.
70. Kim, J. and Manthiram, A. (1998) *J. Electrochem. Soc.*, **145**, L53.
71. Kawai, H., Nagata, M., Takamoto, H., and West, A.R. (1998) *Electrochem. Solid State Lett.*, **1**, 212.
72. Shin, Y. and Manthiram, A. (2003) *Chem. Mater.*, **15**, 2954.
73. Shin, Y. and Manthiram, A. (2003) *Electrochem. Solid-State Lett.*, **6**, A34.
74. Amatucci, G., Pereira, N., Zheng, T., and Tarascon, J.M. (2001) *J. Electrochem. Soc.*, **148**, A171.
75. Kang, Y.J., Kim, J.H., and Sun, Y.K. (2005) *J. Power Sources*, **146**, 237.
76. Oh, S.W., Park, S.H., Kim, J.H., Bae, Y.C., and Sun, Y.K. (2006) *J. Power Sources*, **157**, 464.
77. He, X., Li, J., Cai, Y., Wang, Y., Ying, J., Jiang, C., and Wan, C. (2005) *Solid State Ionics*, **176**, 2571.
78. Choi, W. and Manthiram, A. (2006) *Electrochem. Solid State Lett.*, **9**, A245.
79. Choi, W. and Manthiram, A. (2007) *J. Electrochem. Soc.*, **154**, A792.
80. Luo, Q., Muraliganth, T., and Manthiram, A. (2009) *Solid State Ionics*, **180**, 703.
81. Luo, Q. and Manthiram, A. (2009) *J. Electrochem. Soc.*, **156**, A84.
82. Sun, Y., Hong, K., and Prakash, J. (2003) *J. Electrochem. Soc.*, **150**, A970.
83. Han, J., Myung, S., and Sun, Y. (2006) *J. Electrochem. Soc.*, **153**, A1290.
84. Thackeray, M.M., Johnson, C.S., Kim, J.S., Lauzze, K.C., Vaughey, J.T., Dietz, N., Abraham, D., Hackney, S.A., Zeltner, W., and Anderson, M.A. (2003) *Electrochem. Commun.*, **5**, 752.
85. Amine, K., Tukamoto, H., Yasuda, H., and Fujita, Y. (1996) *J. Electrochem. Soc.*, **143**, 1607.
86. Zhong, Q., Bonakdarpour, A., Zhang, M., Gao, Y., and Dahn, J.R. (1997) *J. Electrochem. Soc.*, **144**, 205.
87. Kim, J.H., Myung, S.T., Yoon, C.S., Oh, I.H., and Suna, Y.K. (2004) *J. Electrochem. Soc.*, **151**, A1911.
88. Arun Kumar, T.A. and Manthiram, A. (2005) *Electrochem. Solid State Lett.*, **8**, A403.
89. Jun, L. and Manthiram, A. (2009) *Chem. Mater.*, **21**, 1695.
90. Murphy, D.W., Cava, R.J., Zahurak, S.M., and Santoro, A. (1983) *Solid State Ionics*, **9&10**, 413.
91. Choi, S. and Manthiram, A. (2002) *J. Solid State Chem.*, **164**, 332.
92. Manthiram, A. and Goodenough, J.B. (1989) *J. Power Sources*, **26**, 403.
93. Manthiram, A. and Goodenough, J.B. (1987) *J. Solid State Chem.*, **71**, 349.
94. Padhi, A.K., Nanjundaswamy, K.S., Masquelier, C., Okada, S., and

Goodenough, J.B. (1997) *J. Electrochem. Soc.*, **144**, 1609.

95. Padhi, A.K., Nanjundaswamy, K.S., Masquelier, C., and Goodenough, J.B. (1997) *J. Electrochem. Soc.*, **144**, 2581.

96. Padhi, A.K., Nanjundaswamy, K.S., and Goodenough, J.B. (1997) *J. Electrochem. Soc.*, **144**, 1188.

97. Ravet, N., Goodenough, J.B., Benser, S., Simoneau, M., Hovington, P., and Armond, M. (1999) The Electrochemical Society and The Electrochemical Society of Japan Meeting Abstracts, Vol. 99-2, Abstract 127, October 17–22, 1999, Honolulu, HI.

98. Huang, H., Yin, S.-C., and Nazar, L.F. (2001) *Electrochem. Solid State Lett.*, **4**, A170.

99. Yamada, A., Chung, S.C., and Hinokuma, K. (2001) *J. Electrochem. Soc.*, **148**, A224.

100. Chung, S.-Y., Bloking, J.T., and Chiang, Y.-M. (2002) *Nat. Mater.*, **1**, 123.

101. Herle, P.S., Ellis, B., Coombs, N., and Nazar, L.F. (2004) *Nat. Mater.*, **3**, 147.

102. Yang, S., Zavalji, P., and Whittingham, M.S. (2001) *Electrochem. Commun.*, **3**, 505.

103. Wang, Y., Wang, J., Yang, J., and Nuli, Y. (2006) *Adv. Funct. Mater.*, **16**, 2135.

104. Dominko, R., Gaberscek, M., Drofenik, J., Bele, M., Pejovnik, S., and Jamnik, J. (2003) *J. Power Sources*, **119**, 770.

105. Ellis, B., Ka, W.H., Makahnou, W.R.M., and Nazar, L.F. (2007) *J. Mater. Chem.*, **17**, 3248.

106. Wang, C.S. and Hong, J. (2003) *Electrochem. Solid State Lett.*, **10**, A65.

107. Doeff, M.M., Hu, Y., McLarnon, F., and Kostecki, R. (2003) *Electrochem. Solid State Lett.*, **6**, A207.

108. Vadivel Murugan, A., Muraliganth, T., and Manthiram, A. (2008) *Electrochem. Commun.*, **10**, 903.

109. Vadivel Murugan, A., Muraliganth, T., and Manthiram, A. (2008) *J. Phys. Chem. C*, **112**, 14665.

110. Muraliganth, T., Vadivel Murugan, A., and Manthiram, A. (2008) *J. Mater. Chem.*, **18**, 5661.

111. Vadivel Murugan, A., Muraliganth, T., and Manthiram, A. (2009) *Inorg. Chem.*, **48**, 946.

112. Yamada, A., Koizumi, H., Nishimura, S.I., Sonoyama, N., Kanno, R., Yonemura, M., Nakamura, T., and Kobayashi, Y. (2006) *Nat. Mater.*, **5**, 357.

113. Delacourt, C., Poizot, P., Tarascon, J.M., and Masquelier, C. (2005) *Nat. Mater.*, **4**, 254.

114. Gibot, P., Cabans, M.C., Laffont, L., Levasseur, S., Carlach, P., Hamelet, S., Tarascon, J.M., and Masqulier, C. (2008) *Nat. Mater.*, **7**, 741.

115. Meethong, N., Carter, W.C., and Chiang, Y.M. (2007) *Electrochem. Solid State Lett.*, **10**, A13.

116. Muraliganth, T. and Manthiram, A. (2010) *J. Phys. Chem. C*, **114**, 15530.

117. Delacourt, C., Laffont, L., Bouchet, R., Wurn, C., Leriche, J.B., Morcrette, M., Tarascon, J.M., and Masquelier, C.J. (2005) *J. Electrochem. Soc.*, **152**, A913.

118. Sauvage, F., Baudrin, E., Gengembre, L., and Tarascon, J.M. (2005) *Solid State Ionics*, **176**, 1869.

119. Delacourt, C., Poizot, P., Morcrette, M., Tarascon, J.M., and Masquelier, C. (2009) *Chem. Mater.*, **16**, 93.

120. Kwon, N.H., Drezen, T., Exnar, I., Teerlinck, I., Isono, M., and Graetzel, M. (2006) *Electrochem. Solid State Lett.*, **9**, A277.

121. Yamada, A., Takei, Y., Koizumi, H., Sonoyama, N., Kanno, R., Itoh, K., Yonemura, M., and Kamiyama, T. (2006) *Chem. Mater.*, **18**, 804.

122. Kim, J.K., Chauhan, G.S., Ahn, J.H., and Ahn, H.J. (2009) *J. Power Sources*, **189**, 391.

123. Zaghib, K., Mauger, A., Gendron, F., Massot, M., and Julien, C.M. (2008) *Ionics*, **14**, 371.

124. Molenda, J., Ojczyk, W., and Marzec, J. (2007) *J. Power Sources*, **174**, 689.

125. Nyten, A., Abouimrane, A., Armand, M., Gustafsson, T., and Thomas, J.O. (2005) *Electrochem. Commun.*, **7**, 156.

126. Nishimura, A.I., Hayese, S., Kanno, R., Yashima, M., Nakayama, N., and Yamada, A. (2008) *J. Am. Chem. Soc.*, **130** (40), 13212.

127. Dominko, R., Bele, M., Gaberšček, M., Meden, A., Remskar, M., and Jamnik, J. (2006) *Electrochem. Commun.*, **8**, 217.

128. Gong, Z.L., Li, Y.X., He, G.N., Li, J., and Yang, Y. (2008) *Electrochem. Solid State Lett.*, **11** (5), A60.
129. Muraliganth, T., Stroukoff, K.R., and Manthiram, A. (2010) *Chem. Mater.*, **22**, 5754.
130. Saidi, M.Y., Barker, J., Huang, H., Sowyer, J.L., and Adamson, G. (2003) *J. Power Sources*, **119**, 266.
131. Yin, S.C., Grond, H., Srobel, P., Huang, H., and Nazar, L.F. (2003) *J. Am. Chem. Soc.*, **125**, 326.
132. Hung, H., Yin, S.C., Kerr, T., Taylor, N., and Nazar, L.F. (2002) *Adv. Mater.*, **14**, 1525.
133. Barker, J., Saidi, M.Y., and Swoyer, J.L. (2003) *J. Electrochem. Soc.*, **150**, A1394.
134. Ren, M.M., Zhou, Z., Gao, X.P., Liu, L., and Peng, W.X. (2008) *J. Phys. Chem. C*, **112**, 13043.
135. Recham, N., Chotard, J.-N., Dupont, L., Delacourt, C., Walker, W., Armand, M., and Tarascon, J.-M. (2009) *Nat. Mater.*, **9**, 68.

13
Rechargeable Lithium Anodes
Jun-ichi Yamaki and Shin-ichi Tobishima

13.1
Introduction

The need to increase the energy density of rechargeable cells has become more urgent as a result of the recent rapid development of new applications, such as electric vehicles, load leveling, and various types of portable equipment, including personal computers, cellular phones, and camcorders. For these practical applications, lithium ion cells have been used since 1991. Many commercial lithium ion cells are composed of a carbon anode and an $LiCoO_2$ cathode with nonaqueous electrolyte solutions. However, the capacity of carbon anodes is now getting closer to the theoretical value (372 mAh g^{-1}). Moreover, new anode materials having higher energy density than carbon have been studied. A lithium-metal anode is an attractive way of delivering the high energy density from such cells. The lithium-metal anode has a very large theoretical capacity of 3860 mAh g^{-1}, in contrast to the value of 372 mAh g^{-1} for an LiC_6 carbon anode. This high energy encourages an attempt to realize a practical lithium-metal anode cell. As an example of these applications requiring supremely high energy density cells, there is a project of New Energy and the Industrial Technology Development Organization (NEDO) in Japan. This project started in 2007. One of the targets of this project is to realize battery packs (not cells) having an energy density exceeding 700 Wh kg^{-1} for future electric vehicles. This value is about 10 times greater than that of present commercial lithium ion cells. In this project, many studies on new high-energy systems have been carried out including lithium metal anode air-rechargeable batteries.

Primary lithium metal cells, including cylindrical and coin-type cells, have been manufactured as high-energy cells since 1973 (Panasonic, Li/polyfluorocarbon cell). In addition, several small rechargeable coin-type cells have appeared on the market (cell capacity < 100 mAh), which are generally categorized as lithium cells. These small rechargeable cells do not employ pure lithium metal as their anode, but have anodes of lithium-metal alloys (Table 13.1). Cells with lithium-ion-inserted compounds have been commercialized more recently (Table 13.2). In spite of their lower energy density, these alternative anodes are used because pure lithium tends

Handbook of Battery Materials, Second Edition. Edited by Claus Daniel and Jürgen O. Besenhard.
© 2011 Wiley-VCH Verlag GmbH & Co. KGaA. Published 2011 by Wiley-VCH Verlag GmbH & Co. KGaA.

Table 13.1 Commercially available rechargeable coin-type cells with lithium-metal alloys.

Anode	Cathode	Cell voltage (V)	Main application	Manufacturer
Pb–Cd–Bi–Sn(–Li)	Carbon	3	Memory backup	Panasonic
Li–Al	c-V_2O_5	3	Memory backup	Panasonic
Li–Al–Mn	$Li_2MnO_3 + \gamma\text{-}\beta\text{-}MnO_2$	3	Memory backup	Sanyo
Li–Al	Polyaniline	3	Memory backup	Seiko/Bridgestone

Abbreviation: c-V_2O_5, crystalline V_2O_5.

Table 13.2 Commercially available rechargeable coin-type cells with lithium-ion inserted anodes.

Anode	Cathode	Cell voltage (V)	Main application	Manufacturer
$Li_xNb_2O_5$	c-V_2O_5	1.5	Watches	Panasonic
Li_xTiO_2	Li_xMnO_2	1.5	Watches	Panasonic
Carbon (–Li)	c-V_2O_5	3	Memory backup	Toshiba
Polyacenic semiconductor (–Li)	Polyacenic semiconductor (–Li)	2	Memory backup	Kanebo
Carbon (–Li)	$Li_xTi_yO_z$	1.5	Memory backup	Hitachi Maxell

Abbreviation: c-V_2O_5, crystalline V_2O_5.

to be deposited on the anode in dendritic form when the cell is charged. These dendrites may cause an internal short as well as a decrease in lithium cycling efficiency.

The alternative alloy anodes, which exhibit good cycle life in coin cells (Table 13.1), are not applied to cylindrical cells. This is because they are brittle, and these alloy anodes turn into fine particles after cycling when the anode is spirally wound in the cylindrical cell. Cylindrical rechargeable lithium-metal cells, such as AA-size cells, are not yet commercially available. Several prototype AA cells with pure lithium anodes have been developed since late 1980 (Table 13.3). However, their cycle life depends on the discharge and charge currents. This problem results from the low cycling efficiency of lithium anodes. Another big problem is the safety of lithium-metal cells. Reasons for their poor thermal stability include the high reactivity and low melting point (180 °C) of lithium.

Nippon Telegraph and Telephone Corporation (NTT) has developed a prototype AA-size lithium-metal anode cell [1] with an amorphous V_2O_5 cathode. The energy of this cell is 2 Wh, which is higher than the value of 1.8 Wh for a lithium-ion cell with an $LiCoO_2$ cathode. (The estimated value for an AA-size lithium-metal cell

Table 13.3 Prototype AA-size rechargeable lithium-metal cells.

Cathode	Voltage (V)	Weight (g)	Capacity (Ah)	Energy (Wh)	Energy density (Wh kg^{-1})^a	(Wh L^{-1})^a	Cycle life^b	Organization
NbSe$_3$	2	20.5	1.1	2.2	107	293	200–400	ATT
TiS$_2$	2.1	15.6	1.0	2.1	135	280	200–250	Duracel
Li$_x$MnO$_2$	2.8	16.3	0.7	2.0	125	267	200–400	Sony
Li$_x$MnO$_2$	2.8	17	0.75	2.1	124	280	100–250	Tadiran
MoS$_2$^{c}	1.8	22	0.8	1.4	64	187	200	Moli Energy
a-V$_2$O$_5$	2.3	18	0.9	2.0	110	267	150–300	NTT

^{a}Assumed cell volume $= 7.5$ cm^3.
^{b}100% depth of discharge; cycle life depends on cycling current.
^{c}B-type cell.
Abbreviation: ATT, American Telephone and Telegraph Corporation.

with an LiCoO$_2$ cathode is about 3 Wh.) However, this metal-anode cell has a cycle life of 150 cycles, and its thermal stability is 130 °C at a high discharge rate. At a low discharge rate, it has a cycle life of 50 cycles and its thermal stability is 125 °C. These values are poor compared with those for lithium-ion cells, whose corresponding values are 500 cycles and >130 °C. This poor performance is explained mainly by the characteristics of the lithium-metal anode, and specifically its low cycling efficiency.

Many studies have been undertaken with a view to improving lithium anode performance to obtain a practical cell. This section will describe recent progress in the study of lithium-metal anodes and the corresponding cells. Sections 13.2–13.7 describe studies on the surfaces of uncycled lithium and lithium coupled with electrolytes, methods for measuring the cycling efficiency of lithium, the morphology of deposited lithium, the mechanism of lithium deposition and dissolution, the amount of dead lithium, the improvement of cycling efficiency, and alternatives to the lithium-metal anode. Section 13.8 describes the safety of rechargeable lithium-metal cells.

13.2
Surface of Uncycled Lithium Foil

Lithium foil is commercially available. Its surface is covered with a 'native film' consisting of various lithium compounds (LiOH, Li$_2$O, Li$_3$N, (Li$_2$O–CO$_2$) adduct, or Li$_2$CO$_3$). These compounds are produced by the reaction of lithium with O$_2$, H$_2$O, CO$_2$, or N$_2$. These compounds can be detected by electron spectroscopy for chemical analysis (ESCA) [2]. As mentioned below, the surface film is closely related to the cycling efficiency.

Lithium foil is made by extruding a lithium ingot through a slit. A study of the influence of the extrusion atmosphere on the kind of native film produced showed that lithium covered with Li_2CO_3 is superior both in terms of storage and discharge because of its stability and because a lithium anode has a low impedance [3, 4].

13.3
Surface of Lithium Coupled with Electrolytes

Lithium metal is chemically very active and reacts thermodynamically with any organic electrolyte. However, in practice, lithium metal can be dissolved and deposited electrochemically in some organic electrolytes [5]. It is generally believed that a protective film is formed on the lithium anode which prevents further reaction [6, 7]. This film strongly affects the lithium cycling efficiency.

According to the solid electrolyte interphase (SEI) model presented by Peled [8], the reaction products of the lithium and the electrolyte form a thin protective film on the lithium anode. This film is a lithium-ion conductor and an electronic insulator, whose nature prevents any further chemical reaction. Aurbach *et al.* and many other research workers have tried to identify the chemical products composing the protection film [9–20] using Fourier transform infrared spectroscopy (FTIR), X-ray photoelectron spectroscopy (XPS), and Raman spectroscopy. The protective films differ depending on the kind of electrolyte, and mainly consist of Li_2CO_3, LiX (X: halogen), ROLi, and ROCOOLi (R: alkyl group). $LiPF_6$ solute forms LiF, and $LiAsF_6$ solute forms some As compounds as the protective films. Chemical composition and physical structure of the surface films are affected by H_2O content of electrolyte solutions and remaining water inside the cell. Ethylene carbonate (EC) and propylene carbonate (PC) are electrolyte solvents with very similar chemical structures but providing different lithium cycling efficiencies. Aurbach *et al.* have reported differences between the lithium surface films in EC and PC, namely that $CH_3CH(OCO_2Li)$ CH_2OCO_2Li and $(CH_2OCO_2Li)_2$ are detected in the lithium surface film with dry PC and EC, respectively [21].

The reaction of the electrolyte with lithium and the resulting film properties affect the cycle life of the lithium cell. Shen *et al.* [22] have examined the stability (reactivity) of the electrolytes by open-circuit storage tests for the Li/TiS_2 cell system by microcalorimetry and alternate current (AC) impedance spectroscopy. They used tetrahydrofuran (THF)-and 2-methyl-tetrahydrofuran (2MeTHF)-based electrolyte, with additives such as 2-methylfuran (2MeF), EC, PC, and 3-methylsulfolane (3MeS), and $LiAsF_6$ as the solute. The heat output of the cells on open circuit for a day (short-term reactivity) or a year (long-term reactivity) is lower for EC/2MeTHF than for 2MeTHF or PC/2MeTHF. Also, the cell with EC/2MeTHF has a lower SEI resistivity of 51 Ωcm^2 than that with 2MeTHF (119 Ωcm^2) or PC/2MeTHF (214 Ωcm^2). The cycle life increases with decreases in heat output and resistivity. They indicate that these measurements are effective in determining electrolyte stability.

13.4
Cycling Efficiency of Lithium Anode

13.4.1
Measurement Methods

Lithium deposited on an anode during a charge is chemically active and reacts with organic electrolytes after deposition. Thus, the lithium is consumed during cycling. The cycling efficiency (percent) of a lithium anode (Eff) is basically defined by Equation 13.1 [23], where Q_p is the amount of electricity needed to plate lithium and Q_s is the amount of electricity needed to strip all the plated lithium. As Eff is less than 100%, an excess of lithium is included in a practical rechargeable cell to compensate for the consumed lithium.

$$\text{Eff} = 100 \times Q_s/Q_p \tag{13.1}$$

The figure of merit (FOM) for lithium cycling efficiency [24] is also often used to evaluate the cyclability of a lithium cell. The *FOM* is defined as the number of cycles completed by one atom of lithium before it becomes electrochemically inactive. Equation 13.2 is derived from the above definition.

$$\text{FOM} = \frac{\text{sum of each discharge capacity to the end of cycle life}}{\text{capacity of lithium anode cell}} \tag{13.2}$$

We can calculate the FOM from Eff, using Equation 13.3 [25].

$$\text{FOM} = \frac{1}{1 - \text{Eff}/100} \tag{13.3}$$

The value of Eff is affected by many experimental conditions other than the electrolyte and anode materials. The experimental conditions include such factors as the cell configuration, electrode orientation, electrode surface area, working electrode substrate, charge–discharge currents, charge quantity, and amount of electrolyte.

When Al, Pt, Ni, or Cu is used as the substrate of lithium plating with 1 mol L^{-1} LiClO$_4$–PC/1,2-dimethoxyethane (DME), Eff decreases in the order is Al > Pt > Ni > Cu [26]. Lithium is easily alloyed electrochemically with many metals [27]; the Eff values measured in these experiments could include those of lithium alloys.

Lithium cycling on a lithium substrate (Li-on-Li cycling) is another frequently used Li half-cell test [28], in which an excess of lithium (Q_{ex} is plated on a metal working electrode, and then constant-capacity cycling (Q_{ps}); Q_{ps} is smaller than Q_{ex}) is continued until all the excess lithium is consumed. The FOM can be evaluated as shown in Equation 13.4.

$$\text{FOM} = \frac{(\text{cycling life}) \times Q_{ps}}{Q_{ex}} \tag{13.4}$$

The influence of the amounts of lithium deposited in Li-on-Li cycling has been examined by Foos *et al.* [29]. They used a cell (I) with a Q_{ex} of 3.4 C cm^{-2} and a Q_{ps} of 1.1 C cm^{-2} and a cell (II) with a Q_{ex} of 18–23 C cm^{-2} and a Q_{ps} of 5–10 C cm^{-2}

with LiAsF$_6$–THF-based electrolytes. The cell (II) experiment provides a more predictable result for the cycle life of the Li/TiS$_2$ full cell it because minimizes the effect of trace impurities.

13.4.2
Reasons for the Decrease in Lithium Cycling Efficiency

The reasons why lithium cycling efficiency is not 100% are generally considered to be as follows;

1) Lithium is consumed by reaction with the electrolyte, which forms a protective film [6]. During the deposition and stripping of lithium, the surface shape changes and a fresh lithium surface is formed with a new protection film on it; lithium is consumed in the process.
2) Lithium is isolated in a protective film [8]. During the deposition of lithium, the protective film may be heated locally by ion transport in the film itself. As a result of this local heating, part of the protective film (SEI) becomes an electronic conductor, and therefore lithium metal is deposited in the film. If local heating does not occur during stripping, the isolated lithium becomes electrochemically inactive.
3) Deposited lithium is isolated from the base anode [30, 31]. When a cell is charged, lithium is deposited on the lithium substrate of the anode. Sometimes, the plated lithium is not flat but fiber-like. When the cell is discharged, the lithium anode dissolves, and sometimes the fiber-like lithium is cut and becomes isolated from the anode substrate [31]. This isolated lithium is called '*dead lithium*,' and it is electochemically inactive but chemically active. During cycling, this dead lithium accumulates on the anode.

We believe that item 3 above is the main reason for the low cycling efficiency. The thermal stability of lithium-metal cells decreases with cycling [30], and the dead lithium may be the cause of this reduction. This indicates that the cycling efficiency is strongly affected by the morphology of the lithium surface.

13.5
Morphology of Deposited Lithium

There have been many reports on the morphology of the lithium that is electrochemically deposited in various kinds of organic electrolyte [32–39].

Figure 13.1 shows a typical lithium deposition morphology. Here, the lithium is deposited on stainless steel at 3 mA cm^{-2} for 1 h with 1.5 mol L^{-1} LiAsF$_6$–EC/2MeTHF (1 : 1, v/v).

Koshina *et al.* have reported that there are three kinds of morphology [40]: dendritic, granular, and mossy. Mossy lithium is formed when the deposition current is small and the salt concentration is high. This mossy lithium provides a high cycling efficiency.

13.5 Morphology of Deposited Lithium

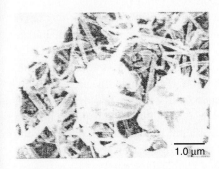

Figure 13.1 Morphology of lithium deposited on stainless steel, 3 mA cm^{-2}, 3 mAh cm^{-2}, 1.5 mol L^{-1} LiAsF$_6$–EC/2MeTHF (1:1), v/v.

A possible mechanism for lithium deposition based on our observations of lithium morphology in LiAsF$_6$–EC/2MeTHF electrolyte is described below [31]. Figure 13.2a shows our image of the mechanism.

1) Lithium is deposited on a lithium anode under the protective film without serious damage to the film.
2) The deposition points on the lithium electrode are the points at which the protective film has a higher lithium-ion conductivity. One example of these

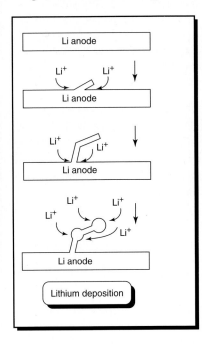 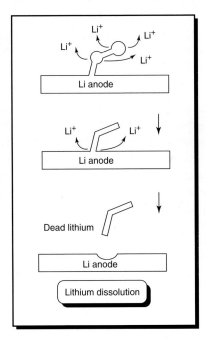

Figure 13.2 A possible mechanism for lithium deposition (a) based on dissolution (b) lithium morphology observations in LiAsF$_6$–EC/2MeTHF electrolyte.

deposition points is the pits on the lithium anode caused by discharge. Crystalline defects and the grain boundaries in lithium may also initiate deposition.

3) As lithium does not deposit uniformly for the reason mentioned above, mechanical stress is created in the lithium electrode under the protective film.
4) The stress causes lithium-atom transport, which means that deformation of the lithium occurs to release the stress in it. The lithium transport is not free but is conditioned by a force created by lithium surface tension (including the surface tension caused by the protective film) at a curved surface, and it may also be affected by crystalline defects and grain boundaries.
5) The protective film is broken in certain places on the lithium surface by the stress. Fiber-like lithium grows, like an extrusion of lithium, through these broken holes in the film. If the deposition current is small enough and the stress is therefore small, the protective film will probably not break. In this case, the deposited lithium may be particle-like or amorphous.
6) After the fiber-like lithium has grown, lithium is still deposited on the lithium substrate, that is, not at the tip of the fiber-like lithium. If the deposition continues for a long time, the lithium electrode becomes covered with long, fiber-like lithium. In this situation, lithium-ion transport in the electrolyte to the lithium electrode surface is hindered by the fiber-like lithium. Then, lithium begins to be deposited on the tip and on kinks of the fiber-like lithium, where there are crystalline defects. The morphology of the deposited lithium is particle-like or amorphous. As there are many kinks, the current density of the lithium deposition becomes very low. This low current density may create particle-like, rather than fiber-like, lithium. Thus the morphology of the lithium as a whole becomes mushroom-like [31].

The dissolution process of plated lithium may be the reverse of the plating process (Figure 13.2b). At first, the particle-like lithium on the kinks is dissolved. Then, the fiber-like lithium at the base is dissolved. During this process, fiber-like lithium is sometimes cut from the lithium substrate and becomes dead lithium. There is a large amount of dead lithium when the diameter of the fiber-like lithium is small under conditions of high-rate and/or low-temperature deposition, because the whiskers are easily cut.

A microelectrode has been used by Uchida *et al.* to study lithium deposition in order to minimize the effect of solution resistance [41]. They used a Pt electrode (10–30 μm in diameter) to measure the lithium-ion diffusion coefficient in 1 mol L^{-1} $LiClO_4$/PC electrolyte. The diffusion coefficient was 4.7×10^{-6} cm^2 s^{-1} at 25 °C.

The lithium morphology at the beginning of the deposition was measured by *in-situ* atomic force microscopy (AFM) [42]. When lithium was deposited at 0.6 C cm-2, small particles 200–1000 nm in size were deposited on the thin lines and grain boundaries in $LiClO_4$–PC. Lump-like growth was observed in $LiAsF_6$–PC along the line.

An electrochemical quartz crystal microbalance (EQCM) or quartz crystal microbalance (QCM) can be used to estimate the surface roughness of deposited lithium [43].

13.6
The Amount of Dead Lithium and Cell Performance

From our experimental results [44], the FOM at a low discharge rate is considerably smaller than that at a high discharge rate. The influence of the discharge rate on the specific surface area of a lithium anode was examined [44] using the Brunauer–Emmett–Teller (BET) equation. The surface area (26 m^2 g^{-1}) for low-rate discharge cycles (0.2 mA cm^{-2}) is double that (13 m^2 g^{-1}) for high-rate discharge cycles (3.0 mA cm^{-2}). In addition, the surface area increases with an increase in cycle number. The surface area after the sixth discharge at a low discharge rate was 30 times greater than that before cycling (1 m^2 g^{-1}). The main reason for the increase in the lithium surface area is considered to be the accumulation of dead lithium on the anode surface.

There are four possible ways of explaining [45] why a higher current discharge creates a smaller amount of dead lithium.

1) When the discharge current is high, delocalized pits (small in size but large in number) are formed on a native lithium anode. As lithium is deposited on these pits, the local charge current density becomes low when the discharge current is high, producing thicker, fiber-like lithium that is not easily cut to form dead lithium.
2) When the discharge current is large, delocalized pits formed in the anode are shallow, so the deposited lithium whiskers can easily emerge from the pits, and stack pressure can be applied to them, as mentioned in Section 13.7.3.
3) Isolated lithium near the anode becomes a local cell because of stray current. As the stray current is high when the cell discharge current is high, lithium recombination occurs easily at a high discharge current [46].
4) When the discharge current is high, transport of lithium ions becomes difficult and stripping occurs from the particle-like lithium on the tip and on the kinks of the fiber-like lithium. In this case, the fiber-like lithium rarely breaks and the efficiency increases.

13.7
Improvement in the Cycling Efficiency of a Lithium Anode

There have been many attempts to improve the cycling efficiency of lithium anodes. We describe some of them below, discussing electrolytes, electrolyte additives, the stack pressure on the electrode, composite anodes, and alternatives to the lithium-metal anode anode.

13.7.1
Electrolytes

Lithium cycling efficiency depends on the electrolyte solutions used. A mixture of EC and 2MeTHF with $LiAsF_6$ as the solute exhibits a high lithium cycling efficiency of 97.2% (FOM = 35.7) as revealed by an Li-on-Li half-cell test with 10^{-2} S cm^{-1} conductivity at 25 °C [47, 48]. A Li/a-V_2O_5–P_2O_5 coin-type cell with $LiAsF_6$–EC/2MeTHF has an FOM of 28.2, while cells with 2MeTHF or EC/PC (1:1) have FOMs of 9.4 and 14.8, respectively [48]. (a-V_2O_5 means amorphous V_2O_5.) Lithium cycling efficiency is strongly influenced by impurities in electrolytes. The relationship between total impurity and the FOM of Li in $LiAsF_6$–EC/2MeTHF has been examined [49]. The FOM for EC/2MeTHF increases with decreases in both water and organic impurities. The influence of the impurity depends on the electrolyte system used.

Hayashi et al. [50] investigated the electrolyte materials and their compositions with various carbonates and ethers as solvents in relation to the cycling efficiency of a lithium-metal anode, using cells with an $LiMn_{1.9}Co_{0.1}O_4$ cathode (Figure 13.3). As electrolyte solvents, they used four carbonates (PC, EC, dimethyl carbonate (DMC), and diethyl carbonate (DEC)) and two ethers (DME and 1, 2-diethoxyethane (DEE))). Of the electrolytes used here, 1.0 mol L^{-1} $LiPF_6$–EC/DMC (1:4) provided a high FOM of around 60 and a long cycle life of about 1200 cycles until the discharge capacity became less than 80% of the initial capacity.

3-Propylsydnone (3-PSD) was proposed as a new solvent by Sasaki et al. [51]. The cycling efficiency of lithium on an Ni electrode of the ternary mixed-solvent electrolyte of 3-PSD, 2MeTHF, and 2,5-dimethyltetrahydrofuran with $LiPF_6$ was about 60%, and it was stable with cycling.

An ether, such as 2MeTHF, has the effect of raising the FOM. When an AA Li/a-V_2O_5–P_2O_5 cell with an $LiAsF_6$–EC/PC electrolyte is cycled with a low discharge current of 60 mA (0.1 C discharge rate), the cell shows a shunting

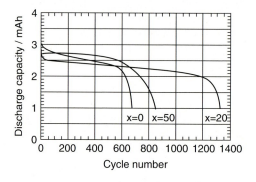

Figure 13.3 Charge–discharge cycling characteristics of an Li/$LiMn_{1.9}Co_{0.1}O_4$ coin-type cell (thickness 2 mm, diameter 23 mm). Charge: 4.3 V, 1 mA cm^{-2}; discharge: 3.3 V, 3 mA cm^{-2}; 1 mol L^{-1} $LiPF_6$–EC/DMC (x:100 − x).

tendency (a partial internal short) near the end of its cycle life [52]. However, the addition of 2MeTHF to EC/PC causes this 'soft shorting' to decrease dramatically.

From a safety aspect of utilizing lithium-metal anodes, various fluorinated solvents have been studied as the components of electrolyte solutions for lithium-metal cells. A typical example of these solvents is methyl difluoroacetate (MFA). MFA exhibited better thermal stability and lithium cycling efficiencies among various fluorinated carboxylic acid esters [53].

Another influence that electrolyte materials have on the cycle life of a practical lithium cell is due to the evolution of gas as a result of solvent reduction by lithium. For example, EC and PC give rise to [54] evolution of ethylene and propylene gas, respectively. In a practical sealed-structure cell, the existence of gas causes irregular lithium deposition. This is because the gas acts as an electronic insulator and lithium is not deposited on an anode surface where gas has been absorbed. As a result, the lithium cycling efficiency is reduced and shunting occurs.

13.7.2
Electrolyte Additives

There have been many studies with the goal of improving lithium cycling efficiency by the use of electrolyte additives. These additives can be classified into three types:

1) stable additives which cover the lithium to limit any chemical reaction between the electrolyte and lithium,
2) additives which modify the state of solvation of lithium ions, and
3) reactive additives used to make a better protective film.

Some of the studies on additives based on this classification will now be described.

13.7.2.1 Stable Additives Limiting Chemical Reaction between the Electrolyte and Lithium

Besenhard et al. [55] studied ways to protect lithium anodes from corrosion by adding saturated hydrocarbons to electrolytes. They considered saturated hydrocarbons to be chemically stable, and thus able to delay the irreversible reduction of organic electrolytes by lithium. They found that the deposited lithium was particle-shaped when *cis*- or *trans*-decalin was added to $LiClO_4$–PC electrolyte. Although there was no change in the cycling efficiency, the long-term storage characteristics improved.

Naoi and co-workers [56], with a QCM, studied lithium deposition and dissolution processes in the presence of polymer surfactants in an attempt to obtain the uniform current distribution at the electrode surface and hence smooth surface morphology of the deposited lithium. The polymer surfactants they used were polyethyleneglycol dimethyl ether (molecular weight $\approx$ 446), or a copolymer of dimethylsilicone (circa 25 wt%) and propylene oxide (circa 75 wt%) (molecular weight $\approx$ 3000) in $LiClO_4$–EC/DMC (3:2, v/v).

Yoshio and co-workers [57, 58] tried using aromatic compounds of benzene, toluene, or 4,4-dipyridyl as additives and found them to be effective.

Later, Saito et al. [59] studied anodes with a layered structure consisting of Li/protective film/additive/protective film/Li/ protective film/additive/−. They made the anode by dropping the additive on a lithium sheet, folding the lithium sheet, and then compressing the folded lithium with an oil press. They repeated this process more than 10 times. The FOM in LiAsF$_6$–EC/2MeTHF electrolyte was 7.41, 13.5, and 37.0 for a lithium anode without additives, a lithium anode with toluene in the electrolyte, and a layered-structure lithium anode containing toluene, respectively.

Other interesting examples of these less reactive additives to improve lithium cycling efficiency are siloxanes [60, 61]. The influence of four poly-ether modified siloxanes as electrolyte additives on charge–discharge cycling properties of lithium was examined. As siloxanes, diethylene glycol methyl-(3-dimethyl(trimethyl siloxy)silyl propyl)ether (sample A), diethylene glycol methyl-(3-dinethyl(trimethyl siloxy)silyl propyl)-2-methylpropyl ether (sample B), diethylene glycol methyl-(3-bis(trimethylsiloxy)silyl propyl)ether (sample C), and diethylene glycol-3-methyl-bis(trimethylsiloxy)silyl-2-methyl propyl)ether (sample D) were investigated. The chemical structures of samples B and D are shown in Figure 13.4. As a base electrolyte solution, 1 M LiPF$_6$- EC/ethylmethyl carbonate (EMC) (mixing volume ratio = 3 : 7) was used (EM). Lithium cycling efficiencies of lithium-metal anodes improved, and an impedance of anode/electrolyte interface decreased on adding poly-ether modified siloxanes. Among these siloxanes, sample B and sample D exhibited much better performance. Figure 13.5 shows the Cole-Cole plot of Li/Li cells after first charge (plating Li on SUS). In this figure, R$_1$ corresponds to the impedance of electrolyte solution. R$_2$ and R$_3$ correspond to the impedance of the interface between Li and the electrolyte solution, that is, between SEI and electrolyte solution. Two components of impedance (R$_2$ and R$_3$) are also observed in EM alone. Two components of impedance may arise from the double layer structure of SEI, for example, layers of organic compounds such as lithium alkyl carbonate and inorganic compounds such as LiF [62]. After first charge, the impedance of the electrolyte/Li interface in EM + sample D (10 vol%) was smaller than that in EM alone. This result suggests that the impedance of the surface film or layer of lithium in EM + siloxanes is smaller than that in EM alone. Figure 13.6 shows the proposed models for the mechanism of the enhancement of lithium cycling efficiency (Eff) by adding siloxanes. Just after charging (lithium deposition), freshly deposited lithium is chemically active. On the lithium surface, EM is chemically reduced by lithium and produces the surface film. The reduction

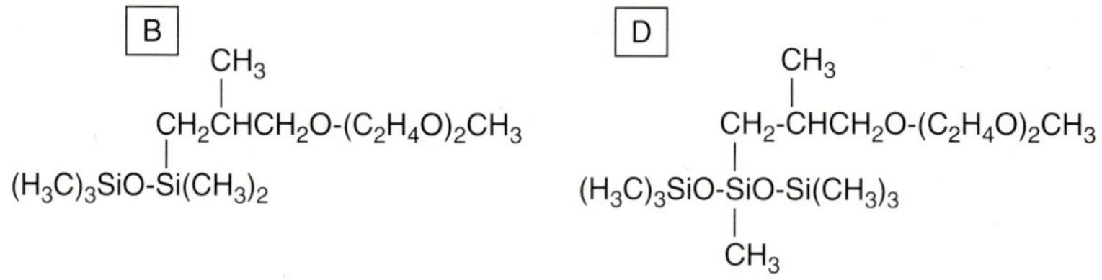

B
$$\underset{(H_3C)_3SiO-Si(CH_3)_2}{CH_2CH\underset{|}{CHCH_2O-(C_2H_4O)_2CH_3}}\;\;\overset{CH_3}{|}$$

D
$$\underset{\underset{CH_3}{\overset{|}{(H_3C)_3SiO-SiO-Si(CH_3)_3}}}{CH_2-CH\underset{|}{CHCH_2O-(C_2H_4O)_2CH_3}}\;\;\overset{CH_3}{|}$$

Figure 13.4 Chemical structure of polyether-modified siloxanes.

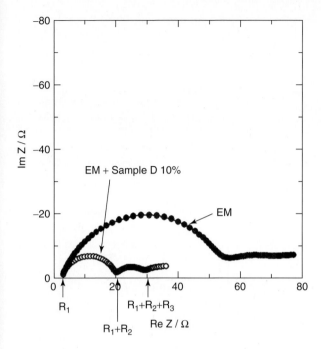

Figure 13.5 Cole–Cole plot of Li/electrolyte interface, Li/Li cells, EM + sample D (10 vol%) after first charge at 25 °C, $I_{ps} = 0.5$ mA cm^{-2}, $Q_p = 0.5$ mAh cm^{-2}, charge–discharge cut-off voltages: -2.0 and 1.0 V vs. Li/Li$^+$.

products were reported to be solid compounds and gas compounds [62]. The solid compounds remaining on the lithium produce the SEI. Figure 13.6a: if there is no addition of siloxane, the surface film is composed of reduction products of EM. Figure 13.6b: after a small addition of siloxane, the surface film is mainly composed of reduction products of EM. Small amounts of siloxane are involved in the SEI, or it adsorbs on the SEI surface. Siloxanes are less reactive than EM. Figure 13.6c: after addition of medium amounts of siloxane (10–20 vol%), the surface film is thin and is composed of both the reduction product of EM and siloxane. Figure 13.6d: by adding larger amounts of siloxane, the excess, a thick siloxane layer, exists around the lithium electrode. This excess siloxane could be resistive in the smooth charge–discharge of lithium (lithium ion diffusion). The cycling efficiency improves on adding siloxanes and shows its maximum value against the added amounts of siloxanes.

13.7.2.2 Additives Modifying the State of Solvation of Lithium Ions

Compounds which produce a complex with Li$^+$ ions have been investigated. The compounds examined were N,N,N',N'-tetramethylethylenediamine (TMEDA), ethylenediamine, crown ethers, cryptand [211], ethylenediamine tetraacetic acid (EDTA), and EDTA-Li$^+$ n ($n = 1, 2, 3$) complexes [63]. The cycling efficiency was improved by adding TMEDA to 1 M LiClO$_4$-PC.

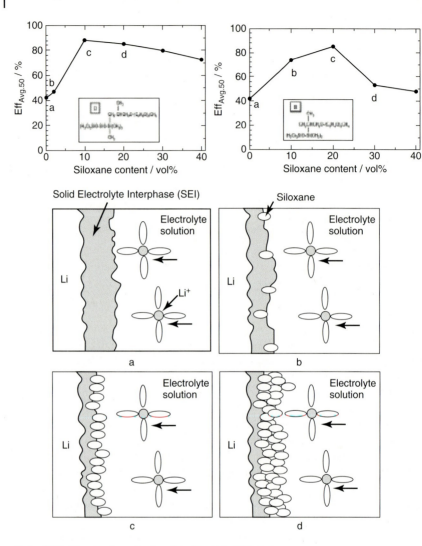

Figure 13.6 Proposed mechanism for change in lithium cycling efficiency in EM + siloxanes.

Smaller-molecular-weight poly(ethylene glycol) dimethyl ethers (($CH_3O(CH_2CH_2O)_nCH_3$, $n = 1,2,3,$ and 4)) are generally known as 'glymes.' Based on a report [64] about the properties of 1 M $LiPF_6$-ternary mixed solvent consisting of n-glyme, EC, and EMC, the glyme solutions exhibited higher conductivity and higher lithium cycling efficiency than EC/MEC (methyl ethyl carbonate). The conductivity tended to increase with decreases in ethylene oxide chain number (n) and solution viscosity. The decrease in the solution viscosity resulted from the selective solvation of the glymes with respect to lithium ions clearly demonstrated by ^{13}C-NMR measurements. The lithium cycling efficiency value depended on the charge–discharge current (I_{ps}). When n increased there was an increase in lithium cycling efficiency at a low I_{ps} and a decrease in the reduction potential of the glymes. When the conductivities, including those at low temperature (below 0 °C), and charge–discharge cycling at a high current are taken into account, diglyme or triglyme is superior to the other glymes.

13.7.2.3 Reactive Additives Used to Make a Better Protective Film

The effect of 'precursors' was examined by Brummer et al. [65] with the aim of producing a thin and Li^+-ion conductive film which was impermeable to solvent molecules. As precursors, they tested CS_2, $PSCl_3$, $PSBr_3$, $POBr_3$, $PNBr_2$, $POCl_3$, CH_3SO_2, $MoOCl_4$, $VOCl_2$, CO_2, N_2O, and SO_2 with 1 mol L^{-1} $LiClO_4$–PC. The concentration of additive ranged from 0.01 to 0.36 mol L^{-1}. A maximum efficiency of 85.1% was obtained by the addition 0.01 mol L^{-1} $POCl_3$, whereas the base electrolyte without any additive showed an average efficiency of 40%. However, without additives, 1 mol L^{-1} $LiAsF_6$–PC has an efficiency of 85.2% and the addition of $POCl_3$ to $LiAsF_6$ provides 75.8% efficiency. These results indicate that the use of $LiAsF_6$ provides a better film for cycling Li than those formed by the precursors. However, we believe that it is still worth attempting to find new precursors.

Also, the influence of adding O_2, N_2, Ar, or CO_2 to $LiAsF_6$–THF on lithium cycling efficiency has been examined [66]. Lithium was cycled on an Ni electrode with $Q_p = 1.125$ C cm^{-2} and cycling currents of 5 mA cm^{-2}. Oxygen and N_2 helped to maintain the cycle life relative to Ar, while CO_2 and ungasified electrolyte did not. The addition of O_2 showed the highest lithium cycling efficiency, which resulted from the formation of an Li_2O film. However, the lithium cycling efficiency rapidly degraded beyond the 10th cycle. On the basis of these results, Dominey et al. [67] examined the effect of adding KOH to ether-based electrolytes such as THF, 2MeTHF, or 1,3-dioxolane. They showed that the presence of the hydroxide modifies the surface film formation. The anode-related heat output was reduced three to fourfold in cyclic-ether electrolytes containing approximately 100 ppm of OH^-. The Li/TiS_2 cell with THF to which 2MeF, KOH, and 12-crown-4 had been added has been reported to show excellent cycling efficiency. Further improvement in the lithium deposition morphology is still needed, however. $LiAsF_6$–2MeTHF has a good cycling efficiency. Abraham et al. [68] showed that the high cycling efficiency is caused by 2MeF, which is naturally contained in 2MeTHF as in impurity.

Quinoneimine dyes, aromatic nitro compounds, and triphenylmethane compounds have been studied [69]. These compounds are highly reactive with lithium. If the lithium cell includes these compounds as the cathode, it will exhibit cell voltages of 2–3 V. The cycling efficiency was improved by adding quinoneimine dyes. However, this effect depends on the charge capacity and the duration of charge–discharge cycling. The effect of hexamethylphosphoric triamide (HMPA) has also been examined. HMPA has an extremely high solvation power for cations whose donor number (DN) is 38 [70]. A unique characteristic of HMPA is that it produces solvated electrons in contact with alkali metals when there is a large

Table 13.4 Influence of HMPA addition on cycle life of Li/Fe-phthalocyanine (FePc) cell[a].

HMPA (vol%)	Cycle life
0.5	240
1.0	220
10.0	80
No additives	55

[a] Electrolyte = 2 mol L^{-1} $LiClO_4$–PC; charge–discharge currents = 0.3 mA cm^{-2}; cycling capacity = 200 mAh g^{-1} (4.3 Li/FePc).

excess of HMPA. A lithium cycling efficiency of 86.6% was obtained by the addition of 0.5 vol% HMPA to 1 mol L^{-1} $LiClO_4$–PC, which exhibits 67.0% efficiency [71]. In addition, Li cells with an organic cathode, Fe phthalocyanine (FePc), containing 1 mol L^{-1} (M) $LiClO_4$–PC with HMPA (0.5–10 vol%) completed 80–240 cycles, whereas a cell without PC completed only 55 cycles [71] (Table 13.4).

Carbon dioxide has been proposed as an additive to improve the performance of lithium batteries [72]. Aurbach et al. [73] studied the film formed on lithium in electrolytes saturated with CO_2, and using in-situ FTIR he found that Li_2CO_3 is a major surface species. This means that the formation of a stable Li_2CO_3 film on the lithium surface could improve cyclability [74]. Osaka and co-workers [75] also studied the dependence of the lithium efficiency on the plating substrate in $LiClO_4$–PC. The addition of CO_2 resulted in an increase in the efficiency when the substrate was Ni or Ti, but no effect was observed with Ag or Cu substrates.

Tekehara and co-workers [76] tried to modify the native film of lithium by an acid–base reaction. HF, HI, H_3PO_4, and HCl were selected as acids because of the possibility of their reacting with the Li_2CO_3, LiOH, and Li_2O, which compose the lithium native film, to form LiA (HA = acid). LiF was observed, by XPS, in the film treated with HF. HF treatment changed the deposition morphology from dendritic to particle-like in $LiPF_6$–PC electrolyte. XPS showed that after HF treatment the lithium surface was composed of two layers (LiF and Li_2O), whereas the native surface was composed of three layers (Li_2CO_3, LiOH, and Li_2O). The impedance of the lithium was reduced by this treatment. The cycling efficiencies [77] in $LiPF_6$–PC were 57 and 70% for as-received and HF-treated lithium, respectively. We have also confirmed the above results reported by Takehara et al. Figures 13.7 and 13.8 show our results, which reveal that HF treatment changed the deposition morphology from dendritic to particle-like in $LiPF_6$–PC electrolyte.

Sulfur is known to be easily reducible in nonaqueous solvents, and its reduction products exist at various levels of reduction of polysulfide radical anions ($S_n^{\cdot-}$) and dianions ($S_m^{\cdot 2-}$) [78]. Recently Besenhard and co-workers [79] have examined the effect of the addition of polysulfide to $LiClO_4$–PC. Lithium is cycled on an Ni substrate with $Q_p = 2.7$ C cm^{-2} and cycling currents of 1 mA cm^{-2}. The cycling efficiency in PC with polysulfide is higher than that without an additive.

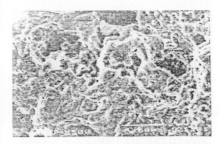

Figure 13.7 Morphology of deposited lithium on lithium, after five cycles with 1 mA cm^{-2}, 2 mAh cm^{-2} in 1 mol L^{-1} LiPF$_6$–PC.

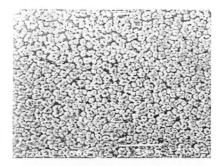
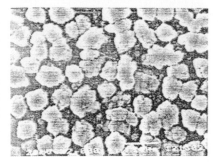

Figure 13.8 Morphology of deposited lithium on lithium after immersion of lithium in 1 mol l^{-1} LiPF$_6$–PC with HF (3 vol%) for three days: five cycles with 1 mAh cm^{-2} in 1 mol L^{-1} LiPF$_6$–PC.

The lithium deposition morphology is compact and smooth in PC with added polysulfide, whereas it is dendritic in PC alone.

Matsuda and co-workers [80, 81] examined LiI, SnI$_2$, AlI$_3$, and 2MeF as additives in LiClO$_4$–PC or LiClO$_4$–PC/DMC electrolyte. They measured the cycling efficiency of lithium on an Ni electrode. All the additives increased the efficiency, the best additive being a combination of AlI$_3$ and 2MeF. They attributed the improvement to the formation of Li–Al alloy on the surface by AlI$_3$ or to a more protective film formed by 2MeF.

We have examined the effects of adding metal chloride (MCl$_x$; CuCl, CuCl$_2$, AlCl$_3$, and NiCl$_2$) on the lithium cycling efficiency in 1 mol L^{-1} LiClO$_4$–PC. The results are shown in Table 13.5.

These compounds may reduce the reactivity of lithium and make the lithium deposition morphology smoother as a result of the spontaneous electrochemical alloy formation during the charging of lithium on the anode. The cling efficiency of the lithium plated on was improved by addition of metal chlorides. The cycling efficiencies were in the order Al > Ni > Cu. β-Li-Al was detected by X-ray diffraction in the surface of the lithium anode after charge–discharge cycling.

In order to improve the lithium battery performance, many reactive additives to electrolyte solutions have been proposed [82]. Vinylene carbonate (VC) and fluoro-ethylene carbonate (F-EC) are well-known additives. VC was developed by SAFT in the first place [83]. Aurbach *et al.* proposed that VC forms polymeric surface species which enhance SEI stability [84]. However, these additives were mainly investigated in carbonate-based electrolytes. Little is known about their suitability to GBL electrolytes. γ-butyrolactone (GBL) has a high boiling point, a low freezing point, a high flash point, a high dielectric constant, and a low viscosity. GBL is a much preferred solvent for lithium batteries. However, GBL readily undergoes reductive decomposition on the surface of the negative electrodes, and it forms SEI with a large resistance and causes deterioration of battery performances. Then, the effects of cyclic carbonates as additives to GBL electrolytes were investigated [85]. The carbonates, EC, PC, VC, vinylethylene carbonate (VEC) [86, 87], and

Table 13.5 Lithium cycling efficiency on Pt in 1 mol L^{-1} LiClO$_4$–PC with 0.1 mol L^{-1} metal halides added.[a]

Metal halide	Eff. 10[b] (%) alloying	Electrochemical alloying efficiency of metals with lithium [27]
AlCl$_3$	91.2	92
CuCl	88.7	42
NiCl$_2$	84.8	50
CuCl$_2$[a]	72.0	–
No additives	65.0	–

[a]Cycling current = 0.5 mA cm^{-2}, plating capacity = 0.6 C cm^{-2}.
[b]Eff, 10 = average cycling efficiency from 1st to the 10th cycle.

13.7 Improvement in the Cycling Efficiency of a Lithium Anode

phenylethylene carbonate (PhEC) [88] were investigated. VC, VEC, and PhEC were effective in suppressing the excessive reductive decomposition of GBL.

In order to increase energy density of the lithium cells, high-voltage cells have been studied. 4 V lithium-ion cells have been commercially applied for electronic portable equipment such as cellular phones and notebook-type personal computers. However, lithium cells are finding new uses, such as for electric vehicles and power storage batteries. These need higher-voltage cells (e.g., 5–6 V) as well as a higher charge–discharge capacity and a higher energy density. To increase the cell voltage, not only is the development of new electrode materials important but also the development of new electrolytes having higher anodic stability than conventional electrolytes using solvents such as EC, PC, DEC, DMC, and EMC. There have been many studies to develop new electrolytes for high-voltage lithium cells [89–93]. For example, fluorinated carbonate solvents exhibit higher anodic stability, relative permittivity, and viscosity [89, 90]. Nitrile solvents show high anodic stability and low viscosity [91, 92]. Ionic liquids are also known to show higher anodic stability, noncombustibility, and high ionic conductivity. Especially, aliphatic ammonium bis(trifluoromethanesulfone)imide shows superior anodic stability [5]. However, charge–discharge properties, cathodic stability, and compatibility of these solvents with a lithium anode have not been fully investigated.

Sulfones are investigated for high voltage cells because of their anodic stability [94–96]. Sulfolane (SL) is a common solvent known to show high anodic stability, high relative permittivity, and low toxicity. However, SL is solid at room temperature, and its viscosity is too high at liquid phase. Ethyl acetate (EA) is also very common as an organic solvent. It has good anodic stability and low viscosity, but its relative permittivity is only 6.02 at 25 °C [97]. This is too low to dissociate supporting salts sufficiently. Then, sulfone–ester mixed solvent electrolytes were examined for 5 V-class high-voltage rechargeable lithium cells [98]. As the base-electrolyte, SL–EA mixed solvent containing $LiBF_4$ solute was investigated. $LiBF_4$ is used here because the tolerance of $LiBF_4$ toward oxidation is reported to be higher than that of $LiPF_6$ [99]. Charge–discharge cycling efficiency of a lithium anode in SL–EA electrolyte was poor, due to its poor tolerance for reduction. To improve lithium charge–discharge cycling efficiency in SL–EA electrolytes, the following three trials were carried out; (i) improvement of the cathodic stability of electrolyte solutions by change in polarization through modification of solvent structure; isopropyl methyl sulfone and methyl isobutyrate were investigated as alternatives to SL and EA, respectively, (ii) suppression of the reaction between lithium and electrolyte solutions by addition to SL–EA electrolytes of low-reactivity surfactants derived from cycloalkanes (decalin and adamantane) or triethylene glycol derivatives (triglyme, 1,8-bis(*tert*-butyldimethylsilyloxy)-3,6-dioxaoctane and triethylene glycol di(methanesulfonate)), and (iii) change in surface film by addition of surface film formation agent VC to SL–EA electrolytes. These trials made lithium cycling behavior better. Of these additives, the addition of VC was the most effective for improvement of lithium cycling efficiency. A stable surface film is formed on the lithium anode by adding VC, and the resistance between anode/electrolyte interfaces showed a constant value with an increase in cycle number. In the electrolyte

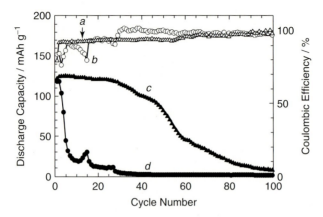

Figure 13.9 Coulombic efficiency and discharge capacity of Li/LiNi$_{0.5}$Mn$_{1.5}$O$_4$ cell (supporting electrolyte: LiBF$_4$, current density: 0.5 mA cm^{-2}, cut-off potential: 3–5 V), a(△): SL–EA(1:1) + 2 vol% VC (Coulombic efficiency), b(○): SL–EA(1:1) (no additive) (Coulombic efficiency), c(▲): SL–EA(1:1) + 2 vol% VC (discharge capacity), d(●): SL–EA(1:1) (no additive) (Coulombic efficiency).

solutions without VC, the interfacial resistance increased with an increase in cycle number. VC addition to SL–EA was effective not only for a Li/LiCoO$_2$ cell with a charge cut-off voltage of 4.5 V but also for Li/LiNi$_{0.5}$Mn$_{1.5}$O$_4$ cells even with the high charge cut-off voltage of 5 V in Li/LiNi$_{0.5}$Mn$_{1.5}$O$_4$ cells (Figure 13.9).

13.7.3
Stack Pressure on Electrodes

Wilkinson *et al.* [100] examined the effect of stack pressure on the lithium turnover (FOM for lithium cycling efficiency) in Li/MoS$_2$ prismatic cells containing 1 mol L^{-1} LiAsF$_6$–PC. The cycle life for spirally wound AA-size Li/MoS$_2$ cells showed that when the electrode assembly is housed tightly in the cell the cycle life is better than with loosely housed cells.

FOMs were also measured [101] for coin-type cells with an amorphous V$_2$O$_5$-P$_2$O$_5$ cathode and lithium anode 20 μm thick (Figure 13.10). LiAsF$_6$–EC/2MeTHF electrolyte had an FOM of 80 at 125 kg cm^{-2}, which was almost four times the value without compression. An scanning electron microscopy (SEM) image of lithium deposited under stack pressure showed that it was densely packed, which reduces the amount of lithium that was isolated from the anode substrate, resulting in a high cycling efficiency.

13.7.4
Composite Lithium Anode

Desjardins and MacLean [102] studied a composite of lithium and Li$_3$N named 'Linode.' Their research cell showed improvements in cycle life, shelf life, and electrode morphology after cycling.

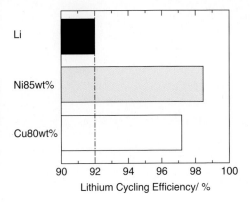

Figure 13.10 Lithium cycling efficiency of composite lithium anodes in 2032 Li/a-V_2O_5 coin-type cell (thickness 2 mm, diameter 23 mm) with 1.5 mol L $LiAsF_6$–EC/2MeTHF.

A lithium anode mixed with conductive particles of Cu or Ni was studied by Saito et al.; they obtained an improvement in the cycling efficiency [103]. Their idea is based on the recombination of dead lithium and formation of many active sites for deposition.

We believe that the advantage of these composite anodes is that they result in a uniform lithium deposition at the boundaries of two components that may improve the cycling efficiency.

13.7.5
Influence of Cathode on Lithium Surface Film

It is generally considered that the lithium surface film is produced by a reaction between the lithium and the electrolyte materials. However, by XPS we have detected vanadium on a lithium anode surface cycled for Li/a-V_2O_5–P_2O_5 cells with EC-based electrolytes [81]. Vanadium comes from the partially dissolved discharge product, $Li_xV_2O_5$, in the electrolyte. Then, with a Li/a-V_2O_5 cell, the cathode also affects the chemical composition of the surface film [104].

Figure 13.11 shows the FOM of an AA cell and the PC content in EC/PC binary mixed-solvent electrolytes. With an increase in PC content, the lithium cycling efficiency (Eff) obtained with Li cycling on a stainless steel substrate increases. However, the FOM of the AA cell reaches its maximum value at EC/PC $= 1 : 9$ [105]. This result arises from the interaction between EC and the a-V_2O_5–P_2O_5 cathode.

13.7.6
An Alternative to the Lithium-Metal Anode (Lithium-Ion Inserted Anodes)

Lithium-ion inserted compounds have been investigated as new anodes. These compounds have the possibility of exhibiting a larger energy density than carbon materials and have anode properties similar to those of lithium metal.

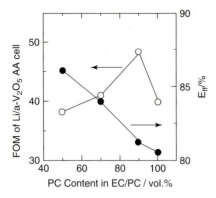

Figure 13.11 Relationship between FOM of AA cells, lithium cycling efficiency (Eff) on stainless steel, and PC content in 1 mol L^{-1} –LiAsF$_6$–EC/PC.

Nishijima et al. [106] reported that lithium ternary nitrides of Li$_3$FeN$_2$, Li$_7$MnN$_4$, and Li$_{2.5}$Co$_{05}$N perform as anodes. These materials exhibit a specific capacity of 200–480 mAh g^{-1}, which is as high as that of carbon. Shodai et al. have found that the capacity of the Li$_{3-x}$Co$_x$N ($x = 0.2$–0.6) system can be increased substantially by extracting lithium ions from the matrix. Li$_{2.6}$Co$_{0.4}$N exhibits a high specific capacity of 760 mAh g^{-1} in the 0–1.4 V vs. Li/Li$^+$ range. Shodai et al. [84] also described the performance of an Li$_{1.6}$Co$_{0.4}$N/LiNiO$_2$ lithium-ion cell, which was designed so that the Li$_{1.6}$Co$_{0.4}$N anode operated at 0–1.0 V and the LiNiO$_2$ cathode operated at 4.2–3.5 V vs. Li/Li$^+$. This cell shows a good reversibility of more than 150 cycles.

Idota et al. have demonstrated [107] that amorphous material based on tin oxide has capacities of 800 mAh g^{-1} and 3200 mAh cm^{-3}, which are respectively two and four times higher than those of carbon. An 18 650-size cell with an LiCoO$_2$ cathode has a capacity of 1850 mAh, which is higher than the value of 1350 mAh for the commercial cells.

Sigala et al. [108] also examined Li$_x$MVO$_4$ (M = Zn, Co, Ni, Cd) as anode materials. The best compounds (M = Zn, Ni) deliver capacities of about 700 mAh g^{-1} after 200 cycles. The search for new anode compounds will prove to be a fruitful area in the future.

13.8
Safety of Rechargeable Lithium Metal Cells

We have developed a prototype AA-size cell which consists of an amorphous (a-) V$_2$O$_5$–P$_2$O$_5$ (95:5, molar ratio) cathode and a lithium anode (Li/a-V$_2$O$_5$ cell) [1]. In this section, we describe safety test results for AA Li/a-V$_2$O$_5$ cells. The AA cell we fabricated has a pressure vent, a Poly-switch (PS, Raychem Co.), a thermal and current fuse composed of a spirally wound cathode sheet, a metallic Li-based anode sheet, and a polyethylene (PE) separator [109].

The basic considerations regarding the cell safety and the test results are described briefly below.

13.8.1
Considerations Regarding Cell Safety

The basic problem in regard to the safety of rechargeable metal cells is how to manage the heat generated in a cell when it is abused. The temperature of a cell is determined by the balance between the amount of heat generated in the cell and the heat dissipated outside the cell. Heat is generated in a cell by thermal decomposition and/or the reaction of materials in the cell, as listed below:

1) by a reaction between an electrolyte and an anode;
2) by the thermal decomposition of an electrolyte;
3) by a reaction between an electrolyte and a cathode;
4) by the thermal decomposition of an anode;
5) by the thermal decomposition of an cathode;
6) by an entropy change in a cathode active material (and an anode inactive material);
7) by current passing through a cell with electric resistance.

When a cell is heated by some trigger (for example, an internal short, application of a high current, or overcharge), heat will be generated if the cell temperature is high enough to cause decomposition and/or a reaction. This situation leads to the thermal runaway of the cell (Figure 13.12). In the worst case, the cell ignites. If the additional heat generation is small, the cell temperature does not increase so much, and the cell is safer.

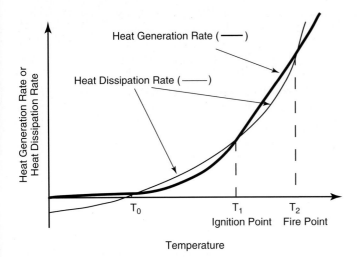

Figure 13.12 Imagination scheme of thermal runaway of lithium cells.

13.8.2
Safety Test Results

13.8.2.1 External Short
We experienced no safety problems during the external short tests because of the Polyswitch inside the cell. We confirmed that even if the Polyswitch fails to operate, the short-circuit current stops flowing before thermal runaway occurs because the micropores are closed by the polyethylene separator, which melts at 125 °C ('separator shutdown').

13.8.2.2 Overcharge
In the overcharge tests we carried out, there was no fire or explosion. The cell impedance increased suddenly in every test. This was due to the oxidation of the electrolyte with a low charging current, or to the separator melting with a high charging current. In practical applications, an electronic device should be used to provide overcharge protection and ensure complete safety.

13.8.2.3 Nail Penetration
The nail penetration test is very important and is considered to simulate an internal short in a cell. No electronic device can protect against an internal short, so the cell itself should pass this test. Our cell did not ignite or explode.

13.8.2.4 Crush
The cell should also be able to survive a crush test because an electronic device cannot provide protection in this case either. Our test cell remained safe in crush tests, both with a bar and with a flat plate.

13.8.2.5 Heating
The heating test is carried out by increasing the temperature at a rate of 5 °C min^{-1} and then holding it constant at least until the maximum cell temperature induced by the internal exothermic reactions starts to decrease. If the thermal stability decreases after cycling, we have to be careful when estimating the safety. The thermal stability of our cell is defined by the maximum temperature at which it can be ensured that no fire will occur. For our cell, this is 130 °C before cycling. The thermal stability limit becomes even higher after cycling. These results are considered to be closely related to the increase in the thermal stability of a lithium anode with an increase in the number of charge–discharge cycles as the result of the formation of a special lithium surface film containing vanadium.

13.9
Conclusion

It is worthwhile attempting to develop a rechargeable lithium metal anode. This anode should have a high lithium cycling efficiency and be very safe. These properties can be realized by reducing the dead lithium. Practical levels of lithium cycling efficiency and safety could be achieved simultaneously by the same technical breakthrough. This will be realized by a wholehearted effort to develop a method of anode construction, a new electrolyte, and a new cell structure. Another interesting area of study is the investigation of new anode materials whose energy density is as close as possible to that of pure lithium metal.

References

1. Tobishima, S., Sakurai, Y., and Yamaki, J. (1997) *J. Power Sources*, **68**, 455.
2. Yen, S.P.S., Dhen, D., Vansquéz, R.P., Grunthaner, F.J., and Somoano, R.B. (1981) *J. Electrochem. Soc.*, **128**, 1434.
3. Hirayama, S., Hiraga, H., Otsuka, K., Ikeda, N., and Sasaki, M. (1993) Extended Abstracts of the 34th Battery Symposium in Japan, Abstract No. 1A06, p. 13.
4. Yamamoto, N., Saito, K., Ishibashi, T., Honjo, M., Fujieda, T., and Higuchi, S. (1993) Extended Abstracts of the 34th Battery Symposium in Japan, Abstract No.1A07, p. 15.
5. Harris, W.R. (1958) Ph. D. thesis, University of California, Berkeley.
6. Dey, A.N. and Rudd, E.J. (1974) *J. Electrochem. Soc.*, **121**, 1294.
7. Rauh, R.D. and Brummer, S.B. (1977) *Electrochim. Acta*, **22**, 75.
8. Peled, E. (1979) *J. Electrochem. Soc.*, **126**, 2047.
9. Dey, A.N. (1977) *Thin Solid Films*, **43**, 131.
10. Nazri, G. and Muller, R.H. (1985) *J. Electrochem. Soc.*, **132**, 2050.
11. Aurbach, D., Daroux, M.L., Faguy, P.W., and Yeager, E. (1987) *J. Electrochem. Soc.*, **134**, 1611.
12. Abraham, K.M. and Chaudhri, S.M. (1986) *J. Electrochem. Soc.*, **133**, 1307.
13. Goldman, J.L., Mank, R.M., Young, J.H., and Koch, V.R. (1980) *J. Electrochem. Soc.*, **127**, 1461.
14. Aurbach, D., Daroux, M.L., Faguy, P.W., and Yeager, E. (1986) *J. Electrochem. Soc.*, **135**, 1307.
15. Odziemkowski, M., Krell, M., and Irish, D.E. (1992) *J. Electrochem. Soc.*, **139**, 3052.
16. Yen, S.P.S., Shen, D., Vasquez, R.P., Grunthaner, F.J., and Samoano, R.B. (1981) *J. Electrochem. Soc.*, **128**, 1434.
17. Aurbach, D. (1989) *J. Electrochem. Soc.*, **136**, 1606.
18. Aurbach, D., Youngman, O., Gofer, Y., and Meitav, A. (1990) *Electrochim. Acta*, **35**, 625.
19. Ely, Y.E. and Aurbach, D. (1992) *Proceedings of the Symposium on High Power Ambient Temperature Lithium Batteries*, The Electrochemical Society, p. 157.
20. Itoh, T., Nishina, T., Matsue, T., and Uchida, I. (1995) Extended Abstracts of the 36th Battery Symposium in Japan, Abstract No. 2B19, p. 171.
21. Aurbach, D., Zaban, A., Gofer, Y., Ely, Y.E., Weissman, I., Chusid, O., and Abramson, O. (1995) *J. Power Sources*, **54**, 76.
22. Shen, D.H., Subbarao, S., Deligiannis, F., and Halpert, G. (1989) *Proceedings of the Symposium on Materials and Processes for Lithium Batteries*, The Electrochemical Society, p. 223.
23. Selim, R. and Bro, P. (1974) *J. Electrochem. Soc.*, **121**, 1457.
24. (a) Klemann, L.P. and Newman, G.H. (1981) *Proceedings of the Symposium on Lithium Batteries*, vol. 81-4, The Electrochemical Society Inc., p. 189; (b) Abraham, K.M., Goldman, J.L., and Natwig, D.L. (1982) *J. Electrochem. Soc.*, **129**, 2404.
25. Yamaki, J., Arakawa, M. Tobishima, S., and Hirai, T. (1987) *Proceedings of the Symposium on Lithium Batteries*, vol. 87-1, The Electrochemical Society Inc., p. 266.
26. Tobishima, S., Yamaki, J., Yamaji, A., and Okada, T. (1984) *J. Power Sources*, **13**, 261.

27. Dey, A.N. (1971) *J. Electrochem. Soc.*, **118**, 1547.
28. Rauh, D., Reise, T.F., and Brummer, S.B. (1983) *J. Electrochem. Soc.*, **130**, 101.
29. Foos, J.S. and Rembetsy, L.M. (1983) Extended Abstracts of Electrochemical Society Fall Meeting, Abstract No. 73, p. 117.
30. Laman, F.C. and Brandt, K. (1988) *J. Power Sources*, **24**, 195.
31. (a) Yoshimatsu, I., Hirai, T., and Yamaki, J. (1988) *J. Electrochem. Soc.*, **135**, 2422; (b) Arakawa, M., Tobishima, S., Nemoto, Y., Ichimura, M., and Yamaki, J. (1993) *J. Power Sources*, **43–44**, 27.
32. Kanamura, K., Shiraishi, S., and Takehara, Z. (1994) *J. Electrochem. Soc.*, **141**, L108.
33. Besenhard, J.O. and Eichinger, G. (1976) *J. Electroanal. Chem*, **68**, 1.
34. Tobishima, S., Yamaki, J., Yamaji, A., and Okada, T. (1984) *J. Power Sources*, **13**, 261.
35. Tobishima, S. and Okada, T. (1985) *Electrochim. Acta*, **30**, 1715.
36. Tobishima, S. and Okada, T. (1985) *J. Appl. Electrochem.*, **15**, 317.
37. Tobishima, S., Yamaki, J., and Okada, T. (1985) *Denki Kagaku*, **53**, 173.
38. Hirai, T., Yoshimatsu, I., and Yamaki, J. (1994) *J. Electrochem. Soc.*, **141**, 611.
39. Fringant, C., Tranchant, A., and Messina, R. (1995) *Electrochim. Acta*, **40**, 513.
40. Koshina, H., Eda, N., and Morita, A. (1989) Extended Abstracts of the 30th Battery Symposium in Japan, Abstract No. 1B11, p. 49.
41. Uchida, I., Wang, X., and Nishina, T. (1994) Extended Abstracts of the 35th Battery Symposium in Japan, Abstract No. 3B04, p. 73.
42. Morigaki, K., Kabuto, N., Yoshino, K., and Ohta, A. (1994) Extended Abstracts of the 35th Battery Symposium in Japan, Abstract No. 3B09, p. 83.
43. Mori, M., Narukawa, Y., Naoi, K., and Futeux, D. (1998) *J. Electrochem. Soc.*, **145**, 2340.
44. Saito, K., Arakawa, M., Tobishima, S., and Yamaki, J. (1994) *Denki Kagaku*, **62**, 888.
45. Hirayama, S., Hiraga, H., Otsuka, K., Ikeda, N., and Sasaki, M. (1993) Extended Abstracts of the 34th Battery Symposium in Japan, Abstract No. 1A06, p. 13.
46. Arakawa, M., Tobishima, S., Nemoto, Y., Ichimura, M., and Yamaki, J. (1993) *J. Power Sources*, **43–44**, 27.
47. Arakawa, M., Tobishima, S., Hirai, T., and Yamaki, J. (1986) *J. Electrochem. Soc.*, **133**, 1527.
48. Tobishima, S., Arakawa, M., Hirai, T., and Yamaki, J. (1987) *J. Power Sources*, **20**, 293.
49. Yamaki, J., Arakawa, M., Tobishima, S., and Hirai, T. (1987) *Proceedings on Lithium Batteries*, vol. 87-1, The Electrochem. Soc. Inc., p. 266.
50. Hayashi, K., Nemoto, Y., Tobishima, S., and Yamaki, J. (1999) *Electrochim.Acta*, **44**, 2344.
51. Sasaki, Y., Kaido, H., Ohashi, H., Minato, T., Handa, M., and Chiba, N. (1993) Extended Abstracts of the 34th Battery Symposium in Japan, Abstract No. 2D02, p. 305.
52. Tobishima, S., Hayashi, K., Saito, K., Shodai, T., and Yamaki, J. (1997) *Electrochim. Acta*, **42**, 119.
53. Sato, K., Yamazaki, I., Okada, S., and Yamaki, J. (2002) *Solid Stae Ionics*, **148**, 463.
54. Stiles, J.A. and Fouchard, D.T. (1988) *Proceedings of the Symposium on Primary and Secondary Ambient Temperature Lithium Batteries*, vol. PV-88, The Electrochemical Society, p. 422.
55. Besenhard, J.O., Guertler, J., and Komenda, P. (1987) *J. Power Sources*, **20**, 253.
56. Mori, M., Kakuta, Y., Naoi, K., and Autcux, D.F. (1998) *J. Electrochem. Soc.*, **145**, 2340.
57. Yoshio, M., Nakamura, H., Isono, K., Itoh, S., and Holzleithner, K. (1988) *Progr. Batt. Sol. Cells*, **7**, 271.
58. Wang, C., Nakamura, H., and Yoshio, M. (1996) Extended Abstracts of the 37th Battery Symposium in Japan, Abstract No. 1A07.

59. Saito, K., Nemoto, Y. Tobishima, S., and Yamaki, J. (1995) Extended Abstracts of the 36lh Battery Symposium in Japan, Abstract No. 2B09, p. 151.
60. Inose, T., Tada, S., Morimoto, H., and Tobishima, S. (2006) *J. Power Sources*, **161**, 550.
61. Takeuchi, T., Noguchi, S., Morimoto, H., and Tobishima, S. (2010) *J. Power Sources*, **195**, 580.
62. Aurbach, D. and Granot, E. (1997) *Electrochim. Acta*, **42**, 697.
63. Tobishima, S., Arakawa, M., Hayashi, K., and Yamaki, J. (1993) Extended Abstracts of the 34th Battery Symposium in Japan, Abstract No. 2D03, p. 307.
64. Tobishima, S., Morimoto, H., Aoki, M., Saito, Y., Inose, T., Fukumoto, T., and Kuryu, T. (2004) *Electrochim. Acta*, **49**, 979.
65. Brummer, S.B., Koch, V.R., and Rauh, R.D. (1980) in *Materials for Advanced Batteries* (eds D.W. Murphy, J. Broadhead, and B.C.H. Steel), Plenum Press, New York, p. 123.
66. Koch, V.R. and Young, J.H. (1978) *J. Electrochem. Soc.*, **125**, 1371.
67. Dominey, L.A., Goldman, J.L., and Koch, V.R. (1989) *Proceedings of the Symposium on Materials und Processes for Lithium Batteries*, The Electrochemical Society, p. 213.
68. Abraham, K.M., Foos, J.S., and Goldman, J.L. (1984) *J. Electrochem. Soc.*, **131**, 2197.
69. Tobishima, S. and Okada, T. (1985) *J. Appl. Electrochem.*, **15**, 901.
70. Gutman, V. (1978) *The Donor–Acceptor Approach to Molecular Interactions*, Chapter 3, Plenum Press, New York.
71. Okada, S., Tobishima, S., and Yamaki, J. (1987) Extended Abstracts of Spring Meeting of Electrochemical Society of Japan, p. D132.
72. Salomon, M. (1989) *J. Power Sources*, **9**, 26.
73. Aurbach, D. and Chusid (Youngman), O. (1993) *J. Electrochem. Soc.*, **140**, L155.
74. Osaka, T., Momma, T., Tajima, T., and Matsumoto, Y. (1994) *Deknki Kagaku*, **62**, 451.
75. Osaka, T., Momma, T., Tajima, T., and Matsumoto, Y. (1995) *J. Electrochem. Soc.*, **142**, 1057.
76. Shiraishi, S., Kanamura, K., and Takehara, Z. (1995) *J. Appl. Electrochem.*, **25**, 584.
77. Shiraishi, S., Kanamura, K., and Takehara, Z. (1999) *J. Appl. Electrochem.*, **29**, 869.
78. Tobishima, S., Yamamoto, H., and Matsuda, M. (1997) *Electrochim. Acta*, **42**, 1019.
79. Wagner, M.W., Liebenow, C., and Besenhard, J.O. (1996) Extended Abstracts of 8th Internationa Meeting on Lithium Batteries, Abstract No. I B12.
80. Ishikawa, M., Yoshitake, S., Morita, M., and Matsuda, Y. (1994) *J. Electrochem. Soc.*, **141**, L159.
81. Uraoka, U., Ishikawa, M., Kishi, K., Morita, M., and Matsuda, Y. (1996) Extended Abstracts of the 37th Battery Symposium of Japan, Abstract No. IA29, p. 103.
82. Ue, M. (2004) Extended abstracts of the Battery and Fuel Cell Materials Symposium, Graz, p. 53.
83. Simon, B. and Boeuve, J.P. (1997) US Patent 5, 626,981.
84. Aurbach, D., Gamoisky, K., Maarkovsku, B., Gofer, Y., Schmidt, M., and Heider, U. (2002) *Electrochim. Acta*, **47**, 1423.
85. Kinoshita, S., Kotato, M., Sakata, Y., Ue, M., Watanabe, Y., Morimoto, H., and Tobishima, S. (2008) *J. Power Sources*, **183**, 755.
86. Kotato, M., Fujii, T., Shima, N., and Suzuki, H. (2000) WO Patent 00/79632.
87. Hu, Y., Kong, W., Li, H., Huang, X., and Chen, L. (2004) *Electrochem. Commun.*, **6**, 126.
88. Suzuki, H., Sato, T., Kotato, M., Ota, H., and Sato, H. (2003) US Patent 6, 664,008.
89. Takehara, M., Ebara, R., Nanbu, N., Ue, M., and Sasaki, Y. (2003) *Electrochemistry*, **71**, 1172.
90. Takehara, M., Tukimori, N., Nanbu, N., Ue, M., and Sasaki, Y. (2003) *Electrochemistry*, **71**, 1201.
91. Ue, M., Ida, K., and Mori, S. (1994) *J. Electrochem. Soc.*, **141**, 2989.

92. Wang, Q., Zakeeruddin, S.M., Exnar, I., and Grätzel, M. (2004) *J. Electrochem. Soc.*, **151**, A1598.
93. Watanabe, M., Kataoka, T., Fuchigami, T., Matsumoto, H., Hayase, S., Murai, S., Satou, T., and Ohno, H. (2004) *Ionic liquids (in Jananese)*, NTS Inc., Tokyo, p. 67.
94. Xu, K. and Angell, C.A. (2002) *J. Electrochem. Soc.*, **149**, A920.
95. Sun, X. and Angell, C.A. (2004) *Solid State Ionics*, **175**, 257.
96. Sun, X. and Angell, C.A. (2005) *Electrochem. Commun.*, **7**, 261.
97. Asahara, T., Tokura, N., Okawara, M., Kumanotani, J., and Seno, M. (1976) *Solvent Handbook (in Japanese)*, Kodansya Scientific Inc., Tokyo, pp. 569, 625, 627, 768.
98. Wataanabe, Y., Kinoshita, S., Morimoto, H., and Tobishima, S. (2008) *J. Power Sources*, **179**, 770.
99. Auborn, J.J. and Ciemieki, K.T. (1983) Fall Meeting of ECS, Vol. 83-2, Extended Abstract, p. 111.
100. Wilkinson, D.P., Blom, H., Brandt, K., and Wainwright, D. (1991) *J. Power Sources*, **36**, 517.
101. Hirai, T., Yoshimatsu, I., and Yamaki, J. (1994) *J. Electrochem. Soc.*, **141**, 611.
102. Desjardins, C.D. and MacLean, G.K. (1989) Extended Abstracts of the Electrochemical Society Meeting, Hollywood, Abstract No. 52.
103. Saito, K., Arakawa, M., Tobishima, S., and Yamaki, J. (1995) Extended Abstracts of the Electrochemical Society Meeting, Reno, Nevada, Abstract No. 14.
104. Arakawa, M., Nemoto, Y., Tobishima, S., Ichimura, M., and Yamaki, J. (1993) *J. Power Sources*, **43–44**, 517.
105. Tobishima, S., Hayashi, K., Nemoto, Y., and Yamaki, J. (1996) *Denki Kagaku*, **64**, 1000.
106. Nishijima, M., Kagohashi, T., Imanishi, N., Takeda, Y., Yamamoto, O., and Kondo, S. (1996) *Solid State Ionics*, **83**, 107.
107. Idota, Y., Mineo, Y., Matsufuji, A., and Miyasaka, T. (1997) *Denki Kagaku*, **65**, 717.
108. Sigala, C., La Salle, A.L., Guyomard, D., and Piffard, Y. (1996) Extended Abstracts of 8th International Meeting on Lithium Batteries, Abstract No. II B63.
109. Tobishima, S. and Yamaki, J. (1999) *J. Power Sources*, **81-82**, 882.

Further Reading

Shodai, T., Okada, S., Tobishima, S., and Yamaki, J. (1996) *Solid State Ionics*, **86-88**, 785.

Yamaki, J., Tobishima, S., Sakurai, Y., Saito, K., and Hayashi, K. (1998) *J. Appl. Electrochem.*, **28**, 135.

14
Lithium Alloy Anodes
Robert A. Huggins

14.1
Introduction

The interest in ever-higher energy content has caused the development of cells with relatively high voltages to receive much attention in the lithium battery research community in recent years. This has led to the exploration of a number of positive electrode materials that operate at potentials of about 4 V, or even more, positive of the potential of elemental lithium.

However, this is only part of the story, for the voltage of a cell is determined by the difference between the potentials of the negative and positive electrodes. The highest voltages are obtained by the use of elemental lithium in the negative electrode. The use of negative electrode reactants with lithium activities of less than unity results in electrodes with more positive potentials, thus reducing the cell voltage.

The voltage is only one of the important parameters of batteries, and other considerations also are often important in practical systems. One that has received increasing attention in recent years is the question of safety. Batteries that store large amounts of energy can be very dangerous if that energy is suddenly released, and there have been a number of accidents involving lithium batteries. It is now recognized that these safety problems generally relate to phenomena at the negative electrode. Local heating to high temperatures, especially above the melting point of lithium when elemental lithium is used, can lead to serious disasters.

In addition, the cycling behavior of lithium cells is often limited by negative electrode problems. These may include gradually increasing impedance, which is observed as decreasing output voltage. In some cases there is a macroscopic shape change. If elemental lithium is used (below its melting point), there may be dendrite growth, or a tendency for filamentary or whisker formation. This may lead to disconnection and electrical isolation of active material, resulting in loss of capacity. It may also result in potentially dangerous electrical shorting between electrodes.

Whereas there had been a significant amount of work on the properties of lithium alloys in the research community for a number of years, this alternative

Handbook of Battery Materials, Second Edition. Edited by Claus Daniel and Jürgen O. Besenhard.
© 2011 Wiley-VCH Verlag GmbH & Co. KGaA. Published 2011 by Wiley-VCH Verlag GmbH & Co. KGaA.

did not receive much attention in the commercial world until about 1990, when Sony began producing batteries with lithium–carbon negative electrodes. Since then, there has been a large amount of work on the preparation, structure, and properties of various carbons in lithium cells.

Another aspect is now beginning to receive attention, also on the basis of commercial development rather than arising directly from activities in the public research community. This is the development by Fuji Photo Film Co. of the use of materials based upon tin oxide as negative electrodes. As will be discussed later, this involves the formation of alloys by the *in-situ* conversion of the oxide.

14.2
Problems with the Rechargeability of Elemental Electrodes

In the case of an electrochemical cell with a negative electrode consisting of an elemental metal, the process of recharging is apparently very simple, for it merely involves the electrodeposition of the metal. There are problems, however.

One of these is the 'shape change' phenomenon, in which the location of the electrodeposit is not the same as that of the discharge (deplating) process. Thus, upon cycling, the electrode metal is preferentially transferred to new locations. For the most part, this is a problem of current distribution and hydrodynamics rather than being a materials issue, therefore it will not be discussed further here.

A second type of problem relates to the inherent instability of a flat interface upon electrodeposition [1]. This is analogous to the problems of the interface evolution during electropolishing and the morphology development during the growth of an oxide layer upon a solid solution alloy, problems that were discussed by Wagner [2, 3] some time ago.

Another analogous situation is present during the crystallization of the solute phase from liquid metal solutions. This leads to the production of protuberances on the growth interface; these gradually become exaggerated, and then develop into dendrites. A general characteristic of dendrites is a tree-and-branches type of morphology, which often has very distinct geometric and crystallographic characteristics due to the orientation dependence of surface energy or growth velocity. The current distribution near the front of a protrusion develops a three-dimensional (3-D) character, leading to faster growth than that at the main electrode's surface, where the mass transport is essentially one-dimensional (1-D). In relatively low-concentration solutions, this leads to a runaway type of process, so that the dendrites consume most of the solute and grow farther and farther ahead of the main, or bulk, interface.

A third type of problem, which is often mistakenly confused with dendrite formation, is due to the presence of a reaction–product layer upon the growth interface if the electrode and electrolyte are not stable in the presence of each other. This leads to filamentary or hairy growth, and the deposit often appears to have a spongy character. During a subsequent discharge step, the filaments often become disconnected from the underlying metal, so that they cannot participate in the electrochemical reaction, and the rechargeable capacity of the electrode is reduced.

This is a common problem when using elemental lithium negative electrodes in contact with electrolytes containing organic cationic groups, regardless of whether the electrolyte is an organic liquid or a polymer [4].

In order to achieve good rechargeability, one has to maintain a consistent geometry on both the macro and micro scales and avoid electrical disconnection of the electroactive species.

14.3
Lithium Alloys as an Alternative

Attention has been given for some time to the use of lithium alloys as an alternative to elemental lithium. Groups working on batteries with molten salt electrolytes that operate at temperatures of 400–450 °C, well above the melting point of lithium, were especially interested in this possibility. Two major directions evolved. One involved the use of lithium–aluminum alloys [5, 6], while the other was concerned with lithium–silicon alloys [7–9].

Whereas this approach can avoid the problems related to lithium melting, as well as the others mentioned above, there are always at least two disadvantages related to the use of alloys. Because they reduce the activity of the lithium, they necessarily reduce the cell voltage. In addition, the presence of additional species that are not directly involved in the electrochemical reaction always brings additional weight, and generally, volume. Thus the maximum theoretical values of the specific energy and the energy density are always reduced in comparison with what might be attained with pure lithium.

In practical cases, however, the excess weight and volume due to the use of alloys may not be very far from those required with pure lithium electrodes, for one generally has to operate with a large amount of excess lithium in rechargeable cells in order to make up for the capacity loss related to the filament growth problem upon cycling.

Lithium alloys have been used for a number of years in the high-temperature 'thermal batteries' that are produced commercially for military purposes. These devices are designed to be stored for long periods at ambient temperatures, where their self-discharge kinetic behavior is very slow, before they are used. They must be heated to high temperatures when their energy output is desired. An example is the Li alloy/FeS_2 battery system that employs a molten chloride electrolyte. In order to operate, the temperature must be raised to above the melting point of the electrolyte. This type of cell typically uses either Li–Si or Li–Al alloys in the negative electrode.

The first use of lithium alloys as negative electrodes in commercial batteries to operate at ambient temperatures was the employment of Wood's metal alloys in lithium-conducting button-type cells by Matsushita in Japan. Development work on the use of these alloys started in 1983 [10], and they became commercially available somewhat later.

It was also shown in 1983 [11] that lithium can be reversibly inserted into graphite at room temperatures when a polymeric electrolyte is used. Prior experiments with liquid electrolytes were unsuccessful due to co-intercalation of species from the organic electrolytes that were used at that time. This problem has been subsequently solved by the use of other electrolytes.

There has been a large amount of work on the development of graphites and related carbon-containing materials for use as negative electrode materials in lithium batteries in recent years, due in large part to the successful development by Sony of commercial rechargeable batteries containing negative electrodes based upon materials of this family.

Lithium–carbon materials are, in principle, no different from other lithium-containing alloys. However, since this topic is treated in more detail in Part II, Chapter 13, only a few points will be briefly discussed here.

One is that the behavior of these materials is very dependent upon the details of both the nanostructure and the microstructure. Therefore, the composition, as well as thermal and mechanical treatments, play especially important roles in determining the resulting thermodynamic and kinetic properties. Materials with a more graphitic structure have more negative potentials, whereas those with less well-organized structures typically operate over much wider potential ranges, resulting in a cell voltage that is both lower and more dependent on the state-of-charge.

Another important consideration in the use of carbonaceous materials as negative electrodes in lithium cells is the common observation of a considerable loss of capacity during the first charge–discharge cycle due to irreversible lithium absorption into the structure. This has the distinct disadvantage that it requires an additional amount of lithium to be initially present in the cell. If this irreversible lithium is supplied by the positive electrode, this means that an extra amount of the positive electrode reactant material must be put into the cell during its fabrication. As the positive electrode reactant materials often have relatively low specific capacities, for example, around 140 mAh g^{-1}, this irreversible capacity in the negative electrode leads to a requirement for an appreciable amount of extra material weight and volume in the total cell.

There are some other matters that should be considered when comparing metallic lithium alloys with the lithium–carbons. The specific volume of some of the metallic alloys can be considerably lower than that of the carbonaceous materials. As will be seen later, it is possible by selection among the metallic materials to find good kinetics and electrode potentials that are sufficiently far from that of pure lithium for there to be a much lower possibility of the potentially dangerous formation of dendrites or filamentary deposits under rapid recharge conditions.

It has been shown that there is a significant advantage in the use of very small particles in cases in which there is a substantial change in specific volume upon charging and discharging electrode reactants [12]. Since the absolute magnitude of the local dimensional changes is proportional to the particle size, smaller particles lead to fewer problems with the decrepitation, or 'crumbling', of electrode microstructure that often leads to loss of electrical contact, and thus capacity loss, as well as macroscopic dimensional problems.

14.4
Alloys Formed *In situ* from Convertible Oxides

A renewed interest in noncarbonaceous lithium alloy electrodes arose recently as the result of the announcement by Fuji Photo Film Co. of the development of a new generation of lithium batteries based upon the use of an amorphous tin-based composite oxide in the negative electrode [13]. It is claimed that these electrodes have a volumetric capacity of 3200 $Ah\,L^{-1}$, which is four times that commonly achieved with carbonaceous negative electrodes, and a specific capacity of 800 $mAh\,g^{-1}$, twice that generally found in carbon-containing negative electrodes.

According to the public announcement, a new company, Fujifilm Celltec Co., has been formed to produce products based upon this approach. It was reported that some 200 patents have been applied for in this connection. Unfortunately, there is little yet available in the standard literature concerning these matters. To date, there are only references to some of the patents [14–17]. However, what must be happening seems rather obvious.

If, as an example, we make the assumption that the electrode initially has the composition SnO, if we introduce lithium into it there will be a displacement reaction in which Li_2O will be formed at the expense of the SnO due to the difference in the values of their Gibbs free energies of formation ($-562.1\,kJ\,mol^{-1}$ for Li_2O and $-256.8\,kJ\,mol^{-1}$ in the case of SnO). This is equivalent to a driving force of 1.58 V. The other product will be elemental Sn, and as additional Li is brought into the electrode this will react further to form the various Li-Sn alloys that are discussed in some detail later in this section. This simplified picture is consistent with what has been found in experiments of this general type [18].

14.5
Thermodynamic Basis for Electrode Potentials and Capacities under Conditions in which Complete Equilibrium can be Assumed

The general thermodynamic treatment of binary systems which involve the incorporation of an electroactive species into a solid alloy electrode under the assumption of complete equilibrium was presented by Weppner and Huggins [19–21]. Under these conditions the Gibbs Phase Rule specifies that the electrochemical potential varies with composition in the single-phase regions of a binary phase diagram and is composition-independent in two-phase regions if the temperature and total pressure are kept constant.

Thus the variations of the electrode potential during discharge and charge, as well as the phases present and the charge capacity of the electrode, directly reflect the thermodynamics of the alloy system.

A series of experiments have been undertaken to evaluate the relevant thermodynamic properties of a number of binary lithium alloy systems. The early work was directed toward determination of their behavior at about 400 °C because of interest in their potential use as components in molten salt batteries operating in that general temperature range. Data for a number of binary lithium alloy systems at about 400 °C are presented in Table 14.1. These were mostly obtained by the use of an experimental arrangement employing the LiCl–KCl eutectic molten salt as a lithium-conducting electrolyte.

It was shown some time ago that one can also use a similar thermodynamic approach to explain and/or predict the composition dependence of the potential of

Table 14.1 Plateau potentials and composition ranges of some binary lithium alloys Li_yM at 400 °C.

Voltage vs Li	M	Range of y	References
0.047	Si	3.25–4.4	[22]
0.058	Cd	1.65–2.33	[23]
0.080	In	2.08–2.67	[24]
0.089	Pb	3.8–4.4	[25]
0.091	Ga	1.53–1.93	[26]
0.122	Ga	1.28–1.48	[26]
0.145	In	1.74–1.92	[24]
0.156	Si	2.67–3.25	[22]
0.170	Sn	3.5–4.4	[27]
0.237	Pb	3.0–3.5	[25]
0.271	Pb	2.67–3.0	[25]
0.283	Si	2–2.67	[22]
0.283	Sn	2.6–3.5	[27]
0.300	Al	0.08–0.9	[28]
0.332	Si	0–2	[22]
0.373	Cd	0.33–0.45	[23]
0.375	Pb	1.1–2.67	[25]
0.387	Sn	2.5–2.6	[27]
0.430	Sn	2.33–2.5	[27]
0.455	Sn	1.0–2.33	[27]
0.495	In	1.2–0.86	[24]
0.507	Pb	0–1.0	[25]
0.558	Cd	0.12–0.21	[23]
0.565	Ga	0.15–0.82	[26]
0.570	Sn	0.57–1.0	[27]
0.750	Bi	1.0–2.82	[21]
0.875	Sb	2.0–3.0	[21]
0.910	Sb	0–2.0	[21]

electrodes in ternary systems [29–32]. This followed from the development of the analysis methodology for the determination of the stability windows of electrolyte phases in ternary systems [33]. In these cases, one uses isothermal sections of ternary phase diagrams, the so-called Gibbs triangles, upon which to plot compositions. In ternary systems, the Gibbs Phase Rule tells us that three-phase equilibria will have composition-independent intensive properties, that is, activities and potentials. Thus compositional ranges that span three-phase regions will lead to potential plateaus at constant temperature and pressure.

Estimated data on a number of ternary lithium systems theoretically investigated as extensions of the Li–Si binary system are included in Table 14.2. Also included are comparable data for the binary Li–Si alloy that are currently being used in commercial thermal batteries.

This thermodynamically based methodology provides predictions of the lithium capacities in addition to the electrode potentials of the various three-phase equilibria under conditions of complete equilibrium. This information is included as the last column in Table 14.2, in terms of the number of moles of lithium per kilogram total alloy weight.

From a practical standpoint, the most useful compositions would be those with quite negative potentials (so as to give high cell voltages) that also have large capacities for lithium. However, it must be recognized that the materials with the most negative potentials, and thus the highest lithium activities, will be the most reactive, and thus will be more difficult to handle than those whose potentials are somewhat farther from that of pure lithium.

As recently pointed out [32], several of these ternary systems appear to have potentials and capacities that should make them quite interesting for practical

Table 14.2 Estimated data relating to lithium–silicon-based ternary systems at 400 °C.

System	Starting composition	Phases in equilibrium	Voltage (mV) vs Li	Li (mol kg^{-1})
Li–Si–Mo	Mo_5Si_3	$Mo_5Si_3–Mo_3Si–Li_{22}Si_5$	3	9.7
Li–Si–Ca	$CaSi$	$CaSi–Ca_2Si–Li_{22}Si_5$	13	26.4
Li–Si–Mn	Mn_3Si	$Mn_3Si–Mn–Li_{22}Si_5$	43	19.7
Li–Si–Mn	Mn_5Si_3	$Mn_5Si_3–Mn_3Si–Li_{13}Si_4$	45	11.1
Li–Si–Mg	Mg_2Si	$Mg_2Si–Mg–Li_{13}Si_4$	60	32.7
Li–Si–Mo	$MoSi_2$	$MoSi_2–Mo_5Si_3–Li_{13}Si_4$	120	24.8
Li–Si–Cr	Cr_5Si_3	$Cr_5S_{13}–Cr_3Si–Li_{13}Si_4$	138	11.6
Li–Si	Li_7Si_3	$Li_7Si_3–Li_{13}Si_4$	158	18.1
Li–Si–Mn	$MnSi$	$MnSi–Mn_5Si_3–Li_7Si_{13}$	163	10.4
Li–Si–Ti	$TiSi$	$TiSi–Ti_5Si_3–Li_7Si_3$	182	11.3
Li–Si–Nb	$NbSi_2$	$NbSi_2–Nb_5Si_3–Li_7Si_3$	184	19.0
Li–Si–V	VSi_2	$VSi_2–V_5Si_3–Li_7Si_3$	191	25.2
Li–Si–Cr	$CrSi$	$CrSi–Cr_5Si_3–Li_7Si_3$	205	10.8
Li–Si–Ta	$TaSi_2$	$TaSi_2–Ta_5Si_3–Li_7Si_3$	211	12.6
Li–Si–Cr	$CrSi_2$	$CrSi_2–CrSi–Li_7Si_3$	223	18.8
Li–Si–Ni	Ni_7Si_{13}	$Li_7Si_{13}–Nisi–Li_{12}Si_7$	316	12.1

applications. Li–Si ternary systems with Mg, Ca, and Mo seem especially interesting from the standpoint of their potentials and capacities. As an example, if one assumes that a positive electrode is used that has a potential 2.0 V positive of elemental lithium and a capacity of 1 mol of lithium per 60 g of active component, these negative electrode materials provide a maximum theoretical specific energy of 574, 544, and 502 Wh kg^{-1}, respectively, whereas the binary Li–Si alloy currently used in thermal batteries would have a maximum value of 428 Wh kg^{-1}. Confirmatory experimental information on the Li–Mg–Si system [34] was recently presented [35].

14.6
Crystallographic Aspects and the Possibility of Selective Equilibrium

If we look at the mechanistic and crystallographic aspects of the operation of poly-component electrodes, we see that the incorporation of electroactive species such as lithium into a crystalline electrode can occur in two basic ways. In the examples discussed above, and in which complete equilibrium is assumed, the introduction of the guest species can either involve a simple change in the composition of an existing phase by solid solution or it can result in the formation of new phases with different crystal structures from that of the initial host material. When the identity and/or amounts of phases present in the electrode change, the process is described as a reconstitution reaction. That is, the micro-structure is reconstituted.

In the simple case of a reconstitution reaction in which the incorporation of additional electroactive species occurs by the nucleation and growth of a new phase, the relative amount of this new phase with a higher solute content increases. If the initial phase and the new phase are in local equilibrium, the respective compositions at their joint interface do not change with the extent of the reaction. The amounts of the phases, determined by the motion of the interfaces between these phases, are related to the lengths of the two-phase constant-potential plateaus in binary systems and the three-phase constant-potential plateaus in ternary systems, and these, in turn, are determined by the extent of the corresponding regions in the relevant phase diagrams.

In many systems, both single-phase and polyphase behaviors are found in different composition ranges. Intermediate, as well as terminal, phases often have been found to have quite wide ranges of composition. Examples are the broad Zintl phases found in several of the binary lithium systems studied by Wen [23].

The second way in which an electroactive species such as lithium can be incorporated into the structure of an electrode is by a topotactic insertion reaction. In this case the guest species is relatively mobile and enters the crystal structure of the host phase so that no significant change in the structural configuration of the host lattice occurs.

Thus the result is the formation of a single-phase solid solution. The insertion of additional guest species involves only a change in the overall (and thus also the local) composition of the solid solution, rather than the formation of additional phases.

From a thermodynamic viewpoint, there is selective, rather than complete, equilibrium under conditions in which this type of reaction occurs. We can assume equilibrium in the sublattice of the mobile solute species, but not in the host substructure, as strong bonding makes atomic rearrangements relatively sluggish in that part of the crystal structure.

In general, equilibrium within the guest species sublattice results in their being randomly arranged among the various interstitial locations within the host structure. There are, however, a number of cases in which the guest species are distributed among their possible sites within the host structure in an ordered, rather than random, manner. There can be different sets of these ordered sites, each having the thermodynamic characteristics of a separate phase. Thus, as the concentration of guest species is changed, such materials can appear thermodynamically to go through a series of phase changes, even though the host structure is relatively stable. This type of behavior was demonstrated for the case of lithium insertion into a potassium tungsten oxide [36].

The thermodynamic properties of topotactic insertion reaction materials with selective equilibrium are quite different from those of materials in which complete equilibrium can be assumed, and reconstitution reactions take place. Instead of flat plateaus related to polyphase equilibria, the composition-dependence of the potential generally has a flat S-type form.

Under near-equilibrium conditions the shape of this curve is related to two contributions: the compositional dependence of the configurational entropy of the guest ions and the contribution to the chemical potential from the electron gas [37].

The configurational entropy of the mobile guest ions, assuming random mixing and a concentration x, residing in x° lattice sites of equal energy, is

$$S = -R \ln[x/(x^\circ - x)] \tag{14.1}$$

There is also a small contribution from thermal entropy, but this can be neglected.

If we can assume that the electrode material is a good metal and the electronic gas is fully degenerate, the chemical potential of the electrons is given by the Fermi level, E_F, which can be written as

$$E_F = [\text{Constant}][(x)^{2/3}/m^*] \tag{14.2}$$

where m^* is the effective mass of the electrons.

14.7
Kinetic Aspects

In addition to the questions of the potentials and capacities of electrodes, which are essentially thermodynamic considerations, practical utilization of alloys as electrodes also requires attractive kinetic properties.

The primary question is the rate at which the mobile guest species can be added to, or deleted from, the host microstructure. In many situations the critical problem is the transport within a particular phase under the influence of gradients in chemical composition rather than kinetic phenomena at the electrolyte/electrode interface. In this case, the governing parameter is the chemical diffusion coefficient of the mobile species, which relates to transport in a chemical concentration gradient.

Diffusion has often been measured in metals by the use of radioactive tracers. The resulting parameter, D_T, is related to the self-diffusion coefficient by a correlation factor f that is dependent upon the details of the crystal structure and jump geometry. The relation between D_T and the self-diffusion coefficient D_{self} is thus simply

$$D_T = D_{self} * f \tag{14.3}$$

Whereas in many metals with relatively simple and isotropic crystal structures the parameter f has values between 0.5 and 1, it can have much more extreme values in materials in which the mobile species move through much less isotropic structures with one-dimensional (1-D) or two-dimensional (2-D) channels, as is often the case with insertion reaction electrode materials. As a result, radiotracer experiments can provide misleading information about self-diffusion kinetics in such cases.

More importantly, the chemical diffusion coefficient D_{chem}, instead of D_{self}, is the parameter that is relevant to the behavior of electrode materials. They are related by

$$D_{chem} = D_{self} * W \tag{14.4}$$

where W is an enhancement factor. This is sometimes called the 'thermodynamic factor,' and can be written as

$$W = d\ln a_i / d\ln c_i \tag{14.5}$$

in which a_i and c_i are the activity and concentration of the neutral mobile species i, respectively. Experimental data have shown that the value of W can be very large in some cases. An example is the phase Li_3Sb, in which it has a value of 70 000 at 360 °C [38].

It is thus much better to measure the chemical diffusion coefficient directly. Descriptions of electrochemical methods for doing this, as well as the relevant theoretical background, can be found in the literature [39, 40]. Available data on the chemical diffusion coefficient in a number of lithium alloys are included in Table 14.3.

14.8
Examples of Lithium Alloy Systems

14.8.1
Lithium–Aluminum System

Because of the interest in its use in elevated-temperature molten salt electrolyte batteries, one of the first binary alloy systems studied in detail was the

14.8 Examples of Lithium Alloy Systems

Table 14.3 Data on chemical diffusion in lithium alloy phases.

Composition		Max. D_{chem} (cm² s⁻¹)	Max. W	Temperature (°C)	References
Nominal	Range (%Li)				
LiAl	16.4	1.2×10^{-4}	70	415	[28, 40]
Li_3Sb	0.05	7.0×10^{-5}	70 000	360	[38]
Li_3Bi	1.37	2.0×10^{-4}	370	380	[41]
$Li_{12}Si_7$	0.54	8.1×10^{-5}	160	415	[22]
Li_7Si_3	3.0	4.4×10^{-5}	111	415	[22]
$Li_{13}Si_4$	1.0	9.3×10^{-5}	325	415	[22]
$Li_{22}Si_5$	0.4	7.2×10^{-5}	232	415	[22]
LiSn	1.9	4.1×10^{-6}	185	415	[42]
Li_7Sn_3	0.5	4.1×10^{-5}	110	415	[42]
Li_5Sn_2	1.0	5.9×10^{-5}	99	415	[42]
$Li_{13}Sn_5$	0.5	7.6×10^{-4}	1150	415	[42]
Li_7Sn_2	1.4	7.8×10^{-5}	196	415	[42]
$Li_{22}Sn_5$	1.2	1.9×10^{-4}	335	415	[42]
LiGa	22.0	6.8×10^{-5}	56	415	[26]
LiIn	33.0	4.0×10^{-5}	52	415	[24]
LiCd	63.0	3.0×10^{-6}	7	415	[23]

lithium–aluminum system. As shown in Figure 14.1, the potential–composition behavior shows a long plateau between the lithium-saturated terminal solid solution and the intermediate β phase 'LiAl,' and a shorter one between the composition limits of the β and γ phases, as well as composition-dependent values in the single-phase regions [28]. This is as expected for a binary system with complete equilibrium. The potential of the first plateau varies linearly with temperature, as shown in Figure 14.2.

Chemical diffusion in the β phase determines the kinetic behavior of these electrodes when lithium is added, so this was investigated in detail using four different electrochemical techniques [28, 40]. It was found that chemical diffusion is remarkably fast in this phase, and that the activation energy attains very low values on the lithium-poor side of the composition range. These data are shown in Figure 14.3.

In addition to this work on the β phase, both the thermodynamic and kinetic properties of the terminal solid-solution region, which extends to about 9 atom% lithium at 423 °C, were also investigated in detail [43].

14.8.2
Lithium–Silicon System

The lithium–silicon system has also been of interest for use in the negative electrodes of elevated-temperature molten salt electrolyte lithium batteries. A

14 Lithium Alloy Anodes

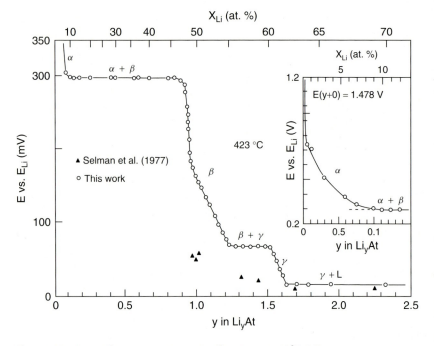

Figure 14.1 Potential vs composition in Li–Al system at 423 °C [28].

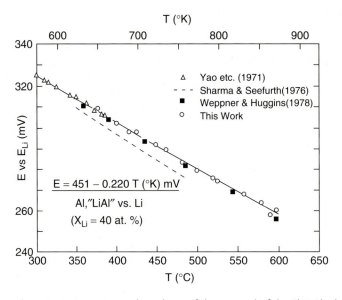

Figure 14.2 Temperature dependence of the potential of the Al/'LiAl' plateau [28].

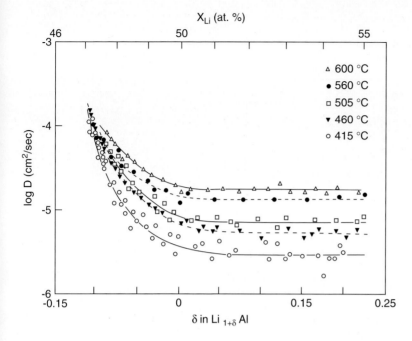

Figure 14.3 Variation of the chemical diffusion coefficient with composition in the 'LiAl' phase at different temperatures [28].

composition containing 44 wt% Li, where Li/Si = 3.18, has been used in commercial thermal batteries developed for military purposes.

Experiments have been performed to study both the thermodynamic and kinetic properties of compositions in this system [22], and the composition dependence of the equilibrium potential at 415 °C is shown in Figure 14.4. It is seen that the commercial electrode composition sits upon a two-phase plateau at a potential 158 mV positive of pure lithium. As lithium is removed the overall composition will first follow that plateau. The potential will then become more positive as it traverses the other plateaus at 288 and 332 mV versus lithium, in that order.

As a result, the cell voltage will decrease during the discharge, regardless of the behavior of the positive electrode.

14.8.3
Lithium–Tin System

The lithium–tin binary system is somewhat more complicated, as there are six intermediate phases, as shown in the phase diagram in Figure 14.5. A thorough study of the thermodynamic properties of this system was undertaken [27]. The composition dependence of the potential at 415 °C is shown in Figure 14.6.

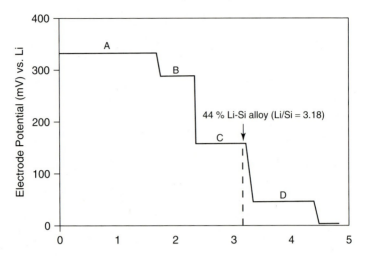

Figure 14.4 Potential vs composition in the Li–Si system at 415 °C [22].

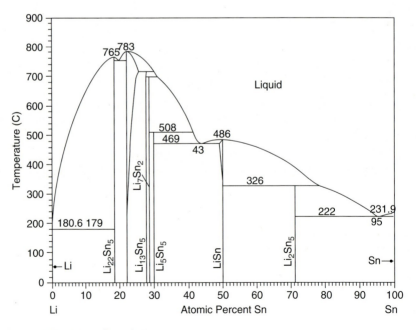

Figure 14.5 Li–Sn phase diagram.

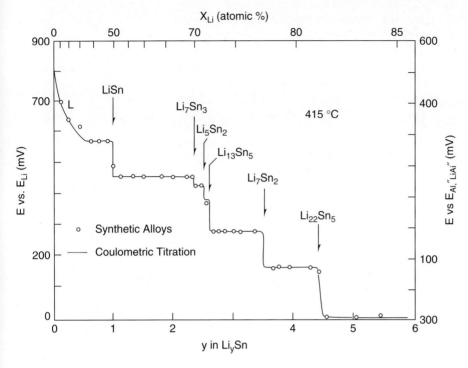

Figure 14.6 Potential vs composition in the Li–Sn system at 415 °C [27].

Measurements were also made of the potential–composition behavior, as well as the chemical diffusion coefficient and its composition dependence, in each of the intermediate phases in the Li–Sn system at 415 °C [42].

It was found that chemical diffusion is reasonably fast in all of the intermediate phases in this system. The self-diffusion coefficients are all high and of the same order of magnitude. However, due to its large value of thermodynamic enhancement factor W, the chemical diffusion coefficient in the phase $Li_{13}Sn_5$ is extremely high, approaching 7.6×10^{-4} cm^2 s^{-1}, which is about 2 orders of magnitude higher than that in typical liquids. These data are included in Table 14.3.

14.9
Lithium Alloys at Lower Temperatures

A smaller number of binary lithium systems have also been investigated at lower temperatures. This has involved measurements using $LiNO_3$–KNO_3 molten salts at about 150 °C [44] as well as experiments with organic solvent-based electrolytes at ambient temperatures [45, 46]. Data on these are included in Table 14.4.

14 Lithium Alloy Anodes

Table 14.4 Plateau potentials and composition ranges of lithium alloys Li_yM at 25 °C.

Voltage vs Li	M	Range of y	References
0.005	Zn	1–1.5	[46]
0.055	Cd	1.5–2.9	[46]
0.157	Zn	0.67–1	[46]
0.219	Zn	0.5–0.67	[46]
0.256	Zn	0.4–0.5	[46]
0.292	Pb	3.2–4.5	[46]
0.352	Cd	0.3–0.6	[46]
0.374	Pb	3.0–3.2	[46]
0.380	Sn	3.5–4.4	[45]
0.420	Sn	2.6–3.5	[45]
0.449	Pb	1–3.0	[46]
0.485	Sn	2.33–2.63	[45]
0.530	Sn	0.7–2.33	[45]
0.601	Pb	0–1	[46]
0.660	Sn	0.4–0.7	[45]
0.680	Cd	0–0.3	[46]
0.810	Bi	1–3	[45]
0.828	Bi	0–1	[45]
0.948	Sb	2–3	[45]
0.956	Sb	1–2	[45]

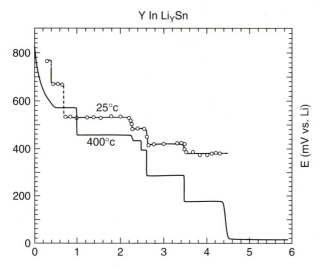

Figure 14.7 Potential vs. composition in the Li–Sn system at 25 °C compared with data at 400 °C [45].

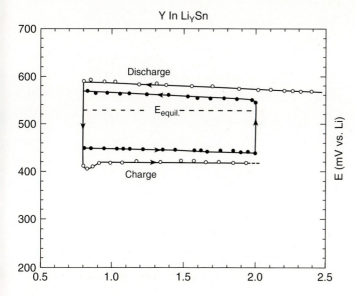

Figure 14.8 Charge–discharge curves for Li$_x$Sn ($x = 0.8$–2.5) at ambient temperature. Solid points are at a current density of 0.24 mA cm^{-2}, and open points at a current density of 0.5 mA cm^{-2}. The equilibrium potential is also shown [45].

The lithium–tin system has been investigated at room temperature and the influence of temperature upon the composition dependence of the potential is shown in Figure 14.7. It is seen that five constant-potential plateaus are found at 25 °C. Their potentials are listed in Table 14.4. It was also shown that the kinetics on the longest plateau, from $x = 0.8$ to 2 in Li$_x$Sn, are quite favorable, even at quite high currents (see Figure 14.8).

The composition dependence of the potential of the Li$_{4.4}$Sn phase was determined, as shown in Figure 14.9.

The chemical diffusion coefficient in that phase was also evaluated and found to be quite high, as can be seen in Figure 14.10 [47]. The chemical diffusion coefficient was also measured in two other Li–Sn phases; these data are all included in Table 14.5.

Comparable information on the Li–Bi and Li–Sb systems was also obtained, and their room-temperature potentials are also included in Table 14.4. The temperature dependence of the potentials of the different two-phase plateaus is shown in Figure 14.11.

This work was extended to the investigation of the Li–Zn, Li–Cd, and Li–Pb alloy systems [46, 47]. The potentials of the various plateaus found in these systems are included in Table 14.4, and are summarized in Figure 14.12.

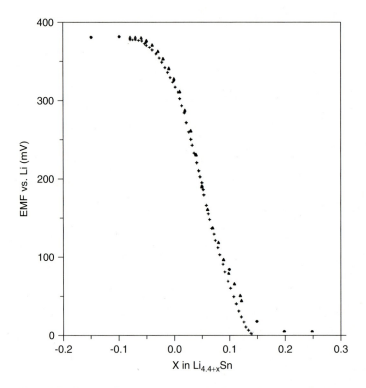

Figure 14.9 Potential vs composition in the Li$_{4.4}$Sn phase at ambient temperature [47].

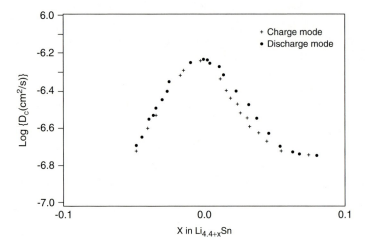

Figure 14.10 Composition dependence of the chemical diffusion coefficient in the Li$_{4.4}$Sn phase at ambient temperature [47].

Table 14.5 Chemical diffusion data for lithium–tin phases at 25 °C.

Phase	Volts vs. Li	D_{chem} (cm^2 s^{-1})
$Li_{0.7}Sn$	0.560	$(6-8) \times 10^{-8}$
$Li_{2.33}Sn$	0.520	$(3-5) \times 10^{-7}$
$Li_{4.4}Sn$	0.0–0.380	$(1.8-5.9) \times 10^{-7}$

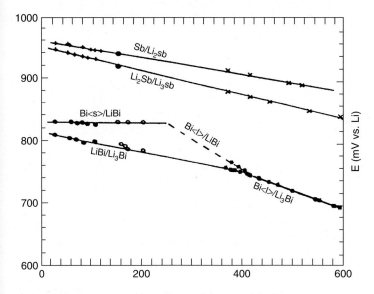

Figure 14.11 Temperature dependence of the potential of the two-phase plateaus in the Li–Sb and Li–Bi systems [45].

14.10
The Mixed-Conductor Matrix Concept

In order to achieve appreciable macroscopic current densities while maintaining low local microscopic charge and particle flux densities, many battery electrodes that are used in conjunction with liquid electrolytes are produced with porous micro-structures containing very fine particles of the solid reactant materials. This porous structure of high reactant surface area is permeated with the electrolyte.

This porous fine-particle approach has several characteristic disadvantages, among which are difficulties in producing uniform and reproducible microstructures and limited mechanical strength when the structure is highly porous. In addition, these systems often suffer Ostwald ripening, sintering, or other time-dependent changes in both microstructure and properties during cyclic operation.

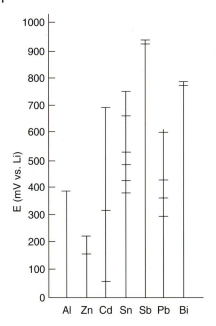

Figure 14.12 Plateau potentials of seven lithium alloy systems at ambient temperature [46].

A quite different approach was introduced in the early 1980s [48–50], in which a dense solid electrode is fabricated which has a composite microstructure in which particles of the reactant phase are finely dispersed within a solid, electronically conducting matrix in which the electroactive species is also mobile. There is thus a large internal reactant/mixed-conductor matrix interfacial area. The electroactive species is transported through the solid matrix to this interfacial region, where it undergoes the chemical part of the electrode reaction. Since the matrix material is also an electronic conductor, it can also act as the electrode's current collector. The electrochemical part of the reaction takes place on the outer surface of the composite electrode.

When such an electrode is discharged by removal of the electroactive species, the residual particles of the reactant phase remain as relics in the microstructure. This provides fixed permanent locations for the reaction to take place during subsequent cycles, when the electroactive species again enters the structure. Thus this type of configuration can provide a mechanism for the achievement of true microstructural reversibility.

In order for this concept to be applicable, the matrix and the reactant phase must be thermodynamically stable in contact with each other. One can evaluate this possibility if one has information about the relevant phase diagram – which typically involves a ternary system – as well as the titration curves of the component binary systems. In a ternary system, the two materials must lie at corners of the same constant-potential tie-triangle in the relevant isothermal ternary phase diagram in order to not interact. The potential of the tie-triangle determines the electrode reaction potential, of course.

An additional requirement is that the reactant material must have two phases present in the tie-triangle, but the matrix phase only one. This is another way of saying that the stability window of the matrix phase must span the reaction potential, but that the binary titration curve of the reactant material must have a plateau at the tie-triangle potential. It has been shown that one can evaluate the possibility that these conditions are met from knowledge of the binary titration curves, without having to perform a large number of ternary experiments.

The kinetic requirements for a successful application of this concept are readily understandable. The primary issue is the rate at which the electroactive species can reach the matrix/reactant interfaces. The critical parameter is the chemical diffusion coefficient of the electroactive species in the matrix phase. This can be determined by various techniques, as discussed above.

The first example that was demonstrated was the use of the phase with the nominal composition $Li_{13}Sn_5$ as the matrix, in conjunction with reactant phases in the lithium–silicon system at temperatures near 400 °C. This is an especially favorable case, due to the high chemical diffusion coefficient or lithium in the Li_3Sn_5 phase.

The relationship between the potential–composition data for these two systems under equilibrium conditions is shown in Figure 14.13. It is seen that the phase $Li_{2.6}Sn$ ($Li_{13}Sn_5$) is stable over a potential range that includes the upper two-phase reconstitution reaction plateau in the lithium–silicon system. Therefore, lithium can react with Si to form the phase $Li_{1.7}Si$ ($Li_{12}Si_7$) inside an all-solid composite electrode containing the $Li_{2.6}Sn$ phase, which acts as a lithium transporting, but electrochemically inert, matrix. Figure 14.14 shows the relatively small polarization

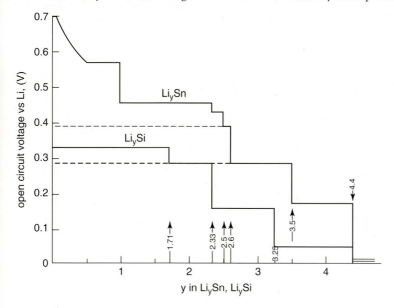

Figure 14.13 Composition dependence of the potential in the Li–Sn and Li–Si systems at 415 °C [48].

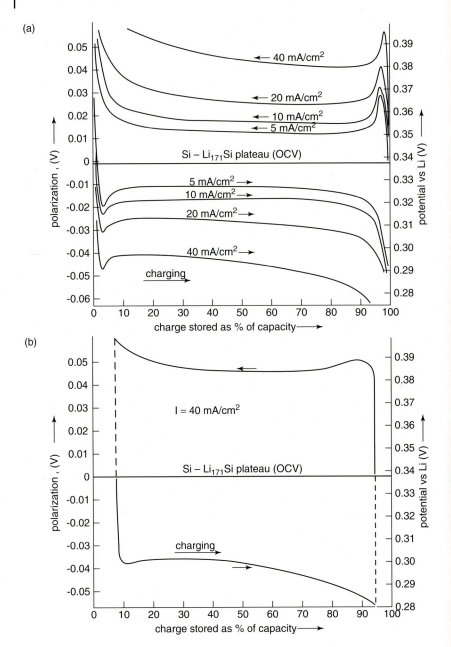

Figure 14.14 Charge-discharge curves for the upper plateau in the Li_xSi system inside a matrix of the $Li_{2.6}Sn \rightarrow$ phase at 415 °C. (a) Effect of current density. (b) The potential overshoot related to the nucleation of the second phase is mostly eliminated if the electrode is not cycled to the ends of the plateau [48].

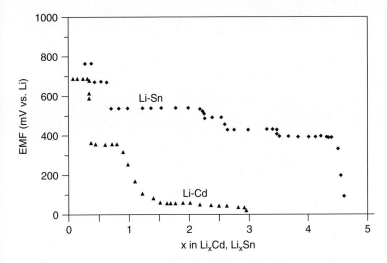

Figure 14.15 Potential vs composition for the Li–Sn and Li–Cd systems at ambient temperature [52].

that was observed during the charge and discharge of this electrode, even at relatively high current densities. It is seen that there is a potential overshoot due to the free energy involved in the nucleation of a new second phase if the reaction goes to completion in each direction. On the other hand, if the composition is not driven quite so far, this nucleation-related potential overshoot does not appear.

This concept has also been demonstrated at ambient temperature in the case of the Li–Sn–Cd system [51, 52]. The composition dependences of the potentials in the two binary systems at ambient temperatures are shown in Figure 14.15, and the calculated phase stability diagram for this ternary system is shown in Figure 14.16. It was shown that the phase $Li_{4.4}Sn$, which has fast chemical diffusion for lithium, is stable at the potentials of two of the Li–Cd reconstitution reaction plateaus, and therefore can be used as a matrix phase.

The behavior of this composite electrode, in which Li reacts with the Cd phases inside of the Li–Sn phase, is shown in Figure 14.17.

In order to achieve good reversibility, the composite electrode microstructure must have the ability to accommodate any volume changes that might result from the reaction that takes place internally. This can be taken care of by clever microstructural design and alloy fabrication techniques.

14.11
Solid Electrolyte Matrix Electrode Structures

In solid-state systems it is often advantageous to have some of the electrolyte material mixed in with the reactant. There are two general advantages that result

428 | 14 Lithium Alloy Anodes

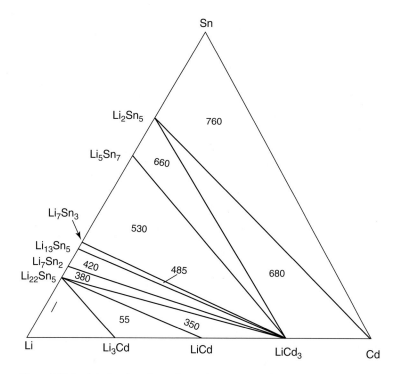

Figure 14.16 Calculated isothermal Li–Cd–Sn ternary phase stability diagram at ambient temperature [52].

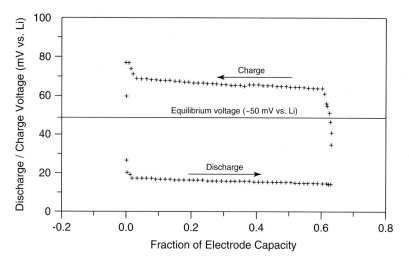

Figure 14.17 Sixth charge-discharge curve of a composite Li–Sn/Li–Cd electrode at a current density of 0.1 mA cm^{-2} at ambient temperature [52].

from doing this. One is that the contact area between the electrolyte phase and the electrode phase (the electrochemical interface) is greatly increased. The other is that the presence of the electrolyte material changes the thermal expansion characteristics of the electrode structure so as to be closer to that of the pure electrolyte. By doing so, the stresses that arise as the result of a difference in the expansion coefficients of the two adjacent phases that can use mechanical separation of the interface are reduced.

It is interesting to note that the recently announced Fujifilm development of convertible oxide electrodes results in the formation of a microstructure containing fine dispersions of both Li–Sn alloys and Li_2O. The latter is known to be a lithium-transporting solid electrolyte. Thus these electrodes can be thought of as having a composite microstructure with an electrolyte as well as the reactant phase.

14.12
What about the Future?

The recent development of the convertible oxide materials at Fuji Photo Film Co. will surely cause much more attention to be given to alternative lithium alloy negative electrode materials in the near future from both scientific and technological standpoints. This work has shown that it may pay not only to consider different known materials, but also to think about various strategies that might be used to form attractive materials *in situ* inside the electrochemical cell.

References

1. Huggins, R.A. and Elwell, D. (1977) *J. Crystal Growth*, **37**, 159.
2. Wagner, C. (1954) *J. Electrochem. Soc.*, **101**, 225.
3. Wagner, C. (1956) *J. Electrochem. Soc.*, **103**, 571.
4. Deublein, G. and Huggins, R.A. (1986) *Solid State Ionics*, **18–19**, 1110.
5. Yao, N.P., Heredy, L.A., and Saunders, R.C. (1971) *J. Electrochem. Soc.*, **118**, 1039.
6. Gay, E.C. (1976) *J. Electrochem. Soc.*, **123**, 1591.
7. Lai, S.C. (1976) *J. Electrochem. Soc.*, **123**, 1196.
8. Sharma, R.A. and Seefurth, R.N. (1976) *J. Electrochem. Soc.*, **123**, 1763.
9. Seefurth, R.N. and Sharma, R.A. (1977) *J. Electrochem. Soc.*, **124**, 1207.
10. Ogawa, H. (1984) *Proceedings of the 2nd International Meeting on Lithium Batteries*, Elsevier Sequoia, S. A. Lausanne, p. 259.
11. Yazami, R. and Touzain, P. (1983) *J. Power Sources*, **9**, 365.
12. Besenhard, J.O., Yang, J. and Winter, M. (1997) *J. Power Sources*, **68**, 87.
13. (1996) Internet: http://www.fujifilm.co.jp/eng/news_e/nr079.html.
14. Kubota, T. and Tanaka, M. (1994) Jpn. Kokai Tokkyo Koho, JP 94-55614940325.
15. Funatsu, E. (1994) Jpn. Kokai Tokkyo Koho, JP 94-2592940114.
16. Idota, Y., Nishima, M., Miyaki, Y., Kubota, T., and Miyasaka, T. (1996) European Patent Application EP 651450 A1 950503.

17. Idota, Y., Nishima, M., Miyaki, Y., Kubota, T., and Miyasaka, T. (1994) Canadian Patent Application 21134053.
18. Courtney, I.A. and Dahn, J.R. (1997) *J. Electrochem. Soc.*, **144**, 2045.
19. Weppner, W. and Huggins, R.A. (1977) in *Proceedings of the Symposium on Electrode Materials and Processes for Energy Conversion and Storage* (eds J.D.E. Mc Intyre, S. Srinivasan, and F.G. Will), The Electrochemical Society, Pennington, NJ, p. 833.
20. Weppner, W. and Huggins, R.A. (1977) *Z. Phys. Chem. N.F.*, **108**, 105.
21. Weppner, W. and Huggins, R.A. (1978) *J. Electrochem. Soc.*, **125**, 7.
22. Wen, C.J. and Huggins, R.A. (1981) *J. Solid State Chem.*, **37**, 271.
23. Wen, C.J. (1980) Ph.D. Dissertation, Stanford University.
24. Wen, C.J. and Huggins, R.A. (1980) *Mater. Res. Bull.*, **15**, 1225.
25. Saboungi, M.L., Marr, J.J., Anderson, K., and Vissers, D.R. (1979) *J. Electrochem. Soc.*, **126**, 322.
26. Wen, C.J. and Huggins, R.A. (1981) *J. Electrochem. Soc.*, **128**, 1636.
27. Wen, C.J. and Huggins, R.A. (1981) *J. Electrochem. Soc.*, **128**, 1181.
28. Wen, C.J., Boukamp, B.A., Huggins, R.A., and Weppner, W. (1979) *J. Electrochem. Soc.*, **126**, 2258.
29. Luedecke, C.M., Doench, J.P., and Huggins, R.A. (1983) in *Proceedings of the Symposium on High Temperature Materials Chemistry* (eds Z.A. Munir and D. Cubicciotti), The Electrochemical Society, Pennington, NJ, p. 105.
30. Doench, J.P. and Huggins, R.A. (1983) in *Proceedings of the Symposium on High Temperature Materials Chemistry* (eds Z.A. Munir and D. Cubicciotti), The Electrochemical Society, Pennington, NJ, p. 115.
31. Anani, A. and Huggins, R.A. (1988) in *Proceedings of the Symposium on Primary arid Secondary Ambient Temperature Lithium Batteries* (eds J.-P. Gabano, Z. Takehara, and P. Bro), The Electrochemical Society, Pennington, NJ, p. 635.
32. Anani, A. and Huggins, R.A. (1992) *J. Power Sources*, **38**, 351.
33. Huggins, R.A. (1979) in *Fast Ion Transport in Solids* (eds P. Vashishta, J.N. Mundy, and G.K. Shenoy), North-Holland, Amsterdam, p. 53.
34. Huggins, R.A. and Anani, A.A. (1990) US Patent 4950566.
35. Anani, A. and Huggins, R.A. (1992) *J. Power Sources*, **38**, 363.
36. Raistrick, I.D. and Huggins, R.A. (1983) *Mater. Res. Bull.*, **18**, 337.
37. Raistrick, I.D., Mark, A.J., and Huggins, R.A. (1981) *Solid State Ionics*, **5**, 351.
38. Weppner, W. and Huggins, R.A. (1977) *J. Electrochem. Soc.*, **124**, 1569.
39. Weppner, W. and Huggins, R.A. (1978) *Annu. Rev. Mater. Sci.*, **8**, 269.
40. Wen, C.J., Ho, C., Boukamp, B.A., Raistrick, I.D., Weppner, W., and Huggins, R.A. (1981) *Int. Metals Rev.*, **5**, 253.
41. Weppner, W. and Huggins, R.A. (1977) *J. Solid State Chem.*, **22**, 297.
42. Wen, C.J. and Huggins, R.A. (1980) *J. Solid State Chem.*, **35**, 376.
43. Wen, C.J., Weppner, W., Boukamp, B.A., and Huggins, R.A. (1980) *Met. Trans. B*, **11**, 131.
44. Doench, J.P. and Huggins, R.A. (1982) *J. Electrochem. Soc.*, **129**, 341.
45. Wang, J., Raistrick, I.D., and Huggins, R.A. (1986) *J. Electrochem. Soc.*, **133**, 457.
46. Wang, J., King, P., and Huggins, R.A. (1986) *Solid State Ionics*, **20**, 185.
47. Anani, A., Crouch-Baker, S., and Huggins, R.A. (1987) in *Proceedings of the Symposium on Lithium Batteries* (ed. A.N. Dey), The Electrochemical Society, Pennington, NJ, p. 365.
48. Boukamp, B.A., Lesh, G.C., and Huggins, R.A. (1981) *J. Electrochem. Soc.*, **128**, 725.
49. Boukamp, B.A., Lesh, G.C., and Huggins, R.A. (1981) in *Proceedings of the Symposium on Lithium Batteries* (ed. H.V. Venkatasetty), The Electrochemical Society, Pennington, NJ, p. 467.
50. Huggins, R.A. and Boukamp, B.A. (1984) US Patent 4436796.
51. Anani, A., Crouch-Baker, S., and Huggins, R.A. (1987) in *Proceedings of the Symposium on Lithium Batteries*

(ed. A.N. Dey), The Electrochemical Society, Pennington, NJ, p. 382.

52. Anani, A., Crouch-Baker, S., and Huggins, R.A. (1988) *J. Electrochem. Soc.*, **135**, 2103.

Further Reading

Park, C.M., Kim, J.H., Kim, H., and Sohn, H.J. (2010) *Chem. Soc. Rev.*, **39** (8), 3115–3141.

Pottgen, R. (2006) *Z. Naturforsch. B, J. Chem. Sci.*, **61** (6), 677–698.

Sun, H., He, X.M., Ren, J.G., Li, J.J., Jiang, C.Y., and Wan, C.R. (2007) *Rare Met. Mater. Eng.*, **36** (7), 1313–1316.

Tirado, J.L. (2003) *Mater. Sci. Eng. R-Reports*, **40** (3), 103–136.

Winter, M. and Besenhard, J.O. (1999) *Electrochim. Acta*, **45**, 31–50.

15
Lithiated Carbons

Martin Winter and Jürgen Otto Besenhard†

15.1
Introduction

The rapid proliferation of new technologies, such as portable consumer electronics and electric vehicles, has generated the need for batteries that provide both high energy density and multiple rechargeability. In order to accomplish such high energy density batteries, the use of electrode materials with high charge-storage capacity is inevitable. Considering thermodynamic reasons for the selection of an anode material, light metals M, such as Li, Na, K, or Mg, are favored as they combine outstanding negative standard redox potentials with low equivalent weights. However, a realization of batteries using these metals as active anode materials is in most cases not possible because the strong reducing power of the metals results in a spontaneous reaction in contact with an electrolyte.

Among the light metals M, only metallic lithium shows a chemical and electrochemical behavior which favors its use in high energy-density batteries [1, 2]. In suitable nonaqueous electrolytes 'passivating' films of Li^+-containing electrolyte decomposition products, spontaneously formed upon immersion in the electrolyte, protect the lithium surfaces. These films act as a 'sieve,' being selectively permeable to the electrochemically active charge carrier, the Li^+ cation, but impermeable to any other electrolyte component that would react with lithium, that is, they behave as an electronically insulating solid/electrolyte interphase (SEI) [3–5].

The composition, structure, and formation process of the SEI on metallic lithium depend on the nature of the electrolyte. The variety of possible electrolyte components makes this topic very complex; it is reviewed by Peled, Golodnitsky, and Penciner in Part III, Chapter 16 of this handbook. The types and properties of liquid nonaqueous electrolytes that are commonly used in lithium cells are reviewed by Barthel and Gores in Part III, Chapter 17.

The observation of the kinetic stability of lithium in a number of nonaqueous electrolytes was the foundation of the research on 'lithium batteries' in the 1950s, and the commercialization of primary (not rechargeable) lithium batteries followed quickly in the late 1960s and early 1970s [2, 6–12]. Today, primary metallic lithium systems have found a variety of applications, for example, military, consumer

Handbook of Battery Materials, Second Edition. Edited by Claus Daniel and Jürgen O. Besenhard.
© 2011 Wiley-VCH Verlag GmbH & Co. KGaA. Published 2011 by Wiley-VCH Verlag GmbH & Co. KGaA.

and medical, and commercial interest is still growing. However, apart from the rechargeable Li/MnO$_2$ cell commercialized by Tadiran (Israel) [13–15], the commercial breakthrough of rechargeable secondary batteries based on metallic lithium anodes has not been achieved so far. Upon recharge of the anode, lithium plating occurs simultaneously with lithium corrosion and 'passivation' (i.e., formation of SEI). Thus, lithium is deposited as highly dispersed, highly reactive metal particles. These dendrites are covered with SET films, and therefore are partially electrochemically inactive. This reduces the efficiency of the lithium deposition/dissolution process. Moreover, the dendrites grow to filaments upon cycling, which may short circuit, overheat the cell locally, and cause a disastrous thermal runaway due to the low melting point of Li (~180 °C) [10, 16–19]. In contrast, the lithium insertion materials used for the cathode exhibited sufficient cyclability and safety.

Beginning in the early 1980s [20, 21] metallic lithium was replaced by lithium insertion materials having a lower standard redox potential than the positive insertion electrode; this resulted in a 'Li-ion' or 'rocking-chair' cell with both negative and positive electrodes capable of reversible lithium insertion (see recommended papers and review papers [7, 10, 22–28]). Various insertion materials have been proposed for the anode of rechargeable lithium batteries, for example, transition metal oxides and chalcogenides, carbons, lithium alloys, lithium transition metal nitrides, and several polymers. In general, both the specific charges and the charge densities of lithium insertion materials are theoretically lower than those of metallic lithium, because the use of an electrochemically inactive lithium insertion host is associated with additional weight and volume (Figure 15.1). However, as the lithium is stored in the host in ionic and not in atomic form, the packing densities and thus the charge densities of several lithium insertion materials, for example, Sn (Figure 15.1) and others [29], are close to those of Li. Considering, moreover, that in practical cases the cycling efficiency of metallic lithium is ≤99%, one has to employ a large excess of lithium [10, 19, 30, 31] to reach a reasonable cycle

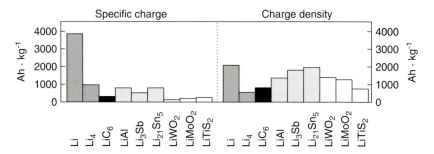

Figure 15.1 Specific charges and charge densities of several lithiated anode materials for lithium batteries, calculated by using data from Refs [10, 32–35], Li$_4$ denotes a fourfold excess of lithium, which is necessary to attain a sufficient cycle life.

life. Therefore, the practical specific charge and the charge density of a secondary lithium metal electrode are much lower than the theoretical values, almost in the same order as those of graphite. (More information on the properties of the metallic lithium anode is given in Part III, Chapter 14.)

From a thermodynamic point of view, apart from charge density and specific charge, the redox potential of lithium insertion into/removal from the electrode materials has to be considered also. For instance, the redox potential of many Li alloys is between ~0.3 and ~1.0 V vs Li/Li$^+$, whereas it is only ~0.1 V vs Li/Li$^+$ for lithiated graphite (Figure 15.2). From the point of view of energy density, the use of anodes with highly negative potential, yielding high cell voltages, would be advantageous. However, the materials with the potential closest to that

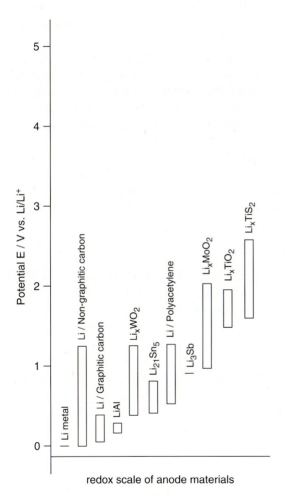

Figure 15.2 Redox potentials for lithium insertion into/removal from several anode materials for lithium cells.

of metallic lithium may be the most reactive (although due to kinetic effects they are not necessarily so), and thus they can cause safety problems as well as handling difficulties. This will be discussed further in the next section of this chapter.

15.1.1
Why Lithiated Carbons?

Among the mentioned lithium insertion materials above, lithiated carbons (Li_xC_n) are considered to be the most promising at present. Carbonaceous materials exhibit higher lithium storage capacities and more negative redox potentials versus the cathode than polymers, metal oxides, or chalcogenides. Furthermore, they show long-term cycling performance superior to Li alloys due to their better dimensional stability. In addition, most carbons suitable as anodes for lithium-ion cells are cheap and abundant compared with the other materials.

Though considerable safety improvements were the major driving force for the introduction of lithiated carbons into rechargeable lithium cells, it has to be kept in mind that the lithium activity of lithium-rich carbons is similar to that of metallic lithium. Thus the redox potential vs Li/Li^+ is quite close to 0 V (Figure 15.2) and the reactivity is high. Additionally, the particle size of Li_xC_n in practical electrodes is only in the order of 10 µm, that is, the reactive surface area is large. Moreover, *ex situ* investigations after cycling have shown that cycling of graphite electrodes increases the specific surface area of Li_xC_n by a factor of 5 [36]. Recent differential scanning calorimetry studies on polymer-bonded lithiated carbons reveal that the SEI films degrade at temperatures of approx. 120–140 °C, then undergo a reaction with the electrolyte *and* the binder material at temperatures above 200 °C. The degradation reactions are proportional to the surface area of the carbon [37], and furthermore can be expected to depend on the SEI films formed, that is, the electrolytes used. The tendency of the SEI film to peel off the carbon anode is assumed to be suppressed (the adherence between carbon and SEI is supposed to be improved) by proper surface pre-treatment of the carbon [38].

However, the reaction rate of Li_xC_n depends on the lithium concentration at the surface of the carbon particles, which is limited by the rather slow transport kinetics of lithium from the bulk to the surface [17–19].[1] As the melting point of metallic lithium is low (~180 °C) there is some risk of melting of lithium under abuse conditions such as short-circuiting, followed by a sudden breakdown of the SEI and a violent reaction of liquid lithium with the other cell components. In contrast, there is no melting of lithiated carbons.

---

1) Lithiated carbons are mostly multiphase systems. Hence, the determination of chemical diffusion coefficients for Li^+ causes experimental problems because the propagation of a reaction front has to be considered.

15.1.2
Electrochemical Formation of Lithiated Carbons

The electroinsertion reaction of mobile lithium ions into a solid carbon host proceeds according to the general reaction scheme

$$\text{Li}_x\text{C}_n \underset{\text{charge}}{\overset{\text{discharge}}{\rightleftarrows}} x\text{Li}^+ + xe^- + \text{C}_n \qquad (15.1)$$

During electrochemical reduction (charge) of the carbon host, lithium cations from the electrolyte penetrate into the carbon and form a lithiated carbon Li_xC_n. The corresponding negative charges are accepted by the carbon host lattice. As in any other electrochemical insertion process, the prerequisite for the formation of lithiated carbons is a host material that exhibits mixed (electronic and ionic) conductance.

The reversibility of this so-called 'intercalation' reaction can be demonstrated by a subsequent electrochemical oxidation (discharge) of Li_xC_n, that is, the de-intercalation of Li^+. This is considered to be a special type of intercalation in that, unusually, a layer of guest ions slides between the sheets of a layered host matrix, while the host broadly retains its structural integrity. This occurs in the case of the insertion of lithium ions into graphite. In most cases, however, a strict differentiation between insertion and intercalation is a formal question and both terms are used inter-changeably. Following historical conventions, the terms '*intercalation*' and '*lithium/carbon intercalation compounds*' will be used in this review, even though only a small fraction of layered structure units may be present in a specific carbon material (see also Refs [2, 6]).

15.2
Graphitic and Nongraphitic Carbons

The electrochemical performance of lithiated carbons depends basically on the electrolyte, the parent carbonaceous material, and the interaction between the two (see also Part III, Chapter 17). As far as the lithium intercalation process is concerned, interactions with the electrolyte, which limit the suitability of an electrolyte system, will be discussed in Sections 15.2.2.3, 15.2.3 and 15.2.4 of this chapter. First, several properties of the carbonaceous materials will be described.

The quality and quantity of sites which are capable of *reversible* lithium accommodation depend in a complex manner on the crystallinity, the texture, the (micro)structure, and the (micro)morphology of the carbonaceous host material [7, 19, 22, 39–56]. The type of carbon determines the current/potential characteristics of the electrochemical intercalation reaction and also potential side-reactions. Carbonaceous materials suitable for lithium intercalation are commercially available in many types and qualities [19, 42, 57–60]. Many exotic carbons have been specially synthesized on a laboratory scale by pyrolysis of various precursors, for example, carbons with a remarkably high lithium storage capacity (see Sections 15.2.4 and 15.2.5), and tailored carbons, which were prepared by the use of inorganic templates [61, 62]. It has to be emphasized that the assumed suitability of a carbonaceous material for a lithium intercalation host depends strongly on the method of its evaluation, and quite a few carbons may have been rejected as anode materials due to an inadequate evaluation method. As a consequence, sometimes the classification of a carbon as 'good' or 'poor' anode material can be only preliminary. For instance, though in principle electrochemical intercalation in graphite was already observed in the mid-1970s [63, 64], it took about 15 years for an appropriate electrolyte allowing the highly reversible operation of a graphite anode to be found [65].

15.2.1
Carbons: Classification, Synthesis, and Structures

Because of the variety of available carbons, classification is inevitable. Most carbonaceous materials which are capable of reversible lithium intercalation can be classified roughly as graphitic and nongraphitic (disordered).

Graphitic carbons basically comprise sp^2-hybridized carbon atoms which are arranged in a planar hexagonal ('honeycomb-like') network such that a so-called 'graphene layer' is formed. Van der Waals forces provide a weak cohesion of the graphene layers, leading to the well-known *layered* graphite structure. From a strictly crystallographic point of view the term '*graphite*' is, however, only applicable for carbons having a layered lattice structure with a perfect stacking order of graphene layers, either the prevalent AB (hexagonal graphite, Figure 15.3) or the

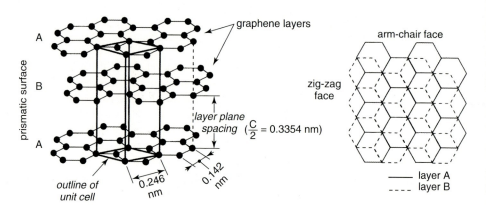

Figure 15.3 (a) Schematic drawing of the crystal structure of hexagonal graphite showing the AB graphene layer stacking sequence and the unit cell. (b) View perpendicular to the basal plane of hexagonal graphite. Prismatic surfaces can be subdivided into armchair and zig-zag faces. Modified and redrawn from Ref. [2].

less common ABC (rhombohedral graphite). Due to the small amount of required energy for transformation of AB into ABC stacking (and vice versa), perfectly stacked graphite crystals are not readily available. For instance, typically about 5% of the graphene layers in natural graphite are arranged rhombohedrally. Therefore, the term '*graphite*' is often used regardless of stacking order.

The actual structure of practical carbonaceous materials deviates more or less from the ideal graphite structure. Moreover, carbonaceous materials consisting of aggregates of graphite crystallites are called *graphites* as well. For instance, the terms '*natural graphite*', '*artificial*' or '*synthetic graphite*', and '*pyrolytic graphite*' are commonly used, although the materials are polycrystalline [66]. The crystallites may vary considerably in size, ranging from the order of nanometers to micrometers. In some carbons, the aggregates are large and relatively free of defects, for example, in highly oriented pyrolytic graphite (HOPG). Furthermore, texture effects can be observed as the crystallites may be differently oriented to each other. In addition to essentially graphitic crystallites, carbons may also include crystallites containing carbon layers (or packages of stacked carbon layers) with significant, randomly distributed misfits and misorientation angles of the stacked segments to each other (turbostratic orientation or turbostratic disorder [67]). The latter disorder can be identified from a nonuniform, and on average increased, interlayer spacing compared with graphite [66, 68].

When the disorder in the structure becomes more dominant among the crystallites, the carbonaceous material can no longer be considered graphitic but must be regarded as a nongraphitic carbon. For carbon samples that contain both characteristic graphitic *and* nongraphitic structure units, the classification into graphitic and nongraphitic types can be somewhat arbitrary and in many cases is only made for the sake of convenience.

In the case of nongraphitic (disordered) carbons, most of the carbon atoms are arranged in a planar hexagonal network, too. Though layered structure segments are probable, there is actually no far-reaching crystallographic order in the *c*-direction. The structure of these carbons is characterized by amorphous areas embedding and partially crosslinking more graphitic (layered) structure segments [69–71] (Figure 15.4). The number and the size of the areas vary, and depend both on the precursor material and on the manufacturing process, for example, on the manufacturing temperature and pressure. Using a simple model [19, 42] the complex X-ray diffraction (XRD) patterns of nongraphitic carbons can be correlated with the probability of finding unorganized (randomly oriented and amorphous) and organized (layered) areas. As a result the lithium storage capacity of a specific nongraphitic carbon material can be predicted approximately.

Most nongraphitic carbons are prepared by pyrolysis of organic polymer or hydrocarbon precursors at temperatures below $\sim$1500 °C. Further heat treatment of most nongraphitic carbons at temperatures from $\sim$1500 to $\sim$3000 °C makes it possible to distinguish between two different types of carbons.

Graphitizing carbons develop the graphite structure continuously during the heating process. The carbon layers are mobile enough to form graphite-like crystallites as crosslinking between the layers is weak. Nongraphitizing carbons

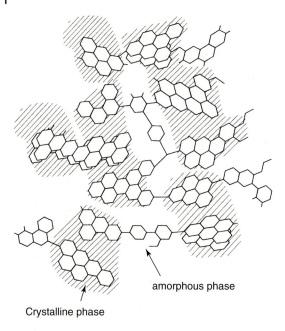

Figure 15.4 Schematic drawing of a nongraphitic (disordered) carbon [2].

exhibit no true development of the graphite structure, not even at high temperatures (2500–3000 °C), since the carbon layers are immobilized by strong crosslinking. Since nongraphitizing carbons are mechanically harder than graphitizing ones, it is common to divide the nongraphitic carbons into 'soft' and 'hard' carbons [69]. The precursors and – at least to some extent – the preparation and assumed structure of the hard carbons resemble those of glassy carbon [72, 73]. Franklin [69] reported that, compared with graphitizing carbons, nongraphitizing carbons exhibit a considerably more extensive fine-structure porosity (nanoporosity). Models for only partially graphitizing carbons are also discussed [69, 74].

The mobility of the carbon structure units, which determines the degree of microstructural ordering as well as the texture of the carbonaceous material, depends on the state of aggregation of the intermediate phase during pyrolysis, which can be solid, liquid, or gaseous [71]. Nongraphitizing carbons are usually products of solid-phase pyrolysis, whereas graphitizing carbons are commonly produced by liquid- or gas-phase pyrolysis. Examples of products of solid-phase pyrolysis are chars and glassy (vitreous) carbon, which are produced from crosslinked polymers. Because of small crystal size and a high structural disorder of the polymers, the ability of these carbons to graphitize is low. Pyrolysis of thermally stabilized polyacrylonitrile or pitch, which are the precursors for carbon fibers, also yields solid intermediate phases, but stretching of the fibrous material during the manufacturing process produces an ordered microstructure [71]. The synthesis of petroleum coke, which is the most important raw material for the manufacture of carbons and graphites, is an example of liquid-phase pyrolysis. Petroleum coke is produced by the pyrolysis of petroleum pitch, which is the residue from the distillation of petroleum fractions. Cokes are also products from pyrolysis of coal tar pitch and aromatic hydrocarbons at 300–500 °C. Carbon black, pyrocarbon, and carbon films are examples of gas-phase pyrolysis products, that is, products of thermal cracking of gaseous hydrocarbon compounds which are deposited as carbon on a substrate [66, 71].

The ability to graphitize also depends on the pre-ordering and pre-texture of the respective precursor. For example, the graphitization ability is higher (i) if the precursor material comprises highly condensed aromatic hydrocarbons which can be considered to have a graphene-like structure and (ii) if neighboring graphene layers or graphitic crystallites are suitably orientated to each other.

Apart from manifold structures, carbons can have various shapes, forms, and textures, including powders with different particle size distributions, foams, whiskers, foils, felts, papers, fibers [75, 76], spherical particles [75] such as mesocarbon microbeads (MCMBs) (Nakagawa, Y. Osaka Gas Chemicals Co., Ltd., personal communication), and so on. Comprehensive overviews are given, for example, in Refs [66, 70, 71]. Further information on the synthesis and structures of carbonaceous materials can be found in Refs [66, 69, 71, 74, 77]. Details of the surface composition and surface chemistry of carbons are reviewed in Part II, Chapter 11, and in Part III, Chapter 17, of this handbook. Some aspects of surface chemistry of lithiated carbons will also be discussed in Section 15.2.2.3.

15.2.2
Lithiated Graphitic Carbons (Li_xC_n)

15.2.2.1 In-Plane Structures

The first lithiated graphitic carbons were lithium–graphite intercalation compounds, abbreviated as Li–GICs (Li_xC_n), and were obtained by chemical synthesis in the mid-1950s [78, 79]. At ambient pressure, a maximum lithium content of one Li guest atom per six carbon host atoms can be reached for highly crystalline graphite ($n \geq 6$ in LiC_n or $x \leq 1$ in Li_xC_6). The intercalation reaction proceeds via the prismatic surfaces (armchair and zig-zag faces). Through the basal plane, intercalation is possible at defect sites only. During intercalation the stacking order of the graphene layers shifts to AA. Thus, two neighboring graphene layers in LiC_6 directly face each other (Figure 15.5a). The energetically favored AA stacking sequence of LiC_6 has been proved by *ab initio* studies [80]. Due to the hosted lithium, the interlayer distance between the graphene layers increases moderately (10.3% has been calculated for LiC_6 [35, 81]). The stacking order of the lithium interlayers is $\alpha\alpha$ (a $Li-C_6-Li-C_6-Li$ chain exists along the *c*-axis) [82, 83]. In LiC_6 the lithium is distributed in-plane in such a manner that the occupation of the nearest-neighbor sites is avoided (Figure 15.5b).

A higher lithium in-plane density by occupation of nearest-neighbor sites is obtained in the phases LiC_2-LiC_4, that is, $x = 2-3$ in Li_xC_6, which are prepared chemically from graphitic carbon under high pressure (~60 kbar) and high temperature (~300 °C) conditions [43, 84–87]. The close Li-Li distance in LiC_2

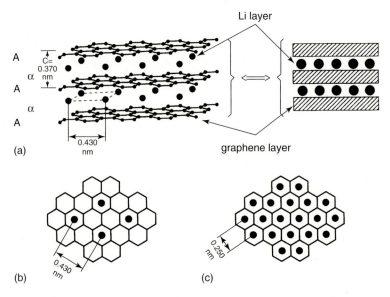

Figure 15.5 Structure of LiC$_6$ (a) Left: schematic drawing showing the AA layer stacking sequence and the $\alpha\alpha$ interlayer ordering of the intercalated lithium. Right: Simplified representation [2]. (b) In-plane distribution of Li in LiC$_6$. (c) In-plane distribution of Li in LiC$_2$.

(Figure 15.5c) results in a higher chemical activity of lithium than that of lithium metal (Li–Li bond length (20 °C) = 0.304 nm [34]). Under ambient conditions LiC$_2$ decomposes slowly via various metastable intermediate Li/C phases to LiC$_6$ and metallic lithium [43, 86]. A preliminary study of the electrochemical behavior of LiC$_2$ can be found in Ref. [43]. For more comprehensive details on the chemical synthesis of Li$_x$C$_6$ see the literature [41, 43, 79, 88–90]. Selected general reviews on GICs are cited in Ref. [6].

15.2.2.2 Stage Formation

A general feature of intercalation into graphite is the formation of a periodic array of unoccupied layer gaps at low concentrations of guest species, called *stage formation* [78, 79, 90–96]. This stepwise process can be described by the stage index *s* which is equal to the number of graphene layers between two nearest-guest layers. Staging is a thermodynamic phenomenon related to the energy required to 'open' the van der Waals gap between two graphene layers for the guests entering the hosts. The repulsive coulombic interactions between the guest ions are less effective. As a consequence, only a few (but highly occupied) van der Waals gaps are energetically favored over a random distribution of guests.

Staging phenomena as well as the degree of intercalation can be easily observed during the electrochemical reduction of carbons in Li$^+$-containing electrolytes. Figure 15.6a shows a schematic potential/composition curve for the galvanostatic (constant current) reduction (= charge) of graphite to LiC$_6$ corresponding to

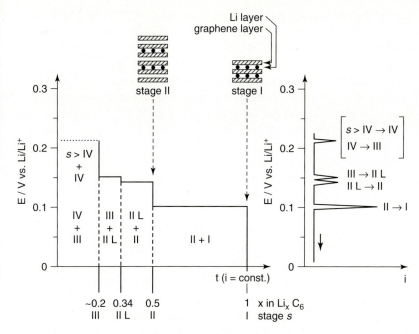

Figure 15.6 Simplified scheme showing the stage formation during electrochemical formation of lithiated graphite. (a) Schematic galvanostatic curve. (b) Schematic voltammetric curve. Prepared with data from Refs [90, 98, 102, 103, 108]. For a more detailed discussion, see text.

a lithium storage capacity of 372 Ah kg^{-1} with respect to the graphite mass (compare Figure 15.1, where the lithium storage capacities are given with respect to the lithiated hosts, i.e., 339 Ah kg^{-1} for LiC$_6$). The plateaus arising during reduction in the curve indicate two-phase regions (coexistence of two phases) [97, 98]. Under potentiodynamic control (linear potential sweep voltammetry) the two-phase regions are indicated by current peaks (Figure 15.6b). Apart from the stage $s = $ I, other binary phases corresponding to the stages $s = $ IV, III, II L, and II (which can also be obtained by chemical synthesis [78, 79, 92, 93, 97, 99–101]) were identified by electrochemical experiments and confirmed by XRD [35, 97, 98, 100–104] and Raman spectroscopy [105, 106]. Stages higher than $s = $ IV were reported, too [98, 106–108].[2] However, there are some discrepancies in the reported literature concerning the staging process, in particular regarding stages $s > $ II. This is discussed in Ref. [109].

2) During the first charge the SEI layers on the surface of the graphite electrode are not fully developed. Moreover, there is a strong tendency to co-intercalate solvent molecules at low lithium concentrations in graphite. Thus, particularly in the first charge the Li-GICs, which correspond to high stage numbers, might be insufficiently protected from a reaction with the electrolyte, and the intercalated lithium is irreversibly consumed for the formation of films between the graphene layers (see Section 15.2.3).

The splitting of the second stage into two, $s = $ II ($x = 0.5$ in Li_xC_6) and $s = $ II L ($x = 0.33$ in Li_xC_6), is due to different lithium packing densities. It disappears at temperatures below $\sim 10\,°C$ [98]. At temperatures above $700\,°C$ Li_xC_6 ($0.5 \le x \le 1$) is transformed into lithium carbide Li_2C_2 and carbon [94, 110].

Commercial graphites can contain a considerable proportion of rhombohedral structure units. It has been reported that lithium intercalation mechanisms and storage capabilities are similar for both rhombohedral and hexagonal graphite structures [58, 60]. However, the preparation of graphite with a higher proportion of rhombohedrally structured graphene planes and its use as anode material is claimed in a patent [111]. Lithiated graphites can also be prepared from a KC_8 precursor, either (i) chemically by ion exchange reactions [112] or (ii) electrochemically after de-intercalation of potassium [113, 114].

15.2.2.3 Reversible and Irreversible Specific Charge

Experimental constant current charge–discharge curves for Li^+ intercalation/de-intercalation into/out of graphite clearly prove the staging phenomenon (Figure 15.7). Nevertheless, there are no sharp discontinuities between the two-phase regions because (i) the packing density of Li_xC_6 varies slightly (a phase width exists), and (ii) various types of overpotentials cause plateau-sloping in galvanostatic measurements (and peak-broadening in voltammetric measurements). Theoretically, Li^+ intercalation into carbons is fully reversible. In the practical charge–discharge curve, however, the charge consumed in the first cycle significantly exceeds the theoretical specific charge of 372 Ah kg^{-1} for LiC_6 (Figure 15.7). The subsequent de-intercalation of Li^+ recovers only ~ 80–95% of this charge. In the second and subsequent cycles, then, the charge consumption for the Li^+ intercalation half-cycle is lower and the charge recovery is close to 100%.

The excess charge consumed in the first cycle is generally ascribed to SEI formation and corrosion-like reactions of Li_xC_6 [19, 65, 115–117]. Like metallic lithium and Li-rich Li alloys, lithiated graphites, and more generally lithiated carbons, are thermodynamically unstable in all known electrolytes, and therefore the surfaces which are exposed to the electrolyte have to be kinetically protected by SEI films (see Part III, Chapter 17). Nevertheless, there are significant differences in the film formation processes between metallic lithium and lithiated carbons. Simplified, these differences are as follows: Film formation on metallic Li takes place upon contact with the electrolyte. Various electrolyte components decompose spontaneously with low selectivity and some of the decomposition products form the film. When the film grows, the activity of the metallic lithium electrode versus the electrolyte decreases because of an increasing I R drop in the film. At this stage the electrolyte reduction processes become more and more selective as the number of electrolyte components which are still sensitive to reduction versus the (now partially electronically 'passivated') lithium electrode is limited. In contrast, film formation on carbonaceous hosts takes place as a charge-consuming side reaction in the first few Li^+ intercalation/deintercalation cycles, especially during the first reduction of the carbon host material. In this case, the electrolyte components which

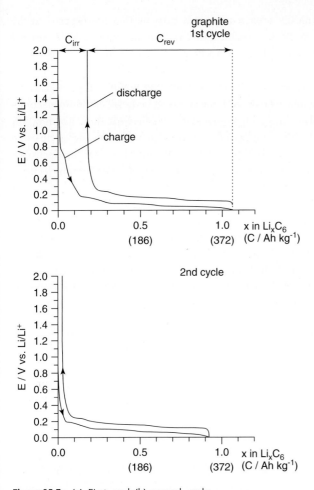

Figure 15.7 (a) First- and (b) second- cycle constant-current charge/discharge curves of graphite Timrex® KS44 in LiN(SO$_2$CF$_3$)$_2$/ethylene carbonate/dimethyl carbonate as the electrolyte (C_{irr} = irreversible specific charge; C_{rev} = reversible specific charge) [2].

are least stable toward reduction are the first to react selectively. As a result, in the case of metallic lithium the decomposition products of electrolyte components which are highly sensitive to reduction will be accumulated in the *external* film regions near the electrolyte solution, whereas in the case of lithiated carbons the *internal* film regions near the carbon surface will contain more decomposition products stemming from these electrolyte components. Moreover, as a consequence of the differences in the reduction behavior, dissimilar intermediate reduction products may be formed, which can initiate different decomposition mechanisms of the electrolyte. From the viewpoint of the above considerations, it should be emphasized that chemically prepared lithiated carbons that are exposed to the electrolyte behave similarly to metallic lithium (and not similarly to lithiated carbons which have been prepared electrochemically, e.g., Figure 15.7) regarding film formation.

Finally, regarding long-term cycling of metallic lithium and of lithiated carbon, respectively, and its influence on the SEI, the surface of a metallic lithium electrode is periodically renewed during cycling, causing irreversible formation of a 'new' SEI in each cycle. Unless the lithium electrode becomes electrochemically inactive due to passivation, this process can be repeated until the lithium and/or the electrolyte are completely consumed. In contrast, the surface of the lithiated carbon electrode is passivated by the SEI throughout the cycling process.

Since film formation on Li_xC_6 is associated with the irreversible consumption of material (lithium and electrolyte), the corresponding charge loss is frequently called 'irreversible specific charge' or 'irreversible capacity.' Reversible lithium intercalation, on the other hand, is called 'reversible specific charge' or 'reversible capacity.' The losses have to be minimized because the losses of charge and of lithium are detrimental to the specific energy of the whole cell and, moreover, increase the material expenses because of the necessary excess of costly cathode material, which is the lithium source in a lithium-ion cell after cell assembly.

The extent of the irreversible charge losses due to film formation depends to a first approximation on the surface area of the lithiated carbon which is wetted by the electrolyte [36, 65, 117–121]. Electrode manufacturing parameters influencing the pore size distribution within the electrode [36, 118, 121, 122] and the coverage of the individual particles by a binder [121, 123] have an additional influence on the carbon electrode surface exposed to the electrolyte. These and other technical aspects which are important in this respect are reviewed in recent papers [2, 6].

Besides the irreversible charge loss caused by electrolyte decomposition, several authors claim that the following reactions are also responsible for (additional) irreversible charge losses:

1) irreversible reduction of impurities such as H_2O or O_2 on the carbon surface,
2) reduction of 'surface complexes' such as 'surface oxides' at the prismatic surfaces of carbon, and
3) irreversible lithium incorporation into the carbon matrix ('formation of residue compounds' [124–126], e.g., by irreversible reduction of 'internal surface groups' at prismatic surfaces of domain boundaries in polycrystalline carbons).

In order to improve the electrochemical performance with respect to lower irreversible capacity losses, several attempts have been made to modify the carbon surface. Here the work of Peled's [38, 127–129] and Takamura's groups [130–135] deserves mention. A more detailed discussion can be found in Part III, Chapter 17.

15.2.3
Li_xC_6 vs $Li_x(solv)_yC_n$

Apart from reactions with the electrolyte at the carbon surface, the irreversible specific charge is furthermore strongly affected by the possible co-intercalation of polar solvent molecules between the graphene layers of highly graphitic matrices [136]. This so-called 'solvated intercalation reaction' depends (i) on the crystallinity and the morphology of the parent carbonaceous material, which will be discussed in Section 15.2.4 and (ii) on the composition of the electrolyte, which is discussed in this section.

Whereas the electrochemical decomposition of propylene carbonate (PC) on graphite electrodes at potentials between ~ 1 and ~ 0.8 V vs Li/Li^+ was already reported in 1970 [137], it took about four years to find out that this reaction is accompanied by a partially reversible electrochemical intercalation of solvated lithium ions, $Li^+(solv)_y$, into the graphite host [63]. In general, the intercalation of Li^+ (and other alkali-metal) ions from electrolytes with organic donor solvents into fairly crystalline graphitic carbons quite often yields solvated (ternary) lithiated graphites, $Li_x(solv)_yC_n$ (Figure 15.8) [7, 24, 26, 64, 65, 138–143].

The co-intercalation of the large solvent molecules is associated with extreme expansion of the graphite matrix (typically in the region above $\sim 100\%$), frequently leading to exfoliation of graphene layers and mechanical disintegration of the electrode. In the 'best' case reduction of charge storage capabilities, and in the worst case complete electrode destruction are the typical results of this reaction.

As long as the content of lithium in the graphitic carbon is low ($x \leq 0.33$ in Li_xC_6), the ternary lithiated graphites are thermodynamically favored over the corresponding binary lithiated graphites Li_xC_6. Hence, the potentials of their electrochemical

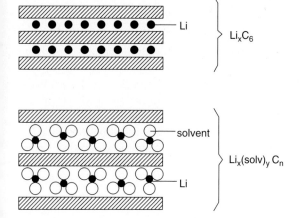

Figure 15.8 Schematic drawing of binary (Li_xC_n) and ternary [$Li_x(solv)_yC_n$] lithiated graphites. Modified and redrawn from Ref. [26].

formation are more positive than those for the formation of the corresponding compound Li_xC_6. At this stage of Li^+ intercalation the coulombic interaction between the lithium guest layer (Li^+) and the balancing negative charge distributed over the graphene layers (C_n^-) is weak, and space to accommodate large solvent molecules is still available [24, 26, 144, 145]. As a further result of the low coulombic interactions, the mobility of the intercalants is high, and therefore the guest distribution among the interlayer spaces is incommensurable or random. By contrast, the lithium guests in the binary phase LiC_6 form a commensurable structure, that is, they are organized according to the honeycomb 'raster' of the graphene layers [90]. The solvated GICs are thermodynamically unstable with respect to the reduction of the co-intercalated solvent molecules [136]. The kinetically controlled reduction depends on the type of co-intercalated solvent. It is slow for, for example, dimethyl sulfoxide, where even staging of solvated GICs can be observed [138, 146], but very much faster, for example, for PC [63, 137, 147–151], where the electrochemical intercalation followed by fast decomposition of the intercalated $Li^+(solv)_y$ can be misunderstood as simple electrolyte decomposition. Anyway, the reduction of $Li^+(solv)_y$ inside the graphite is associated with an increase of irreversible losses of charge and of material.

Effects resulting from solvent co-intercalation, mechanical destruction, and higher irreversible specific charge losses seriously complicate the operation of graphitic anode materials. Since ternary lithiated carbons are thermodynamically favored at low lithium concentrations in the graphite, that is, at the beginning of intercalation, kinetic measures have to be applied to prevent, or at least suppress, solvent co-intercalation. This can be accomplished by using electrolyte components that form effectively protecting SEI films on the external graphite surfaces in the very early stages of the first reduction, at potentials which are positive relative to the potentials of a significant formation of $Li_x(solv)_yC_n$. The first [65], and for a long time the only, effective solvent in this respect seemed to be ethylene carbonate (EC) [136, 152, 153]. Since the viscosity of electrolytes based on pure EC is rather high, mixtures of EC with low viscosity solvents such as dimethyl carbonate (DMC) and diethyl carbonate (DEC) are widely used [154–159]. Several papers which report on investigations and applications of these electrolyte blends are compiled in recent reviews [2, 6].

Although the formation of binary Li–GICs prevails in EC-based electrolytes, many investigations indicate a mechanism of film formation in which solvated lithiated graphites also participate. Film formation in the first cycle during the first reduction of the host material due to electrolyte decomposition is not a simple surface reaction but a rather complex three-dimensional process, taking place basically at a potential of $\sim$1 to $\sim$0.8 V vs Li/Li^+ (Figure 15.7a). *In-situ* dilatometric [152], scanning tunneling microscopy (STM) [160, 161] and Raman [106] methods indicate a (fairly large) expansion of the graphite host corresponding to the (intermediate) formation of $Li_x(solv)_yC_n$ at those potentials. The reduction of $Li_x(solv)_y$ on parts of the *internal* surfaces *between* the graphene layers results in an 'extra' film, which penetrates into the bulk of the graphite host (Figure 15.9). Correspondingly 'extra' irreversible charge losses are observed [152]. Several other

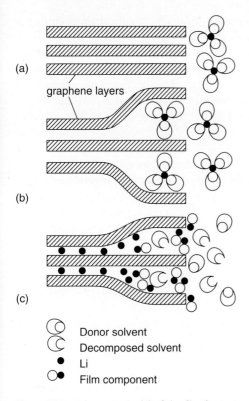

Figure 15.9 Schematic model of the film-formation mechanism on/in graphite: (a) the situation before reaction, (b) formation of ternary lithiated graphite $Li_x(solv)_yC_n$, and (c) film formation due to decomposition of $Li_x(solv)_y$. Prepared with data from Ref. [152].

investigations indicate that the film-formation process on graphite can be even more complex [107, 160–163].

In analogy to unsolvated intercalation reactions, solvent co-intercalation takes place via the prismatic surfaces only; not only the total external surface area but also the ratio between the prismatic and basal surfaces affect the irreversible charge loss [117]. This is in good agreement with the observations of Imanishi *et al.* [164, 165], who found that the tendency for PC co-intercalation into graphitized carbon fibers depends on the fiber texture. Carbon fibers in which the graphite packages are concentrically arranged expose a smaller amount of prismatic surfaces to the electrolyte, that is, they are less sensitive to solvent co-intercalation than fibers with a radial texture.

To take into account the effect of the thickness of the respective graphite flake on the formation of $Li_x(solv)_yC$, several simple models have been suggested recently [117]. The intercalation of all kinds of species into graphite generally requires energy to expand the gaps between the graphene layers held together by van der

Waals forces. Firstly, the 'expansion energy' (related to the threshold intercalation potential) depends on the mechanical flexibility of those graphene layers that are deformed by the intercalation process. The average 'expansion energy' increases with the thickness of the graphite flake, or more precisely with the number of adjacent graphene layers on both sides of a particular gap. Therefore, intercalation typically starts close to the basal planes of the flake, in the gaps adjacent to the end basal plane. Then the intercalation progresses toward internal layer gaps.

Secondly, the 'expansion energy' increases with the size of the guest species. Intercalation of large solvated lithium ions into the outer van der Waals gaps produces a considerable deformation (bending) of the outer graphene layers. Further intercalation into the internal gaps increases the bending angles of the outer graphene layers. However, when the outer graphene layers cannot be bent any more, the intercalation of solvated lithium into the internal van der Waals gaps is hindered. In conclusion, the graphite particle thickness effect should be particularly regarded for the intercalation of solvated lithium ions (Figure 15.10 and [117, 166]). This means, furthermore, that it should be possible to diminish some expansion due to solvent co-intercalation by sufficient external pressure on the electrode, for example, by close packing of the electrodes in the cell.

Thirdly, strong solvent co-intercalation, in particular into *internal* van der Waals gaps, can only be expected for kinetically stable ternary compounds $Li_x(solv)_yC_n$. For example, comparison of DMC and DEC with dimethoxyethane (DME) shows that the kinetic stability of $Li_x(DME)_yC_n$ can be considered to be much higher than that of $Li_x(DMC)_yC_n$ and $Li_x(DEC)_yC_n$ (and of course $Li_x(EC)_yC_n$) [166]. With EC/DME, solvent co-intercalation proceeds on a macroscopic scale, that is, the external van der Waals gaps and some internal ones can participate in the solvent co-intercalation reaction. When DMC or DEC is used as co-solvent, solvent co-intercalation can be expected to take place at the more external gaps only. Instrumentation such as STM with which it is possible to investigate the edge of a basal plane surface can still detect a local expansion [160, 161], whereas instrumentation providing information on a macroscopic scale, such as dilatometry [152] or XRD, cannot.

Numerous research activities have focused on the improvement of the protective films and the suppression of solvent co-intercalation. Beside EC, significant improvements have been achieved with other film-forming electrolyte components such as CO_2 [153, 166–174], N_2O [167, 174], SO_2 [152, 166, 174–176], S_x^{2-} [167, 174, 177, 178], ethyl propyl carbonate [179], ethyl methyl carbonate [180, 181] and other asymmetric alkyl methyl carbonates [182], vinyl PC [183], ethylene sulfite [184], S,S-dialkyl dithiocarbonates [185], vinylene carbonate [186], and chloroethylene carbonate [187–191] (which evolves CO_2 during reduction [192]). In many cases the suppression of solvent co-intercalation is due to the fact that the electrolyte components form effective SEI films already at potentials which are positive relative to the potentials of solvent co-intercalation. An excess of DMC or DEC in the electrolyte inhibits PC co-intercalation into graphite, too [180].

Furthermore, the molecular size of the Li^+-solvating solvents may affect the tendency for solvent co-intercalation. Crown ethers [19, 149–151, 193, 194] and

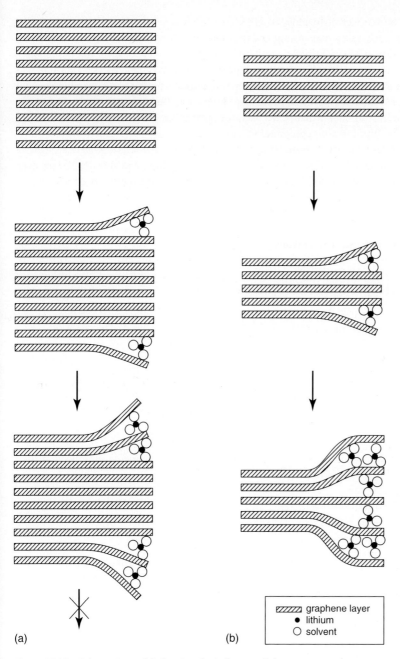

Figure 15.10 Schematic model showing the influence of the thickness of a graphite flake on the extent of co-intercalation of solvent molecules in the internal van der Waals gaps of graphite. (a) Thick graphite flakes and (b) thin graphite flakes. Prepared with data from Ref. [166].

other bulky electrolyte additives [193] are assumed to coordinate Li⁺ ions in solution in such a way that solvent co-intercalation is suppressed. The electrochemical formation of binary lithiated graphites Li_xC_6 was also reported for the reduction of graphite in electrolytes containing high-molecular-mass polymers as solvent. The claimed lithium intercalation, however, proceeds in a potential region where usually solvated lithiated graphites appear [41, 43, 195, 196] (for comparison, see Ref. [197]).

Graphitic anodes which have been 'pre-filmed' in an electrolyte 'A' containing effective film-forming components before they are used in a different electrolyte 'B' with less effective film-forming properties show lower irreversible charge losses and/or a decreased tendency to solvent co-intercalation [152, 198, 199]. However, sufficient insolubility of the pre-formed films in electrolyte 'B' is required to ascertain long-term operation of the anode.

15.2.4
Lithiated Nongraphitic Carbons

The use of nongraphitic (disordered) carbons as anode materials in lithium-ion cells is highly attractive for two reasons:

1) The crosslinking between the graphene layers (or packages of graphene layers) by sp^3-hybridized carbon atoms (Figure 15.4) mechanically suppresses the formation of solvated lithiated graphites, $Li_x(solv)_yC_n$, [19, 26, 65, 152]. As a result the gap between the layers cannot expand very much, and thus there is not enough space for the solvent to co-intercalate. Moreover, these carbons have the advantage that they can operate in EC-free electrolytes. Consequently, the first practically applicable lithiated carbon anodes [200, 201] were based on these nongraphitic carbons and not on graphitic materials. Furthermore, the use of composite carbonaceous materials comprising a 'core' of graphite and a protective 'shell' of nongraphitic carbon is an alternative to inhibit the solvent co-intercalation reaction in graphite [202–205].
2) In comparison with graphite, nongraphitic carbons can provide additional sites for lithium accommodation. As a result, they show a higher capability of reversible lithium storage than graphites, that is, stoichiometries of $x > 1$ in Li_xC_6 are possible.

The latter, so-called 'high specific charge' or 'high capacity,' carbons have received considerable attention in recent research and development. Usually they are synthesized at rather low temperatures, ranging from ~500 to ~1000 °C, and can exhibit reversible specific charges from ~400 to ~2000 Ah kg⁻¹ ($x = $ ~1.2 to ~5 in Li_xC_6), depending on the heat treatment, the organic precursor, and the electrolyte [206].[3] Such materials have been known since the late 1980s,

---

3) Carbons chemically pre-lithiated before the production of the electrode can exhibit specific charges greater than 400 Ah kg⁻¹, too [206].

15.2 Graphitic and Nongraphitic Carbons | 453

when nongraphitic carbons with specific charges of up to ~500 Ah kg^{-1} were synthesized [200]. When lithium is stored in the carbon bulk, one can suppose that a higher specific charge (in ampere-hour per kilogram) requires a correspondingly higher volume (i.e., lower density) of the carbonaceous matrix to accommodate the lithium. As a consequence, the charge densities (in ampere-hour per liter) of the high-specific-charge carbons should be comparable with those of graphite. Several models have been suggested to explain the high specific charge of these lithiated carbons. This may be because the variety of precursor materials and of manufacturing processes leads to carbonaceous host materials with various structures and compositions.

Yazami *et al.* [41, 207–211] proposed the formation of nondendritic metallic (Figure 15.11a) lithium multilayers on external graphene sheets and surfaces. Peled *et al.* [127, 128] suggested that the extra charge gained by mild oxidation of graphite is attributed to the accommodation of lithium at the prismatic surfaces (Figure 15.3) between two adjacent crystallites and in the vicinity of defects. Sato *et al.* [212, 213] suggested that lithium molecules occupy nearest-neighbor sites in intercalated carbons. Yata's and Yamabe's groups discussed the possibility of the formation of LiC$_2$ in carbons with a high interlayer spacing of ~0.400 nm (graphite: 0.335 nm) for their 'polyacenic semiconductor' (Figure 15.11b) [214–219]. In Refs [220–225] it is assumed that carbons with a small particle size can store a considerable amount of lithium on graphite edges and surfaces in addition to the lithium located between the graphene layers (Figure 15.11c). The existence of different 'Li storage sites' (Figure 15.11d) was discussed in Refs [226–228], too. Others [40, 229–232] proposed that additional lithium can be accommodated in nanocavities which are present in the carbon at temperatures below ~800 °C [233] (Figure 15.11e). Kureha Chemical Industry Co. proposes a cluster-like storage of lithium in pores where the electrolyte solvent cannot enter (Figure 15.11f). This carbon, so-called 'Carbotron P' (Morimoto, S. Kureha Chemical Industry Co., Ltd., personal communication) or 'pseudo isotropic carbon (PIC)' [234], was used in the second generation of Sony's lithium-ion cell [235].

The probabilities of the models have been discusses rather controversially [50, 51, 236–238]. There are various attempts by several researchers (in particular, Dahn's group [50, 51, 239–244]) to interpret the behavior of the high-specific-charge carbons systematically. Many graphitizing (soft) and nongraphitizing (hard) carbons which were prepared below approximately 800–900 °C and which show very high specific charges, exhibit a hysteresis [50, 51, 215, 241, 242]: the lithium intercalation occurs close to 0 V vs Li/L$^+$ whereas the lithium de-intercalation occurs at much more positive potentials (Figure 15.12b). The potential hysteresis seems to be proportional to the hydrogen content in the carbon. The ratio of hydrogen to carbon in the material is high (i) when a substantial amount of hydrogen is present during manufacture, either because hydrogen is already incorporated in the precursor material and/or because manufacture takes place under an H$_2$ atmosphere and (ii) when the manufacturing temperature has been so low that hydrogen has not yet been removed. It has been suggested that lithium is bound near H-terminated edge carbon atoms, which induces a partial bond change at the carbon from

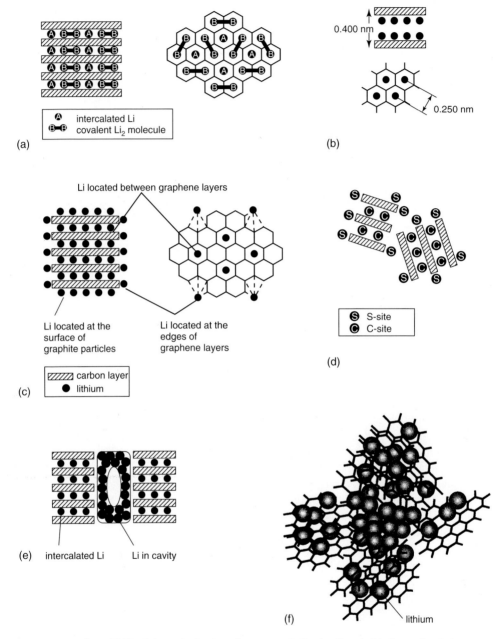

Figure 15.11 Schematic drawing of some mechanisms for reversible lithium storage in 'high-specific-charge' lithiated carbons as proposed in Refs (a) [212], (b) [214], (c) [220], (d) [226], (e) [40], and (f) (Morimoto, S. Kureha Chemical Industry Co., Ltd., personal communication). The last figure has been reproduced with kind permission of Kureha Chemical Industry Co., Ltd.

(a)

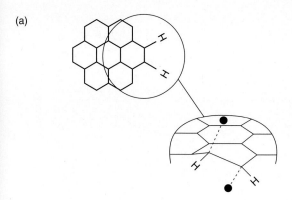

(b)

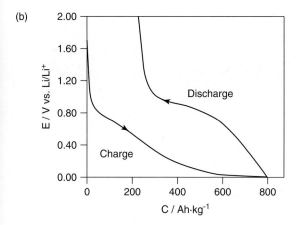

Figure 15.12 (a) Schematic model showing the mechanism of lithium storage in hydrogen-containing carbons as proposed in Ref. [242]. (b) Schematic charge–discharge curve of a hydrogen-containing carbon.

sp^2- to sp^3- hybridization [242]. This bond change requires energy for both the intercalation and removal of lithium, leading to the observed potential hysteresis (Figure 15.12a).[4]

Serious drawbacks of the carbons which are prepared at temperatures below ~800–900 °C, and which exhibit hysteresis, are their very low charge–discharge efficiencies (Figure 15.13) and their low cycle lives (Takamura, T. Tamaki, T. Petoca Ltd., personal communication). With increasing temperature the hydrogen is removed. The specific charges achieved after the removal of hydrogen depend on the structure of the carbonaceous material, whether it is a soft (graphitizing) or a hard (nongraphitizing) carbon [47, 51, 229, 230, 242, 247–252].

4) Other explanations for the increase in specific charge obtained by using hydrogen-containing carbons and for the observation of potential hysteresis were suggested in Refs [245, 246].

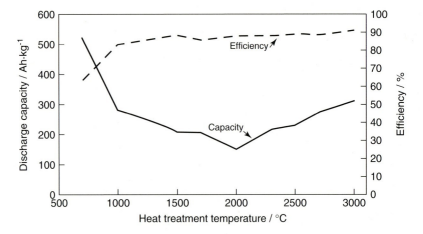

Figure 15.13 First cycle discharge capacities (—) and charge/discharge efficiencies (---) of a soft (graphitizing) carbon (Melblon® carbon fibers (MPCF)) at different heat-treatment temperatures. Reproduced with kind permission of Petoca, Ltd. (Takamura, T. Tamaki, T. Petoca Ltd., personal communication).

Around or above ~1000 °C the graphitizing (soft) carbons develop a structure with 'wrinkled' and 'buckled' structure segments (see Figure 15.4). This structure offers fewer sites for lithium intercalation than graphite [7, 19, 24, 42]. In addition, crosslinking of carbon sheets in disordered carbons hampers the shift to AA stacking, which is necessary for the accommodation of a greater amount of lithium into graphitic sites [253–255]. Correspondingly, rather low specific charges are observed (x in Li_xC_6 is typically between ~0.5 and ~0.7) in soft carbons such as turbostratic carbons [19, 42, 47, 115, 253–255] and more disordered carbons like cokes [19, 42, 47, 256–266] and certain carbon blacks [260, 267–269]. On the other hand the charge/discharge efficiency increases with the temperature during heat treatment (Figure 15.13). A type of soft carbon has been used in the first generation of Sony's lithium-ion cell [235].

Figure 15.14 shows the first Li^+ intercalation/de-intercalation cycle of a coke electrode. The potential profile differs from that of graphite, in the sense that the reversible intercalation of Li^+ begins at a potential above 1 V vs Li/Li^+, and the curve slopes without distinguishable plateaus. This behavior is a consequence of the disordered structure providing electronically and geometrically nonequivalent sites, whereas for a particular intercalation stage in highly crystalline graphite the sites are basically equivalent [19, 26]. With increasing temperature soft carbons develop more graphitic structure segments. The sites for lithium storage which were formerly determined by the disordered structure (see above) change to graphitic sites, where lithium resides in the van der Waals gaps between ordered graphene layers. Finally, at ~3000 °C, the structure and the specific charge (Figure 15.13) of graphite are achieved. The probability of finding disordered and ordered (graphitic)

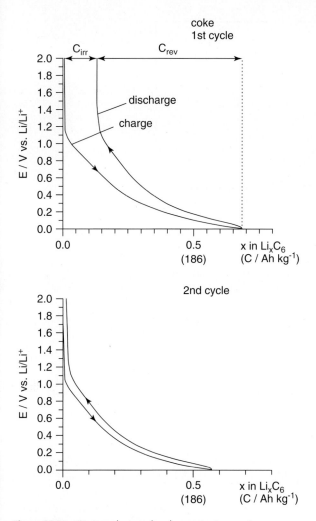

Figure 15.14 First- and second-cycle constant-current charge–discharge curves of coke (Conoco) in LiN(SO$_2$CF$_3$)$_2$/ethylene carbonate/dimethyl carbonate as the electrolyte (C_{irr} = irreversible specific charge; C_{rev} = reversible specific charge) [2].

structure segments for soft carbons obtained at temperatures between ∼1000 and ∼3000 °C has been intensively discussed in a model developed by Dahn's group [19, 42], who correlated the measured XRD data with the experimental lithium storage capacity of a specific carbon material.

In contrast to soft carbons, many nongraphitizing (hard) carbons obtained at temperatures of ∼1000 °C show a high specific charge with little hysteresis (several hundreds of Ah kg^{-1}) which can be reached only at a very low potential of a few

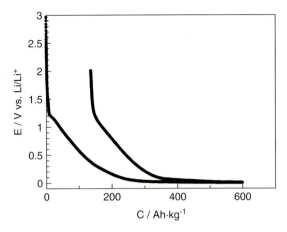

Figure 15.15 First-cycle constant-current charge/discharge curve of hard carbon ('Carbotron P'). The figure has been reproduced with kind permission of Kureha Chemical Industry Co., Ltd. (Morimoto, S. Kureha Chemical Industry Co., Ltd., personal communication.)

millivolts vs Li/Li$^+$ (Figures 15.15 and 15.16b). Compared with graphitic carbons and cokes, which show volume expansions and contractions in the region of ~10% during lithium intercalation and de-intercalation, the hard carbons are claimed not to be subject to dimensional changes during lithium uptake and removal because of the high separation (~0.380 nm) between neighboring carbon layers [235]. In order to explain the high capability for lithium storage in nongraphitizing carbons, Dahn et al. [50, 239, 243, 244, 270, 271] suggested that lithium is 'adsorbed' on both sides of single graphene layer sheets which are arranged like a 'house of cards' [243] or like 'falling cards' [244] (Figure 15.16a). The 'falling cards' model is the advanced form of the 'house of cards' model and also takes into account the storage of lithium in micropores. The accumulation of lithium in the micropores is in line with other mechanisms proposing the storage of lithium clusters or agglomerates in specific carbon 'spaces' (Morimoto, S. Kureha Chemical Industry Co., Ltd., personal communication) [234, 235]. Both the pore size and the pore openings should be small to avoid the reaction of stored lithium with the electrolyte [51, 239, 270, 272]. The proportion of nanopores in the carbons can be expected to increase with the cross-linking density of the precursor material [225, 233, 273]. Recent studies on a hard carbon prepared at 1000 °C reveal that the sizes of the single layers as well as the sizes of the pores are in the region of 1 nm [274]. In addition to microporosity, micro-texture should also be considered [273].

In conclusion, a high proportion of single-layer sheets and nanopores is beneficial for the lithium storage capabilities of hard carbons. On the other hand, heat treatment, above 1000 °C, leads to drastically increased irreversible specific charges of hard carbons [249, 275]. This may be related to the burn-off of material, which causes the opening of the nanopores. Electrolyte can penetrate in, and the sites

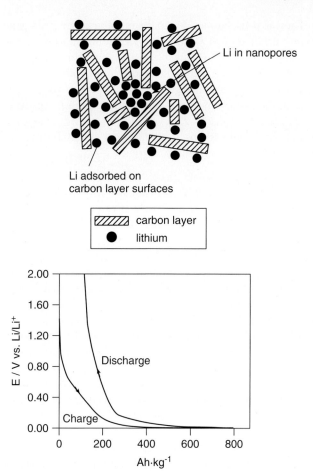

Figure 15.16 (a) Schematic model showing the mechanism of lithium storage in hard (nongraphitizing) carbons as proposed in Refs [243, 244]. (b) Schematic charge–discharge curve of hard carbon.

in the pore are no longer available for lithium storage. In this respect, it should be noted that the sensitivity of nanoporous hard carbons toward strong heat might be a difficulty for surface treatment processes typically involving higher temperatures.

Finally, it should be emphasized that the models proposed above for lithium storage in nongraphitic carbons are based on a limited number of experiments made with a limited number of carbonaceous materials. This means that carbons may be synthesized which do not belong to any of the above categories, that is, hydrogen-containing carbon, soft or hard carbon, and so on. For example, it was reported that milling [276] or strong burn-off [179, 249] of hard carbons in air alters the charge–discharge curve completely, as these carbons show hysteresis, too.

The change in electrochemical behavior may due to the introduction of hydrogen- and/or oxygen-containing surface groups. Others report a lithium storage mechanism for hard carbons prepared at 1000 °C, which considers lithium intercalated between turbostratically disordered graphene layers and lithium accommodation in amorphous hydrogen-containing carbon regions [277]. In recent work much effort has been invested in the evaluation of carbons prepared from inexpensive and abundant precursors [240, 243, 244, 278, 279].

Although the high-specific-charge carbons exhibit several times the specific charge of graphite, there are still some problems to solve:

1) In many cases extremely high irreversible specific charges were observed [44, 50, 51, 214, 216, 226, 229, 239, 280, 281], occasionally also at higher cycle numbers [51, 226, 280–282]. The irreversible capacities can be correlated with the formation of the SEI [51, 229, 239, 280, 283]. However, an 'irreversible lithium incorporation' into the carbon is discussed, too [226, 283, 284]. In analogy to graphites (Section 15.2.2.3) this irreversible reaction can be related to the reaction with surface groups on the carbon. In comparison with highly graphitic carbons with a relatively low number of surface groups, the large fraction of (internal and external) heteroatoms in the nongraphitic carbons, such as hydrogen and oxygen, can bind irreversibly a considerable amount of lithium during reduction of the carbon and thus increase the irreversible specific charge losses. Anyway, any charge losses have to be compensated by an excess of cathode material, as lithium-ion cells are assembled in the discharged state. Therefore, the specific charges calculated for the masses of *both* anode and cathode material put in the cell can be about the same for graphitic (with low irreversible charge losses) and high-specific-charge carbons (with high irreversible charge losses).

2) Carbons exhibiting hysteresis show poor cycling performance, and can be discharged only in a broad potential region of about 1–2 V (Figure 15.13) [40, 50, 51, 214–216, 230–232, 239, 270, 271, 280]. As a result, the energy efficiency of a lithium-ion cell is reduced.

3) The end-of-charge potential of nongraphitic carbons, either hydrogen-containing carbons (Figure 15.12) or cokes (Figure 15.14), but in particular of hard carbons treated at 1000 °C (Figures 15.15 and 15.16b) must be chosen very close to 0 V vs Li/Li^+ in order to obtain the full specific charge available. The narrow 'safety zone' separating this from the potential where metallic lithium is deposited on the carbon surface might give rise to some safety problems for these carbons, in particular if fast charging is required. In some cases [51, 239, 270, 271] the electrode was indeed charged to potentials below 0 V vs Li/Li^+ to achieve the high specific charge.

In contrast, there is a difference of approximately 0.1 V between the potential of graphitic LiC_6 and the potential of lithium deposition (Figure 15.7). This might be why – apart from Sony and Hitachi Maxell – many battery companies, for example, Sanyo (Nishio, K. Sanyo Electric Co., Ltd., personal communication) [285, 286], Nikkiso (Abe, H. Nikkiso Co., Ltd., personal communication) [287],

Matsushita/Panasonic [288], Moli Energy Ltd. [289], Varta [290], and A&T [291] (Satoh, A. Toshiba Corp., personal communication), use graphitic anodes in their prototype or already-commercialized lithium-ion cells. Even Sony is said to use graphitic carbon in their newest generation of lithium-ion cells. Moreover, there is a search for carbons which combine both high specific charge and graphite-like charge/discharge potential characteristics [292].

15.2.5
Lithiated Carbons Containing Heteroatoms

Depending on the precursor and the heat-treatment temperature, the carbonaceous materials discussed so far contain heteroatoms in addition to the prevailing carbon atoms. Even highly crystalline graphite is saturated with heteroatoms at dislocations in the crystallites and at the edges of the graphene layers (prismatic surfaces), which are supposed to affect the formation of residue compounds and the (electro)chemistry at the electrode/electrolyte interface. Substantial amounts of heteroatoms, such as hydrogen, are furthermore believed to change the bulk electrochemical properties of the host material, such as reversible specific charge (see previous section). From this point of view the influence of other 'noncarbonaceous' elements is of great interest. The pyrolysis of precursor materials (i) containing carbon atoms together with (various) heteroatoms and/or (ii) combining a 'carbon precursor' with 'heteroatom precursors' can result in a variety of new anode materials with interesting properties. Several aspects are reviewed briefly in this section.

Boron is assumed to act as an electron acceptor in a carbon host and thus to cause a shift of the intercalation potential to more positive values and an increase in the reversible capacity because there are more electronic sites [19, 42, 293–296]. This can enable a more rapid charging process. Nitrogen presumably functions as an electron donor and should therefore have the opposite effect on the reversible capacity and the intercalation potential [19, 297]. Contrarily, others [298] report that nitrogen-containing carbons can provide high specific charges of ~ 500 Ah kg^{-1}. Whereas the higher lithium storage capability of the above B- and N-containing anode materials is explained by different electronic properties of the host [19, 294], the high specific charge of phosphorus-doped carbons [299–304] is discussed on the basis of steric effects [299, 300]. Anyway, in analogy to 'pure' carbonaceous hosts, the influence of structure should be considered as well. Other substituted carbons, such as layered B/C/N materials [305–311] and C_xS [298], have been intensively evaluated, too.

Recently, several results obtained with high-specific-charge lithium storage materials derived from (composite) silicon-containing precursors have been published [312–318]. In addition to silicon and carbon these materials can contain a substantial proportion of hydrogen and oxygen, which can lead to a large variety of compositions [313, 314]. Very high specific charges of up to 770 Ah kg^{-1} can be reached, which seem to depend on the Si content in the matrix [314, 315]. However, these materials usually exhibit strong potential hysteresis and large irreversible specific charges (>150 Ah kg^{-1}). The interactions of the various heteroatoms with each other and with the carbon neighbors seem to be quite complex and have not been entirely clarified. Also, the nature of the lithium insertion mechanism is not yet certain. In particular, it seems not to be clear if the higher lithium storage capacity is due to the presence of Si or simply due to structural effects and/or the presence of other heteroatoms. However, in earlier papers it was speculated that composite carbon/silicon insertion materials exhibit high reversible capacities because of the high lithium alloying capacity of silicon in addition to the lithium incorporation proceeding independently in disordered carbon regions [319, 320]. Furthermore, in analogy to carbonaceous materials, manufacturing parameters such as the pyrolysis temperature have to be considered [317]. For further details, in particular regarding the synthesis and preparation of the materials discussed in this section, see the literature [2, 6, 19, 42, 321, 322].

15.2.6
Lithiated Fullerenes

Fullerenes C_{60} and C_{70} have been evaluated for use as anode materials by several groups [323–328]. A maximum reversible capacity corresponding to the approximate stoichiometry Li_2C_{60} ($x = 0.2$ in Li_xC_6) is available [323, 326], but is not sufficient for application in high-energy-density batteries. Moreover, fullerenes show some solubility in nonaqueous organic electrolytes [329].

15.3
Lithiated Carbons vs Competing Anode Materials

Despite the fact that currently commercialized lithium-ion cells basically contain anode materials based on lithiated carbons, there are still strong interests in replacing the carbon by other anode materials which show better electrochemical performance in terms of irreversible and reversible specific charges. Hence, recently proposed anode materials with specific charges higher than those of graphites and hard carbons have attracted significant interest. In particular, a lithium-ion cell with the trademark Stalion announced by Fujifilm Celltec Co. [330] has found considerable publicity, because it can provide a higher energy density and specific energy than 'conventional' lithium-ion cells (Figure 15.17). The improved performance is due to the replacement of the carbon anode by an 'amorphous tin-based composite oxide' (abbreviated to TCO or ATCO). The TCO combines both (i) a promising cycle life [330, 331] and (ii) a high specific charge (>600 Ah kg^{-1}, Figure 15.18) and charge density (>2200 Ah l^{-1}) (Idota, Y. Fujifilm Celltech Co. Ltd., personal communication) [332].

The TCO is synthesized from SnO, B_2O_3, $Sn_2P_2O_7$, Al_2O_3, and other precursors. The nature of the insertion mechanism of lithium into the Fuji material has been considered by several groups [29, 333–336]. To date it is quite clear that the lithium is not simply inserted into the TCO as in the case of the dioxides of the transition

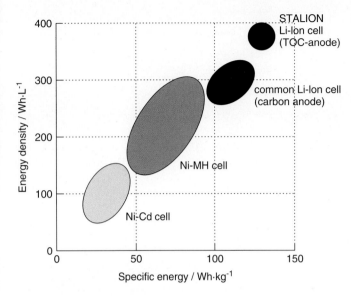

Figure 15.17 Specific energies and energy densities of rechargeable cells. Prepared from data kindly provided by Fujifilm Celltech Co., Ltd. (Idota, Y. Fujifilm Celltech Co. Ltd., personal communication).

metals Mo [32, 337], W [21, 32, 338–342], and Ti [343–345]. Fujifilm Celltech claims that only the Sn(II) compounds in the composite oxide form the electrochemically active centers for Li insertion, whereas the oxides of B, P, or Al are electrochemically inactive. In order to explain the high specific charge, a mechanism is suggested in which the tin oxide reacts to Li_2O and metallic Sn [29, 333–336]. This reaction is associated with large charge losses due to the irreversible formation of Li_2O during the first charge (Figure 15.18). In a second step the Sn then alloys with lithium reversibly. Though Fuji Celltec Co. has stopped its R&D activities on the TCO recently, the idea that the high specific charge of the TCO is due to the alloying of metallic tin has led to a revival of research and development of Li alloys and related materials [135, 334–336, 346–348].

The good cycling stability of the tin in TCO is quite unusual, because the electrochemical cycling of Li_xSn and also of other Li alloy electrodes is commonly associated with large volume changes in the order of 100–300% (Figure 15.19) [2, 7, 22, 24, 26, 349–351]. Moreover, lithium alloys Li_xM have a highly ionic character ('Zintl-Phases,' $Li_x^{x+}M^{x-}$). For this reason they are usually fairly brittle. Mechanical stresses related to the volume changes induce a rapid decay in mechanical properties and, finally, a 'pulverization' of the electrode (see Part III, Chapter 15). In the TCO, however, the Sn is finely distributed within the matrix of the oxides of B, P, and Al. The matrix compounds have glass-forming properties, form a network, and thus stabilize the composite microstructure during charge–discharge cycling [332]. The strategy for the improvement of cycle life by using a composite comprising

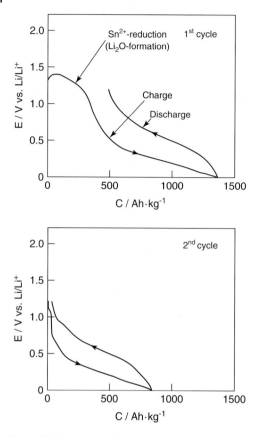

Figure 15.18 First- and second- cycle constant current charge–discharge curves of a tin composite oxide (TCO) electrode. Prepared by using data kindly provided by Fujifilm Celltech Co., Ltd. (Idota, Y. Fujifilm Celltech Co. Ltd., personal communication).

an (amorphous) lithium insertion material and a network-former follows the ideas reported in Refs [341, 352–357].

A strategy to counteract the mechanical degradation of Li alloys without incurring the irreversible lithium losses due to the formation of considerable amounts of Li_2O during the first reduction of the TCO has been suggested alternatively [29, 358]. Using thin layers of materials of small particle size or small grain size ('submicro' or 'nano' materials), relatively large dimensional changes in the crystallites ($\sim$100%) do not cause particle cracking, as the absolute changes in particle dimensions are small. For instance, small particle size, submicro-structured multiphase matrices, such as Sn/SnSb alloys, show a significant improvement in the cycling performance compared with coarse particles of (single-phase) metallic tin (Figure 15.20). The multiphase composition is favorable for the cycling behavior of the electrode as it allows the more reactive domains of the matrix (SnSb) to expand in the

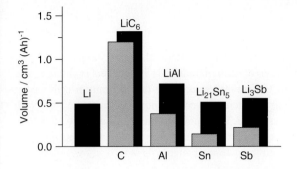

Figure 15.19 The volumes of several anode materials for lithium-ion cells before (gray) and after (black) lithiation.

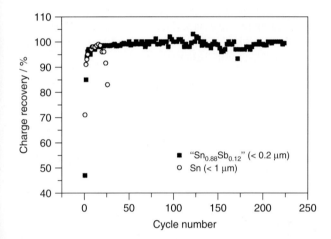

Figure 15.20 Cycling behavior of Sn + SnSb powder (analytical composition 'Sn$_{0.88}$Sb$_{0.12}$;' particle size < 0.2 µm) and Sn powder (particle size 1 µm) in 1 mol L^{-1} LiClO$_4$/propylene carbonate as electrolyte. Constant time charge with a charge input of 1.6 Li/M (~360 Ah kg^{-1}); potential-controlled discharge with a cut-off of 1.6 V vs Li/Li$^+$. $i_c = i_d = 0.3$ mA cm^{-2}. Prepared with data from Ref. [360].

neighborhood of as-yet unreacted and ductile material (Sn) in the first charging (alloying) step [29, 358–360].

Carbons containing dispersed silver [202, 361–363] also show good cycling behavior. Independently of the alloying of lithium with the metal, lithium intercalation into the carbon takes place. Apart from carbons and metal phases, novel oxides such as Li$_x$MVO$_4$ (M = Co, Cd, Ni, Zn; 1 < x ≤ 8) [364] or MnV$_2$O$_{6+\delta}$ (0 < δ ≤ 1) [365] have been proposed as anode materials. Furthermore, lithium/transition metal nitrides Li$_x$M$_y$N$_z$ with M = Co, Fe, Mn [366–370] with high specific charge of up to 760 Ah kg^{-1} are reviewed intensively in Ref. [109].

15.4
Summary

Lithiated carbons are the state-of-the-art anode materials for rechargeable lithium batteries. An immense number of carbon materials have been evaluated, and many more will be tested in the future. Since R&D on lithium-ion cells began in 1985 [235] the specific energy increased continuously because carbons with ever-higher specific charge were used. Until now, graphite and hard carbons have been preferred. At present, the graphitic materials seem to be winning *this* race. However, future trends are directed toward materials which show still-higher specific charges and charge densities while having reduced irreversible specific charges. In this respect, attention should be paid to alternative anode materials, for example, Li alloys and related materials, because their specific charges and charge densities can far exceed those of graphite. Moreover, several important features such as electrode manufacturing technologies, composite anode materials, appropriate bulk material, surface pre-treatments, and safety aspects, which have not been comprehensively reviewed here, will have a great influence on the future selection of anode materials for rechargeable lithium batteries.

Acknowledgments

The authors are grateful to Professor Tsutomu Takamura, Petoca Ltd., for several personal communications and for his support in gathering some of the data presented in this section. The companies and academic organizations which provided data and literature presented and cited in this work are gratefully acknowledged. We thank M. Wachtler and G. H. Wrodnigg for careful reading of this manuscript and for additional technical support. This work was partially supported by the Austrian Science Fund (FWF) in the 'Electroactive Materials' special research program.

References

1. Besenhard, J.O. and Winter, M. (1998) *Pure App. Chem.*, **70**, 603.
2. Winter, M., Besenhard, J.O., Spahr, M.E., and Novák, P. (1998) *Adv. Mater.*, **10**, 725.
3. Peled, E. (1983) in *Lithium Batteries*, Chapter 3 (ed. J.-P. Gabano), Academic Press, London.
4. Peled, E. (1995) in *Rechargeable Lithium and Lithium-Ion Batteries* (eds S. Megahed, B. Barnett, and L. Xie), The Electrochemical Society, Pennington, NJ, PV 94 28, p. 1.
5. Peled, E., Golodnitzky, D., and Ardel, G. (1997) *J. Electrochem. Soc.*, **144**, L208.
6. Winter, M. and Besenhard, J.O. (1998) in *Lithium Ion Batteries*, Chapter 6 (eds M. Wakihara and O. Yamamoto), Kodansha/Wiley-VCH Verlag GmbH, Tokyo, Weinheim.
7. Besenhard, J.O. (1994) in *Progress in Intercalation Research* (eds W. Müller-Warmuth and R. Schöllhorn), Kluwer, Dordrecht, p. 457.
8. Linden, D. (1984) *Handbook of Batteries and Solar Cells*, McGraw-Hill, New York.
9. Linden, D. (1995) *Handbook of Batteries*, McGraw-Hill, New York.
10. Brandt, K. (1994) *Solid State Ionics*, **69**, 173.

11. Besenhard, J.O. and Eichinger, G. (1976) *J. Electroanal. Chem.*, **68**, 1.
12. Eichinger, G. and Besenhard, J.O. (1976) *J. Electroanal. Chem.*, **72**, 1.
13. Dan, P., Mengeritsky, E., Geronov, Y., and Aurbach, D. (1995) *J. Power Sources*, **54**, 143.
14. Dan, P., Mengeritsky, E., Aurbach, D., Weissman, I., and Zinigrad, E. (1997) *J. Power Sources*, **68**, 443.
15. Mengeritsky, E., Dan, P., Weissman, I., Zaban, A., and Aurbach, D. (1996) *J. Electrochem. Soc.*, **143**, 2110.
16. Levy, S.C. and Bro, P. (1994) *Battery Hazards and Accident Prevention*, Plenum Press, New York.
17. von Sacken, U., Nodwell, E., Sundher, A., and Dahn, J.R. (1994) *Solid State Ionics*, **69**, 284.
18. Fouchard, D., Xie, L., Ebner, W., and Megahed, S. (1995) in *Rechargeable Lithium and Lithium-Ion Batteries* (eds S. Megahed, B. Barnett, and L. Xie), The Electrochemical Society, Pennington, NJ, PV 94-28, p. 349.
19. Dahn, J.R., Sleigh, A.K., Shi, H., Way, B.M., Weydanz, W.J., Reimers, J.N., Zhong, Q., and von Sacken, U. (1994) in *Lithium Batteries, New Materials, Developments and Perspectives* (ed. G. Pistoia), Elsevier, Amsterdam, p. 1.
20. Armand, M.B. (1980) in *Materials for Advanced Batteries* (eds D.W. Murphy, J. Broadhead, and B.C.H. Steele), Plenum Press, New York, p. 145.
21. Scrosati, B. (1992) *J. Electrochem. Soc.*, **139**, 2776.
22. Fauteux, D. and Koksbang, R. (1993) *J. Appl. Electrochem.*, **23**, 1.
23. Megahed, S. and Scrosati, B. (1994) *J. Power Sources*, **51**, 79.
24. Besenhard, J.O. (1994) in *Soft Chemistry Routes to New Materials*, Materials Science, Vols. **152–153** (eds J. Rouxel, M. Tournoux, and R. Brec), Trans-Tech Publications, Switzerland, p. 13.
25. Basu, S. (1983) US Patent 4423125.
26. Besenhard, J.O. and Winter, M. (1995) in *Ulmer Elektrochemische Tage, Ladungsspeicherung in der Doppelschicht*, vol. 2 (ed. W. Schmickler), Universitätsverlag Ulm, p. 47.
27. Nagaura, T. (1991) in *Progress in Batteries and Solar Cells*, vol. 10 (eds JEC Press Inc. and IBA Inc.), JEC Press, Brunswick, OH, p. 218.
28. Nagaura, T. and Tozawa, K. (1990) in *Progress in Batteries and Solar Cells*, vol. 9 (eds JEC Press Inc. and IBA Inc.), JEC Press, Brunswick, OH, p. 209.
29. Besenhard, J.O., Yang, J., and Winter, M. (1997) *J. Power Sources*, **68–69**, 87.
30. Brandt, K. (1995) *J. Power Sources*, **54**, 151.
31. Brandt, K. and Kruger, F.J. (1993) *Batteries Int.*, 24.
32. Whittingham, M.S. (1978) *Prog. Solid State Chem.*, **12**, 41.
33. Nesper, R. (1990) *Prog. Solid State Chem.*, **20**, 1.
34. Weast, R.C. (1987) *Handbook of Chemistry and Physics*, 68th edn, CRC Press, Boca Raton, FL.
35. Billaud, D., McRae, E., and Hérold, A. (1979) *Mater. Res. Bull.*, **14**, 857.
36. Manev, V., Naidenov, I., Puresheva, B., and Pistoia, G. (1995) *J. Power Sources*, **57**, 133.
37. Du Pasquier, A., Disma, F., Bowmer, T., Gozdz, A.S., Amatucci, G., and Tarascon, J.-M. (1998) *J. Electrochem. Soc.*, **145**, 472.
38. Menachem, C., Golodnitzky, D., and Peled, E. (1997) in *Batteries for Portable Applications and Electric Vehicles* (eds C.F. Holmes and A.R. Landgrebe), The Electrochemical Society, Pennington, NJ, PV97-18, p. 95.
39. Sawai, K., Iwakoshi, Y., and Ohzuku, T. (1994) *Solid State Ionics*, **69**, 273.
40. Mabuchi, A. (1994) *Tanso*, **165**, 298.
41. Yazami, R. and Munshi, M.Z.A. (1995) in *Handbook of Solid State Batteries and Capacitors* (ed. M.Z.A. Munshi), World Scientific, Singapore, p. 425.
42. Dahn, J.R., Sleigh, A.K., Shi, H., Reimers, J.N., Zhong, Q., and Way, B.M. (1993) *Electrochim. Acta*, **38**, 1179.
43. Yazami, R. (1994) in *Lithium Batteries, New Materials. Developments and Perspectives* (ed. G. Pistoia), Elsevier, Amsterdam, p. 49.
44. Yamamoto, O., Takeda, Y., Imanishi, N., and Kanno, R. (1993)

in *New Sealed Rechargeable Batteries and Supercapacitors* (eds B.M. Barnett, E. Dowgiallo, G. Halpert, Y. Matsuda, and Z. Takehara), The Electrochemical Society, Pennington, NJ, PV 93-23, p. 302.
45. Yazami, R. and Guérard, D. (1993) *J. Power Sources*, **43–44**, 39.
46. Takami, N., Satoh, A., Hara, M., and Ohsaki, T. (1995) *J. Electrochem. Soc.*, **142**, 371.
47. Satoh, A., Takami, N., and Ohsaki, T. (1995) *Solid State Ionics*, **80**, 291.
48. Endo, M., Nakamura, J., Emori, A., Sasabe, Y., Takeuchi, K., and Inagaki, M. (1994) *Mol. Cryst. Liq. Cryst.*, **245**, 171.
49. Yazami, R., Zhaghib, K., and Deschamps, M. (1994) *Mol. Cryst. Liq. Cryst.*, **245**, 165.
50. Zheng, T., Xue, J.S., and Dahn, J.R. (1996) *Chem. Mater.*, **8**, 389.
51. Zheng, T., Liu, Y., Fuller, E.W., Tseng, S., von Sacken, U., and Dahn, J.R. (1995) *J. Electrochem. Soc.*, **142**, 2581.
52. Yamamoto, O., Imanishi, N., Takeda, Y., and Kashiwagi, H. (1995) *J. Power Sources*, **54**, 72.
53. Li, G., Xue, R., Chen, L., and Huang, Y. (1995) *J. Power Sources*, **54**, 271.
54. Yamaura, J., Ohzaki, Y., Morita, A., and Ohta, A. (1993) *J. Power Sources*, **43–44**, 233.
55. Li, G., Lu, Z., Huang, B., Huang, H., Xue, R., and Chen, L. (1995) *Solid State Ionics*, **81**, 15.
56. Takami, N., Sato, A., and Ohsaki, T. (1993) in *Lithium Batteries* (eds S. Surampudi and V.R. Koch), The Electrochemical Society, Pennington, NJ, PV 93-24, p. 44.
57. Tran, T.D., Feikert, J.H., Song, X., and Kinoshita, K. (1995) *J. Electrochem. Soc.*, **142**, 3297.
58. Shi, H., Barker, J., Saiidi, M.Y., Koksbang, R., and Morris, L. (1997) *J. Power Sources*, **68**, 291.
59. Miura, S., Nakamura, H., and Yoshio, M. (1993) in *Progress in Batteries and Battery Materials*, vol. 12 (eds ITE-JEC Press Inc. and IBA Inc.), IEC Press, Brunswick, OH, p. 115.
60. Shi, H., Barker, J., Saidi, M.Y., and Koksbang, R. (1996) *J. Electrochem. Soc.*, **143**, 3466.
61. Winans, R.E. and Carrado, K.A. (1995) *J. Power Sources*, **54**, 11.
62. Sandi, G., Winans, R.E., and Carrado, K.A. (1996) *J. Electrochem. Soc.*, **143**, L95.
63. Besenhard, J.O. and Fritz, H.P. (1974) *J. Electroanal. Chem.*, **53**, 329.
64. Besenhard, J.O. (1976) *Carbon*, **14**, 111.
65. Fong, R., von Sacken, U., and Dahn, J.R. (1990) *J. Electrochem. Soc.*, **137**, 2009.
66. Pierson, H.O. (1993) *Handbook of Carbon, Graphite, Diamond and Fullerenes*, Noyes Publications, Park Ridge, NJ, p. 43.
67. Warren, B.E. (1941) *Phys. Rev.*, **59**, 693.
68. Franklin, R.E. (1951) *Acta Crystallogr.*, **4**, 253.
69. Franklin, R.E. (1951) *Proc. R. Soc. London Ser. A*, **209**, 196.
70. Kinoshita, K. (1987) *Carbon, Electrochemical and Physicochemical Properties*, Wiley-VCH Verlag GmbH, New York.
71. Vohler, O., von Sturm, F., Wege, E., von Kienle, H., Voll, M., and Kleinschmitt, P. (1986) in *Ullmann's Encyclopedia of Industrial Chemistry*, vol. 5A (Executive ed. W. Gerhartz), 5th edn, Wiley-VCH Verlag GmbH, Weinheim, p. 95.
72. Jenkins, G.M. and Kawamura, K. (1976) *Polymeric Carbons, Carbon Fibre, Glass and Char*, Cambridge University Press, Cambridge.
73. Dübgen, R. and Popp, G. (1984) *Z. Werkstofftech.*, **15**, 331.
74. Marsh, H. (1989) *Introduction to Carbon Science*, Butterworth, London.
75. Inagaki, M. (1996) *Solid State Ionics*, **86–88**, 833.
76. Ruland, W. (1990) *Adv. Mater.*, **2**, 528.
77. Kelly, B.T. (1981) *Physics of Graphite*, Applied Science, Englewood, NJ.
78. Hérold, A. (1955) *Bull. Soc. Chim. Fr.*, **187**, 999.
79. Hérold, A. (1987) in *Chemical Physics of Intercalation*, NATO ASI Series, Vol. B172 (eds A.P. Legrand and S.

Flandrois), Plenum Press, New York, p. 3.
80. Hartwigsen, C., Witschel, W., and Spohr, E. (1997) *Ber. Bunsenges. Phys. Chem.*, **101**, 859.
81. Song, X.Y., Kinoshita, K., and Tran, T.D. (1996) *J. Electrochem. Soc.*, **143**, L120.
82. Moret, R. (1986) in *Intercalation in Layered Materials*, NATO ASI Series, Vol. B148 (ed. M.S. Dresselhaus), Plenum Press, New York, p. 185.
83. Rossat-Mignod, J., Fruchart, D., Moran, M.J., Milliken, J.W., and Fisher, J.E. (1980) *Synth. Met.*, **2**, 143.
84. Avdev, V.V., Nalimova, V.A., and Semenenko, K.N. (1990) *High Press. Res.*, **6**, 11.
85. Guerard, D. and Nalimova, V.A. (1995) in *Solid State Ionics IV*, Materials Research Society Symposia Proceeding, Vol. 369 (eds G.-A. Nazri, J.-M. Tarascon, and M. Schreiber), p. 155.
86. Nalimova, V.A., Guerard, G., Lelaurain, M., and Fateev, O.V. (1995) *Carbon*, **33**, 177.
87. Nalimova, V.A., Bindra, C., and Fisher, J.E. (1996) *Solid State Commun.*, **97**, 583.
88. Yazami, R., Cherigui, A., Nalimova, V.A., and Guerard, D. (1993) in *Lithium Batteries* (eds S. Surampudi and V.R. Koch), The Electrochemical Society, Pennington, NJ, PV 93–24, p. 379.
89. Mizutani, Y., Abe, T., Ikeda, K., Ihara, E., Asano, M., Harada, T., Inaba, M., and Ogumi, Z. (1997) *Carbon*, **35**, 61.
90. Besenhard, J.O. and Fritz, H.P. (1983) *Angew. Chem. Int. Ed. Engl.*, **95**, 950.
91. Rüdorff, W. and Hofmann, U. (1938) *Z. Anorg. Allg. Chem.*, **238**, 1.
92. Daumas, N. and Hérold, A. (1969) *C. R. Acad. Sci. Paris Ser. C*, **268**, 373.
93. Daumas, N. and Hérold, A. (1971) *Bull. Soc. Chim.*, **5**, 1598.
94. Ebert, L.B. (1976) *Annu. Rev. Mater. Sci. Fr.*, **6**, 181.
95. Rüdorff, W. (1959) in *Advances in Inorganic Chemistry and Radiochemistry*, vol. 1 (eds H.J. Eméleus and A.G. Sharpe), Academic Press, New York, p. 223.
96. Schlögl, R. (1994) in *Progress in Intercalation Research*, (eds W. Müller-Warmuth and R. Schöllhorn), Kluwer, Dordrecht, p. 83.
97. Fisher, J.E. (1987) in *Chemical Physics of Intercalation*, NATO ASI Series, Vol. B172 (eds A.P. Legrand and S. Flandrois), Plenum Press, New York, p. 59.
98. Dahn, J.R. (1991) *Phys. Rev. B*, **44**, 9170.
99. Bagouin, M., Guerard, D., and Hérold, A. (1966) *C. R. Acad. Sci. Paris Ser. C*, **262**, 557.
100. Guerard, D. and Hérold, A. (1975) *Carbon*, **13**, 337.
101. Pfluger, P., Geiser, V., Stolz, S., and Güntherodt, H.-J. (1981) *Synth. Met.*, **3**, 27.
102. Billaud, D., Henry, F.X., and Willmann, P. (1993) *Mater. Res. Bull.*, **28**, 477.
103. Billaud, D., Henry, F., and Willmann, P. (1994) *Mol. Cryst. Liq. Cryst.*, **245**, 159.
104. Whitehead, A.H., Edström, K., Rao, N., and Owen, J.R. (1996) *J. Power Sources*, **63**, 41.
105. Inaba, M., Yoshida, H., Ogumi, Z., Abe, T., Mizutani, Y., and Asano, M. (1995) *J. Electrochem. Soc.*, **142**, 20.
106. Huang, W. and Frech, R. (1998) *J. Electrochem. Soc.*, **145**, 765.
107. Mori, S., Asahina, H., Suzuki, H., Yonei, A., and Yasukawa, E. (1997) *J. Power Sources*, **68**, 59.
108. Ohzuku, T., Iwakoshi, Y., and Sawai, K. (1993) *J. Electrochem. Soc.*, **140**, 2490.
109. Imanishi, N., Takeda, Y., and Yamamoto, O. (1998) in *Lithium Ion Batteries*, Chapter 5 (eds M. Wakihara and O. Yamamoto), Kodansha/Wiley-VCH Verlag GmbH, Tokyo, Weinheim.
110. Avdeev, V.V., Savchenko, A.P., Monyakina, L.A., Nikol'skaya, I.V., and Khvostov, A.V. (1996) *J. Phys. Chem. Solid*, **57**, 947.
111. Flandrois, S., Fevrier, A., Biensan, P., and Simon, B. (1996) US Patent 5, 554, 462.

112. Isaev, Y.V., Lemenko, N.D., Gumileva, L.V., Buyanovskaya, A.G., Novikov, Y.N., and Stumpp, E. (1997) *Carbon*, **35**, 563.
113. Tossici, R., Berrettoni, M., Marassi, R., and Scosati, B. (1996) *J. Electrochem. Soc.*, **143**, L65.
114. Tossici, R., Berrettoni, M., Rosolen, M., Marassi, R., and Scosati, B. (1997) *J. Electrochem. Soc.*, **144**, 186.
115. Yokoyama, K. and Nagawa, N. (1993) in *New Sealed Rechargeable Batteries and Supercapacitors* (eds B.M. Barnett, E. Dowgiallo, G. Halpert, Y. Matsuda, and Z. Takehara), The Electrochemical Society, Pennington, NJ, PV 93-23, p. 270.
116. Kanno, R., Kawamoto, Y., Takeda, Y., Ohashi, S., Imanishi, N., and Yamamoto, O. (1992) *J. Electrochem. Soc.*, **139**, 3397.
117. Winter, M., Novák, P., and Monnier, A. (1998) *J. Electrochem. Soc.*, **145**, 428.
118. Manev, V., Naidenov, I., Puresheva, B., and Pistoia, G. (1995) *J. Power Sources*, **57**, 211.
119. Imhof, R. and Novák, P. (1997) in *Batteries for Portable Applications and Electric Vehicles* (eds C.F. Holmes and A.R. Landgrebe), The Electrochemical Society, Pennington, NJ, PV97-18, p. 313.
120. Imhof, R. and Novák, P. (1998) *J. Electrochem. Soc.*, **145**, 1081.
121. Novák, P., Scheifele, W., Winter, M., and Haas, O. (1997) *J. Power Sources*, **68**, 267.
122. Rosolen, J.M. and Decker, F. (1996) *J. Electrochem. Soc.*, **143**, 2417.
123. Hirasawa, K.A., Nishioka, K., Sato, T., Yamaguchi, S., and Mori, S. (1997) *J. Power Sources*, **69**, 97.
124. Bittihn, R., Herr, R., and Hoge, D. (1993) *J. Power Sources*, **43–44**, 409.
125. Hennig, G.R. (1959) *Prog. Inorg. Chem.*, **1**, 125.
126. Hooley, J.G. (1969) *Chem. Phys. Carbon*, **5**, 321.
127. Peled, E., Menachem, C., Bar-Tow, D., and Melman, A. (1996) *J. Electrochem. Soc.*, **143**, L4.
128. Menachem, C., Peled, E., Burstein, L., and Rosenberg, Y. (1997) *J. Power Sources*, **68**, 277.
129. Bar-Tow, D., Peled, E., and Burstein, L. (1997) in *Batteries for Portable Applications and Electric Vehicles* (eds C.F. Holmes and A.R. Landgrebe), The Electrochemical Society, Pennington, NJ, PV97-18, p. 324.
130. Takamura, T., Awano, H., Ura, T., and Ikezawa, Y. (1995) *Anal. Sci. Technol.*, **8**, 583.
131. Kikuchi, M., Ikezawa, Y., and Takamura, T. (1995) *J. Electroanal. Chem.*, **396**, 451.
132. Takamura, T., Kikuchi, M., Ebana, J., Nagashima, M., and Ikezawa, Y. (1993) in *New Sealed Rechargeable Batteries and Supercapacitors* (eds B.M. Barnett, E. Dowgiallo, G. Halpert, Y. Matsuda, and Z. Takehara), The Electrochemical Society, Pennington, NJ, PV 93 23, p. 229.
133. Takamura, T., Kikuchi, M., and Ikezawa, Y. (1995) *Rechargeable Lithium and Lithium-Ion Batteries* (eds S. Megahed, B. Barnett, and L. Xie), The Electrochemical Society, Pennington, NJ, PV 94-28, p. 213.
134. Takamura, T., Awano, H., Ura, T., and Sumiya, K. (1997) *J. Power Sources*, **68**, 114.
135. Sumiya, K., Sekine, K., and Takamura, T. (1997) in *Batteries for Portable Applications and Electric Vehicles* (eds C.F. Holmes and A.R. Landgrebe), The Electrochemical Society, Pennington, NJ, PV97-18, p. 523.
136. Winter, M., Besenhard, J.O., and Novák, P. (1996) *GDCh Monogr.*, **3**, 438.
137. Dey, A.N. and Sullivan, B.P. (1970) *J. Electrochem. Soc.*, **117**, 222.
138. Besenhard, J.O., Möhwald, H., and Nickl, J.J. (1980) *Carbon*, **18**, 399.
139. Rose, M., Nacchache, C., and Golé, J. (1968) *C.R. Acad. Sci. Paris Ser. C*, **266**, 421.
140. Due, C.-M., Rose, M., and Pascault, J.-P. (1970) *C.R. Acad. Sci. Paris Ser. C*, **270**, 569.
141. Beguin, F., Setton, R., Hamwi, A., and Touzain, P. (1979) *Mater. Sci. Eng.*, **40**, 167.
142. Besenhard, J.O., Witty, H., and Klein, H.-F. (1984) *Carbon*, **22**, 98.

143. Besenhard, J.O., Klein, H.-F., Möhwald, H., and Nickl, J.J. (1981) *Synth. Met.*, **4**, 51.
144. Besenhard, J.O., Kain, I., Klein, H.-F., Möhwald, H., and Witty, H. (1983) in *Intercalated Graphite* (eds M.S. Dresselhaus, G. Dresselhaus, J.E. Fischer, and M.J. Moran), North-Holland, New York, p. 221.
145. Marcus, B. and Touzain, P. (1988) *J. Solid State Chem.*, **77**, 223.
146. Schoderböck, P. and Boehm, H.P. (1991) *Synth. Met.*, **44**, 239.
147. Arakawa, M. and Yamaki, J. (1987) *J. Electroanal. Chem.*, **219**, 273.
148. Eichinger, G. (1976) *J. Electroanal. Chem.*, **74**, 183.
149. Shu, Z.X., McMillan, R.S., and Murray, J.J. (1993) *J. Electrochem. Soc.*, **140**, 922.
150. Shu, Z.X., McMillan, R.S., and Murray, J.J. (1993) *J. Electrochem. Soc.*, **140**, L101.
151. Shu, Z.X., McMillan, R.S., and Murray, J.J. (1993) in *New Sealed Rechargeable Batteries and Supercapacitors* (eds B.M. Barnett, E. Dowgiallo, G. Halpert, Y. Matsuda, and Z. Takehara), The Electrochemical Society, Pennington, NJ, PV 93-23, p. 238.
152. Besenhard, J.O., Winter, M., Yang, J., and Biberacher, W. (1995) *J. Power Sources*, **54**, 228.
153. Aurbach, D., Ein-Eli, Y., Chusid, O., Carmeli, Y., Babai, M., and Yamin, H. (1994) *J. Electrochem. Soc.*, **141**, 603.
154. Guyomard, D. and Tarascon, J.M. (1995) *J. Power Sources*, **54**, 92.
155. Guyomard, D. and Tarascon, J.M. (1994) *Solid State Ionics*, **69**, 222.
156. Guyomard, D. and Tarascon, J.M. (1993) US Patent 5, 192, 629.
157. Aurbach, D., Markovsky, B., Shechter, A., Ein-Eli, Y., and Cohen, H. (1996) *J. Electrochem. Soc.*, **143**, 3809.
158. Aurbach, D., Zaban, A., Schechter, A., Ein-Eli, Y., Zinigrad, E., and Markovsky, B. (1995) *J. Electrochem. Soc.*, **142**, 2873.
159. Ohta, A., Koshina, H., Okuno, H., and Murai, H. (1995) *J. Power Sources*, **54**, 6.
160. Inaba, M., Siroma, Z., Funabiki, A., and Ogumi, Z. (1996) *Langmuir*, **12**, 1535.
161. Inaba, M., Siroma, Z., Kawadate, Y., Funabiki, A., and Ogumi, Z. (1997) *J. Power Sources*, **68**, 221.
162. Hirasawa, K.A., Sato, T., Asahina, H., Yamaguchi, S., and Mori, S. (1997) *J. Electrochem. Soc.*, **144**, L81.
163. Aurbach, D. and Ein-Eli, Y. (1995) *J. Electrochem. Soc.*, **142**, 1746.
164. Imanishi, N., Kashiwagi, H., Ichikawa, T., Takeda, Y., Yamamoto, O., and Inagaki, M. (1993) *J. Electrochem. Soc.*, **140**, 315.
165. Imanishi, N., Kashiwagi, H., Takeda, Y., Yamamoto, O., and Inagaki, M. (1992) in *High Power, Ambient Temperature Lithium Batteries* (eds S. Megahed, B. Barnett, and L. Xie), The Electrochemical Society, Pennington, NJ, PV 92-15, p. 80.
166. Winter, M. (1995) Doctoral thesis, University of Münster, Germany.
167. Besenhard, J.O., Wagner, M.W., Winter, M., Jannakoudakis, A.D., Jannakoudakis, P.D., and Theodoridou, E. (1993) *J. Power Sources*, **43–44**, 413.
168. Ein-Eli, Y., Markovsky, B., Aurbach, D., Carmeli, Y., Yamin, H., and Luski, S. (1994) *Electrochim. Acta*, **39**, 2559.
169. Aurbach, D., Ein-Eli, Y., Markovsky, B., Zaban, A., Lusky, S., Carmeli, Y., and Yamin, H. (1995) *J. Electrochem. Soc.*, **142**, 2882.
170. Besenhard, J.O., Castella, P., and Wagner, M.W. (1992) *Mater. Sci. Forum*, **91–93**, 647.
171. Simon, B., Boeuve, J.P., and Broussely, M. (1993) *J. Power Sources*, **43–44**, 65.
172. Chusid, O., Ein-Eli, Y., and Aurbach, D. (1993) *J. Power Sources*, **43–44**, 47.
173. Besenhard, J.O., Winter, M., and Yang, J. (1995) Poster presented at the International Workshop on Advanced Batteries (Lithium Batteries), Osaka, p. 129.
174. Winter, M. (1993) Diploma thesis, University of Münster, Germany.
175. Ein-Eli, Y., Thomas, S.R., and Koch, V.R. (1997) *J. Electrochem. Soc.*, **144**, 1159.

176. Ein-Eli, Y., Thomas, S.R., and Koch, V.R. (1997) *J. Electrochem. Soc.*, **144**, L195.
177. Wagner, M.W., Liebenow, C., and Besenhard, J.O. (1997) *J. Power Sources*, **68**, 328.
178. Wagner, M.W., Besenhard, J.O., and Liebenow, C. (1997) in *Batteries for Portable Applications and Electric Vehicles* (eds C.F. Holmes and A.R. Landgrebe), The Electrochemical Society, Pennington, NJ, PV97-18, p. 26.
179. Ein-Eli, Y., McDevitt, S., Aurbach, D., Markovsky, B., and Schechter, A. (1997) *J. Electrochem. Soc.*, **144**, L180.
180. Nakamura, H., Komatsu, H., and Yoshio, M. (1996) *J. Power Sources*, **62**, 219.
181. Ein-Eli, Y., Thomas, S.R., Koch, V., Aurbach, D., Markovsky, B., and Schechter, A. (1996) *J. Electrochem. Soc.*, **143**, L273.
182. Ein-Eli, Y., McDevitt, S.F., and Laura, R. (1998) *J. Electrochem. Soc.*, **145**, L1.
183. Jehoulet, C., Biensan, P., Bodet, J.M., Broussely, M., Moteau, C., and Tessier-Lescourret, C. (1997) in *Batteries for Portable Applications and Electric Vehicles* (eds C.F. Holmes and A.R. Landgrebe), The Electrochemical Society, Pennington, NJ, PV97-18, p. 974.
184. Wrodnigg, G.H., Besenhard, J.O., and Winter, M. HCD to *Electrochem. Solid-State Lett.*, submitted.
185. Ein-Eli, Y. and McDevitt, S.F. (1997) *J. Solid State Electrochem.*, **1**, 227.
186. Naruse, Y. Fujita, S., and Omaru, A. (1998) US Patent 5, 714, 281.
187. Shu, Z.X., McMillan, R.S., and Murray, J.J. (1995) *J. Electrochem. Soc.*, **142**, L161.
188. Shu, Z.X., McMillan, R.S., and Murray, J.J. (1995) in *Rechargeable Lithium and Lithium-Ion Batteries* (eds S. Megahed, B.M. Barnett, and L. Xie), The Electrochemical Society, Pennington, NJ, PV 94-28, p. 431.
189. Shu, Z.X., McMillan, R.S., Murray, J.J., and Davidson, I.J. (1996) *J. Electrochem. Soc.*, **143**, 2230.
190. Shu, Z.X. McMillan, R.S., and Murray, J.J. (1996) US Patent 5, 529, 859.
191. Shu, Z.X. McMillan, R.S., and Murray, J.J. (1996) US Patent 5, 571, 635.
192. Winter, M. and Novàk, P. (1998) *J. Electrochem. Soc.*, **145**, L27.
193. Wilkinson, D.P. and Dahn, J.R. (1992) US Patent 5, 130, 211.
194. Roh, Y.B., Tada, K., Kawai, T., Araki, H., Yoshino, K., Takase, M., and Suzuki, T. (1995) *Synth. Met.*, **69**, 601.
195. Yazami, R. and Touzain, P. (1995) *J. Power Sources*, **9**, 365.
196. Yazami, R. and Guerard, D. (1993) *J. Power Sources*, **43–44**, 65.
197. Lemont, S. and Billaud, D. (1995) *J. Power Sources*, **54**, 340.
198. Billaud, D., Naji, A., and Willmann, P. (1995) *J. Chem. Soc. Chem. Commun.*, 1867.
199. Naji, A., Ghanbaja, J., Willmann, P., and Billaud, D. (1997) *Carbon*, **35**, 845.
200. Kanno, R., Takeda, Y., Ichikawa, T., Nakanishi, K., and Yamamoto, O. (1989) *J. Power Sources*, **26**, 535.
201. Mohri, M., Yanagisawa, N., Tajima, Y., Tanaka, H., Mitate, T., Nakajima, S., Yoshida, M., Yoshimoto, Y., Suzuki, T., and Wada, H. (1989) *J. Power Sources*, **26**, 545.
202. Momose, H., Funahashi, A., Aragane, J., Matsui, K., Yoshitake, S., Mitsuishi, I., Awata, H., and Iwahori, T. (1997) in *Batteries for Portable Applications and Electric Vehicles* (eds C.F. Holmes and A.R. Landgrebe), The Electrochemical Society, Pennington, NJ, PV97-18, p. 376.
203. Shu, X., Schmidt, L.D., and Smyrl, W.H. (1995) in *Rechargeable Lithium and Lithium-Ion Batteries* (eds S. Megahed, B. Barnett, and L. Xie), The Electrochemical Society, Pennington, NJ, PV 94-28, p. 196.
204. Kuribayashi, I., Yokoyama, M., and Yamashita, M. (1995) *J. Power Sources*, **54**, 1.
205. Kurokawa, H., Maeda, T., Nakanishi, N., Nohma, T., and Nishio, K. (1996) Poster presented at the 8th International Meeting on Lithium Batteries, Nagoya, p. 222. Extended Abstract.
206. Yazami, R. and Moreau, M. (1996) US Patent No. 5543021.

207. Yazami, R. and Deschamps, M. (1995) in *Advances in Porous Materials* (eds S. Komarneni, D.M. Smith, and J.S. Beck), Materials Research Society Symposia Proceedings, Vol. 369, Materials Research Society, p. 165.
208. Yazami, R. and Deschamps, M. (1995) *J. Power Sources*, **54**, 411.
209. Yazami, K. and Deschamps, M. (1995) in *Rechargeable Lithium and Lithium-Ion Batteries* (eds S. Megahed, B. Barnett, and L. Xie), The Electrochemical Society, Pennington, NJ, PV 94-28, p. 183.
210. Deschamps, M. and Yazami, R. (1997) *J. Power Sources*, **68**, 236.
211. Yazami, R. and Deschamps, M. (1996) in *Progress in Batteries and Battery Materials*, vol. 15 (eds ITE-JEC Press Inc./IBA Inc.), JEC Press, Brunswick, OH, p. 161.
212. Sato, K., Noguchi, M., Demachi, A., Oki, N., and Endo, M. (1994) *Science*, **264**, 556.
213. Sato, K., Noguchi, M., Demachi, A., Oki, N., Endo, M., and Sasabe, Y. (1995) Poster presented at the International Workshop on Advanced Batteries (Lithium Batteries), Osaka, p. 219.
214. Yata, S., Kinoshita, H., Komori, M., Ando, N., Kashiwamura, T., Harada, T., Tanaka, K., and Yamabe, T. (1994) *Synth. Met.*, **62**, 153.
215. Yata, S., Sakurai, K., Osaki, T., Inoue, Y., and Yamaguchi, K. (1990) *Synth. Met.*, **33**, 177.
216. Yata, S., Hato, Y., Kinoshita, H., Ando, N., Anekawa, A., Hashimoto, T., Yamaguchi, M., Tanaka, K., and Yamabe, T. (1995) *Synth. Met.*, **73**, 273.
217. Yata, S. (1995) Poster presented at the International Workshop on Advanced Batteries (Lithium Batteries), Osaka, p. 204.
218. Yata, S., Kinoshita, H., Komori, M., Ando, N., Kashiwamura, T., Harada, T., Tanaka, K., and Yamabe, T. (1993) in *New Sealed Rechargeable Batteries and Supercapacitors* (eds B.M. Barnett, E. Dowgiallo, G. Halpert, and Y. Matsuda), The Electrochemical Society, Pennington, NJ, PV 93-23, p. 502.
219. Yata, S. (1997) *Denki Kakagu*, **9**, 65.
220. Wang, S., Matsumura, Y., and Maeda, T. (1995) *Synth. Met.*, **71**, 1759.
221. Matsumura, Y., Wang, S., and Mondori, J. (1995) *Carbon*, **33**, 1457.
222. Matsumura, Y., Wang, S., Kasuh, T., and Maeda, T. (1995) *Synth. Met.*, **71**, 1755.
223. Matsumura, Y., Wang, S., Shinohara, K., and Maeda, T. (1995) *Synth. Met.*, **71**, 1757.
224. Li, W.S., Jiang, L.C., Xie, G.Y., and Jiang, X. (1996) *J. Power Sources*, **58**, 235.
225. Xiang, H., Fang, S., and Jiang, Y. (1997) *J. Electrochem. Soc.*, **144**, L187.
226. Mori, Y., Iriyama, T., Hashimoto, T., Yamazaki, S., Kawakami, F., Shiroki, H., and Yamabe, T. (1995) *J. Power Sources*, **56**, 205.
227. Mori, Y., Iriyama, T., Hashimoto, T., Yamazaki, S., Kawakami, F., and Shiroki, H. (1995) Poster presented at the International Workshop on Advanced Batteries (Lithium Batteries), Osaka, p. 199.
228. Takami, N., Satoh, A., Ohsaka, T., and Kanda, M. (1997) *Electrochim. Acta*, **42**, 2537.
229. Fujimoto, H., Mabuchi, A., Tokumitsu, K., and Kasuh, T. (1995) *J. Power Sources*, **54**, 440.
230. Mabuchi, A., Fujimoto, H., Tokumitsu, K., and Kasuh, T. (1995) *J. Electrochem. Soc.*, **142**, 1041.
231. Tokumitsu, K., Mabuchi, A., Fujimoto, H., and Kasuh, T. (1995) *J. Power Sources*, **54**, 444.
232. Tokumitsu, K., Mabuchi, A., Kasuh, T., and Fujimoto, H. (1996) Poster presented at the 8th International Meeting on Lithium Batteries, Nagoya. Extended Abstract No. 212.
233. Tokumitsu, K., Mabuchi, A., Fujimoto, H., and Kasuh, T. (1996) *J. Electrochem. Soc.*, **143**, 2235.
234. Nagai, A., Ishikawa, M., Masuko, J., Sonobe, N., Chuman, H., and Iwasaki, T. (1995) *Mater. Res. Soc. Symp. Proc.*, **393**, 339.
235. Nishi, Y. (1998) in *Lithium Ion Batteries*, Chapter 8 (eds M. Wakihara and O. Yamamoto), Kodansha/Wiley-VCH Verlag GmbH, Tokyo, Weinheim.

236. Ebert, L.B. (1996) *Carbon*, **34**, 671.
237. Dahn, J.R., Zheng, T., Liu, Y., and Xue, J.S. (1995) *Science*, **270**, 590.
238. Zhou, P., Papanek, P., Bindra, C., Lee, R., and Fischer, J.E. (1997) *J. Power Sources*, **68**, 297.
239. Xue, J.S. and Dahn, J.R. (1995) *J. Electrochem. Soc.*, **142**, 3668.
240. Dahn, J.R., Xue, J.S., Xing, W., Wilson, A.M., and Gibaud, A. (1996) Poster presented at the 8th International Meeting on Lithium Batteries, Nagoya, Extended Abstract, p. 89.
241. Zheng, T. and Dahn, J.R. (1997) *J. Power Sources*, **68**, 201.
242. Zheng, T., McKinnon, W.R., and Dahn, J.R. (1996) *J. Electrochem. Soc.*, **143**, 2137.
243. Xing, W., Xue, J.S., Zheng, T., Gibaud, A., and Dahn, J.R. (1996) *J. Electrochem. Soc.*, **143**, 3482.
244. Dahn, J.R., Xing, W., and Gao, Y. (1997) *Carbon*, **35**, 825.
245. Wang, S., Zhang, Y., Yang, L., and Liu, Q. (1996) *Solid State Ionics*, **86–88**, 919.
246. Takami, N., Satoh, A., Ohsaki, T., and Kanda, M. (1998) *J. Electrochem. Soc.*, **145**, 478.
247. Satoh, A., Takami, N., Ohsaki, T., and Kanda, M. (1995) in *Rechargeable Lithium and Lithium-Ion Batteries* (eds S. Megahed, B. Barnett, and L. Xie), The Electrochemical Society, Pennington, NJ, PV 94-28, p. 143.
248. Tatsumi, K., Mabuchi, A., Iwashita, N., Sakaebe, H., Shioyama, H., Fujimoto, H., and Higuchi, S. (1993) *Batteries and Fuel Cells for Stationary and Electric Vehicle Applications* (eds A.R. Landgrebe and Z. Takehard), Electrochemical Society, Pennington, NJ, PV 93-8, p. 64.
249. Mabuchi, A., Fujimoto, H., Tokumitsu, K., and Kasuh, T. (1995) *J. Electrochem. Soc.*, **142**, 3049.
250. Tokumitsu, K., Mabuchi, A., Fujimoto, H., and Kasuh, T. (1995) in *Rechargeable Lithium and Lithium-Ion Batteries* (eds S. Megahed, B. Barnett, and L. Xie), The Electrochemical Society, Pennington, NJ, PV 94-28, p. 136.
251. Tatsumi, K., Akai, T., Imamura, T., Zaghib, K., Iwashita, N., Higuchi, S., and Sawda, Y. (1996) *J. Electrochem. Soc.*, **143**, 1923.
252. Tatsumi, K., Iwashita, N., Sakaebe, H., Shioyama, H., Higuchi, S., Mabuchi, A., and Fujimoto, H. (1995) *J. Electrochem. Soc.*, **142**, 716.
253. Zheng, T. and Dahn, J.R. (1995) *Synth. Met.*, **73**, 1.
254. Zheng, T. and Dahn, J.R. (1996) *Phys. Rev. B*, **53**, 3061.
255. Zheng, T., Reimers, J.N., and Dahn, J.R. (1995) *Phys. Rev. B*, **51**, 734.
256. Qiu, W., Zhou, R., Yang, L., and Liu, Q. (1996) *Solid State Ionics*, **86–88**, 903.
257. Jean, M., Desnoyer, C., Tranchant, A., and Messina, R. (1995) *J. Electrochem. Soc.*, **142**, 2122.
258. Moshtev, R.V., Zlatilova, P., Puresheva, B., and Manev, V. (1995) *J. Power Sources*, **56**, 137.
259. Chen, J.M., Yao, C.Y., Cheng, C.H., Hurng, W.M., and Kao, T.H. (1995) *J. Power Sources*, **54**, 494.
260. Sleigh, A.K. and von Sacken, U. (1992) *Solid State Ionics*, **57**, 99.
261. Ma, S., Li, J., Jing, X., and Wang, F. (1996) *Solid State Ionics*, **86–88**, 911.
262. Alcántara, R., Jiminéz-Mateos, J.M., Lavela, P., Morales, J., and Tirado, J.L. (1996) *Mater. Sci. Eng.*, **B39**, 216.
263. Kostecki, R., Tran, T., Song, X., Kinoshita, K., and McLarnon, F. (1997) *J. Electrochem. Soc.*, **144**, 3111.
264. Jean, M., Tranchant, A., and Messina, R. (1996) *J. Electrochem. Soc.*, **143**, 391.
265. Jean, M., Chausse, A., and Messina, R. (1997) *J. Power Sources*, **68**, 232.
266. Avery, N.R. and Black, K.J. (1997) *J. Power Sources*, **68**, 191.
267. Takei, K., Terada, N., Kumai, K., Iwahori, T., Uwai, T., and Miura, T. (1995) *J. Power Sources*, **55**, 191.
268. Takei, K., Kumai, K., Kobayashi, Y., Miyashiro, H., Iwahori, T., Uwai, T., and Miura, T. (1995) *J. Power Sources*, **54**, 171.
269. Takei, K., Kumai, K., Iwahori, T., Uwai, T., and Furusyo, M. (1993) *Denki Kakagu*, **4**, 421.
270. Liu, Y., Xue, J.S., Zheng, T., and Dahn, J.R. (1996) *Carbon*, **34**, 193.
271. Zheng, T., Zhong, Q., and Dahn, J.R. (1995) *J. Electrochem. Soc.*, **142**, L211.

272. Ago, H., Tanaka, K., Yamahe, T., Miyoshi, T., Takegoshi, K., Terao, T., Yata, S., Hato, Y., Nagaura, S., and Ando, N. (1997) *Carbon*, **12**, 1781.
273. Sonobe, N., Iwasaki, T., and Masuko, J. (1996) US Patent No. 5, 527, 643.
274. Zhou, P., Papanek, P., Lee, R., Fisher, J.E., and Kamitakahara, W.A. (1997) *J. Electrochem. Soc.*, **144**, 1744.
275. Iomoto, H., Omaru, A., Azuma, A., and Nishi, Y. (1993) in *Lithium Batteries* (eds S. Surampudi and V.R. Koch) The Electrochemical Society, Pennington, NJ, PV 93-24, p. 9.
276. Xing, W., Dunlap, R.A., and Dahn, J.R. (1998) *J. Electrochem. Soc.*, **145**, 62.
277. Dai, Y., Wang, Y., Eshkenazi, V., Peled, E., and Greenbaurn, S.G. (1998) *J. Electrochem. Soc.*, **145**, 1179.
278. Xing, W., Xue, J.S., and Dahn, J.R. (1996) *J. Electrochem. Soc.*, **143**, 3046.
279. Zheng, T., Xing, W., and Dahn, J.R. (1996) *Carbon*, **34**, 1501.
280. Imoto, H., Nagamine, M., and Nishi, Y. (1995) in *Rechargeable, Lithium and Lithium-Ion Batteries* (eds S. Megahed, B. Barnett, and L. Xie), The Electrochemical Society, Pennington, NJ, PV 93-24, p. 43.
281. Alamgir, M., Zuo, Q., and Abraham, K.M. (1994) *J. Electrochem. Soc.*, **141**, L143.
282. Iijima, T., Suzuki, K., and Matsuda, Y. (1995) *Synth. Met.*, **73**, 9.
283. Matsumura, Y., Wang, S., and Mondori, J. (1995) *J. Electrochem. Soc.*, **142**, 2914.
284. Jung, Y., Suh, M.C., Lee, H., Kim, M., Lee, S., Shim, S.C., and Kwak, J. (1997) *J. Electrochem. Soc.*, **144**, 4279.
285. Nishio, K., Fujimoto, M., Shoji, Y., Ohshita, R., Nohma, T., Moriwaki, K., Narukawa, S., and Saito, T. (1996) Poster presented at the 8th International Meeting on Lithium Batteries, Nagoya. Extended Abstract, p. 93.
286. Kurokawa, H., Nohma, T., Fujimoto, M., Maeda, T., Nishio, K., and Saito, T. (1995) Poster presented at the International Workshop on Advanced Batteries (Lithium Batteries), Osaka, p. 332.
287. Zaghib, K., Tatsumi, K., Abe, H., Ohsaki, T., Sawada, Y., and Higuchi, S. (1998) *J. Electrochem. Soc.*, **145**, 210.
288. (1994) *JEC Battery Newsletter*, vol. 5, JEC Press, Brunswick OH, p. 32.
289. (1994) *JEC Battery Newsletter*, vol. 5, JEC Press, Brunswick, OH, p. 33.
290. Koehler, U., Kruger, F.J., Kuempers, J., Maul, M. and Niggemann, E. (1997) Varta Special Report, December, p. 7.
291. Takami, N., Satoh, A., Hara, M., and Ohsaki, T. (1995) *J. Electrochem. Soc.*, **142**, 2564.
292. Kasuh, T., Mabuchi, A., Tokumitsu, K., and Fujimoto, H. (1997) *J. Power Sources*, **68**, 99.
293. Dahn, J.R., Reimers, J.N., Sleigh, A.K., and Tiedje, T. (1992) *Phys. Rev. B*, **45**, 3773.
294. Way, B.M. and Dahn, J.R. (1994) *J. Electrochem. Soc.*, **141**, 907.
295. Tamaki, T. and Tamaki, M. (1996) Poster presented at the 8th International Meeting on Lithium Batteries, Nagoya, Extended Abstract, p. 216.
296. Way, B.M., Dahn, J.R., Tiedje, T., Myrtle, K., and Kasrai, M. (1992) *Phys. Rev. B*, **46**, 1697.
297. Weydanz, W.J., Way, B.M., van Buuren, T., and Dahn, J.R. (1994) *J. Electrochem. Soc.*, **141**, 900.
298. Ito, S., Murata, T., Hasegawa, M., Bito, Y., and Toyoguchi, Y. (1997) *J. Power Sources*, **68**, 245.
299. Tran, T.D., Feikert, J.H., Mayer, S.T., Song, X., and Kinoshita, K. (1995) in *Rechargeable Lithium and Lithium-Ion Batteries* (eds S. Megahed, B. Barnett, and L. Xie), The Electrochemical Society, Pennington, NJ, PV 94-28, p. 110.
300. Schönfelder, H.H., Kitoh, K., and Nemoto, H. (1997) *J. Power Sources*, **68**, 258.
301. Tran, T.D., Mayer, S.T., and Pekala, R.W. (1995) in *Advances in Porous Materials*, Materials Research Society Symposia Proceedings, Vol. 369 (eds S. Komarneni, D.M. Smith, and J.S. Beck), p. 449.
302. Zhong, Q. and von Sacken, U. (1996) Poster presented at the 8th International Meeting on Lithium Batteries, Nagoya, Extended Abstract, p. 202.

303. Omaru, A., Azuma, H., Aoki, M., Kita, A., and Nishi, Y. (1993) in *Lithium Batteries* (eds S. Surampudi and V.R. Koch), The Electrochemical Society, Pennington, NJ, PV 93-24, p. 21.
304. Morrison, R.L. (1996) US Patent 5, 558, 954.
305. Matsuda, Y., Ishikawa, M., Nakamura, T., Morita, M., Tsujioka, S., and Kawashima, T. (1995) in *Rechargeable Lithium and Lithium-Ion Batteries* (eds S. Megahed, B. Barnett, and L. Xie), The Electrochemical Society, Pennington, NJ, PV 94-28, p. 85.
306. Morita, M., Hanada, T., Tsotsumi, H., Matsuda, Y., and Kawaguchi, M. (1992) *J. Electrochem. Soc.*, **139**, 1227.
307. Matsuda, Y., Morita, M., Hanada, T., and Kawaguchi, M. (1993) *J. Power Sources*, **43–44**, 75.
308. Ishikawa, M., Nakamura, T., Morita, M., Matsuda, Y., Tsujioka, S., and Kawashima, T. (1995) *J. Power Sources*, **55**, 127.
309. Ishikawa, M., Nakamura, T., Morita, M., Matsuda, Y., and Kawaguchi, M. (1994) *Denki Kakagu*, **62**, 897.
310. Morita, M., Hanada, T., Tsutsumi, H., and Matsuda, Y. (1992) in *High Power Ambient Temperature Lithium Batteries* (eds W.D. Clark and G. Halpert), The Electrochemical Society, Pennington, NJ, PV 92-15, p. 101.
311. Ishikawa, M., Morita, M., Hanada, T., Matsuda, Y., and Kawaguchi, M. (1993) *Denki Kakagu*, **61**, 1395.
312. Wilson, A.M., Reimers, J.N., Fuller, E.W., and Dahn, J.R. (1995) *Solid State Ionics*, **74**, 249.
313. Wilson, A.M., Zank, G., Eguchi, K., Xing, W., and Dahn, J.R. (1997) *J. Power Sources*, **68**, 195.
314. Xing, W., Wilson, A.M., Eguchi, K., Zank, G., and Dahn, J.R. (1997) *J. Electrochem. Soc.*, **144**, 2410.
315. Wilson, A.M., Zank, G., Eguchi, K., Xing, W., Yates, B., and Dahn, J.R. (1997) *Chem. Mater.*, **9**, 2139.
316. Xue, J.S., Myrtle, K., and Dahn, J.R. (1995) *J. Electrochem. Soc.*, **142**, 2927.
317. Guidotti, R.A. and Johnson, B.J. (1996) Poster presented at the Proceedings of the 37th Power Sources Conference, Cherry Hill, p. 219.
318. Xing, W., Wilson, A.M., Zank, G., and Dahn, J.R. (1997) *Solid State Ionics*, **93**, 239.
319. Wilson, A.M. and Dahn, J.R. (1995) *J. Electrochem Soc.*, **142**, 326.
320. Wilson, A.M. and Dahn, J.R. (1995) in *Rechargeable Lithium and Lithium-Ion Batteries* (eds S. Megahed, B. Barnett, and L. Xie), The Electrochemical Society, Pennington, NJ, PV 94-28, p. 158.
321. Riedel, R. (1994) *Adv. Mater.*, **6**, 549.
322. Kawaguchi, M. (1997) *Adv. Mater.*, **9**, 615.
323. Lemont, S., Ghanbaja, J., and Billaud, D. (1994) *Mol. Cryst. Liq. Cryst.*, **244**, 203.
324. Firlej, L., Zahab, A., Brocard, F., and Bernier, P. (1995) *Synth. Met.*, **70**, 1373.
325. Zho, W., Huang, H., Li, Y., Xue, R., Huang, Y., Zhao, Z., and Chen, L. (1994) Poster presented at the 7th International Meeting on Lithium Batteries, Boston, Extended Abstract, p. 278.
326. Yazami, R. and Cherigui, A. (1994) Poster presented at the 7th International Meeting on Lithium Batteries, Boston, Extended Abstract, p. 255.
327. Yazami, R., Cherigui, A., Nalimova, V.A., and Guerard, D. (1993) in *Lithium Batteries* (eds S. Surampudi and V.R. Koch), The Electrochemical Society, Pennington, NJ, PV 93-24, p. 1.
328. Kamitakahara, W.A. (1996) *J. Phys. Chem. Solids*, **57**, 671.
329. Soucaze-Guillous, B., Kutner, W., Jones, M.T., and Kadish, K.M. (1996) *J. Electrochem. Soc.*, **143**, 550.
330. Nippon Denki Shinbun (1996) March 11.
331. Idota, Y., Mineo, Y., Matsufuji, A., and Miyasaka, T. (1997) *Denki Kakagu*, **65**, 717.
332. Idota, Y., Kubota, T., Matsufuji, A., Maekawa, Y., and Miyasaka, T. (1997) *Science*, **276**, 1395.
333. Székely, M., Eid, B., Caillot, E., Herlem, M., Etcheberry, A., Mathieu, C., and Fahys, B. (1995) *J. Electroanal. Chem.*, **391**, 69.

334. Huggins, R.A. (1997) in *Batteries for Portable Applications and Electric Vehicles* (eds C.F. Holmes and A.R. Landgrebe), The Electrochemical Society, Pennington, NJ, PV 97-18, p. 1.
335. Brousse, T., Retoux, R., Herterich, U., and Schleich, D.M. (1997) in *Batteries for Portable Applications and Electric Vehicles* (eds C.F. Holmes and A.R. Landgrebe), The Electrochemical Society, Pennington, NJ, PV 97-18, p. 34.
336. Courtney, I.A. and Dahn, J.R. (1997) *J. Electrochem Soc.*, **144**, 2045.
337. Auborn, J.J. and Barberio, Y.L. (1987) *J. Electrochem. Soc.*, **134**, 638.
338. Di Pietro, B., Patriarca, M., and Scrosati, B. (1982) *J. Power Sources*, **8**, 289.
339. Di Pietro, B., Patriarca, M., and Scrosati, B. (1982) *Synth. Met.*, **5**, 1.
340. Morzilli, S., Scrosati, B., and Sgarlata, F. (1985) *Electrochim. Acta*, **30**, 1271.
341. Shodai, T., Sakurai, Y., and Okada, S. (1995) in *Rechargeable Lithium and Lithium-Ion Batteries* (eds S. Megahed, B. Barnett, and L. Xie), The Electrochemical Society, Pennington, NJ, PV 94-28, p. 224.
342. Abraham, K.M., Pasquariello, D.M., Willstaedt, E.B., and McAndrews, G.F. (1988) in *Primary and Secondary Ambient Temperature Lithium Batteries* (eds J.-P. Gabano, Z. Takehara, and P. Bro), The Electrochemical Society, Pennington, NJ, PV 88-6, p. 669.
343. Huang, S.Y., Kavan, L., Exnar, I., and Grätzel, M. (1995) *J. Electrochem. Soc.*, **142**, L142.
344. Grätzel, M. (1995) *Chem. Technol.*, **67**, 1300.
345. Kavan, L., Grätzel, M., Rathousky, J., and Zukal, A. (1995) *J. Electrochem. Soc.*, **142**, 394.
346. Sekine, K., Shimoyamada, T., Takagi, R., Sumiya, K., and Takamura, T. (1997) in *Batteries for Portable Applications and Electric Vehicles* (eds C.F. Holmes and A.R. Landgrebe), The Electrochemical Society, Pennington, NJ, PV97-18, p. 92.
347. Brousse, T., Retoux, R., Herterich, U., and Schleich, D.M. (1998) *J. Electrochem. Soc.*, **145**, 1.
348. Liu, W., Huang, X., Wang, Z., Li, H., and Chen, L. (1998) *J. Electrochem. Soc*, **145**, 59.
349. Besenhard, J.O., Hess, M., and Komenda, P. (1990) *Solid State Ionics*, **40–41**, 525.
350. Landolt, H.H. and Börnstein, R. (1971) *Structure Data of the Elements and Intermetallic Phases*, vol. 6, Springer-Verlag, Berlin.
351. Zintl, E. and Bauer, G. (1933) *Z. Phys. Chem.*, **20(B)**, 245.
352. Sakurai, Y., Okada, S., Yamaki, J., and Okada, T. (1987) *J. Power Sources*, **20**, 173.
353. Pagnier, T., Fouletier, M., and Souquet, J.L. (1983) *Solid State Ionics*, **9–10**, 649.
354. Sakurai, Y. and Yamaki, J. (1985) *J. Electrochem. Soc.*, **132**, 512.
355. Arakawa, M., Tobishima, S., Hirai, T., and Yamaki, J. (1986) *J. Electrochem. Soc.*, **133**, 1527.
356. Tobishima, S., Arakawa, M., Hirai, T., and Yamaki, J. (1987) *J. Power Sources*, **20**, 293.
357. Sakurai, Y. and Yamaki, J. (1988) *J. Electrochem. Soc.*, **135**, 791.
358. Yang, J., Winter, M., and Besenhard, J.O. (1996) *Solid State Ionics*, **90**, 281.
359. Yang, J., Besenhard, J.O., and Winter, M. (1997) in *Batteries for Portable Applications and Electric Vehicles* (eds C.F. Holmes and A.R. Landgrebe), The Electrochemical Society, Pennington, NJ, PV 97-18, p. 350.
360. Winter, M., Besenhard, J.O., Albering, J.H., Yang, J., and Wachtler, M. (1998) *Prog. Batt. Batt. Mater.*, **17**, 208.
361. Aragane, J., Matsui, K., Andoh, H., Suzuki, S., Fukada, H., Ikeya, H., Kitaba, K., and Ishikawa, R. (1997) *J. Power Sources*, **68**, 13.
362. Nishimura, K., Honbo, H., Takeuchi, S., Horiba, T., Oda, M., Koseki, M., Muranaka, Y., Kozono, Y., and Miyadera, H. (1997) *J. Power Sources*, **68**, 436.
363. Momose, H., Honbo, H., Takeuchi, S., Nishimura, K., Horiba, T., Muranaka, Y., Kozono, Y., and Miyadera, H. (1997) *J. Power Sources*, **68**, 208.

364. Guyomard, D., Sigala, C., Le Gal La Salle, A., and Piffard, Y. (1997) *J. Power Sources*, **68**, 692.
365. Piffard, Y., Leroux, F., Guyomard, D., Mansot, J.-L., and Tournoux, M. (1997) *J. Power Sources*, **68**, 698.
366. Shodai, T., Okada, S., Tobishima, S., and Yamaki, J. (1997) *J. Power Sources*, **68**, 515.
367. Nishijima, M., Kagohashi, T., Imanishi, M., Tdkeda, Y., Yamamoto, O., and Kondo, S. (1996) *Solid State Ionics*, **83**, 107.
368. Yamamoto, O., Takeda, Y., and Imanishi, N. (1995) Poster presented at the International Workshop on Advanced Batteries (Lithium Batteries), Osaka, p. 189.
369. Nishijima, M., Kagohashi, T., Takeda, Y., Imanishi, N., and Yamamoto, O. (1997) *J. Power Sources*, **68**, 510.
370. Nishijima, M. (1997) *Denki Kakagu*, **65**, 711.

Further Reading

Aurbach, D., Markovsky, B., Weissman, I., Levi, E., and Ein-Eli, Y. (1999) *Electrochim. Acta*, **45**, 67–86.
Flandrois, S. and Simon, B. (1999) *Carbon*, **37** (2), 165–180.
Landi, B.J., Ganter, M.J., Cress, C.D., DiLeo, R.A., and Raffaelle, R.P. (2009) *Energy Environ. Sci.*, **2** (6), 638–654.
Nakajima, T. (2007) *J. Fluor. Chem.*, **128** (4), 277–284.
Noel, M. and Suryanarayanan, V. (2002) *J. Power Sources*, **111** (2), 193–209.
Su, D.S. and Schlogl, R. (2010) *ChemSusChem*, **3** (2), 136–168.

16
The Anode/Electrolyte Interface
Emanuel Peled, Diane Golodnitsky, and Jack Penciner

16.1
Introduction

The anode/electrolyte interface, also referred to as the solid-electrolyte interface (SEI), plays a key role in lithium-metal, lithium-alloy, lithium-ion, and any other alkali-metal and alkaline-earth batteries. In primary batteries it determines the safety, self-discharge (shelf life), power capability, low-temperature performance, and faradaic efficiency. In secondary batteries it determines, in addition, the faradaic efficiency on charge, the cycle life, the morphology of lithium deposits, and the irreversible capacity loss (Q_{IR}) for the first charge cycle of lithium-ion batteries. As its importance is well recognized in the scientific community, special sessions are devoted to it in battery-related meetings such as the International Meetings on Lithium Batteries (IMLBs) International Symposia on Polymer Electrolytes (ISPEs), and in others including meetings of the Electrochemical Society (ECS), the Battery Symposium in Japan, and the Materials Research Society (MRS).

The alkali-metal/electrolyte interphase was named [1] the 'solid/electrolyte interphase' (SEI). A good SEI must have the following properties:

1) electronic transference numbers $t_e = 0$, that is, it must be an electronic resistor, in order to avoid SEI thickening leading to a high internal resistance, self-discharge, and low faradaic efficiency (ε_f);
2) cation transference number $t_+ = 1$, to eliminate concentration polarization and to ease the lithium-deposition process;
3) high conductivity to reduce overvoltage;
4) in the case of the rechargeable lithium battery, uniform morphology and chemical composition for homogeneous current distribution;
5) good adhesion to the anode;
6) mechanical strength and flexibility.

The early literature (until 1982) is summarized in Refs [1] and [2]. Hundreds of papers have been published since then (most of them in since 1994), and it is impossible to summarize all of them here. The Proceedings of the conferences mentioned above are good sources of recent developments though sometimes incomplete. Since the early 1980s new systems have been introduced. The most important of these are lithium-ion batteries (which have lithiated carbonaceous anodes) and polymer–electrolyte (PE) batteries. Until 1991 very little was published on the Li/PE interface [3, 4]. The application of the SEI model to Li–PE batteries is addressed in Refs [5] and [6].

Film-forming chemical reactions and the chemical composition of the film formed on lithium in nonaqueous aprotic liquid electrolytes are reviewed by Dominey [7]. SEI formation on carbon and graphite anodes in liquid electrolytes has been reviewed by Dahn et al. [8]. In addition to the evolution of new systems, new techniques have recently been adapted to the study of the electrode surface and the chemical and physical properties of the SEI. The most important of these are X-ray photoelectron spectroscopy (XPS), scanning electron microscopy (SEM), X-ray diffraction (XRD), Raman spectroscopy, scanning tunneling microscopy (STM), energy-dispersive X-ray spectroscopy (EDS), Fourier transformation infrared spectroscopy (FTIR), nuclear magnetic resonance spectroscopy (NMR), electron paramagnetic resonance spectroscopy (EPR), calorimetry, differential scanning calorimetry (DSC), thermogravimetric analysis (TGA), use of quartz-crystal microbalance (QCMB), and atomic force microscopy (AFM).

It is now well established that in lithium batteries (including lithium-ion batteries) containing either liquid electrolytes or PEs, the anode is always covered by a passivating layer called the *SEI*. However, the chemical and electrochemical formation reactions and properties of this layer are as yet not well understood. In this section we discuss the electrode surface and SEI characterizations, film formation reactions (chemical and electrochemical), and other phenomena taking place at the lithium or lithium-alloy anode and at the Li_xC_6 anode/electrolyte interface in both liquid electrolyte and PE batteries. We focus on the lithium anode, but the theoretical considerations are common to all alkali-metal anodes. We address also the initial electrochemical formation steps of the SEI, the role of the solvated-electron rate constant in the selection of SEI-building materials (precursors), and the correlation between SEI properties and battery quality and performance.

16.2
SEI Formation, Chemical Composition, and Morphology

16.2.1
SEI Formation Processes

Alkali and alkaline-earth metals have the most negative standard reduction potentials; these potentials are (at least in ammonia, amines, and ethers) more negative than that of the solvated-electron electrode. As a result, alkali metals (M) dissolve in these highly purified solvents [9, 10] following reactions (16.1) and (16.2) to give the well-known blue solutions of solvated electrons.

$$M^0 - e \rightarrow M^{M+}_{solv} \tag{16.1}$$

$$e(M) \rightarrow e^-_{solv} \tag{16.2}$$

These reactions proceed to equilibrium when the potential of the solvated-electron electrode equals that of the alkali metal [11]:

$$E_M^0 \frac{RT}{F} \ln a_{M^+} = E_{e^-} - \frac{RT}{F} \ln a_{e^-} \tag{16.3}$$

where E_x^0 is the standard potential of the X electrode and a_x is the activity of X. In ammonia the difference between E_M^0 and E_C^0 is a few hundred millivolts, so the equilibrium concentration of e_{solv}^- is in the range of several millimoles per liter.

This dissolution process takes place in many solvents to an extent governed by Equation 16.3. Solvated electrons can be formed in all solvents by many means. Their kinetics is best studied with the use of pulse radiolysis.

Practical primary or secondary alkali-metal or alkaline-earth batteries can be made only if the dissolution of the anode by reactions (16.1) and (16.2) (and by other corrosion reactions) can be stopped. Since e_{solv}^- attacks both the electrolyte and the cathode, the electrolyte must be designed to contain at least one material that reacts rapidly with lithium (or with the alkali-metal anode) to form an insoluble SEI. On inert electrodes, the SEI is formed by reduction of the electrolyte. This type of electrode (completely covered by SEI), was named [1, 2] the 'SEI electrode.'

In this paper we shall discuss four types of SEI electrodes:

1) inert metal covered by SEI (e.g., anode current collectors or battery case connected to the anode);
2) freshly cut lithium immersed in the electrolyte or lithium plated (deposited) in the electrolyte;
3) lithium covered with native oxide film as received from the manufacturer;
4) carbonaceous and alloy anodes.

When an inert metal (or any electronic conductor) is negatively polarized in the electrolyte (typically from 3 V versus the lithium reference electrode (LiRE) to 0 V versus LiRE), the following reactions take place at different potentials and at varying rates, depending on E^0, on concentration, and on i_0 for each of the following electrochemically active materials: (i) solvents, (ii) anions of the salts, and (iii) impurities such as H_2O, O_2, HF, CO_2, and so on (Figure 16.1). Some of the products, especially at more positive potentials, may be soluble and diffuse away from the electrode, while others will precipitate on the surface of the electrode to form the SEI. At potentials lower than a few hundred millivolts, solvated electrons will be formed. These will also react with impurities, solvents, and salts (e_{sol}^- scavengers) to produce similar products. The lifetime τ, ($\tau = 0.69\ K_e[S]$; K_e = rate constant; [S] = scavenger concentration) of solvated electrons in battery-related electrolytes is expected to be in the range of 10^{-4}–10^{-10} s (for more details see Section 16.2.3). Schematics of these processes can be seen in Figure 16.5 in Section 16.3.2, which deals with SEI formation on carbonaceous electrodes. When the thickness of the SEI is larger than the tunneling range of electrons, or when the electric field in the SEI is smaller than that for dielectric breakdown (typically at a thickness of 2–5 nm), the electrons cease flowing through the SEI and lithium deposition at the electrode/SEI interface begins [1, 2].

$$C_3O_3H_4 \xrightarrow{e^-} (C_3O_3H_4)^- \xrightarrow[2Li^+]{e^-} Li_2CO_3 + CH_2 = CH_2$$

$$2[(C_3O_3H_4)^-] \longrightarrow {}^-OCO_2CH_2CH_2CH_2CH_2OCO_2 \xrightarrow{2Li^+}$$

$$LiOCO_2CH_2CH_2CH_2CH_2OCO_2Li$$

$$CH_2 = CH_2 \xrightarrow{e^-} H_2\dot{C}-CH_2^- \xrightarrow{CH_2=CH_2} [CH_2-CH_2]_n$$

$$AsF_6^- + 3Li^+ \xrightarrow{2e^-} 3LiF + AsF_3 \xrightarrow[3Li^+]{3e^-} 6LiF + As^0$$

$$\tfrac{1}{2}O_2 + 2Li^+ \xrightarrow{2e^-} Li_2O$$

$$H_2O + Li^+ \xrightarrow{e^-} LiOH + \tfrac{1}{2}H_2$$

$$BF_4^- + 4Li \xrightarrow{3e^-} B^0 + 4LiF$$

$$PF_6^- + H_2O \longrightarrow HF + PF + OH^-$$

$$Li + HF \longrightarrow LiF + \tfrac{1}{2}H_2$$

$$Li_2CO_3 + 2HF \longrightarrow 2LiF + H_2CO_3$$

$$LiOH + HF \longrightarrow LiF + H_2O$$

$$Li_2O + 2HF \longrightarrow 2LiF + H_2O$$

Figure 16.1 Electrolyte decomposition reactions.

The charge needed to complete the formation of the SEI (about 10^{-3} mAh cm^{-2} [8, 12]) increases with the real surface area of the electrode and decreases with increase in the current density and with decrease in the electrode potential (below the SEI potential). In practice, it may take from less than a second to some hours to build an SEI 2–5 nm thick. When lithium is cut while immersed in the electrolyte, the SEI forms almost instantaneously (in less than 1 ms [13, 14]). On continuous plating of lithium through the SEI during battery charge, some electrolyte is consumed in each charge cycle in a break-and-repair process of the SEI [1, 2], and this results in a faradaic efficiency lower than 1. When a battery is made with commercial lithium foil, the foil is covered with a native surface film. The composition of this surface film depends on the environment to which the lithium is exposed. It consists of Li_2O, $LiOH$, Li_2CO_3, Li_3N, and other impurities. When this type of lithium is immersed in the electrolyte, the native surface film may react with the solvent, salts, and impurities to form an SEI, whose composition may differ from that of electrodeposited lithium in the same electrolyte. The formation of SEI on carbonaceous anodes is discussed in Section 16.3.

16.2.2
Chemical Composition and Morphology of the SEI

16.2.2.1 Ether-Based Liquid Electrolytes

16.2.2.1.1 Fresh Lithium Surface Little work has been done on bare lithium metal that is well defined and free of surface film [13–22]. Odziemkowski and Irish [13] showed that for carefully purified $LiAsF_6$ tetrahydrofuran (THF) and 2-methyltetrahydrofuran (2Me-THF) electrolytes the exchange-current density and corrosion potential on the lithium surface immediately after cutting *in situ* are primarily determined by two reactions: anodic dissolution of lithium and cathodic reduction of the AsF_6^- anion by bare lithium metal. The SEI was formed in less than 1 ms. As pointed out by Holding *et al.* [15], the reduction of AsF_6^-, PF_6^-, BF_6^-, and ClO_4^-, is thermodynamically favored. All of the electrolytes investigated [13, 14] were unstable with respect to bare lithium metal. The most unstable system was lithium hexafluoroarsenate in THF and 2Me-THF [14]. Odziemkowski and Irish [13] concluded that in THF, 2Me-THF, and propylene carbonate (PC), electrochemical reduction of the anion AsF_6^- overshadows any possible solvent-reduction reaction. This was also pointed out by Campbell *et al.* [16], who observed that films on lithium surfaces are formed in most cases by electrochemical reduction of the electrolyte anions. One of the main products detected in $LiAsF_6$/THF electrolyte by *in-situ* and *ex-situ* Raman microscopy [13] was a polytetrahydrofuran (PTHF). The polymerization reaction is determined by: (i) the ratio of the concentration of nucleophilic impurities like H_2O, O_2, and so on, to the concentration of the pentavalent Lewis acid AsF_5, and (ii) the ratio of the volume of the electrolyte to the active surface area of the lithium electrode. In contrast to THF, 2Me-THF does not polymerize. However, the brown film formed on lithium was observed in both electrolytes. It was found that the film decomposes to arsenious oxide when exposed to air, moisture, and heat, and may not be polymeric [17]. The surface film formed in imide-based electrolytes was found to be a better electronic conductor and more permeable to an oxidizer. Thus it is suggested that the SEI in Li imide electrolytes is less compact than that in $LiClO_4$, $LiBF_4$, and $LiPF_6$ electrolytes regardless of the type and purity of solvent. It also was noted that the SEI in $LiBF_4$/THF electrolyte was the most water-sensitive [13].

According to the depth profile of lithium passivated in $LiAsF_6$/dimethoxyethane (DME), the SEI has a bilayer structure containing lithium methoxide, LiOH, Li_2O, and LiF [19]. The oxide–hydroxide layer is close to the lithium surface, and there are solvent-reduction species in the outer part of the film. The thickness of the surface film formed on lithium freshly immersed in $LiAsF_6$/DME solutions is of the order of 100 Å.

Kanamura *et al.* [20] thoroughly studied the chemical composition of surface films of lithium deposited on a nickel substrate in γ-butyrolactone (γ-BL) and THF electrolytes containing various salts, such as $LiClO_4$, $LiAsF_6$, $LiBF_4$, and $LiPF_6$. They found, with the use of XPS, that the outer and inner layers of the surface film covering lithium in $LiClO_4$/γ-BL involve LiOH or possibly some Li_2CO_3 and Li_2O

as the main products. Chlorine and oxygen were produced uniformly over the entire region of the surface film. The authors suggested that the hydrocarbon observed in the C 1s spectrum is due not only to a hydrocarbon contaminant but also to organic compounds incorporated in the surface film. The chemical composition and the depth profile of the surface film formed in Li AsF$_6$ + γ − BL and LiBF$_4$ + γ − BL electrolytes are very similar to what is observed in the case of LiClO$_4$-based electrolyte. The only difference is the type of the lithium halide present. LiF is formed by the acid–base reaction between basic lithium compounds (LiOH, Li$_2$CO$_3$, and Li$_2$O) and HF, which is present as a contaminant in the electrolyte or as a product of the reaction of Li with LiAsF$_6$ ions. However, the reaction products of BF$_4^-$ ions were not present in the surface film. The Li 1s spectra of LiPF$_6$ + γ − BL electrolyte were completely different from those observed for the other γ − BL electrolytes [20]. The surface film consists of LiF as the main component of the outer layer. The inner layer is mainly Li$_2$O. All the spectra of electrochemically deposited lithium in LiClO$_4$ + γ − BL containing 5×10^{-3} mol L^{-1} HF were very similar to those of LiPF$_6$ + γ − BL electrolyte. The same results were obtained with other electrolyte salts [19] and PC-based electrolytes (see Section 16.2.2.2).

The XPS data of the surface film on lithium deposited in LiClO$_4$ + THF before etching show that the outer surface film consists of LiOH, Li$_2$CO$_3$, LiCl, and organic compounds [19]. The depth profiles of the surface in THF were similar to those obtained in γ−BL, but the distribution of LiOH and Li$_2$O was slightly different. The surface film obtained in LiClO$_4$ + THF probably has a thicker mixed LiOH + Li$_2$CO$_3$ layer than that obtained in LiClO$_4$ + γ − BL [19].

The chemical species in the surface film of lithium deposited in LiBF$_4$ + THF electrolyte are not very different from those in LiBF$_4$ + γ − BL. However, more LiF and other fluoride compounds are formed and chemical species including elemental boron were also observed in the B 1s spectra. This means that BF$_4^-$ ion reacts quite strongly with lithium during electrochemical deposition in THF electrolyte. According to the depth profile, the surface film comprises mainly a mixture of LiF, LiOH, and Li$_2$CO$_3$ [20]. The carbon content in LiASF$_6$ + γ − BL was greater than that in LiAsF$_6$ + THF.

16.2.2.1.2 Lithium Covered by Native Film

The formation of a native surface film on lithium is unavoidable. It arises from storage and laboratory handling of the lithium metal in the gaseous atmosphere of the dry box and from the immediate reaction taking place after immersion of lithium electrodes in an electrolyte. The outer part of the native film, which was analyzed by XPS [21], consists of Li$_2$CO$_3$ or LiOH, and the inner part is Li$_2$O. X-ray micro-analysis of lithium electrodes, covered by native film and treated in fluorine-containing salts such as LiAsF$_6$, LiBF$_4$, LiPF$_6$, Li imide, and Li triflate dissolved in THF, always shows fluorine, oxygen, and carbon peaks, which are characterized by different relative intensities [18].

Among many polar aprotic solvents, including ethers, BL, PC, and ethylene carbonate (EC), methyl formate (MF) seems to be the most reactive toward lithium. It is reduced to lithium formate as a major product, which precipitates on the lithium surface and passivates it [22]. The presence of trace amounts of the two expected contaminants, water and methanol, in MF solutions does not affect the surface chemistry. CO_2 in MF causes the formation of a passive film containing both lithium formate and lithium carbonate.

16.2.2.2 Carbonate-Based Liquid Electrolyte

16.2.2.2.1 Fresh Lithium Surface The structure and composition of the lithium surface layers in carbonate-based electrolytes have been studied extensively by many investigators [17–35]. High reactivity of PC to the bare lithium metal is expected, since its reduction on an ideal polarizable electrode takes place at much more positive potentials compared with THF and 2Me-THF [16]. Thevenin and Muller [27] found that the surface layer in $LiClO_4$/PC electrolyte is a mixture of solid Li_2CO_3 and a polymer they suggested is P(PO)$LiClO_4$. Aurbach and Zaban [23] have proposed that the lithium surface deposited on a nickel electrode in any of several electrolytes is covered with Li_2CO_3, LiOH, Li_2O, LiOR, $LiOCO_2R$ (R = hydrocarbon), and lithium halide. Kanamura et al. [24] showed that LiF is a final product of the reduction of electrolytes that contain HF as a contaminant. When a small amount of HF is added to PC containing 1.0 mol L^{-1} $LiClO_4$, lithium deposited on a nickel substrate is covered with a thin LiF/Li_2O surface film. Without the addition of HF, the surface of electrodeposited lithium is covered with a thick film (mainly LiOH and Li_2O). Using in-situ XRD, Nazri and Muller [28] identified two peaks on the lithium surface during its cathodic deposition on a nickel substrate in 1.5 mol L^{-1} $LiClO_4$/PC electrolyte. The broad peak at low diffraction angle (20°) is characteristic of polymer compounds; the second sharp peak is attributed to lithium carbonate.

16.2.2.2.2 Lithium Covered by Native Film Aurbach and Zaban [23] used special methods to study the lithium (covered by native film)/electrolyte interphase formed in uncontaminated PC solutions. They found the interphase to be mainly a matrix of Li alkylcarbonate (probably of the type $CH_3CH(OCO_2Li)CH_2OCO_2Li$), containing salt-reduction products and trace Li_2CO_3. Lithium carbonate is formed by the reaction of $ROCO_2Li$ with traces of water unavoidably present in solutions. They also suggested the possibility of Li_2CO_3 formation by further reduction of Li alkylcarbonate species close to the lithium surface. FTIR spectra obtained from lithium electrodes stored in PC/LiBr solutions show the formation of Li alkyl carbonates ($ROCO_2Li$) and Li_2CO_3, which are the major surface species formed on lithium in PC [18]. It was shown that in the presence of LiI, PC was reduced to $ROCO_2Li$ and propylene. In Li imide/PC electrolytes these products were not dominant. On lithium surfaces stored in Li imide and Li triflate, both fluorine- and sulfur-containing compounds are present. The surface concentration of sulfur in Li-triflate-based solutions was higher than in Li-imide/PC electrolytes. When the solution contains CO_2, the main surface species is lithium carbonate. In PC-based electrolytes, $LiPF_6$, $LiBF_4$, $LiSO_3CF_3$, and $LiN(SO_2CF_3)$ were found to be more reactive toward lithium than were $LiClO_4$ and $LiAsF_6$ [18]. Aurbach and Gofer

[29] investigated the formation of SEI in mixtures of carbonate- and ether-based solvents, such as PC/THF, PC/2Me-THF, DME/PC, EC/THF, EC/dioxolane, and EC/PC, containing $LiClO_4$ or $LiAsF_6$. They found that lithium electrodes treated in $LiAsF_6$/PC/THF solutions are covered by a surface film containing carbon, oxygen, and fluorine. Both PC and THF contribute to the buildup of surface films, but the F and As peaks in pure THF were much higher. This indicates that the addition of reactive PC to the ether decreases salt reduction by competing with it, and the film becomes more organic in nature, containing less LiF. In the case of EC/PC or EC/ether mixtures, the reduction of EC by lithium seems to be the dominant process, followed by the formation of lithium alkylcarbonates (derivatives of ethylene glycol) [29]. It was suggested [32] that in mixed organic solvent systems, the solvent having higher donicity tends to coordinate preferentially with Li^+ ion and consequently to react at the Li-electrode/electrolyte interface. Matsuda and co-workers [33, 34] showed that some cyclic compounds containing heteroatoms and conjugated double bonds, such as 2-methylthiophene (2MeTp), 2-methylfuran (2MeF), and aromatic compounds like benzene are very effective in electrolyte solutions for rechargeable lithium batteries. On the basis of AC-impedance measurements it was estimated that the reaction between lithium and 2MeTp or 2MeF would result in a thick SEI of uniform composition [34]. Tobishima *et al.* [35] showed that the addition of 2Me-THF improves the cycle life of the Li/EC-PC/V_2O_5–P_2O_5 cell. The interaction between the cathode and electrolyte leads to the formation of a film containing vanadium on the lithium anode surface. Mori *et al.* [36], using a QCMB, demonstrated the smooth surface morphology and almost constant thickness of the lithium film in EC/dimethyl carbonate (DMC) solutions in the presence of surfactants like polyethylene glycol dimethyl ether (PEGDME), and a mixture of dimethyl siloxane and propylene oxide (PO). The morphological properties of the surface layers formed on the lithium electrode (covered by native film) in sulfolane-based electrolytes (SFLs) have been investigated [37]. It was found that the surface layers are essentially homogeneous and consist mainly of waxy degradation material in which some long white microcrystals are present.

Matsuda and co-workers [38–40] proposed the addition of some inorganic ions, such as Mg^{2+}, Zn^{2+}, In^{3+}, Ga^{3+}, Al^{3+}, and Sn^{2+}, to PC-based electrolyte in order to improve cycle life. They observed the formation of thin layers of Li/M alloys on the electrode surface during the cathodic deposition of lithium on charge–discharge cycling. The resulting films suppress the dendritic deposition of lithium [39, 40]. The Li/Al layer exhibited low and stable resistance in the electrolyte, but the resistance of the Li/Sn layer was relatively high and unstable.

16.2.2.3 Polymer (PE), Composite Polymer (CPE), and Gelled Electrolytes

It seems clear that in PEs, especially in the gel types, lithium-passivation phenomena are similar to those commonly occurring in liquid electrolytes. The crucial role played by the nature and composition of the PE in controlling electron transfer has been described by several authors [3–6, 41–43]. It is postulated that at least two separate competitive reactions occur simultaneously to form the passive layer. The first is the reaction of lithium with contaminants. The second reaction is that of lithium metal with salt. The passivation film appears to consist mainly of Li_2O and/or LiOH [44]. Chabagno [45] observed that polyethylene oxide (PEO) and LiF will not form a complex. LiF is therefore insoluble and will remain at the interface between the lithium electrode and the PE. In one publication [46] PEGDME (molecular weight MW = 400) was chosen as a model system for the investigation of the process of passivation of lithium in contact with PEs. The authors showed that SEI formation was apparently complete in just 2–3 min. The increase in the SEI resistance (R_{SEI}) over hours and days is apparently due to the relaxation of the initially formed passivation films or to the continuation of the reaction at a much slower rate. Results obtained with PEGDME electrolytes containing different salts showed that the formation of LiF as a result of the reduction of anions like AsF_6^- or $CF_3SO_3^-$ plays a key role in the lithium passivation mechanism [46].

Finely divided ceramic powders, which have a high affinity for water and other impurities, were initially added in order to improve the mechanical and electrical properties of LiX–PEO PEs [47–49]. However, today there is considerable experimental evidence for higher stability of lithium/composite polymer electrolyte (CPE) interfaces as compared with pure PEs [50–53]. On the basis of standard free energies of reactions of lithium with ceramic fillers, such as CaO, MgO, Al_2O_3, and SiO_2, it was concluded [54] that lithium passivation is unlikely to occur when lithium is in contact with either CaO or MgO. However, passivation is possible in the case of Al_2O_3 and SiO_2. It was shown that the interfacial stability can be significantly enhanced by decreasing the ceramic particle size to the scale of nanometers [54, 55]. The mechanism of the processes leading to improved stability is not well understood, and some explanations include scavenging effects and screening of the electrode with the ceramic phase [54].

The morphology of lithium deposits from 1 to 3 mol L^{-1} $LiClO_4$–EC/PC-ethylene oxide (EO)/PO copolymer electrolytes was investigated [56]. It was found that, as the weight ratio of host polymer to liquid electrolyte increased, fewer lithium dendrites were formed, with no dendrites found in electrolytes containing more than 30% w/w host polymer. The authors emphasized that good contact between the polymer and lithium is also of great importance for the suppression of dendrites. Direct *in-situ* observation of lithium dendritic growth in Li imide $P(EO)_{20}$ PE [57] shows that dendrites grow at a rate close to that of anionic drift.

16.2.3
Reactivity of e_{sol}^- with Electrolyte Components – a Tool for the Selection of Electrolyte Materials

As mentioned in Section 16.2.1, solvated electrons may take part in the early stage of SEI formation and during break-and-repair healing processes during lithium plating and stripping. It is most important that the formation and the healing of the SEI, especially on graphite, in the first intercalation step be a very fast reaction, and that the SEI-building materials have extremely low solubility. The best SEI materials seem to be LiF, Li_2CO_3, and Li_2O. They are insoluble in most of the lithium battery organic-based electrolytes, and LiF and Li_2O are thermodynamically stable with respect to lithium metal and are cationic conductors. LiOR (R is an organic residue), Li_2S, LiCl, and LiBr are suitable materials for some electrolytes. Li_2O cannot serve as the sole SEI material; because of its low equivalent volume, it cannot supply sufficient corrosion protection for a metallic lithium anode. In practice, it is mixed with LiF or covered by a second layer of LiF, Li_2CO_3, semicarbonates, polyolefins, or mixtures of these. The electrolyte must be designed to contain one or more SEI precursors having high E_0 and high exchange-current density (i_0) for reduction (electron scavengers such as CO_2, EC, and $LiAsF_6$) where the reduction products are appropriate SEI-building materials [58]. However, the data bank of i_0 for such reactions is limited. The determination of practical i_0 values for the electrolyte components (solvents, salts, and impurities) is a difficult task for two reasons: (i) the electrodes must be solid – mercury cannot be used and (ii) the electrode is passivated during the cathodic sweep. It is therefore suggested that the data bank be used for the bimolecular rate constant (k_e) for the reaction:

$$e_{aq}^- + S \rightarrow \text{product} \tag{16.4}$$

where e_{aq}^- is a hydrated electron and S is an electron scavenger and a candidate material for a lithium-battery electrolyte. This data bank has information on more than 1500 materials [59, 60]. The reactivity of materials toward e_{aq}^- (in aqueous solutions) is expected to be quite close to that for e_{sol}^- in organic solutions, at least in most simple cases. For example, the rate constants measured for the reaction of solvated electrons in methanol, ethanol, and water with H^+, O_2, $C_6H_5CH_2Cl$, and $C_{10}H_8$ are about the same, that is, no solvent effect is observed [61]. Another factor which must be taken into account is that, in some cases, the heterogeneous reaction at the electrode and the homogeneous reaction in the solution may yield different products.

According to the Marcus theory [62] for outer-sphere reactions, there is good correlation between the heterogeneous (electrode) and homogeneous (solution) rate constants. This is the theoretical basis for the proposed use of hydrated-electron rate constants (k_e) as a criterion for the reactivity of an electrolyte component toward lithium or any electrode at lithium potential. Table 16.1 shows rate-constant values for selected materials that are relevant to SEI formation and to lithium batteries. Although many important materials are missing (such as PC, EC, diethyl carbonate (DEC), $LiPF_6$, etc.), much can be learned from a careful study of this table (and its sources).

The first factor to take into account is that rate constants higher than 10^{10} L mol^{-1} s^{-1} relate to diffusion-controlled reactions, that is, reactions which have almost zero energy of activation and are therefore controlled by the diffusion of the reactants. These reactions are expected to proceed very quickly at the lithium electrode potential. Therefore SEI precursors should be chosen from this group, or at least from the group having rate constants higher than 10^9 L mol^{-1} s^{-1}. On the other hand, 'inert' electrolyte components, which are chosen because of their very slow reaction with lithium (or with the Li_xC_6 anode) must be taken from the group that has the smallest rate constant – preferably smaller than 10^7 (or even 10^5 L mol^{-1} s^{-1}, for example, ethers). The reactions of strong oxidants

Table 16.1 A compilation of specific bimolecular rate constants (k_e) for the reaction of hydrated electrons with Li battery related materials [59, 60].

Reactant	pH	Rate constant, k_e at 15–25 °C (L mol^{-1} s^{-1})
AsF_6^-	7	9×10^9
BF_4	5.8	$<2.3 \times 10^5$
CO_2	7	7.7×10^9
CO_3^{2-}	–	$<3.9 \times 10^6$
ClO_4	–	$<10^6$
CrO_4^2	13	5.4×10^{10}
H^+	–	2.5×10^{10}
HF	–	6×10^7
MnO_4	7	4.4×10^{10}
NO_3^-	–	8.2×10^9
O_2	7	1.9×10^9
$H_2PO_4^-$	6.7	7.7×10^6
SO_4^{2-}	7	$<10^6$
$S_2O_8^{2-}$	7	1×10^{10}
SF_6	7	1.6×10^{10}
Acetone	7	6×10^9
Acetonitrile	–	4×10^7
Acetophenone	–	2.4×10^{10}
Acrylamide	–	2.2×10^{10}
Acrylate ion	9.2	5.3×10^9
Acrylonitrile	–	1.3×10^{10}
Carbon disulfide	–	3.1×10^{10}
Cyclopentanone	–	7.4×10^9
Ethyl acrylate	11	8.7×10^9
Cyclohexene	11	$<10^6$
Diethyl ether	–	$<10^7$
Dimethyl fumarate	9.2	3.3×10^{10}
Dimethyl oxalate	–	2×10^{10}
Ethyl acetate	6.5	5.9×10^7
Furan	8	3×10^6
Propylene glycol carbonate [$CH_2CH(CH_3)OCOO$]	–	7.2×10^7
Naphthalene	11	5.4×10^9
Styrene	12.7	1.1×10^{10}
α,α,α-Trifluorotoluene	11	1.8×10^9

like O_2, CrO_4^{2-}, MnO_4^-, and $S_2O_8^{2-}$ are diffusion-controlled. AsF_6^- and CO_2, which are good SEI precursors [13, 14, 22, 23] have values of k_e that approach those for diffusion-controlled reactions. There is good correlation between the SEI composition and the reactivity of electrolyte components toward e_{aq}^-. For example, LiF and As–F–O species are found in the SEI formed in electrolytes containing LiAsF$_6$ [7, 20, 24–26]. Also, when CO_2 is added to the electrolyte, more Li_2CO_3

is found in the SEI [18, 22]. EC is so far the best SEI-forming precursor: we attribute this to its high i_0, although this still needs to be proved. On the other hand BF_4^- and ClO_4^- are much less reactive toward e_{aq}^- ($k_e < 10^6$) and LiCl and B^0 are rarely found in the SEI in γ-BL solutions [18–20]. However, in ether-based solutions containing $LiBF_4^-$, B^0 was found in the SEI [20, 21]. In some cases, BF_4^- electrolytes contain some HF (or hydrolyze to yield HF) which reacts with lithium or with Li_2O, LiOH, or Li_2CO_3 and covers the anode with LiF [24]. Esters such as ethyl acetate and semi-carbonates like propylene glycol carbonate have moderate rate constants (10^7–10^8 l mol^{-1} s^{-1}). This suggests that semicarbonate is not stable with respect to lithium and that the part of the SEI which is close to the lithium (or the Li_xC_6) anode cannot consist of semicarbonate as already reported [24–26, 63]. In the solid phase, Li_2CO_3 is thermodynamically unstable with respect to lithium (i.e., ΔG for the reaction: $4Li(s) + Li_2CO_3(s) \rightarrow 3Li_2O(s) + C(S)$ is negative). As a result, the Li_2CO_3 particles at the Li/SEI (or Li_xC_6/SEI) interface are expected to be reduced to Li_2O + C. The equivalent volumes of both LiF and Li_2O are too small (9.84 and 7.43 mL, respectively) to provide adequate corrosion protection for lithium metal (the equivalent volume of lithium is 12.98 mL). Thus a second layer of Li_2CO_3 (equivalent volume, 17.5 mL) or other organic materials is required to cover the first layer in order to provide this protection. XPS measurements [20, 24] show that the part of the SEI which is closer to the lithium or carbon electrode is rich in Li_2O and low in Li_2CO_3. The carbon may further react to form lithium intercalation compounds – Li_xC_6 (see also Section 16.3.5.2).

A problem associated with the use of organic-carbonate-based electrolytes is gas evolution during the formation of the SEI (or during cycling of cells with lithium-metal anodes). A possible way to overcome this problem is to consider the use of reactive e_{sol}^- scavengers not based on organic carbonates, such as SF_6, trifluoro-toluene, and so on, which may form an LiF-based SEI instead of an Li_2CO_3-based SEI. We recently found [58] that the addition to DEC of dimethyl oxalate (DMO), which is an excellent e_{sol}^- scavenger, enables cycling graphite in 1.2 mol L^{-1} $LiAsF_6$/DMO-DEC (1 : 2).

16.3
SEI Formation on Carbonaceous Electrodes

Carbonaceous materials such as various types of soft and hard carbons and graphites are currently the focus of greatest interest in the field of battery technology. Lithium-ion batteries with this type of anode show excellent performance in terms of cycle life, energy density, power density, and charge rate. For this reason, these anodes deserve special attention.

16.3.1
Surface Structure and Chemistry of Carbon and Graphite

The physicochemical properties of carbon are highly dependent on its surface structure and chemical composition [64–66]. The type and content of surface species, particle shape and size, pore-size distribution, BET surface area, and pore-opening are of critical importance in the use of carbons as anode material. These properties have a major influence on Q_{IR}, reversible capacity Q_R, and the rate capability and safety of the battery. The surface chemical composition depends on the raw materials (carbon precursors), the production process, and the history of the carbon. Surface groups containing H, O, S, N, P, halogens, and other elements have been identified on carbon blacks [64, 65]. There is also ash on the surface of carbon and this typically contains Ca, Si, Fe, Al, and V. Ash and acidic oxides enhance the adsorption of the more polar compounds and electrolytes [64].

The basic building block of carbon is a planar sheet of carbon atoms arranged in a honeycomb structure (called *graphene* or *basal plane*). These carbon sheets are stacked in an ordered or disordered manner to form crystallites. Each crystallite has two different edge sites (Figure 16.2): the armchair and zig-zag sites. In graphite and other ordered carbons, these edge sites are actually the crystallite planes, while in disordered soft and hard carbons these sites, as a result of turbostratic disorder, may not form large planes (i.e., the crystallite dimensions parallel and perpendicular to the basal plane, 1_a and 1_c, are in the nanometer range). The reactivity of carbon atoms at the edge sites (and near lattice defects and foreign atoms) is much higher than that of carbon atoms in the basal planes [64–66]. Consequently, the physical and chemical properties of carbon vary with the basal-plane to edge-plane area ratio. The surface area of carbon powders varies over a wide range from less than a few square meters per gram for large-particle graphite powders to more than 1000 m^2 g^{-1} for high-surface-area carbons. As a result, the content of surface groups or heteroatoms, measured as the ratio of foreign atoms to C, varies from nearly zero up to 1 : 5 in the case of hydrogen [64–66].

Carbons may have closed and open pores with a large variety of dimensions from a few angstroms to several microns. In terms of structure, the pores in active carbons are divided into three basic classes [64, 67]: macropores, transitional pores, and micropores. Pores are formed during the production of carbon (pyrolysis of its precursors), or can be formed by other means such as oxidation by O_2, air, CO_2, or H_2O [64]. According to Dubinin's classification [64, 67], the radius of a macropore is in the range 500–20 000 Å, and its surface area can vary from 0.5 to 2 m^2 g^{-1}.

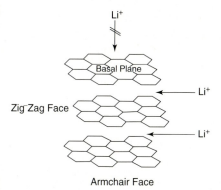

Figure 16.2 The faces of a carbon crystallite.

Transitional pores have an effective radius between 16 and 2000 Å and they can contribute a BET surface area of 20–70 m² g^{-1}. The effective radius of micro-pores is less than 20 Å and can constitute 95% of the total specific BET area. The pore structure has the following pattern: the macropores open out to the external surface of the particle, transitional pores branch off from macropores, and micropores, in turn, branch off from the transitional pores.

Surface species can be present at edge sites inside open and closed pores, and not only on the external surface of the carbon particles. Some pores have very narrow openings and are not accessible to battery electrolyte (solvents or salt), but may be permeable to oxygen and water [68, 69]. Other pores are completely closed and not permeable even to helium. The edge atoms in these closed pores are actually radicals and are said to have a 'dangling' bond [66]. These pores are responsible for 'extra' reversible capacity of disordered carbons [68] and oxidized graphite [70–73]. The surface oxygen species are by far the most important group, influencing the physicochemical properties such as wettability, catalysis, and electrical and chemical bonding to other materials [64–66]. A schematic representation of surface oxygen species that are believed [65] to be present on the carbon surface is given in Figure 16.3. The formation processes and properties of these species are reviewed and widely discussed by Kinoshita and Cookson [64, 65].

It is common practice to classify the surface oxides as basic, neutral, and acidic [64, 65]. Acidic surface oxides (carboxylic groups) are formed when the carbon is exposed to oxygen at temperatures between 200 and 500 °C or by the action of aqueous oxidizing solutions. They start to decompose under vacuum (or inert atmosphere) at about 250 °C and in general are completely desorbed at above 800 °C. The basic surface oxides are formed when the carbon surface is freed of all surface compounds by heating in vacuum or inert atmosphere (above 1200 °C) and then exposed to air or O_2 after cooling [64, 65]. The neutral surface oxides are formed by the irreversible adsorption of oxygen on unsaturated sites [65] to form a –C–O–O–C– bond. These oxides are more stable than the acidic surface oxides and begin to decompose to CO_2 only at 500–600 °C [65]. When graphite or carbon powders are vacuum heated, CO_2, CO, and H_2 begin to desorb at 100, 200–300, and 700 °C respectively, and decomposition is complete at 700, 1000, and >1200 °C (sometimes 1600 °C) respectively [65]. Acidic groups can be titrated by alkali hydroxides, barium hydroxide, and carbonates, whereas basic groups are titrated by HCl and other acids [64, 65]. Various spectroscopic techniques have been used to characterize surface species [64, 65]; these include IR, FTIR, XPS, Raman, and EPR.

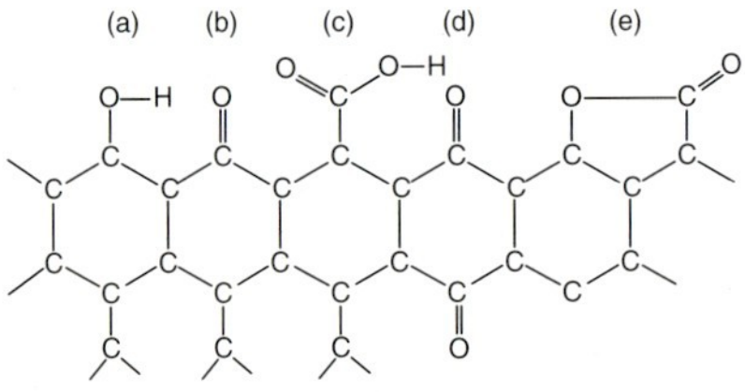

Figure 16.3 Schematic representation of the oxygen functional groups on the carbon surface: (a) phenol; (b) carbonyl; (c) carboxyl; (d) quinone; and (e) lactone [2].

16.3.2
The First Intercalation Step in Carbonaceous Anodes

The first 1.5 charge–discharge cycles of lithium/carbon cells are presented in Figure 16.4 for both graphite (b) and petroleum coke (a) [69]. In both cases, the first intercalation capacity is larger than the first deintercalation capacity.

This difference is the irreversible capacity loss (Q_{IR}). Dahn and co-workers [69] were the first to correlate Q_{IR} with the capacity required for the formation of the SEI. They found that Q_{IR} is proportional to the specific surface area of the carbon electrode and, assuming the formation of an Li_2CO_3 film, calculated an SEI thickness of 45 ± 5 Å on the carbon particles, consistent with the barrier thickness needed to prevent electron tunneling [1, 2]. They concluded [69] that when all the available surface area is coated with a film of the decomposition products, further decomposition ceases.

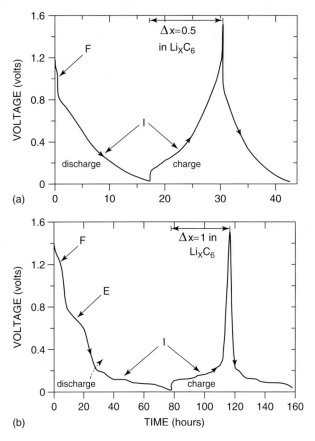

Figure 16.4 (a) The first 1.5 cycles of a lithium/petroleum coke cell and (b) the first 1.5 cycles of a lithium/graphite cell [74].

16 The Anode/Electrolyte Interface

In general, lithium-ion batteries are assembled in the discharged state. That is, the cathode, for example, Li_1CoO_2, is fully intercalated by lithium, while the anode (carbon) is completely empty (not charged by lithium). In the first charge the anode is polarized in the negative direction (electrons are inserted into the carbon), and lithium cations leave the cathode, enter the solution, and are inserted into the carbon anode. This first charge process is very complex. On the basis of many reports it is presented schematically [6, 72, 75] in Figure 16.5. The reactions presented in Figure 16.5 are also discussed in Sections 16.2.1, 16.2.2, and 16.3.5.

At the electrode surface there is competition among many reduction reactions, the rates of which depend on i_0 and overpotential η for each process. Both i_0 and η depend on the concentration of the electro-active materials (and on the catalytic properties of the carbon surface). However, the chemical composition of the SEI is also influenced by the solubility of the reduction products. As a result, the voltage at which the SEI is formed (V_{SEI}) depends on the type of carbon, the catalytic properties of its surface (ash content, type of crystallographic plane, basal to edge planes ratio), temperature, concentrations, and types of solvents, salts and impurities, and current density. For lithium-ion battery electrolytes, V_{SEI} is typically in the range 1.7–0.5 V (Table 16.2) vs LiRE, but it continues to form down to 0 V. In some cases, ε_F is less than 100% in the first few cycles [76]. This means that the completion of SEI formation may take several charge–discharge cycles. Table 16.2

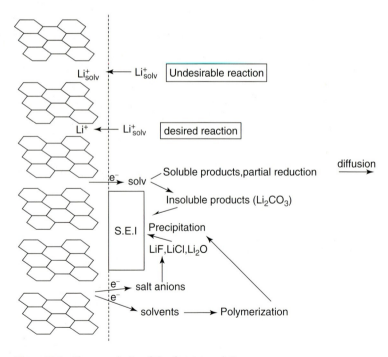

Figure 16.5 The complexity of the first intercalation process into graphite (after Refs [6, 23]).

Table 16.2 Correlation between k_e and V_{SEI}.

Reference	$V_{SEI}{}^a$ (V)	$k_e{}^b$ (l mol^{-1} s^{-1})	Material	Carbon type	Electrolyte
[77]	0.8	$<10^6$	$ClO_4{}^-$	Graphite fibers	$LiClO_4/EC-DEC$
[76]	0.8	$<10^6$	$ClO_4{}^-$	Natural graphite	$LiClO_4/EC-DEC$
[78]	0.8	$<10^6$	$ClO_4{}^-$	Graphite	$LiClO_4/EC-PC$
[79]	1.0	$<10^6$	$ClO_4{}^-$	Graphite	$LiClO_4/EC$
		$10^{10}-10^{9c}$	EC		
[80]	0.65	–	$PF_6{}^-$	Graphite	$LiPF_6/EC-DEC$
[80]	0.7	–	$PF_6{}^-$	Graphite fibers	$LiPF_6/EC-DEC$
[81]	0.6	9×10^9	$AsF_6{}^-$	Graphite	$LiAsF_6/PC$, 12-crown-4
[81]	1.7	9×10^9	$AsF_6{}^-$	Petroleum coke	$LiAsF_6/MF-CO_2$
		7.7×10^9	CO_2		
[82]	1.3	9×10^9	$AsF_6{}^-$	Acetylene black	$LiAsF_6/MF-CO_2$
		7.7×10^9	CO_2		
[82]	1.33	9×10^9	$AsF_6{}^-$	Graphite	$LiAsF_6/MF-CO_2$
		7.7×10^9	CO_2		
[65]	1.2	9×10^9	$AsF_6{}^-$	Graphite	$LiAsF_6/EC-DEC$
Bar-Tow, C. and Menachem, E. Peled, (unpublished results)	1.2	9×10^9	$AsF_6{}^-$	Graphite fibers	$LiAsF_6/EC-DEC$
Bar-Tow, C. and Menachem, E. Peled, (unpublished results)	1.25	9×10^9	$AsF_6{}^-$	Graphite	$LiAsF_6/1,3$-dioxolane–EC
Bar-Tow, C. and Menachem, E. Peled, (unpublished results)	1.2	9×10^9	$AsF_6{}^-$	Graphite	$LiAsF_6/1,3$-dioxolane
Bar-Tow, C. and Menachem, E. Peled, (unpublished results)	1.5	9×10^9	$AsF_6{}^-$	Carbon DB40R	$LiAsF_6/1,3$-dioxolane

(continued overleaf)

Table 16.2 (continued)

Reference	$V_{SEI}{}^a$ (V)	$k_e{}^b$ (l mol^{-1} s^{-1})	Material	Carbon type	Electrolyte
[83]	1.2	9×10^9	$AsF_6{}^-$	Graphite	$LiAsF_6$/EC–PC
[83]	1.1	9×10^9	$AsF_6{}^-$	Petroleum coke	$LiAsF_6$/EC–DEC
[84]	0.75	–	$N(CF_3SO_2)_2{}^-$	Graphite	$LiN(CF_3SO_2)_2$/PC–EC, 12-crown-4
[85]	2.7	10^{10d}	SO_2	Graphite	$LiAsF_6$/MF or PC + SO_2
[86]	2.7	10^{10d}	SO_2	Carbon fiber	LiBr/AN + SO_2

^{a}vs LiRE (or Li/Li$^+$), V peak in dQ/dV curve or inflection point in V vs Q curves.
^{b}From Table 16.1.
^{c}Assumed to be similar to that of other esters (like dimethyl oxalate) in Table 16.1.
^{d}Assumed to equal that of O_2.

shows that V_{SEI} depends on the reactivity of the electrolyte components toward e^-_{aq}; this reactivity correlates with i_0 In the case of reactive components like AsF_6^-, CO_2, and EC, V_{SEI} moves to more positive values, while for more kinetically stable (lower-k_e,) substances like ClO_4^- (and probably PF_6^- and imide), V_{SEI} approaches the Li/Li$^+$ potential, that is, η is higher. It has been reported [71] that if the first intercalation of graphite is not completed (to 0 V vs LiRE) the performance of the graphite anode (as characterized in Li/Li$_x$C$_6$ cells) suffers, that is, x is smaller and there is a higher rate of carbon capacity degradation.

In both graphite and disordered carbons, lithium begins to intercalate in parallel with the formation of the SEI. In the case of disordered carbons, this, in general, does not constitute a problem, as the lithium intercalates as the naked ion, leaving its solvation sheath in the electrolyte. However, in the case of graphite, lithium can intercalate together with some of its solvation molecules, and this leads to the exfoliation of the graphite. Exfoliation may be enhanced in cases where the reduction of the solvated molecules produces gas (Figure 16.1) [69, 78, 83]. In most organic electrolytes this problem is very severe, and the result is complete destruction of the structure of the graphite and almost zero reversible capacity. The only practical electrolyte known today for graphite-anode cells is based on EC [87]. It was reported that graphite can also be cycled with reasonable stability in dioxolane-based electrolytes [81, 82] and in other high-molecular-weight ethers such as 2Me-THF, dimethyl-THF, and 1-methoxybutane [88].

Recently, there have been several very important reports by the groups of Ogumi [89, 90], Farrington [70], Yamaguchi [91], and Besenhard [86], who used STM, AFM, and dilatometry to study the early stages of intercalation into highly oriented pyrolytic graphite (HOPG). Ogumi *et al.* [89, 90], using STM, found that in 1 mol L^{-1} LiClO$_4$/EC–DC, a hill-like structure of about 1 nm height appeared on the HOPG surface (near a step) at 0.95 V vs Li/Li$^+$, and then changed at 0.75 V to an irregular blister-like feature with a maximum height of about 20 nm (see Figure 16.6).

These morphological changes (hill and blister formation) were attributed to the intercalation of solvated lithium ion into graphite interlayers and to the accumulation of its decomposition products (some of them gases), respectively. On the other hand, rapid exfoliation and rupturing of graphite layers were observed in 1 mol L^{-1} LiClO$_4$/PC electrolyte (Figure 16.6), a process which was considered to be responsible for the continuing solvent decomposition when graphite is charged in PC-based electrolytes. This showed that, even in EC-based electrolytes, some degree of solvent co-intercalation exists but does not prevent formation of a stable SEI. It is clear that this phenomenon of co-intercalation must be minimized in practical graphite-anode lithium-ion batteries.

Using dilatometry in parallel with cyclic voltammetry (CV) measurements in 1 mol L^{-1} LiClO$_4$ EC–1,2-DME, Besenhard *et al.* [86] found that over the voltage range of about 0.8–0.3 V (vs Li/Li$^+$), the HOPG crystal expands by up to 150%. Some of this expansion seems to be reversible, as up to 50% contraction due to partial deintercalation of solvated lithium cations was observed on the return step of the CV. It was concluded [86] that film formation occurs via chemical reduction of a solvated graphite intercalation compound (GIC) and that the permselective

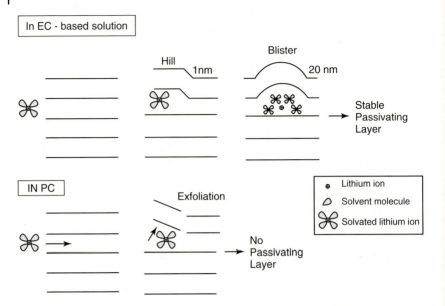

Figure 16.6 Early stages of intercalation into HOPG: (a) in EC-based electrolyte and (b) in PC-based electrolyte [4, 5].

film (SEI) in fact penetrates into the bulk of the HOPG. It is important to repeat the tests conducted by Besenhard *et al.* [86] in other EC-based electrolytes in order to determine the severity of this phenomenon.

Chu *et al.* [70] and Yamaguchi [91] ran AFM measurements on the basal plane of HOPG in 1 mol L^{-1} LiClO$_4$/EC–DMC (1:1) at various potentials versus the Li/Li$^+$ electrode. It was found that at 1.45 V some material begins to form in the vicinity of a step on the basal plane. When the voltage is reduced to 1.17 V, this material expands and becomes about 100 nm thick. Below 0.9 V it covers the basal plane. It seems to be a soft material as the AFM tip drags it. This is in agreement with findings on the formation of polymers as one of the major constituents of the film [13, 28, 75, 92]. It has been found that the addition of crown ethers suppresses the co-intercalation of solvated lithium into the graphite [8, 78, 81]. It was concluded that crown ethers are too large to be intercalated into the graphite layers. The solvated lithium ion in EC or PC is a large cation (not much smaller than the crown ether), containing four to five solvent molecules [93]. It is suggested that the intercalation of Li(solv)$_n^+$ follows a step in which it loses some of its solvated molecules and is adsorbed on the surface of the carbons, losing some of its charge:

$$\text{Li(solv)}_n^{+1} \rightarrow \text{Li(solv)}_{n-m}^{+(1-\partial)} + m\text{(solv)} \tag{16.5}$$

This pattern is similar to that for the formation process of ad-atoms in metal plating [94].

The smaller ion may intercalate faster into the graphite galleries. Reaction (16.5) may be the rate-determining step for the solvent co-intercalation process, and, if so, molecules that form large and stable solvated lithium cations will have a smaller tendency for co-intercalation into the graphite.

In conclusion, it seems that solvents appropriate for lithium-ion batteries employing a graphite anode must have high solvation energy, high E^0, and high i_0 for reduction in order to slow the co-intercalation of the solvated ion, and to enhance the formation of the SEI at the most positive potential (far from the Li/Li^+ potential).

Another phenomenon which has practical importance, especially for large prismatic cells, is gas production during the first charge [95, 96]. There are two sources of gas: (i) the reduction of the electrolyte with the formation of methane or ethane, CO, and hydrogen and (ii) replacement of gases adsorbed on the carbon by SEI-building materials during the formation of the SEI. Gases such as H_2, O_2, CO_2, N_2, and small organic molecules are released from the micropores of the carbon. The second source of gases is very significant if the carbons being used are disordered, but less so if the carbons are graphite-like.

It was concluded [97, 98] that, on long cycling of the lithium-ion battery, the passivating layer on the carbon anode becomes thicker and more resistive, and is responsible, in part, for capacity loss.

16.3.3
Parameters Affecting Q_{IR}

It has been shown [8, 72, 75] that Q_{IR} is consumed mainly in the building of the SEI. However, for a variety of carbons and graphites, Q_{IR} may have other sources [6, 72] (Equation 16.6):

$$Q_{IR} = Q_{SEI} + Q_{SP} + Q_U + Q_T \tag{16.6}$$

where Q_{SEI} is the capacity needed for the formation of the SEI, Q_U is the unused capacity under specified experimental conditions (it is usable at low rates and high potentials), Q_{SP} is the capacity associated with the formation of soluble reduction products [6, 72], and Q_T is the capacity associated with the trapping of lithium inside the structure of the carbon, generally as a result of irreversible reaction with heteroatoms present on the inner surface of closed pores [68].

Q_{IR} depends on the electrolyte type (solvent and salts), the impurity level of the carbon and the electrolyte, the real surface area of the carbon including inner pores which the electrolyte can enter, the surface morphology, and the chemical composition of the carbon. It typically decreases in the order: powders > microbeads > fibers. Impurities such as acids and alcohols, water, or heavy metals may contaminate the SEI, causing side-reactions [1, 2] such as hydrogen evolution and electrolyte reduction; this results in larger Q_{IR} values [99, 100].

Since in many publications the impurity level in the carbons and electrolytes is not specified, it is difficult to correlate Q_{IR} reliably with the type of solvent or salt. However, it seems that in cases where the electrolyte reduction products are Li_2CO_3 and LiF (as in the cases when EC, CO_2, and fluorinated anions are used), Q_{IR} is lowest. We believe that controlled reduction-induced formation of nonconducting polymers (polyolefins) [6, 72] also suppresses Q_{IR}. Polymers have been reported to form on Li_xC_6 and on lithium in several electrolytes [13, 28, 92] (Endo, I., private communication.)

One of the most important factors affecting Q_{SEI} [75, 78, 86] is graphite-anode exfoliation as a result of intercalation of solvated lithium ions. Factors that are reported to decrease Q_{IR} are: increasing the EC content in organic carbonates or dioxolane solutions [101, 102], addition of CO_2 [29, 86, 102] or crown ethers [8, 69, 78], and increasing the current density [71] (this also lowers Q_{SE} [12] as a result of decrease in Q_{SP}).

Q_{SEI} is expected to depend on the morphology of the carbon and should increase with the ratio of cross-section plane area to basal-plane area. This is suggested in view of the recent findings of Besenhard et al. [86], who reported on the penetration of the passivating layer into the graphite galleries through the cross-section planes, and Peled and co-workers [103], who found that the thickness of the SEI at the cross-section planes is greater than that at the basal plane (of an HOPG crystal).

Xing and Dahn recently reported [68] that Q_{IR} for disordered carbon and MCMB 2800 can be markedly reduced from about 180 and 30 mAh g^{-1} to less than 50 and 10 mAh g^{-1} respectively, when the carbon anode and cell assembly are made in an inert atmosphere and never come in contact with air. This indicates that these carbons contain nanopores that are not accessible to the electrolyte but are permeable to O_2, CO_2, and H_2O. The absorption of these gases appears to be the dominant cause of the irreversible loss of capacity [68]. The peaks at about 0.7 and 0.3 V vs Li/Li$^+$ in dQ/dV curves are assigned to electrolyte reduction and reactions with COH and COOH groups respectively.

It is not clear why Q_{IR} is twice as large [6, 75, 100] (or more) in PEs than it is in liquid electrolytes. This may result from larger Q_{SP} and larger Q_{SEI} due to partial exfoliation.

16.3.4
Graphite Modification by Mild Oxidation and Chemically Bonded (CB) SEI

We recently found [6, 71–73, 75] that mild air oxidation (burnoff) of two synthetic graphites and NG7 (natural graphite) improves their performance in Li/Li$_x$C$_6$ cells. The reversible capacity loss of the graphite (Q_R) increased (up to 405 mAh g^{-1} at 4–11% burnoff), its irreversible capacity loss (Q_{IR}) was generally lower, and the degradation rate of the Li$_x$C$_6$ electrode (in three different electrolytes) was much lower. STM images of these modified graphites show nanochannels having openings ranging from a few nanometers up to tens of nanometers (Figure 16.7). It was suggested that these nanochannels are formed at the zig-zag and armchair faces between two adjacent crystallites and in the vicinity of defects and impurities. Performance improvement was attributed to the formation of SEIs chemically bonded (CB) to the surface, carboxylic and oxide groups at the zig-zag and armchair faces (Figure 16.8), better wetting by the electrolyte, and accommodation of extra lithium at the zig-zag, armchair, and other edge sites and nanovoids. This graphite modification, following mild burnoff, was found to make the Li$_x$C$_6$ electrode

16.3 SEI Formation on Carbonaceous Electrodes | 501

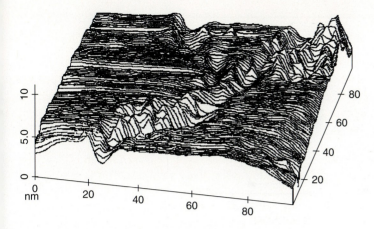

Figure 16.7 An STM image of an oxidized R-LibaD Lonza graphite (3% burnoff, 500 °C, 7.5 h) [23].

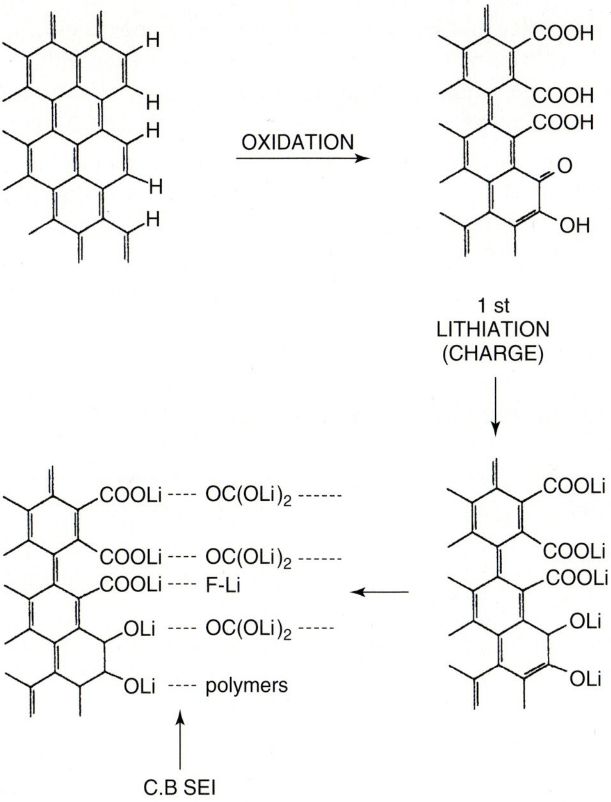

Figure 16.8 The formation of a chemically bonded SEI at the zig-zag and armchair faces (schematic presentation of an organic carbonate-based electrolyte) [23].

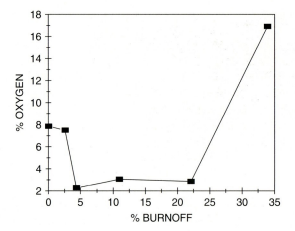

Figure 16.9 Effect of burnoff on surface oxygen content; natural graphite NG7 [38].

performance more reproducible and less sensitive to electrolyte impurities. The increase in capacity at 4–11% burnoff of NG7 was found to be associated with the formation of less than 1% void volume. XPS studies showed [73] that the surface atomic oxygen concentration of NG7 has a broad minimum at 4–22% burnoff (Figure 16.9).

The oxygen peak maximum shifted monotonically with burnoff time, rising from 531.05 eV for pristine NG7 to 534.0 eV for a 34% burnoff sample. The analysis of XPS spectra by curve fitting is presented in Table 16.3.

Table 16.3 Binding energies and fitting parameters of the C 1s curves for the pristine and 34% burnt samples [73].

Peak no.	Sample	Peak position (eV)	Peak shift (eV)	FWHM (eV)	Percentage of total	Peak assignment
1	Pristine	281.48	2.8	1.5	5.2	Hydrocarbons
	34%	281.5	2.6	1.6	4.1	
2	Pristine	283.0	1.3	2.1	15.4	Hydrocarbons
	34%	283.0	1.1	1.8	7	
3	Pristine	0	0.68	53	5.2	Aromatic carbon, C–C
	34%	0	0.63	13.4	4.1	
4	Pristine	285.2	0.9	2.1	19.5	C–OH
	34%	284.9	0.8	1.9	26.6	
5	Pristine	286.9	2.6	2.1	4.8	C=O
	34%	286.7	2.6	1.9	33.1	
6	Pristine	–	–	–	–	O=C–OH
	34%	288.2	4.1	1.9	8.9	
7	Pristine	290.2	5.9	2.2	2.1	Shake-up
	34%	290.1	6.0	2.2	6.8	Satellite

It can be seen that pristine NG7 surface contains mostly (53%) aromatic carbon, about 20% each of CH and COH groups – only 4.8% of CO groups, and no COOH groups. The 34% burnt sample consists of mostly CO groups (33%), C–OH groups (26.6%), and 8.9% of COOH groups.

It was recently found [104] that chemical oxidation of graphite powder by strong oxidizing agents such as ammonium per-oxysulfate and hot concentrated nitric acid gave similar results, that is, it suppresses Q_{IR} and enhances Q_R to 410–430 mAh g^{-1}. Following this wet oxidation, carboxyl groups were identified on the surface of the graphite. Takamura *et al.* found that heat treatment at 700 °C in the presence of acetylene black improved the performance of the graphite-fiber anode [105]. HOPG was used as a model electrode for studying separately the oxidation processes taking place on the basal and on the edge planes [103]. The mechanism of oxidation of the basal plane and that of the cross-section are entirely different (Figure 16.10). Oxygen content on the cross-section rises with oxidation, while that on the basal plane drops from about 10 to 1%. This may correlate with the decrease in the ratio of edge planes to basal planes due to selective burning of the edge planes.

16.3.5
Chemical Composition and Morphology of the SEI

16.3.5.1 Carbons and Graphites

The chemical composition of the SEI formed on carbonaceous anodes is, in general, similar to that formed on metallic lithium or inert electrodes. However, some differences are expected as a result of the variety of chemical compositions and morphologies of carbon surfaces, each of which can affect the i_0 value for the various reduction reactions differently. Another factor, when dealing with graphite, is solvent co-intercalation. Assuming Li_2CO_3 to be a major SEI building material, the thickness of the SEI was estimated to be about 45 Å [69].

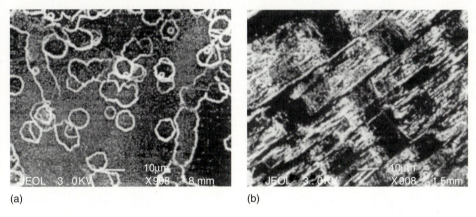

(a)   (b)

Figure 16.10 SEM micrographs of HOPG after burning at 6.50 °C; magnification, ×900. (a) Basal plane 2% burnoff and (b) cross-section 4% burnoff [106].

The anodic behavior of carbon materials, such as acetylene black, activated carbon, and vapor-grown carbon fiber (VGCF), in LiClO$_4$/PC solution was studied by Yamamoto et al. [107]. Irreversible reactions, including gas evolution and disintegration, were mainly observed on that part of the surface occupied by the edge planes of the graphite. XRD measurements indicated that these reactions were the decomposition of the electrolyte, leading to the formation of Li$_2$CO$_3$. The surface reactions on the carbon-fiber electrodes in LiX/PC–DME solution have been investigated by XRD, XPS, and differential scanning calorimetry–thermogravimetric analysis (DSC–TGA) methods [108]. According to this work, the reaction during the first charge, which includes solvent co-intercalation, can proceed by more than two mechanisms, with different kinetics. The mechanism was found to depend on the electrolyte composition and discharge-current density. Aurbach and co-workers [81, 82] carried out an extensive electrochemical and spectroscopic study of carbon electrodes in lithium-battery systems. The carbons investigated included carbon black, graphite, and carbon fibers. The solvents MF, PC, EC, THF, DME, 1,3-dioxolane, and their mixtures were used. The salts tested were LiClO$_4$, LiAsF$_6$, and LiBF$_4$. It was found that the first charging of carbon with lithium is accompanied by irreversible solvent and salt reduction, and this is followed by coating of the carbon surface with passivating films. These films are similar in their chemical structure to those formed on lithium in the same solutions. Thus, PC is reduced on carbon to ROCO$_2$Li, ethers are reduced to alkoxides, and MF to lithium formate. LiAsF$_6$ is reduced to LiF and AsF$_3$, and further to insoluble Li$_x$AsF$_y$ (Figure 16.1). IR spectra of graphite-EPDM electrodes cycled in LiClO$_4$–MF solution seem to prove the existence of LiClO$_3$, LiClO$_2$, or LiClO. CO$_2$ reacts with Li$_x$C$_6$ to form Li$_2$CO$_2$ (and probably CO). Because of the high surface area of graphite particles as compared with the lithium-metal electrode, the role of contaminants, such as HF in LiPF$_6$-and LiBF$_4$-based electrolytes, is much less pronounced [109]. Disordered or graphitized carbons with turbostratic structure were shown to be less sensitive to the solution composition. Aurbach and co-workers [81, 82] emphasized that the most important aspect of the optimization of lithium-ion batteries is the modification of the surface chemistry of carbon by the proper electrolyte additives (e.g., CO$_2$, crown ethers) which form better passivating layers and/or prevent solvent intercalation. The beneficial effect of inorganic additives, such as CO$_2$, N$_2$O, S$_x^{2-}$, and so on, on the formation of SEI on carbons was also emphasized by Besenhard et al. [110]. Tibbets et al. [111] showed that oxidative pretreatment of VGCFs can reduce the capacity of SEI building in LiClO$_4$/PC electrolyte by an order of magnitude. Their experiments confirm the idea that air etching removes the more active carbon atoms – those capable of decomposing the electrolyte – and completely alters the fiber morphology.

Ein-Eli et al. [112] showed that the use of SO$_2$ as an additive to LiAsF$_6$/MF or LiAsF$_6$/PC–DEC–DMC solutions offers the advantage of forming fully developed passive films on graphite at a potential much higher (2.7 V vs Li/Li$^+$) than that of electrolyte reduction (<2 V vs Li/Li$^+$) or of lithium intercalation (0.3–0 V vs Li/Li$^+$). They claimed that the major surface species are organic lithium alkylcarbonates (ROCO$_2$Li) and inorganic lithium salts (Li$_x$AsF$_y$, Li$_2$CO$_3$, Li$_2$SO$_2$O$_4$,

Li_2SO_3, $Li_2S_2O_5$, and Li_2S). The predominating surface reactions in $LiAsF_6/MF-EC$ electrolytes contaminated with water [85] result in the formation of insoluble lithium alkylcarbonates, Li_2O, and LiOH (Figure 16.1). EDX analysis of the surface film formed on mesocarbon-derived carbon fibers in $LiBF_4/EC-PC-DME$ solution [113] indicated that the film is composed of C, F, B, and O. SEM measurements showed that the lithiated carbon fibers cohere as a result of the formation of a passivating film [113]. An unstable passivating layer on petroleum coke in Li triflate/EC–PC–DMC, followed by interaction between the electrolyte and the intercalated lithium , was observed by Jean et al. [114]. The increased stability of lithium-carbon electrodes in EC-containing electrolytes [86] was related to inorganic films formed via secondary chemical decomposition of electrochemically formed EC–graphite-intercalation compounds. Using CV, Inaba et al. [89] found that, for graphite electrodes, an EC–DEC solvent mixture is preferred over EC–DME with respect to the formation of a stable passivating film. When graphite electrodes are charged in PC-based solutions, the solvent decomposes at about 1 V, and this makes SEI formation difficult. It was shown [76] that $LiBF_4$ is more reactive than $LiPF_6$ toward an Li_xC_6 anode. A lithium-ion battery based on $LiPF_6/EC$-DEC (7 : 3) electrolyte [115] underwent more than 300 stable charge–discharge cycles. However an increase of cell resistance from 1.5 to 3.5 kΩ cm^2 was observed on cycling, and this was attributed to the decomposition of electrolyte.

Lithium intercalation into graphite was studied by Morita et al. [116], who used XRD and electrochemical QCMB techniques. The XRD pattern of the graphite electrode after cathodic polarization in $LiClO_4/EC-DMC$ solution shows a spectrum that is more complicated than that for an electrode polarized in EC–PC mixture. The diffraction angles observed do not correspond exactly to the values expected from any idealized stage structure of Li_xC_6. Changes in the resonance frequency of the graphite-coated quartz crystal showed that the cathodic intercalation of lithium is accompanied by electrochemical decomposition of the electrolyte The mass change per coulomb over the potential range of 0.0–0.2 V vs Li/Li$^+$ was higher in EC–DMC than in EC–PC, indicating different surface reactions.

Lithium carbonate and hydrocarbon were identified in XPS spectra of graphite electrodes after the first cycle in $LiPF_6/EC-DMC$ electrolyte [109]. Electrochemical QCMB experiments in $LiAsF_6/EC-DEC$ solution [102] clearly indicated the formation of a surface film at about 1.5 V vs (Li/Li$^+$). However, the values of mass accumulation per mole of electrons transferred (m.p.e), calculated for the surface species, were smaller than those of the expected surface compounds (mainly $(CH_2OCO_2Li)_2$). This was attributed to the low stability of the SEI and its partial dissolution.

16.3.5.2 HOPG

HOPG was used as a model electrode to study separately the formation of the SEI on the basal and cross-section planes [103]. The cross-section planes of HOPG consist of both zig-zag and armchair planes. Carbon atoms on these two planes are considered to be much more active than carbon atoms on the basal plane

Table 16.4 XPS measurements of HOPG after one cycle in 1.2 mol L^{-1} LiAsF$_6$/EC–DEC electrolyte.

Element	Cross-section (%)	Basal (%)
C	40.27	63.51
O	23.40	17.16
F	14.46	4.71
Li	19.52	12.10
As	2.35	0.87

[64, 66, 72, 73]. The SEI compositions on these planes are therefore expected to be different.

Table 16.4 presents the elemental composition of SEI on both basal and cross-section planes as measured by XPS its after formation in 1.2 mol L^{-1} LiAsF$_6$/EC : DEC (1 : 2). For light atoms, the XPS detection depth is about 5–10 atomic layers, so Table 16.4 reflects the composition of the SEI which was exposed to the electrolyte. It is clear that the SEI on the cross section is rich in inorganic components (Li, F, O, As), while that on the basal plane is rich in carbon compounds. Detailed analysis of the XPS spectra for C, O, As, F, and Li at different sputtering times, together with atomic accounting, leads [103] to the formation of an estimated depth profile for the SEI materials (Figure 16.11). As the vacuum in the XPS instrument is about 10^{-10} Torr, it was concluded that the hydrocarbons found in the SEI are actually polymers (polyolefins). We believe that the hydrocarbons reported by Takehara to be present in the SEI lithium [24–26] are also polymers (for the same reason). This conclusion is in agreement with a number of authors [6, 13, 28, 72, 92] who reported on the formation of polymers on the surface of lithium or lithium amalgam. It was concluded [103] that at the basal plane, reduction of the solvents is more pronounced, and this causes the surface of the SEI to contain about 60% polymers, while at the cross-section the SEI is formed by the reduction products of the salt anion (AsF$_6^-$). The thickness of the SEI on the cross-section planes is larger than that on the basal plane. This is in agreement with Besenhard's [86] conclusion that the film penetrates into the graphite galleries as a result of co-intercalation of solvent molecules. Additional support for this is our finding of much carbonate material on the cross-section plane (after 8 min of sputtering), but none on the basal plane (Figure 16.11). The bulk of the SEI on both planes consists mainly of LiF (60–80%) and carbonates (more carbonates on the cross-section planes) (Figure 16.11) [103]. The SEI composition is in agreement with thermodynamics: at the carbon/SEI interphase we found fully reduced anions (Figure 16.11) [103], such as Li$_2$O, LiF, and As, whereas on the solution side we found partially reduced materials: semicarbonates, polymers, AsO$_x$, and AsF$_3$ (as well as LiF). The cross-section SEI contains very few Li–O–C groups, but on the basal plane they constitute up to 40% of the surface. The cross-section SEI has a very strong As XPS peak at 48 eV which is absent from the basal plane (it may be

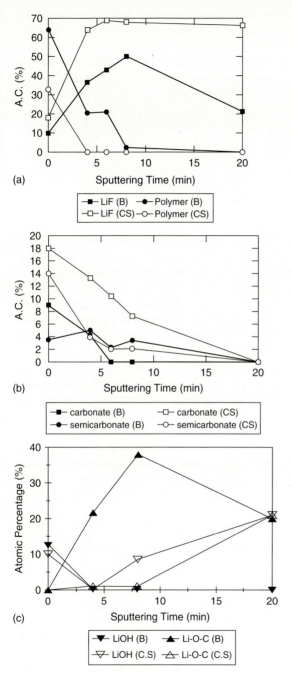

Figure 16.11 Depth profile of estimated SEI composition on the basal (B) and cross section (CS) plane of HOPG, 1.2 mol L^{-1} LiAsF$_6$/EC–DEC (1:2) (Part III Chapter 14).

assigned to some AsO_xF_y compounds). Considering the sputtering depth profiles, it must be noted that the sputtering yield is not the same for all materials; organic and thermally unstable materials may have higher sputtering rates. These findings are in agreement with EDS measurements of the SEI on NG7 formed in the same electrolyte (Bar-Tow, C. and Menachem, E. Peled, unpublished results). The SEI on the basal plane of a single graphite flake consists of less F, As, and O compared with the content of a large sample area (200 μm × 200 μm) which contains both basal and cross-section planes. The SEI on oxidized HOPG seems to be thinner (Bar-Tow, C. and Menachem, E. Peled, unpublished results) than that on pristine HOPG.

16.3.6
SEI Formation on Alloys

The processes taking place in the first intercalation of lithium into an alloy anode in a lithium-ion battery assembled in the discharged state are expected to be very similar to those in a disordered-carbon anode. The intercalation of lithium into the alloy proceeds in parallel with the reduction of the electrolytes and the building of the SEI. However, because of the dependence of i_0 on the catalytic nature of the alloy, the chemical composition and the morphology of the SEI may vary from alloy to alloy. A problem unique to lithium-alloy anodes is the high degree of expansion and contraction during charge–discharge cycles. This may result in shorter cycle life and lower faradaic efficiency as a result of the formation of cracks in both the alloy and the SEI. Therefore, in this case the flexibility of the SEI is highly important. Besenhard et al. [117] recently showed that when thin alloy layers are used, a longer cycle life can be achieved. Yoshio et al. [118] applied for a patent which covers tin oxide-based materials, which on the first intercalation are likely to turn into lithium–tin alloys [119]. It seems that lithium alloys in the form of (probably disordered) small particles may be potential candidates for high-capacity anodes (as a replacement for carbon anodes) in lithium-ion batteries.

16.4
Models for SEI Electrodes

16.4.1
Liquid Electrolytes

In the first papers dealing with SEI electrodes it was suggested that the passivating layer consists of one or two layers [1, 2]. The first one (the SEI) is thin and compact; the second (if it exists), on top of the SEI, is a more porous, or structurally open, layer that suppresses the mass transport of ions in the electrolyte filling the pores of this layer.

According to this model, the SEI is made of ordered or disordered crystals that are thermodynamically stable with respect to lithium. The grain boundaries

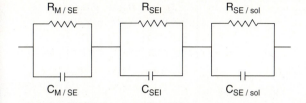

Figure 16.12 Equivalent circuit for the SEI electrode [1, 2].

(parallel to the current lines) of these crystals make a significant contribution to the conduction of ions in the SEI [1, 2]. It was suggested that the equivalent circuit for the SEI consists of three parallel RC circuits in series combination (Figure 16.12). Later, Thevenin and Muller [27] suggested several modifications to the SEI model:

1) the polymer electrolyte interphase (PEI) model, in which the lithium in PC electrolyte is covered with a PET which consists of a mixture of Li_2CO_3, $P(PO)_x$, and $LiClO_4$, where $P(PO)_x$ is polypropylene oxide, formed by reduction-induced polymerization of PC;
2) the solid-polymer–layer (SPL) model, where the surface layer is assumed to consist of solid compounds dispersed in the PE;
3) the compact-stratified-layer (CSL) model, in which the surface layer is assumed to consist of two sub layers; the first layer on the electrode surface is the SEI, while the second layer is either the SEI or the PEI.

The first two models are irrelevant to lithium-battery systems since the PEIs are not thermodynamically stable with respect to lithium. Perchlorate (and other anions but not halides) were found to be reduced to LiCl [13, 14, 20–25]. It is commonly accepted that in lithium batteries the anode is covered by SEI which consists of thermodynamically stable anions (such as O^{2-}, S^{2-}, halides). Recently, Aurbach and Zaban [23] suggested an SEI which consists of five different consecutive layers. They represented this model by a series of five parallel RC circuits representing the capacitance and resistance of each layer. Some of these layers have a thickness of only a few ångstroms, a fact which makes it difficult to assign physical properties such as dielectric constant E, ionic conductivity, energy of activation, and so on. In addition, between each two adjacent layers there is an interface which must be represented by another RC circuit. Thus, a model which consists of three different layers with two interfaces seems to be more appropriate to their AC data.

It is well known today that the SEI on both lithium and carbonaceous electrodes consists of many different materials including LiF, Li_2CO_3, $LiCO_2R$, Li_2O, lithium alkoxides, nonconductive polymers, and more. These materials form simultaneously and precipitate on the electrode as a mosaic of microphases [5, 6]. These phases may, under certain conditions, form separate layers, but in general it is more appropriate to treat them as heteropolymicrophases. We believe that Figure 16.13a is the most accurate representation of the SEI.

The equivalent circuit of a section of this SEI is presented in Figure 16.13b. It was recently found [120, 121] that at temperatures lower than 90 °C, the grain

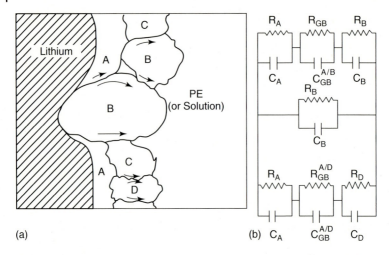

Figure 16.13 Schematic presentation of a small segment of polyheteromicrophase SEI (a) and its equivalent circuit (b): A, native oxide film; B, LiF or LiCl; C, nonconducting polymer; D, Li_2CO_3, or $LiCO_3$ R; GB, grain boundary. R_A, R_B, R_D, ionic resistance of microphase A, B, D. $R_{GB}^{A/B}$, $R_{GB}^{A/D}$ charge-transfer resistances at the grain boundary of A to B or A to D, respectively. C_A, C_B, C_D SEI capacitance for each of the particles A to D. $C_{GB}^{A/B}$, $C_{GB}^{A/D}$, grain boundary (GB) capacitance for A/B and A/D interfaces.

boundary resistance of CPEs and composite solid electrolytes based on $LiI–Al_2O_3$ is many times larger than their ionic resistance. At 30 °C R_{GB} is several orders of magnitude larger than R_B (the ionic resistance), and for 100 μm-thick CPE foils or $LiI–Al_2O_3$ pellets it reaches [122] $10^5–10^6$ Ω cm^2 (depending on CPE composition).

A value of R_{GB} for an SEI 10 nm thick can be estimated from its values for CPE and CSE by assuming that these solid electrolytes consist of nanometer-sized particles. Thus the expected value for R_{GB} at 30 °C for a 10 nm SEI is in the range 10–100 Ω cm^2, that is, it cannot be neglected. In some cases it may be larger than the ionic (bulk) resistance of the SEI. This calculation leads us to the conclusion that R_{GB} and C_{GB} must be included in the equivalent circuits of the SEI, for both metallic lithium and for Li_xC_6 electrodes. The equivalent circuit for a mosaic-type SEI electrode is extremely complex and must be represented by a very large number of series and parallel distributions of parallel RC elements (Figure 16.13b). Since the exact composition, size, and distribution of these particles are generally unknown, we prefer to make the following approximations [5, 6]: the contributions R_B of all particles (in the same sublayer (SL)) are combined in one term – the apparent SEI ionic resistance, R_{SEI} – and the values R_{GB} of all the particles in the same SL are combined into another term – the apparent R_{GB} of the SL [122]. The same is done for C_{SEI} and C_{GB}, and the result is the equivalent circuit shown in Figure 16.14 representing a two-SL SEI [122]. $C_{E/SE}$, $C_{1/2}$, and $C_{SE/sol}$ are the double-layer capacitances between, respectively, the following pairs of phases: electrode/solid electrolyte, SL-1/SL-2, and solid electrolyte/solution (or battery electrolyte). R_{CT1}, $R_{1/2}$, R_{CT2} are the charge-transfer resistances between the pairs of phases mentioned above. W represents the Warburg impedance on the solution side of the SEI electrode.

In many cases, the Nyquist plot for SEI electrodes consists of only one, almost perfect, semicircle whose diameter increases with storage time (and a Warburg section at low frequencies). For these cases the following can be concluded: the SEI consists of only one SL, R_{CT1}, R_{GB}, and $R_{CT2} \ll R_{SEI}$; C_{GB}, $C_{E/sol}$, and $C_{E/SE} \gg C_{SEI}$. Under these conditions the SEI can be represented by a single RC element-R_{SEI} and C_{SEI} (and the Warburg element). In other cases, aside from the Warburg section, the Nyquist plot can consist of two semicircles [123–125], many semicircles [18, 23, 124], or a shallow arc [126]. For these cases, the equivalent circuit of Figure 16.14 or a similar one should be considered.

16.4.2
Polymer Electrolytes

The major differences between PEs and liquid electrolytes result from the physical stiffness of the PE. PEs are either hard-to-soft solids, or a combination of solid and molten in phases equilibrium. As a result, wetting and contact problems are to be expected at the Li/PE interface. In addition, the replacement of the native oxide layer covering the lithium, under the OCV conditions, by a newly formed SEI is expected to be a slow process. The SEI is necessary in PE systems in order to prevent the entry of solvated electrons to the electrolyte and to minimize the direct reaction between the lithium anode and the electrolyte. SEI-free Li/PE batteries are not practical. The SEI cannot be a pure polymer, but must consist of thermodynamically stable inorganic reduction products of PE and its impurities.

Figure 16.15 [6] schematically represents the Li/PE interphase. Solid PEs have a rough surface, so when they are in contact with lithium, some spikes, like '2' in Figure 16.15, penetrate the oxide layer and the lithium metal, and a fresh SEI is formed at the Li/PE interface. In other parts of the interface, softer contacts between the PE and lithium are formed ('1' and '3' in Figure 16.15). Here the fresh SEI forms on the native oxide layer or, as a result of the retreat of lithium during its corrosion, the native oxide layer breaks and the gap is filled by a fresh SEI ('1' in Figure 16.15). The net result is that only a fraction (θ) of the lithium surface is in intimate contact with the PE. The situation in composite solid electrolytes (CPEs) is more severe because of their greater stiffness. This complex morphology of the Li/PE and Li/CPE interfaces causes difficulties in measuring SEI and PE

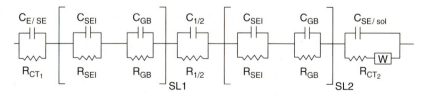

Figure 16.14 Equivalent circuit for two sublayer poly-heteromicrophase SEI (for notation, see text) [122].

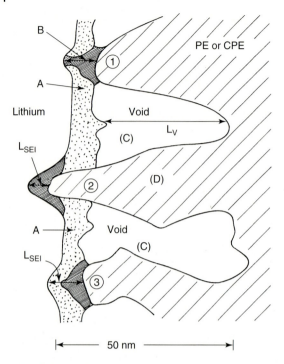

Figure 16.15 Schematic presentation of the Li/Pe interphase [5, 6]; A, native oxide film; B, freshly formed SEI; C, void; and D, PE (solid).

properties such as conductivity and energy of activation for conduction, and also in the interpretation of the results. As θ changes with temperature, stack pressure, the preparation process for the PE, and the morphology of the native lithium surface film, there is often disagreement between measurements taken in different laboratories for the same PE.

For simplicity we assume here that $R_{GB} = 0$ and there is only one SL in the SEI. In this case, the SEI resistance (R_{SEI}) is given by $R_{SEI} = R_{SEI}^0/\theta$ where R_{SEI}^0 is the value for $\theta = 1$. The SEI capacitance (C_{SEI}) is given, to a first approximation, by $C_{SEI} = \theta C_{SEI}^0$ where C_{SEI}^0 is the capacitance for $\theta = 1$.

For $\theta = 1$ the apparent thickness of the SEI (L_{SEI}^0) can be calculated from Equation 16.7,

$$L_{SEI}^0 = \frac{\varepsilon \varepsilon_0 A}{C_{SEI}^0} \tag{16.7}$$

where A is the electrode area, ε is the dielectric constant of the SEI, and ε_0 is the dielectric constant of a vacuum. Substituting the experimental C_{SEI} values gives $L_{SEI} = L_{SEI}^0/\theta$. However, it is difficult to determine both θ and L_{SEI}^0 simultaneously.

At temperatures above or near the eutectic temperature of the polymer phase, C_{SEI} values are typically in the range of 0.1–$2\,\mu F\,cm^{-2}$ [5]. However, for stiff CPEs or below this temperature, C_{SEI} can be as low as $0.001\,\mu F\,cm^{-2}$ (Figure 16.16).

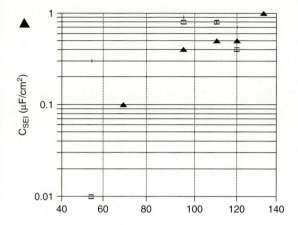

Figure 16.16 Temperature dependence of C_{SEI} in Li/CPE cell [5]. LiI/P(EO)$_n$–9% Al$_2$O$_3$: 1, (+) $n = 9$,; 2, (□) $n = 2.5$; 3, (▲) $n = 6$.

When a CPE is cooled from 100 to 50 °C, the C_{SEI} falls by a factor of 2–3, and on reheating to 100 °C it returns to its previous value. This is an indication of void formation at the Li/CPE interface. As a result, the apparent energy of activation for ionic conduction in the SEI cannot be calculated from Arrhenius plots of $1/R_{SEI}$ but rather from Arrhenius plots of σ_{SEI} (the apparent conductivity of the SEI), which, to a first approximation, does not depend on θ. This approximation is based on the assumption that the heights of the voids (L_v in Figure 16.15) are much greater than the thickness of the SEI:

$$\sigma_{SEI} = (L/R)_{SEI} = ((\varepsilon_0 \varepsilon A / RC^0 \theta))_{SEI} = \varepsilon_0 \varepsilon A / C^0_{SEI} R^0_{SEI} \tag{16.8}$$

Similar contact problems are expected at the PE/carbon interface, where the wetting of the carbon by the PE is of crucial importance [124, 127].

16.4.3
Effect of Electrolyte Composition on SEI Properties

16.4.3.1 Lithium Electrode
The properties of SEI electrodes, the growth rate of the SEI, the mechanism of dissolution and deposition, and the effects of various factors on SEI conductivity have been addressed elsewhere [1, 2]; space limitations do not permit their repetition here.

A comparison of the SEI properties on bare lithium in four electrolytes, LiClO$_4$/PC, LiClO$_4$/PC–DME, LiAsF$_6$/EC–2Me-THF, and LiAsF$_6$/THF–2Me-THF, was made by Montesperelli et al. [30] using impedance spectroscopy. The resistance of the passivation film in LiAsF$_6$-based solutions was found to be twice as large as in LiClO$_4$ electrolyte after 10 days of storage. High values of R_{film} (∼45Ω cm^2) in THF-containing electrolyte were explained by the high reactivity of this solvent toward lithium, followed by the formation of a thick (∼220 Å) surface film. The impedance spectra of the lithium electrode in LiPF$_6$ electrolyte were analyzed by Takami and Ohsaki [123]. The two semicircles in the impedance spectra have been interpreted as resulting from two kinds of passivating film. The formation of the first film is considered to be closely related to decomposition of the PF$_6^-$ anion, which should lead to a thick outer passivation film, probably consisting of LiF [20]. It was found that the most effective solvent for decreasing the film resistance is an EC–2Me-THF mixed solvent. This may result from enrichment of the SEI with Li$_2$CO$_3$. It was found [20] that the SEI formed in LiPF$_6$/γ-BL electrolyte is much thinner than those formed in LiAsF$_6$, LiClO$_4$, and LiBF$_4$/γ-BL-based electrolytes. The SEI thickness was found to be less than a few tens of ångstroms in LiPF$_6$/γ-BL, while for other electrolytes it exceeds 200 Å. Moreover, the film formed in the LiPF$_6$-containing electrolyte was very uniform and sufficiently compact. The thickness of the lithium surface layer in a lithium perchlorate/PC solution, calculated from the apparent resistance according to the CSL interface model, was found to increase exponentially with storage time from 100 to 1000 Å [27]. The values are in good agreement with those deduced from ellipsometric measurements [31].

The order of the interfacial resistance of the SEI on lithium covered by native film in 1 mol L^{-1} LiX/PC solutions was determined by Aurbach and Zaban [18, 23] from Nyquist plots. For the different salts, the order of R_{SEI} was: LiPF$_6$ $\gg$ LiBF$_4$ > LiSO$_3$CF$_3$ $\gg$ LiAsF$_6$ > LiN(SO$_2$CF$_3$)$_2$ > LiBr, LiClO$_4$ [18]. The values for LiPF$_6$/PC and LiN(SO$_2$CF$_3$)$_2$/PC were about 800 and 23 Ω cm^2, respectively. The resistivity of the film was found to be directly proportional to the salt concentration, and the presence of CO$_2$ in solutions considerably reduced the interfacial resistance.

In PC-based electrolytes, inorganic ions like Mg^{2+}, Zn^{2+}, In^{3+}, and Ga^{3+} form thin layers of lithium alloys at the electrode surface during cathodic deposition of lithium, and the resulting thin films suppress the dendritic deposition of lithium that causes the lowering of the coulombic efficiencies in the charge–discharge cycles [34, 38–40]. The Li–Sn electrode shows the greatest increase in interfacial resistance with immersion time and has a double-layer capacitance, C_{dl}, between 0.03 and 0.08 mF [38]. The most stable and lowest interfacial resistance (80–100 Ω cm^2) was observed with the Li-3% (w/w) Al alloy electrode. The SEI resistance decreased in the order: no additive > LiI > SnI$_2$ > AlI$_3$ $\cong$ AlI$_3$-2MeF. For systems containing AlI$_3$, in particular, the film resistance was low (5 Ω), almost constant, and independent of the cycle number.

The interfacial phenomena in LiX/PE systems were studied extensively by Scrosati and co-workers [3, 47, 128]. They found that the high-frequency semicircle in the impedance spectrum of LiClO$_4$/P(EO)$_8$ electrolyte (EO), which is attributed to the interfacial resistance, is often irregularly shaped and seems to contain an additional arc. The authors suggested that this impedance response is based on more than one relaxation phenomenon. The resistance of the passivation film was found to increase continuously upon storage, reaching a value 3 orders of magnitude higher than the initial resistance (10^5 Ω). In some cases, film growth leads to the blocking of lithium ion transport and to the almost complete inactivity of the polymer cell. The progressive decay of capacitance from 0.65 to 0.5 µF in the initial stage of film evolution during 100 h of storage was associated with the increase of film thickness. Hiratani et al. [129] postulated that the interfacial-impedance semicircle in the lithium/solid electrolyte system corresponds to two main processes: the ionic conduction of the interphase film and the charge-transfer process ($Li^+ + e \rightarrow Li$). The increase of interfacial resistance throughout the temperature cycle (heating–cooling) was time-independent and explained by a reduction in the contact area. This is in good agreement with other results [5, 6]. A study was made of the passivation of metallic lithium in contact with PEO-based PEs as a function of the nature and concentration of the salt ($LiClO_4$, $LiCF_3SO_3$, $LiAsF_6$, LiI), time, temperature, and current density [44, 45]. The $LiCF_3SO_3$ $P(EO)_n$-based electrolytes, with $\infty > n > 4$, show similar behavior, where the increase in R_{SEI} is proportional to the growth of the film with time. However, at low salt concentrations for single-phase PE composition, the apparent activation energy of ionic conduction in SEI ($E_{A,SEI}$) did not exceed 0.65 eV, while at high salt concentrations ($n < 20$) for the two-phase electrolyte, $E_{A,SEI}$ rose to 0.78 eV. Sloop and Lerner [130] showed that SEI formation can be affected by treatment of the cross-linked polymer, poly–[oxymethylene oligo(oxyethylene)] (PEM) with an alkylating agent. Cross-linked films of PEM do not form a stable interface with lithium; however, upon treatment with methyl iodide, R_{SEI} stabilizes at 2000 Ω cm^{-1}. Such an SEI is characterized by low conductivity, from 10^{-12} to 10^{-8} Ω^{-1} cm^2, which is linear over the temperature range of 25–85 °C.

CPEs, containing various inorganic fillers, show a trend of impedance behavior quantitatively similar to that of pure LiX–PEO electrolytes. However, the growth rate of the passive film and the capacitance changes were found to be considerably lower in CPEs. Addition of $LiAlO_2$ [128], Al_2O_3, MgO, SiO_2 [47, 50, 51, 131], Li_3N, or zeolite [132] improves SEI stability. Kumar et al. [52] reported suppression of the charge-transfer resistance by a factor of 3 when glass powder (composition: $0.4B_2O_3$, $4Li_2O$, $0.2Li_2SO_4$) was added to the $LiBF_4/P(EO)_n$ electrolyte. It was shown [133] that glass-polymer composite (GPC) electrolytes, prepared by mixing and grinding 87% (v/v), of $0.56Li_2S$, $0.19B_2S_3$, 0.25 LiI glass powder with 13% (v/v) Li imide/$P(EO)_6$, appear to be stable with respect to lithium and Li_xC_6 electrodes, since the interfacial impedances are relatively constant (60 and 30 Ω, respectively) at 70 °C for up to 375 h. The lithium/PE interface is extremely sensitive to the amount of water absorbed in the $LiClO_4$/PEM electrolyte [134].

An extensive study of the fundamental processes taking place in Li/CPE interphases and the properties of the SEI was made by Peled and co-workers [5, 6, 50, 122, 135, 136]. It was found that the use of a thermodynamically stable anion like I^- or Br^- and fine Al_2O_3 or MgO powders resulted in very stable Li/CPE ($n > 3$) and Li/CSE ($n \leq 3$) interphases. The maximum value of R_{SEI} in LiI/P $(EO)_9P(MMA)_{0.5}$ EC_1 9% Al_2O_3 or MgO electrolytes was 8 Ω cm^2 at 120 °C. The apparent thickness (L_{SEI}) of the SEI did not exceed 120 Å, and remained constant or decreased slightly during 1800 h of storage at 120 °C. The SEI conductivity in some cases was stable and in others decreased with storage at elevated temperatures. This was explained by composition changes or recrystallization of the SEI particles to yield a more ordered and less conductive film, as found previously [1, 2] for nonaqueous solutions. In another test, an Li/CPE/Li cell was stored for over 3000 h with almost no change in R_{SEI}. At temperatures above or near the eutectic temperature of the PE, C_{SEI} values were in the range $0.1-2\,\mu\text{F cm}^{-2}$. However, below this temperature or for CSEs, which are stiffer than CPEs, the SEI capacitance could be as low as $0.001\,\mu\text{F cm}^{-2}$. The conductivity of the SEI (σ_{ESI}) was found to be three to four orders of magnitude lower than that of the CPE and did not change much with the salt concentration [5]. The replacement of Al_2O_3 by MgO, or LiI by LiBr, had little effect on σ_{SEI}. However, both changes result in a severe decrease in C_{SEI}, probably due to the stiffness of these CPEs. The apparent energy for ionic conduction in the SEI measured at $130°C > T > 60°C$ was found to be $7-11\,\text{kcal mol}^{-1}$. The addition of copolymers such as poly(butyl acrylate) (PBA) and poly(methyl acrylate) (PMA) and the variation of the EO:Li ratio from 6:1 to 10:1 were found to have some effect on the SEI properties. In general, C_{SEI} increases and R_{SEI} decreases with increasing organic content of the CPE. The stiffer CPE, which contained PBA, had the highest SEI resistance (due to low θ values), and the lowest conductivity.

Figure 16.17 presents the effect of EC and PEGMDE on the Li/CPE interphase. The addition of EC was followed by a decrease by 1 order of magnitude in R_{SEI} and L_{SEI}. Both R_{SEI} and L_{SEI} were stable for over 500 h of storage at 120 °C. Similar behavior was observed in CPEs containing PEGDME with a MW of 500 and 2000. High SEI stability was achieved in these electrolytes for 1200 h of storage. The positive effect of plasticizers may result from better wetting of the lithium metal by the PE, that is, an increase in the contact area. In addition, the formation of a thin and stable SEI composed mainly of lithium carbonate is expected in CPE containing EC.

The interfacial properties of gel electrolytes containing EC immobilized in a polyacrylonitrile (PAN) matrix with a lithium (bis)trifluoromethane sulfonimide (LiTFSI) salt have been studied [137]. SEI stability appeared to be strongly dependent on the LiTFSI concentration. A minimum value of R_{SEI} of about $1000\,\Omega\,\text{cm}^2$ was obtained after 200 h of storage of an electrolyte containing 14% salt. This value was

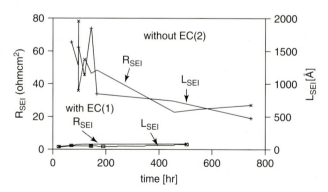

Figure 16.17 Effect of EC on the apparent thickness and stability of SEI [6]. CPE composition: (1) LiI/P(EO)$_9$P(MMA)$_{0.5}$(EC)$_2$ + 6% (v/v)Al$_2$O$_3$. (2) LiI/P(EO)$_9$P(MMA)$_{0.5}$ + 6%(v/v)Al$_2$O$_3$.

doubled after 1000 h of storage; however, for 9 and 18% LiTFSI electrolytes, a sixfold increase of R_{SEI} was observed. The increase of interfacial resistance in poly(methyl methacrylate) (PMMA) gel electrolyte with storage time was described by Osaka et al. [138]. Croce et al. [128] emphasized that the passivation of lithium in LiClO$_4$ PC/EC–PAN electrolytes is very severe and induces the growth at the interface of a layer having a resistance orders of magnitude higher than the bulk resistance of the electrolyte itself. Fan and Fedkiw [53] found that in gel-like composite electrolytes, based on fumed silica, PEGDME (PEO, oligomer), and Li imide or Li triflate, the interfacial stability and conductivity are significantly improved by the addition (10 or 20%) of fumed-silica R805 (Degussa). Nagasubramanian and Boone [139] found that saturated cyclic compounds with functional groups decrease the interfacial impedance of LiPF$_6$–PVDF EC/PC gel electrolyte, especially at low temperature.

16.4.3.2 Li$_x$C$_6$ Electrode

Since this is a new field, little has been published on the Li$_x$C$_6$/electrolyte interface. However, there is much similarity between the SEIs on lithium and on Li$_x$C$_6$ electrodes. The mechanism of formation of the passivation film at the interface between lithiated carbon and a liquid electrolyte or PE was studied by AC impedance [124, 125]. Two semicircles observed in AC-impedance spectra of LiAsF$_6$/EC–2Me-THF electrolytes at 0.8 V vs Li/Li$^+$ [125] were attributed to the formation of a surface film during the first charge cycle. However, in the cases of LiClO$_4$ or LiBF$_4$/EC–PC–DME, only one high-frequency distorted semicircle was found in the impedance spectra [124]. Yazami et al. [124] explained the complicated arc shape by surface-film formation followed by electrode gassing during the decomposition of the electrolyte. This phenomenon is less pronounced in Li triflate, Li imide, and lithium hexafluorophosphate. However, we believe that the depressed high-frequency arc may be due to the overlapping of two, or even more, arcs and may be associated with grain-boundary resistance in the SEI (see Sections 16.4.1 and 16.4.2). In another investigation [125] it was found that the interfacial resistance of graphite electrodes in LiPF$_6$ and LiBF$_4$/EC–DMC solutions is about 1 order of magnitude higher than that of LiAsF$_6$-based electrolytes and increases considerably upon storage. This is explained by different surface chemistry, namely, by the increased resistance of a passive film containing LiF.

Yazami et al. [124, 127] studied the mechanism of electrolyte reduction on the carbon electrode in PEs. Carbonaceous materials, such as cokes from coal pitch and spherical mesophase and synthetic and natural graphites, were used. The change in R_{film} with composition on Li$_x$C$_6$ electrodes was studied for three ranges of x in an Li/POE–LiX/carbon cell [124]. The first step in the lithium intercalation ($0 < x < 0.5$) is characterized by a sharp increase in R_{film} and is attributed to the formation of a bond between lithiated coke and POE. Such intercalated lithium is irreversible in the 1.5–0.5 V range. In the second step, ($\Delta x \approx 1$), lithium intercalates mainly into the coke, and the film does not grow significantly, so a slow increase in R_{film} is observed. In the third step, excess lithium is formed on the surface of the coke, and this induces a further increase in the film thickness and its resistance.

16.5
Summary and Conclusions

The anode/electrolyte interphase (the SEI) plays a key role in lithium-metal, lithium-alloy, and lithium-ion batteries. Close to the lithium side, it consists of fully reduced (thermodynamically stable) anions such as F^-, O^{2-}, S^{2-}, and other elements such as As, B, C (or their lithiated compounds). The equivalent volumes of both LiF and Li_2O are too small (9.84 and 7.43 mL, respectively) to provide adequate corrosion protection for lithium metal. Thus a second layer of Li_2CO_3 (equivalent volume, 17.5 mL) or other organic materials is required to cover the first layer in order to provide this protection. The outer part of the SEI (near the solution) consists of partially reduced materials such as polyolefins, poly-THF, Li_2CO_3, $LiRCO_3$, ROLi, LiOH, and LiF, LiCl, Li_2O, and so on. Often, polymers are the major constituent of the outer part of the SEI. It has been shown that the rate constants of the reactions of solvated electrons with electrolyte and solvent components (and impurities) are a good measure of the stability of these substances toward lithium. Use of the rate constants (k_e) for these reactions is suggested as a tool for the selection of electrolyte components. Good correlation was found between k_e and SEI formation voltage and composition.

The SEI is formed by parallel and competing reduction reactions, and its composition thus depends on i_0, η, and the concentrations of each of the electroactive materials. For carbon anodes, i_0 also depends on the surface properties of the electrode (ash content, surface chemistry, and surface morphology). Thus, SEI composition on the basal plane is different from that on the cross-section planes. Mild oxidation of graphite was found to improve anode performance. Improvement was attributed to the formation of an SEI chemically bonded to the surface carboxylic and oxide groups at the zig-zag and armchair faces, better wetting by the electrolyte, and accommodation of extra lithium at the zig-zag, armchair, and other edge sites and nanovoids. Since the SEI consists of a mosaic of heteropolymicrophases, its equivalent circuit is extremely complex and must be represented by a very large number of series and parallel distributions of RC elements representing bulk ionic conductivity and grain boundary phenomena aside from the Warburg element. In some cases it can be reduced to simpler equivalent circuits.

In lithium-ion batteries, with carbonaceous anodes, Q_{IR} can be lowered by decreasing the true surface area of the carbon, using pure carbon and electrolyte, applying high current density at the beginning of the first charge, and using appropriate electrolyte combinations.

Today we have some understanding of the first lithium intercalation step into carbon and of the processes taking place on the lithium metal anode. A combination of a variety of analytical tools including dilatometry, STM, AFM, XPS, EDS, SEM, XRD, QCMB, FTIR, NMR, EPR, Raman spectroscopy, and DSC is needed in order to understand better the processes occurring at the anode/electrolyte interphase. This understanding is crucial for the development of safer and better lithium-based batteries.

References

1. Peled, E. (1979) *J. Electrochem. Soc.*, **126**, 2047.
2. Peled, E. (1983) in *Lithium Batteries* (ed. J.P. Gabano), Academic Press, p. 43.
3. Scrosati, B. (1987) in *Polymer Electrolyte Reviews*, vol. 1 (eds J.R. MacCallum and C.A. Vincent), Elsevier Applied Science, London, p. 315.
4. Gray, F.M. (1991) *Solid Polymer Electrolytes*, Chapter 10, Wiley-VCH Verlag GmbH, Weinheim, p. 215.
5. Peled, E., Golodnitsky, D., Ardel, G., and Eshkenazy, V. (1995) *Electrochim. Acta*, **40**, 2197.
6. Peled, B., Golodnitsky, D., Ardel, G., Menachem, C., Bar-Tow, D., and Eshkenazy, V. (1995) *Mater. Res. Soc. Symp.*, **393**, 209.
7. Dominey, L.A. (1994) in *Lithium Batteries, New Materials, Development and Perspectives* (ed. G. Pistoia), Elsevier, New York, p. 137.
8. Dahn, J.R. et al. (1994) in *Lithium Batteries, New Materials, Development and Perspectives* (ed. G. Pistoia), Elsevier, New York, p. 22.
9. Mann, C.K. (1969) in *Electroanalytical Chemistry*, vol. 3 (ed. A.J. Bard), Marcel Dekker, New York.
10. Yoshio, M. and Ishibashi, N. (1973) *J. Appl. Electrochem.*, **3**, 321.
11. Makishima, S. (1938) *J. Fac. Eng., Tokyo Univ.*, **21**, 115.
12. Peled, E. and Yamin, H. (1981) in *Power Sources* (ed. S. Thompson), Academic Press, New York, p. 101.
13. Odziemkowski, M. and Irish, D.E. (1994) *J. Electrochem. Soc.*, **140** (6), 1546.
14. Odziemkowski, M. and Irish, D.E. (1992) *J. Electrochem. Soc.*, **139**, 363.
15. Holding, A.D., Pletcher, D., and Jones, R.V.H. (1989) *Electrochim. Acta*, **34**, 1529.
16. Campbell, S.A., Bowes, C., and McMillan, R.S. (1990) *J. Electroanal. Chem.*, **284**, 195.
17. Odziemkowski, M., Krell, M., and Irish, D.E. (1992) *J. Electrochem. Soc.*, **139**, 3063.
18. Aurbach, D., Weissman, I., Zaban, A., and Chuzid (Youngman), O. (1994) *Electrochim. Acta*, **39**, 51.
19. Aurbach, D., Daroux, M., McDougal, G., and Yeager, E.B. (1993) *J. Electroanal. Chem.*, **358**, 63.
20. Kanamura, K., Tamura, H., Shiraishi, S., and Takehara, Z.-I. (1995) *J. Electroanal. Chem.*, **394**, 49.
21. Kanamura, K., Shiraishi, S., and Takehara, Z.-I. (1995) *Chem. Lett.*, 209.
22. Ely, Y.E. and Aurbach, D. (1992) *Langmuir*, **8**, 1845.
23. Aurbach, D. and Zaban, A. (1993) *J. Electroanal. Chem.*, **348**, 155.
24. Kanamura, K., Shiraishi, S., and Takehara, Z.-I. (1996) *J. Electrochem. Soc.*, **143** (7), 2187.
25. Kanamura, K., Shiraishi, S., and Takehara, Z.-I. (1994) *J. Electrochem. Soc.*, **141** (9), L108–L110.
26. Kanamura, K., Tamura, H., and Takehara, Z. (1992) *J. Electroanal. Chem.*, **333**, 127.
27. Thevenin, J.G. and Muller, R.H. (1987) *J. Electrochem. Soc.*, **134** (2), 273.
28. Nazri, G. and Muller, R.H. (1985) *J. Electrochem. Soc.*, **132** (6), 1385.
29. Aurbach, D. and Gofer, Y. (1991) *J. Electrochem. Soc.*, **138** (12), 3529.
30. Montesperelli, G., Nunziante, P., Pasquali, M., and Pistoia, G. (1990) *Solid State Ionics*, **37**, 149.
31. Schwager, F., Geronov, Y., and Muller, R.H. (1985) *J. Electrochem. Soc.*, **132**, 285.
32. Matsuda, Y. (1992) in *Electrochemistry in Transition*, Chapter 14 (eds O.J. Murphy, S. Srinivasan, and B.E. Conway), Plenum, New York.
33. Morita, M., Aoki, S., and Matsuda, Y. (1992) *Electrochim. Acta*, **37** (1), 119.
34. Matsuda, Y. (1993) *J. Power Sources*, **43**, 1.
35. Tobishima, S., Hayashi, K., Nemoto, Y., and Yamaki, J. (1996) 37th Battery Symposium, Tokyo.
36. Mori, M., Kakuta, Y., Naoi, K., and Fauteux, D. (1996) 37th Battery Symposium, Tokyo.

37. Uraoka, Y., Ishikawa, M., Kishi, K., Morita, M., and Matsuda, Y. (1996) 37th Battery Symposium, Tokyo.
38. Matsuda, Y., Ishikawa, M., Yoshitake, S., and Morita, M. (1995) *J. Power Sources*, **54**, 301.
39. Ishikawa, M., Yoshitake, S., Morita, M., and Matsuda, Y. (1994) *J. Electrochem. Soc.*, **141** (12), L159.
40. Ishikawa, M., Otani, K.-Y., Morita, M., and Matsuda, Y. (1996) *Electrochim. Acta*, **41** (7/8), 1253.
41. Fauteux, D. (1988) *J. Electrochem. Soc.*, **135** (9), 2231.
42. Takehara, Z., Ogumi, Z., Uchimoto, Y., and Endo, E. (1991) Proceedings of the Symposium on High Power, Ambient Temperature Lithium Batteries, Phoenix, p. 176.
43. Xue, R., Huang, H., Menetrier, M., and Chen, L. (1993) *J. Power Sources*, **44**, 431.
44. Fauteux, D. (1985) *Solid State Ionics*, **17**, 133.
45. Chabagno, J.M. (1980) Thesis, INP–Grenoble.
46. Xu, J.J. and Farrington, G.C. (1996) Abstract of Electrochemical Society Meeting, Sari Antonio, October, p. 74.
47. Scrosati, B. (1990) *J. Electrochem. Soc.*, **136**, 2774.
48. Capuano, F., Croce, F., and Scrosati, B. (1991) *J. Electrochem. Soc.*, **138**, 1918.
49. Croce, F. and Scrosati, B. (1993) *J. Power Sources*, **43–44**, 9.
50. Golodnitsky, D., Ardel, G., and Peled, E. (1996) *Solid State Ionics*, **85**, 231.
51. Krawiec, W.T., Fellner, J.P., Giannelis, E.P., and Scanlon, L.G. (1994) *Diffus. Defect Data, Part B: Solid State Phenom.*, **39**, 175.
52. Kumar, B., Schaffer, J.D., Munichandraiah, N., and Scanlon, L.G. (1994) *J. Power Sources*, **47**, 63.
53. Fan, J. and Fedkiw, P.S. (1996) Abstract of 190th Electrochemical Society Meeting, San Antonio, Vol. 96-2, p. 89.
54. Kumar, B. and Scanlon, L.G. (1994) Proceedings of 36th Power Sources Conference, Cherry Hill, p. 236.
55. Nagasubramanian, G., Attia, A.I., Halpert, G., and Peled, E. (1993) *Solid State Ionics*, **67**, 51.
56. Matsui, T. and Takeyama, K. (1995) *Electrochim. Acta*, **40** (13–14), 2165.
57. Brissot, C., Rosso, M., and Chazalviel, J.-N. (1996) *Electrochim. Acta*, **43** (10–11), 1569.
58. Peled, E. and Golodnitsky, D. (1997) Abstract 192nd Electrochemical Society Meeting, Paris.
59. Anbar, M. and Neta, P. (1967) *Int. J. Appl. Radiat. Isotopes*, **18**, 493.
60. Buxton, G.V., Greenstock, C.L., Helman, W.P., and Ross, A.B. (1988) *J. Phys. Chem. Ref. Data*, **17** (2), 513.
61. Matheson, M.S. and Dorfman, L.M. (1969) *Pulse Rudiolysis*, MIT Press, Cambridge, MA, p. 173.
62. Marcus, R.A. (1968) *Electrochim. Acta*, **13**, 995.
63. Koch, V.R. (1979) *J. Electrochem. Soc.*, **126**, 181.
64. Cookson, J.T. (1978) in *Carbon Adsorption Handbook* (eds P.N. Cheremisinoff and F. Ellerbush), Ann Arbor Science, Ann Arbor, MI, p. 241.
65. Kinoshita, K. (1987) *Carbon, Electrochemical and Physicochemical Properties*, Wiley-Inter Science, New York.
66. Pierson, H.P. (1993) *Handbook of Carbon., Graphite, Diamond and Fullerenes*, Noyes, Park Ridge, NJ.
67. Dubinin, M.M. (1966) in *Chemistry and Physics of Carbon*, Chapter 2. (ed. P.L. Walker Jr.), Marcel Dekker, New York, p. 251.
68. Xing, W. and Dahn, J.R. (1996) Abstract 190th Electrochemical Society Meeting, Sail Antonio, Vol. 96-2, p. 1055.
69. Fong, R., Sacken, U.V., and Dahn, J.R. (1990) *J. Electrochem. Soc.*, **137**, 2009.
70. Chu, A.C., Josefowicz, J.X., Fischer, J.E., and Farrington, G.C. (1996) Abstract 190th Electrochemical Society Meeting, San Antonio, Vol. 96-2, p. 1062.
71. Menachem, C., Peled, E., and Burstein, L. (1996) Proceedings of 37th Power Sources Conference, Cherry Hill, p. 208.
72. Peled, E., Menachem, C., Bar-Tow, D., and Melman, A. (1996) *J. Electrochem. Soc.*, **143**, L4.
73. Menachem, C., Peled, E., Burstein, L., and Rosenberg, Y. (1996) Abstract 8th

International Meeting on Li Batteries, Nagoya, p. 224; idem, *J. Power Sources*, **68**, 277.
74. Arima, Y. and Aoyagui, S. (1979) *Israel J. Chem.*, **18**, 81.
75. Peled, E. (1995) in *Rechargeable Lithium and Lithium-Ion Batteries*, vol. 94-28 (eds S. Megahed and B.M. BarnettwarningXie), The Electrochemical Society, Pernnington, NJ, p. 1.
76. Zaghib, K., Tatsumi, K., Abe, H., Sakaebi, H., Higuchi, S., Ohsaki, T., and Sawada, Y. (1994) Proceedings of Electrochemical Society Meeting, Vol. 94-1, San Fransisco, Abstract No. 581.
77. Tatsumi, K., Zaghib, K., and Sawada, Y. (1995) *J. Electrochem. Soc.*, **142**, 1090.
78. Shu, Z.X., McMillan, R.S., and Murray, J. (1993) *J. Electrochem. Soc.*, **140**, 992.
79. Billaud, D., Naji, A., and Willmann, P. (1995) *J. Chem. Soc. Chem. Commun.*, 1867.
80. Takami, N., Satah, A., Hara, M., and Ohsaki, T. (1995) *J. Electrochem. Soc*, **142**, 8.
81. Aurbach, D., Ein-Eli, Y., Chusid, O., Carmeli, Y., Babai, M., and Yamin, H. (1994) *J. Electrochem. Soc.*, **141**, 603.
82. Chusid, O., Ein-Eli, Y., and Aurbach, D. (1993) *J. Power Sources*, **43**, 47.
83. Arakawa, M., Yamaki, J., and Okada, T. (1984) *J. Electrochem. Soc.*, **131** (11), 2605.
84. Dahn, J.R. (1991) *Phys. Rev. B*, **44**, 17.
85. Ein-Eli, Y., Thomas, S.R., and Koch, V.R. (1996) Abstract 190th Electrochemical Society Meeting, San Antonio, Vol. 96-2, p. 1046.
86. Besenhard, J.O., Winter, M., Yang, J., and Biberacher, W. (1995) *J. Power Sources*, **54**, 228.
87. Johnson, B.A. and Write, R.E. (1996) Abstract 190th Electrochemical Society Meeting, San Antonio, Vol. 96-2, p. 170.
88. Abe, T., Mizutani, Y., Ikada, K., Inaba, M., and Ogumi, Z. (1996) 37th Battery Symposium, Tokyo, p. 53.
89. Inaba, M., Siroma, Z., Kawatate, Y., Funabiki, A., and Ogumi, Z. (1997) *J. Power Sources*, **68**, 221.
90. Ogumi, Z., Inaba, M., Siroma, Z., Kawatate, Y., and Funabiki, A. (1996) Abstract 190th Electrochemical Society Meeting, San Antonio, Vol. 96-2, p. 1045.
91. Yamaguchi, S., Hirasawa, K.A., and Mori, S. (1996) 37th Battery Symposium, Tokyo, p. 49.
92. Peled, E., Bar-Tow, D., and Eshkenazy, V. (1996) 37th Battery Symposium, Tokyo, Abstract no. 3102.
93. Blint, R. (1994) in *Proceedings of Symposium on Lithium Batteries*, vol. 94-4 (eds N. Doddapaneni and A.R. Langrebe), The Electrochemical Society, Pennington, NJ, p. 1.
94. Bockris, J.O.M. and Reddy, A.K.N. (1970) *Modern Electrochemistry*, Plenum Press, New York, p. 1176.
95. Fujimoto, M., Kida, Y., Nohma, T., Takahashi, M., Nishio, K., and Saito, T. (1996) *J. Power Sources*, **63**, 127.
96. Yoshida, H., Fukunaga, T., Hazama, T., Terasaki, M., Mizutani, M., and Yamachi, M. (1996) Abstract 8th International Meeting on Li Batteries, Nagoya, p. 252.
97. Tanaka, K., Itabashi, M., Aoki, M., Hiraka, S., Kataoka, M., Fujita, S., Sekai, K., and Ozawa, K. (1993) 184th Electrochemical Society Meeting, New Orleans, Vol. 93-2, Abstract No. 21.
98. Huang, C.K., Smart, M., Davies, E., Cortez, R., and Surampudi, S. (1996) Abstract 190th Electrochemical Society Meeting, San Antonio, Vol. 96-2, p. 171.
99. Peled, E., Bar-Tow, D., Melman, A., Gerenrot, E., Lavi, Y., and Rosenberg, Y. (1994) *Proceedings of Symposium on Lithium Batteries*, vol. 94-4, The Electrochemical Society, Penninton, NJ, p. 177.
100. Zaghib, K., Choquette, Y., Guerbi, A., Simoneau, M., Belanger, A., and Gauthier, M. (1996) Abstract 8th International Meeting on Li Batteries, Nagoya, p. 298.
101. Guyomard, D. and Tarascon, J.M. (1993) US Patent 5192629.
102. Huang, C.K., Surampudi, S., Shen, D.H., and Halpert, G. (1993) Abstract 184th Electrochemical Society Meeting, San Antonio, Vol. 93-2, p. 26.

103. Bar-Tow, D., Peled, E., and Burstein, L. (1996) Abstract 190th Electrochemical Society Meeting, San Antonio, Vol. 96-2, Paper No. 1028.
104. Ein-Eli, Y. and Koch, V.R. (1997) *J. Electrochem. Soc.*, **144**, 2968.
105. Takamura, T., Kikuchi, M., and Ikezawa, Y. (1995) in *Rechargeable Lithium-Ion Batteries*, vol. 94-28 (eds S. Megahed, B.M. Barnett, and L. Xie), The Electrochemical Society, Pennington, NJ, p. 213.
106. Jortner, J. and Kestner, N.R. (eds) (1973) in *Electrons in Fluids*, Springer-Verlag, Berlin.
107. Yamamoto, O., Takeda, Y., and Imanishi, N. (1993) Proceedings of Electrochemical Society Meeting, Honolulu.
108. Kanno, R., Kawamoto, Y., Takeda, Y., Ohashi, S., Imanishi, N., and Yamamoto, O. (1992) *J. Electrochem. Soc*, **139** (12), 3397.
109. Momose, H., Honbo, H., Takeuchi, S., Nishimura, K., Horiba, T., Muranaka, Y., Kozono, Y., and Miadera, H. (1996) Abstract 8th International Meeting on Li Batteries, Nagoya, p. 172.
110. Besenhard, J.O., Wagner, M.W., Winter, M., Jannakoudakis, A.D., Jannakoudakis, P.D., and Theodoridou, E. (1993) *J. Power Sources*, **43–44**, 413.
111. Tibbets, G.G., Nazri, G.-A., and Howie, B.J. (1996) Abstract 190th Electrochemical Society Meeting, San Antonio, Vol. 96-2, p. 117.
112. Ein-Eli, Y., Thomas, S.R., and Koch, V.R. (1996) *J. Electrochem. Soc.*, **143** (9), L195.
113. Zaghib, K., Yazami, R., and Broussly, M. (1996) Abstract 8th International Meeting on Li Batteries, Nagoya, p. 192.
114. Jean, M., Chausse, A., and Messina, R. (1996) Abstract 8th International Meeting on Li Batteries, Nagoya, p. 186.
115. Tada, T., Yoshida, H., Yagasaki, E., Hanafusa, K., and Tanaka, G. (1995) Extended Abstract 36th Battery Symposium, Kyoto, p. 195.
116. Morita, M., Ichimura, T., Ishikawa, M., and Matsuda, Y. (1996) *J. Electrochem. Soc.*, **143** (2), L26.
117. Besenhard, J.O., Yang, J., and Winter, M. (1996) Abstract 8th International Meeting on Li Batteries, Nagoya, p. 96.
118. Yoshio, I., Masayuki, M., Yukio, M., Tadahiko, K., and Tsutomu, M. (1994) European Patent Application 94116643-1, October 21, 1994.
119. Courtney, I.A., Wilson, A.M., Xing, W., and Dahn, J.R. (1996) Abstract 190th Electrochemical Society Meeting, Sun Antonio, Vol. 96-2, Paper No. 66.
120. Ardel, G., Golodnitsky, D., Strauss, E., and Peled, E. (1996) Proceedings of the 37th Power Sources Conference (ARL), NJ, p. 283.
121. Ardel, G., Golodnitsky, D., Strauss, E., and Peled, E. (1996) Abstract 190th Electrochemical Society Meeting, San Antonio, Vol. 96-2, Paper No. 104a.
122. Ardel, G., Golodnitsky, D., and Peled, E. (1997) *J. Electrochem. Soc.*, **144** (8), L208.
123. Takami, N. and Ohsaki, T. (1992) *J. Electrochem. Soc.*, **139** (7), 1849.
124. Yazami, R., Deschamps, M., Genies, S., Frison, J.C., and Ledran, J. (1996) Abstract 8th International Meeting on Li Batteries, Nagoya, p. 107.
125. Aurbach, D., Markovsky, B., and Shechter, A. (1996) Abstract 190th Electrochemical Society Meeting, San Antonio, Vol. 96-2, Paper No. 164.
126. Fouache-Ayoub, S., Garreau, M., Prabhu, P.V.S.S., and Thevenin, J. (1990) *J. Electrochem. Soc.*, **137** (6), 1659.
127. Yazami, R. and Deschamps, M. (1996) Abstract 8th International Meeting on Li Batteries, Nagoya, p. 190.
128. Croce, F., Panero, S., Passerini, S., and Scrosati, B. (1994) *Electrochim. Acta*, **39** (2), 255.
129. Hiratani, M., Miyauchi, K., and Kudo, T. (1988) *Solid State Ionics*, **28–30**, 1431.
130. Sloop, S.E. and Lerner, M.M. (1996) *Solid State Ionics*, **85**, 251.
131. Peled, E., Golodnitsky, D., Menachem, C., and Ardel, G. (1994) Abstract 7th International Meeting on Lithium Batteries, Boston, Paper no. 21.
132. Munichandraiah, N., Scanlon, L.G., Marsh, R.A., Kumar, B., and Sircar,

A.K. (1995) *Appl. Electrochem.*, **25** (9), 857.

133. Cho, J. and Liu, M. (1996) Abstract Electrochemical Society Meeting, San Antonio, Vol. 96-2, Paper no. 104.
134. Ismail, I.M., Kadiroglu, U., Gray, N.D., and Owen, J.R. (1996) Abstract Electrochemical Society Meeting, San Antonio, Vol. 96-2, p. 84.
135. Peled, E., Golodnitsky, D., Menachem, C., Ardel, G., and Lavi, Y. (1993) The Electrochemical Society Fall Meeting, New Orleans, Vol. 93-2, Abstract no. 504.
136. Peled, E., Golodnitsky, D., Ardel, G., Lang, J., and Lavi, Y. (1994) in *Proceedings of 11th International Seminar on Primary and Secondary Battery Technology and Applications* (eds S.P. Wolsky and N. Marincic), Florida.
137. du Pasquier, A., Sarrazin, C., Andrieu, X., and Fauvarque, J.-F. (1996) Abstract 190th Electrochemical Society Meeting, San Antonio, Vol. 96-2, Paper no. 82.
138. Osaka, T., Momma, T., and Ito, H. (1996) Abstract Electrochemical Society Meeting, San Antonio, Vol. 96-2, Paper no. 87.
139. Nagasubramanian, G. and Boone, D.R. (1996) Abstract 190th Electrochemical Society Meeting, San Antonio, Vol. 96-2, Paper no. 166.

Further Reading

Aurbach, D., Zaban, A., and Moshkovich, M. (1996) Abstract 8th International Meeting on Li Batteries, Nagoya, p. 246.

Fu, L.J., Liu, H., Li, C., Wu, Y.P., Rahm, E., Holze, R., and Wu, H.Q. (2006) *Solid State Sci.*, **8** (2), 113–128.

Goodenough, J.B. and Kim, Y. (2010) *Chem. Mater.*, **22** (3), 587–603.

Verma, P., Maire, P., and Novak, P. (2010) *Electrochim. Acta*, **55** (22), 6332–6341.

Zhang, S.S. (2006) *J. Power Sources*, **162** (2), 1379–1394.